TRIBOLOGY OF ABRASIVE MACHINING PROCESSES

TRIBOLOGY OF ABRASIVE MACHINING PROCESSES

by

Ioan D. Marinescu
University of Toledo
Toledo, Ohio, USA

W. Brian Rowe
Liverpool John Moores University
Liverpool, United Kingdom

Boris Dimitrov
Institute for Precision Engineering
Bucharest, Romania

Ichiro Inasaki
Keio University
Yokohama-shi, Japan

William Andrew
publishing

Cover Art © 2004 by Brent Beckley / William Andrew, Inc.

Library of Congress Catalog Card Number: 2004002376
ISBN: 0-8155-1490-5
Printed in the United States

Published in the United States of America by
William Andrew, Inc.
13 Eaton Avenue
Norwich, NY 13815
1-800-932-7045
www.williamandrew.com
www.knovel.com

10 9 8 7 6 5 4 3 2 1

This book may be purchased in quantity at discounts for education, business, or sales promotional use by
contacting the Publisher.

NOTICE

To the best of our knowledge the information in this publication is accurate; however the
Publisher does not assume any responsibility or liability for the accuracy or complete-
ness of, or consequences arising from, such information. This book is intended for
informational purposes only. Mention of trade names or commercial products does not
constitute endorsement or recommendation for use by the Publisher. Final determination
of the suitability of any information or product for any use, and the manner of that use,
is the sole responsibility of the user. Anyone intending to rely upon any recommendation
of materials or procedures mentioned in this publication should be independently
satisfied as to such suitability, and must meet all applicable safety and health standards.

Library of Congress Cataloging-in-Publication Data

Marinescu, Ioan D.
Tribology of abrasive machining processes / Ioan D. Marinescu ... [et al.].
p. cm.
ISBN 0-8155-1490-5 (alk. paper)
1. Grinding and polishing. 2. Tribology. I. Marinescu, Ioan D.
TJ1280.T68 2004
621.8'9--dc22

2004002376

Related Titles

ADVANCED CERAMIC PROCESSING AND TECHNOLOGY, Volume 1: edited by Jon G. P. Binner (ISBN: 0-8155-1256-2)

CARBON-CARBON MATERIALS AND COMPOSITES: by John Buckley and Dan Edie (ISBN: 0-8155-1324-0)

CRYSTAL GROWTH TECHNOLOGY: by K. Byrappa (ISBN: 0-8155-1453-0)

CEMENTED TUNGSTEN CARBIDES: by Gopal S. Upadhyaya (ISBN: 0-8155-1417-4)

CERAMIC CUTTING TOOLS: edited by E. Dow Whitney (ISBN: 0-8155-1355-0)

CERAMIC FILMS AND COATINGS: edited by John B. Wachtman and Richard A. Haber (ISBN: 0-8155-1318-6)

CERAMIC TECHNOLOGY AND PROCESSING: by Alan G. King (ISBN: 0-8155-1443-3)

CORROSION OF GLASS, CERAMICS AND CERAMIC SUPERCONDUCTORS: edited by David E. Clark and Bruce K. Zoitos (ISBN: 0-8155-1283-X)

DUPLEX STAINLESS STEELS: by R. Gunn (ISBN: 1-884207-61-8)

FIBER REINFORCED CERAMIC COMPOSITES: edited by K. S. Mazdiyasni (ISBN: 0-8155-1233-3)

FLUIDIZATION, SOLIDS HANDLING, AND PROCESSING: by Wen-Ching Yang (ISBN: 0-8155-1427-1)

GUIDE TO WEAR PROBLEMS AND TESTING FOR INDUSTRY: by Michael Neale and Mark Gee (ISBN: 0-8155-1471-9)

HANDBOOK OF CERAMICS GRINDING AND POLISHING: edited by Ioan D. Marinescu, Hans K. Tonshoff, and Ichiro Inasaki (ISBN: 0-8155-1424-7)

HANDBOOK OF CARBON, GRAPHITE, DIAMOND AND FULLERENES: by Hugh O. Pierson (ISBN: 0-8155-1339-9)

HANDBOOK OF ELLIPSOMETRY: edited by Harland G. Tompkins and Eugene A. Irene (ISBN: 0-8155-1499-9)

HANDBOOK OF ENVIRONMENTAL DEGRADATION OF MATERIALS: edited by Myer Kutz (ISBN: 0-8155-1500-6)

HANDBOOK OF FILLERS, Second Edition: edited by George Wypych (ISBN: 1-884207-69-3)

HANDBOOK OF HARD COATINGS: edited by Rointan F. Bunshah (ISBN: 0-8155-1438-7)

HANDBOOK OF HYDROTHERMAL TECHNOLOGY: edited by K. Byrappa and Masahiro Yoshimura (ISBN: 0-8155-1445-x)

HANDBOOK OF INDUSTRIAL REFRACTORIES TECHNOLOGY: by Stephen C. Carniglia and Gordon L. Barna (ISBN: 0-8155-1304-6)

HANDBOOK OF MATERIAL WEATHERING, Third Edition: edited by George Wypych (ISBN: 0-8155-1478-6)

HANDBOOK OF MULTILEVEL METALLIZATION FOR INTEGRATED CIRCUITS: edited by Syd R. Wilson, Clarence J. Tracy, and John L. Freeman, Jr. (ISBN: 0-8155-1340-2)

HANDBOOK OF PLASTICIZERS: edited by George Wypych (ISBN: 0-8155-1496-4)

HANDBOOK OF REFRACTORY CARBIDES AND NITRIDES: by Hugh O. Pierson (ISBN: 0-8155-1392-5)

HANDBOOK OF SOLVENTS: edited by George Wypych (ISBN: 0-8155-1458-1)

INDUSTRIAL MINERALS AND THEIR USES: by Peter Ciullo (ISBN: 0-8155-1408-5)

MECHANICAL ALLOYING FOR FABRICATION OF ADVANCED ENGINEERING MATERIALS: by M. Sherif El-Eskandarany (ISBN: 0-8155-1462-X)

MEMS: A PRACTICAL GUIDE TO DESIGN, ANALYSIS, AND APPLICATIONS: edited by Oliver Paul and Jan Korvink (ISBN: 0-8155-1497-2)

NANOSTRUCTURED MATERIALS: edited by Carl C. Koch (ISBN: 0-8155-1451-4)

SEMICONDUCTOR MATERIALS AND PROCESS TECHNOLOGY HANDBOOK: edited by Gary E. McGuire (ISBN: 0-8155-1150-7)

SOL-GEL TECHNOLOGY FOR THIN FILMS, FIBERS, PREFORMS, ELECTRONICS AND SPECIALTY SHAPES: edited by Lisa C. Klein (ISBN: 0-8155-1154-X)

SOL-GEL SILICA: by Larry L. Hench (ISBN: 0-8155-1419-0)

SUPERCRITICAL FLUID CLEANING: edited by John McHardy and Samuel P. Sawan (ISBN: 0-8155-1416-6)

TRIBOLOGY OF ABRASIVE MACHINING PROCESSES: by Ioan Marinescu, Brian Rowe, Boris Dimitrov, Ichiro Inasaki (ISBN: 0-8155-1490-5)

ULTRA-FINE PARTICLES: edited by Chikara Hayashi, Ryozi Uyeda and Akira Tasaki (ISBN: 0-8155-1404-2)

WEATHERING OF PLASTICS: edited by George Wypych (ISBN: 1-884207-75-8)

Preface

In 1966, a report published by the UK Department of Education and Science introduced the concept of *tribology*, which was defined as the science of interacting surfaces in relative motion. Tribology, as a "new science," studies friction, wear, and lubrication. These three processes affect each other with interacting causes and effects: tribology is the study of them as they interact.

Abrasive machining processes are part of the large field of "manufacturing processes" and include grinding, superfinishing, honing, lapping, polishing, etc. The common characteristic of these processes is the fact that the main stock removal mechanism is the abrasive process. But, different from classic tribology, in this case the "abrasive wear" is a useful process, helping remove unwanted material from workpieces. At the same time, "abrasive wear" is a negative term when applied to abrasive tools which, during all abrasive machining processes, exhibit predominant abrasive wear.

Most abrasive machining processes have been studied very little and most of the studies were deterministic, based on experiments and experience and less on science. It is often said that abrasive machining processes are more an art rather than a science.

This book is an attempt to introduce science into the study of abrasive machining processes. The authors considered that the "marriage" between tribology and abrasive machining processes is a good match because tribology provides the right tool to study these processes. The main characteristic of abrasive machining processes is that these processes are more random than many manufacturing processes. Tribology offers a good approach for describing abrasive machining processes and offers the ability to predict some of the outputs of the processes.

This book aims to bring attention back to tribology. Tribology was fashionable in the sixties, but today academia and the research community in the U.S., Japan, and to some extent in Europe consider the field of less importance. I asked Dr. Ernest Rabinovicz, one of my mentors in the field of tribology, why this happened. He answered that the concept of tribology was oversold in the mid-1970s and very soon other topics became more fashionable: robotics, manufacturing systems, etc.

This book considers knowledge in the field of tribology as a tool for the study of abrasive machining processes. This will help to explain scientifically each process as well as to quantify the relationships between the main parameters involved in abrasive machining processes.

In this book's chapters, we will consider the important elements of the abrasive machining system and the tribological factors which control the efficiency and quality of the processes.

Since grinding is by far the most commonly employed abrasive machining process, it will be given detailed consideration. By understanding the tribological principles, it is possible to propose process improvements and solutions to many commonly experienced industrial problems such as poor accuracy, poor surface quality, rapid wheel wear, vibrations, workpiece burn, and high process costs.

The chapter on kinematics examines factors affecting the size, shape and variability of the material in the process of being removed. These factors influence the wear of the abrasive tool, the surface roughness of the machined parts, the process forces and energy, and the surface integrity of the workpiece. Whereas traditional texts assume a uniform distribution of abrasive grains, this analysis addresses the effects of variable grain distributions.

Traditionally, it is assumed that the machining energy is distributed over an area represented by the geometry of the undeformed contact between the abrasive tool and the workpiece. The chapter on contact mechanics reveals that this assumption is quite misleading for many cases of precision machining and provides a more practical analysis. Realistically, the power density in the contact zone may be less than half the value given by traditional calculations.

The study of friction, forces, and energy explores the importance of the various factors which govern the stresses and deformations of abrasion. The effects of grain shape, depth of penetration, and lubrication on the process forces are explored.

With a knowledge of contact mechanics and process energy, the principles of heat transfer can be applied to explore the effect of process variables on surface temperatures and surface quality. It is also shown how thermal processes define permissible domains of operation ranging from creep grinding with deep cuts at low workspeeds to conventional grinding with shallow cuts at higher workspeeds to high-efficiency deep grinding with deep cuts at extremely high removal rates.

Process fluids play important roles in cooling and lubrication. New information is provided on the requirements for effective fluid delivery.

Further chapters deal with important aspects such as assessment of the workpiece surface, the grinding wheel topography, the nature of abrasive tools, dressing processes, free abrasive processes, lubricants and process fluids, tribochemical processes, and the characteristics of work-piece materials. The final chapter examines some fundamental information about deformation processes revealed through molecular dynamics simulation.

The reader should not have to read every chapter consecutively to understand the development of the subject. The material is presented in a logical order so that a reader can refer back to earlier passages to fully understand some aspects.

The reader can use this text as a reference work to directly access a topic of interest. It is designed to make the material accessible to technicians, undergraduates, and graduates. The book could well form the basis for a series of lectures for a specialized course at any of these levels.

The idea for this book came to me in the early 1980s when, in a CIRP paper, I published a table which contained the analogy between technologic parameters and tribological parameters for the grinding process. I have had the privilege to spend many years in Europe, many years in the US and a lot of time in Japan during the last five years. I was able to observe the evolution of tribology in these parts of the world and to get in touch with the professors and researchers in this field. I was very encouraged to continue to look at abrasive machining processes by my colleagues from Japan (Drs. Koji Kato and Takeo Nakagawa,) from Europe (Drs. W. Konig, Hans Kurt Toenshoff, Tom Childs, and Trevor Howes), as well as from my colleagues from the US (Drs. Nam Suh, Ken Ludema, Said Jahanamir, Steven Malkin, and Steven Danyluk). I would like to thank to all of them for their encouragement and for their help.

I would like to express special thanks to Dr. R. Rentsch for the help with Ch. 7 regarding molecular dynamics simulation of abrasive processes, and my former student, Dr. C. Spanu, for his help with Chs. 9 and 11.

The authors of this book cover some regions of the world where tribology was initiated and developed in the last forty years: Western and Eastern Europe, the US, and Japan. I would like to thank my co-authors for taking time from their busy schedules to write these chapters and to review each others' chapters. This means that the book is unitary with integrated chapters, not just a collection of disparate chapters.

My special thanks to my wife Jocelyn for encouraging me, checking my English, and putting up with my long working days punctuated by more working days on the weekends. She minimizes the friction and wear in my life.

Ioan Marinescu   December 2003
Toledo, Ohio

Table of Contents

1

Introduction

1.1 ABRASIVE PROCESSES

Abrasive machining processes are manufacturing techniques which employ very hard granular particles in machining, abrading, or polishing to modify the shape and surface texture of manufactured parts.

A wide range of such processes is mostly used to produce high quality parts to high accuracy and to close tolerances. Examples range from very large parts such as machine slideways to small parts such as contact lenses, needles, electronic components, silicon wafers, and ball bearings.

While accuracy and surface texture requirements are common reasons for selecting abrasive processes, there is another common reason. Abrasive processes are the natural choice for machining and finishing hard materials and hardened surfaces.

Most abrasive processes may be categorized into one of four groups: (*i*) grinding, (*ii*) honing, (*iii*) lapping, (*iv*) polishing.

This is not a completely inclusive list, but the four groups cover a wide range of processes and are a sufficient representation for a study of fundamental characteristics of such processes. These four groups are illustrated in Fig. 1.1. Grinding and honing are processes which employ bonded or fixed abrasives within the abrasive tool, whereas lapping and polishing employ free abrasive particles, often suspended in a liquid or wax medium.

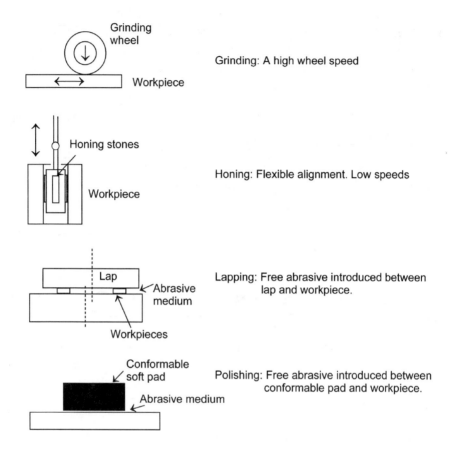

Figure 1.1. Basic principles of grinding, honing, lapping, and polishing.

1.1.1 Grinding

In grinding, the abrasive tool is a grinding wheel which moves at a high surface speed compared to other machining processes such as milling and turning. Surface speeds are typically in the range of 20 m/s (4,000 ft/min) to 45 m/s (9,000 ft/min) in conventional grinding. In high-speed grinding, the wheel moves at speeds up to 140 m/s with wheels especially designed to withstand the high bursting stresses. Speeds greatly in excess of 140 m/s may be employed, but the proportion of applications at such speeds is small due to the expense and sophistication of the machines and techniques involved.

Although grinding can take place without lubrication, wet grinding is preferred wherever possible due to the reduced frictional losses and improved quality of the surfaces produced. Commonly used lubricants include oil in water emulsions and neat oils.

1.1.2 Honing

In honing, the abrasive particles, or *grains* as they are commonly known, are fixed in a bonded tool as in grinding. The honing process is mainly used to achieve a finished surface in the bore of a cylinder. The honing stones are pressurized radially outwards against the bore. Honing is different than grinding in two ways.

First, in honing, the abrasive tool moves at a low speed relative to the workpiece. Typically, the surface speed is 0.2 m/s to 2m/s. Combined rotation and oscillation movements of the tool are designed to average out the removal of material over the surface of the workpiece and produce a characteristic "cross-hatch" pattern favored for oil retention in engine cylinder bores.

Another difference between honing and grinding is that a honing tool is flexibly aligned to the surface of the workpiece. This means that eccentricity of the bore relative to an outside diameter cannot be corrected.

1.1.3 Lapping

In lapping, free abrasive is introduced between a lap, which may be a cast iron plate, and the workpiece surface. The free abrasive is usually suspended in a liquid medium, such as oil, providing lubrication and helping to transport the abrasive. The lap and the abrasive are both subject to wear. To maintain the required geometry of the lap and of the workpiece surface, it is necessary to pay careful attention to the nature of the motions involved to average out the wear across the surface of the lap. Several laps may be employed and periodically interchanged to assist this process.

1.1.4 Polishing

Polishing, like lapping, also employs free abrasive. In this case, pressure is applied on the abrasive through a conformable pad or soft cloth. This allows the abrasive to follow the contours of the workpiece surface and limits the penetration of individual grains into the surface. Polishing with a fine abrasive is a very gentle abrasive action between the grains and the workpiece, thus ensuring a very small scratch depth.

The main purpose of polishing is to modify the surface texture rather than the shape. Highly reflective mirror surfaces can be produced by polishing. Material is removed at a very low rate. Consequently, the geometry of the surface needs to be very close to the correct shape before polishing is commenced.

1.2 ABRASIVES

In all four classes of abrasive machining processes, the abrasive grain must be harder than the workpiece at the point of interaction. This means that the grain must be harder than the workpiece at the temperature of the interaction. Since these temperatures of short duration can be very high, the abrasive grains must retain their hardness even when hot. This is true in all abrasive processes, without exception, since if the workpiece is harder than the grain, it is the grain that will suffer the most wear.

Some typical hardness values of abrasive grains are given in Table 1.1 based on data published by de Beers.[1] A value for a typical M2 tool steel is given for comparison. The values given are approximate since variations can arise due to the particular form and composition of the abrasive.

Table 1.1. Typical Hardness Values of Abrasive Grain Materials at Ambient Temperatures

UNITS	GPA
Diamond	56–102
Cubic boron nitride (CBN)	42–46
Silicon carbide	~ 24
Aluminium oxide	~ 21
M2 tool steel (double tempered)	~ 0.81

The hardness of the abrasive is substantially reduced at typical contact temperatures between a grain and a workpiece. At 1,000°C, the hardness of most abrasives is approximately halved. Cubic boron nitride (CBN) retains its hardness better than most abrasives which makes it a wear resistant material. Fortunately, the hardness of the workpiece is also reduced. As can be seen from the table, the abrasive grains are at least one order of magnitude harder than a hardened steel.

The most common abrasives are aluminium oxide and silicon carbide. These abrasives are available in a number of different compositions, are inexpensive, and widely available.

Diamond and CBN abrasives are much more expensive, but are finding increasing applications due to their greater hardness and wear resistance.

1.3   TRIBOLOGICAL PRINCIPLES

The scientific principles underlying abrasive machining processes lie within the domain of *tribology*. Tribology is defined as the science and technology of interacting surfaces in relative motion.[2] Tribology is primarily concerned with the study of friction, lubrication, and wear.

In machining, material removal from the workpiece is referred to rather than "wear." The idea that material is cleanly cut away from the workpiece is preferred to material being rubbed away. However, cutting and rubbing are merely two aspects of abrasion as will be discussed in Ch. 5, "Friction, Forces, and Energy." Whereas in a bearing, the objective may be to minimize wear of a critical machine element, in abrasive machining the objective is more likely to be to maximize removal rate.

In abrasive machining, the main objectives are usually to minimize friction and wear of the abrasive while maximizing abrasive wear of the workpiece. Other objectives are concerned with the quality of the workpiece, including the achievement of a specified surface texture and avoidance of thermal damage.

In tribological terms, grinding and honing involve 2-body abrasion while lapping and polishing may be considered 3-body abrasion processes. These terms are illustrated in Fig. 1.2.

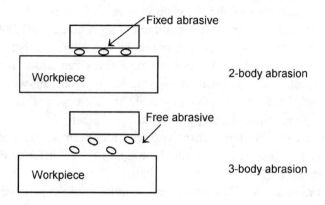

Figure 1.2. Two-body and three-body abrasive processes.

1.3.1 Two-body Abrasion

In 2-body abrasion, the abrasive particles are assumed to be constrained by the tool. The relative motion between the abrasive and the workpiece is usually considered to be pure sliding.

1.3.2 Three-body Abrasion

In 3-body abrasion, the abrasive grains are free to rotate and slide, experiencing collisions both with the workpiece and with the pad and other abrasive grains. From an energy viewpoint, this is obviously a less efficient process since each collision leads to energy dissipation. However, an advantage of the 3-body process is that as the grains rotate, new cutting edges can be brought into action.

In practice, a 2-body abrasive process involves an element of 3-body abrasion, since abraded material from the workpiece and fractured abrasive particles from the grains can form a 3-body action in grinding and honing. In general, 3-body action in 2-body processes is an effect which causes quality problems, since the loose material can become embedded in the workpiece surface. Embedded particles detract from the surface texture and create an abrasive finished surface on the workpiece which can damage other surfaces with which the part comes into contact.

1.4　A TYPICAL GRINDING PROCESS

Figure 1.3 illustrates a typical reciprocating grinding operation. The five main elements are the grinding wheel, the workpiece, the grinding fluid, the atmosphere, and the grinding swarf. The grinding wheel performs the machining of the workpiece, although there is also an inevitable reverse process. The workpiece wears the grinding wheel.

The grinding swarf includes chips cut from the workpiece mixed with a residue of grinding fluid and worn particles from the abrasive grains of the wheel. The swarf can be considered an undesirable outcome of the process, although it is not necessarily valueless.

The grinding fluid serves three main objectives:

- Lubrication and reduction of friction between the abrasive grains, the chips, and the workpiece in the contact zone.

- Cooling the workpiece and reducing temperature rise of the bulk of the workpiece material within and outside of the contact zone.

- Flushing away the grinding swarf to minimize 3-body abrasion.

Although it may not be immediately obvious, the atmosphere also plays an important role. Most metals, when machined, experience increased chemical reactivity due to two effects:

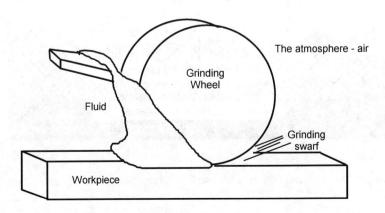

Figure 1.3. The reciprocating horizontal spindle surface grinding process.

• Nascent surfaces created in the cutting process are much more highly reactive than an already oxidized surface.

• High temperatures at the interfaces between the grains and the workpiece, and the grains and the chip, also increase the speed of reaction.

The result is that oxides or other compounds are formed very rapidly on the underside of the chips and on the new surfaces of the workpiece.

Oxides of low shear strength assist the lubrication of the process and reduce friction at conventional grinding speeds. The lubrication effect of the oxides lessens with increasing grinding speeds.

The surface of the workpiece as it enters the contact zone can be considered as several layers extending from the atmosphere down to the parent workpiece material. This is illustrated schematically in Fig. 1.4 for grinding with a fluid.

The two uppermost layers are the air in the atmosphere and the fluid. The fluid is separated from the parent workpiece material, at least partially, by the oxide layer and adsorbed contaminants from the environment.

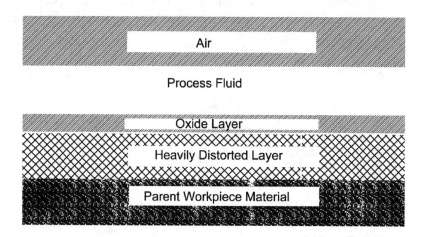

Figure 1.4. The interacting layers at a workpiece surface as it enters the contact zone.

On a freshly ground workpiece surface, the oxide is less than 0.0001 mm thick. There is possibly a transition layer between the oxide layer and the layer heavily distorted from the previous machining operation. In a polished surface, this transition layer consisting of oxides and heavily distorted material is known as a *Beilby layer*. The thickness of the Beilby layer and the heavily distorted layer underneath are dependent on factors such as the amount of working of the surface, the ductility of the material, and the temperature of the previous process which formed the surface.

Although it is tempting to emphasize the importance of physical aspects of abrasive processes, it is clear that chemical and thermal aspects play an important role.

Understanding the tribological principles of abrasive processes is crucial to discovering how advances may be made to improve accuracy, production rate, and surface quality in manufacturing. The purpose of this book is to systemize and explore the nature of the key tribological elements of abrasive machining processes, including their effects on process efficiency and product quality.

In broad terms, the main elements of an abrasive machining system are:

- The workpiece material: shape, hardness, speed, stiffness, and thermal and chemical properties.
- The abrasive tool: structure, hardness, speed, stiffness, thermal and chemical properties, grain size, and bonding.
- The geometry and motions governing the engagement between the abrasive tool and the workpiece (kinematics).
- The process fluid: flowrate, velocity, pressure, and physical, chemical, and thermal properties.
- The atmospheric environment.
- The machine: accuracy, stiffness, temperature stability, and vibrations.

The properties of the machine only indirectly affect the abrasive machining process that takes place in the contact zone between the tool and the workpiece. However, the machine has a very important role in providing static and dynamic constraints on displacements between the tool and the workpiece. There is some evidence to show that increased vibrations between the tool and the workpiece leads to increased fracture wear of the grains.

1.5   A TRIBOLOGICAL SYSTEM

The systematic investigation of a tribological system requires consideration of the inputs and outputs.[3] Figure 1.5 shows the nature of the inputs and outputs to be considered.

The inputs and outputs can be broken down into motions, materials, energy, and information. In a detailed analysis, each of these categories is examined to determine its influences on the process. In addition, there are disturbances to the process such as vibrations which may, in a few cases, be controllable, but not always avoidable.

There are also outputs from the process which may be considered as losses. These include frictional losses and wear products.

Some of the factors to be considered in an abrasive machining process are illustrated in Fig. 1.6.

In the following chapters, the important elements of the abrasive machining system and the tribological factors that control the efficiency and quality of the process are considered. Since grinding is by far the most commonly employed abrasive machining process, it is described in detail.

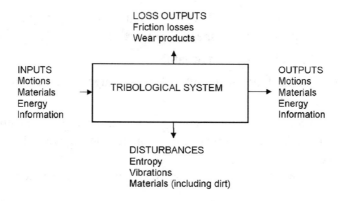

Figure 1.5. Inputs and outputs of a tribological system.[3]

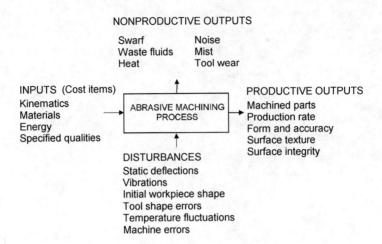

Figure 1.6. Inputs and outputs of abrasive machining processes.

REFERENCES

1. De Beers, Diamond and CBN Grit Products.

2. HMSO, Lubrication (Tribology) Education and Research, DES (Jost) Report, London (1966)

3. Czichos, H., Tribology—A Systems Approach to the Science and Technology of Friction, Lubrication and Wear, Elsevier Press (1978)

2

Tribosystems of Abrasive Machining Processes

2.1 INTRODUCTION

The purpose of a system approach is to systematically identify the parameters that can influence the course of a process and to identify the strength of each influence and its interaction with other parameters. With this knowledge of the system, it is possible to approach the optimization of a process for a desired set of outcomes. In abrasive machining, there are numerous interactions between many variables governing the elasto-plastic and microcutting processes that occur at numerous cutting edges. The analysis of abrasive machining is, therefore, best carried out within the context of a system concept.[1]

A system may be considered as a *black box* with inputs and outputs. The inputs are a group of elements {X} that are transformed within the system into the second group of elements, called outputs {Y}. The relationship between the two groups defines the transfer function of the system:[2]

Eq. (2.1) $\{X\} \rightarrow \{Y\}$

Abrasive machining processes are mainly open systems and are open to the environment. They are unprotected and dissipative. The inputs can be subdivided into useful inputs and disturbances. The outputs can be subdivided into useful outputs and loss outputs (Fig. 2.1).

13

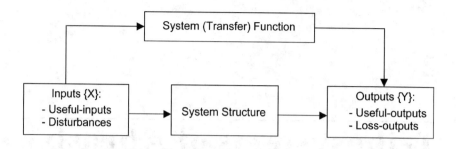

Figure 2.1. Tribosystem representation.

Briefly, a system is represented by elements linked to each other by a transfer function and a system structure. The transfer function of the abrasive processing system defines the influence of the inputs on the outputs of the process.

The following information about an abrasive machining system must be listed in order to establish the nature of the transfer function:

- The capacity of the machine-tool and its auxiliary units.

- The kinematic and dynamic movements of the tool and workpiece including the force and speed conditions.

- The properties of the raw material and the finished workpiece, the cooling system, and the temperature of the workpiece in the processing area.

- The energy required for the process and the duration of processing.

This information is summarized in Table 2.1.

It is relatively easy to establish a transfer function of a system by mathematical modeling, relating inputs and results obtained from industrial practice. However, understanding the processes taking place inside the black box, that is, inside the structure of the system, involves solving complex situations. Understanding the system structure requires the multidisciplinary cooperation of process technologists with other specialists in physics, chemistry, and tribology. The use of concepts and methods specific to the science of tribology leads to the concept of the tribosystem.[3]

Table 2.1a. Functional External Elements of the System

GROUPS	SUBGROUPS	EXTERNAL ELEMENTS
Inputs {X}:	Useful Inputs:	- Machine-tools working parameters - Form of motion - Display of motion - Auxiliary processes (cooling & dressing) - Processing time - Raw workpiece material - Energy
	Disturbances:	- Heat from the environment - Vibrations - Materials from the environment
Outputs {Y}:	Useful Outputs:	- Processed workpiece
	Loss outputs:	- Heat - Noise & vibrations - Swarf - Tool wear

Table 2.1b. Main Useful Input Parameters

GROUP	SYMBOL	PARAMETER	UNIT
Machine-tool	a_d	Dressing depth	mm
	a_e	Depth of cut	mm
	f_d	Dressing infeed	mm
	n_s	Tool rotational speed	rpm
	n_w	Workpiece rotational speed	rpm
	v_s	Peripheral speed of the tool	m/s
	v_f	Delivery speed	mm/min
	v_w	Workpiece speed	m/min
	P_c	Processing power	W
	p_k	Process fluid pressure	bars
	Q'_k	Specific process fluid rate	l/mm · min
	t_s	Effective processing time	min
Workpiece	a_w	Stock removal of workpiece	mm
	R_a	Mean roughness of raw workpiece	μm

Table 2.1c. Type and Form of Relative Motion

Type of relative motion	- uni-directional
	- opposite-directional
	- oscillating
	- intermittent
	- impulsive
	- etc.
Form of relative motion	- sliding
	- rolling

2.2 STRUCTURE OF TRIBOMECHANICAL PROCESSING

The structure of abrasive processing technology may be investigated by correlating the system concept with knowledge of friction, wear, and lubrication procedures for abrasive processes.[4][5] On this basis, as shown in Tables 2.1b, and 2.1c, the components of the tribosystem structure emerge.

2.2.1 The Interconnecting Elements in the Tribosystem Structure

The interconnecting elements make up a *couple* with 2 or 3 parts (i.e., a 2- or 3-body system). The couple includes the abrasive tool and the workpiece. An abrasive process may also involve free abrasive particles in the contact area and a process fluid. The process also involves the active agents of the environment (Table 2.2). Group A defines typical elements, where

Eq. (2.2) $A = \{a_1, a_2, a_n\}$ and $2 \le n \le 5$

2.2.2 The Optimum Properties of the Tribosystem

After defining the nature of the interconnecting elements, the properties of those elements must be optimized. Optimum properties of each of the interconnecting elements are usually established after several successive attempts, on the basis of criteria that depend on the required process outputs. The criteria will concern such aspects as removal rate, size and shape tolerances, and surface integrity. The properties of Group B of the tribosystem are

Eq. (2.3) $B = \{B_{(a,I)}\}$

Table 2.2. Tribosystem Structural Elements

NOTATION	SYSTEM ELEMENTS	ELEMENT SHAPE AND STATE
a_1	Bonded abrasive tool	Wheel, ring, disc, cone, segment, point, etc. (solid state)
a_2	Workpiece	Various shapes (solid state)
a_3	Loose-abrasive	Powder, paste, spray, liquid suspension, etc.
a_4	Process fluid	Solution, emulsion, suspension (liquid state)
a_5	Materials from environment	Air, water, dust, fog, corrosive effluvia, accidental swarf, etc. (suspension)

The elements that make up the couple have properties that need to be described as follows:

- *Bonded abrasive tool:* A bonded-abrasive tool can be defined by shape, dimension, and specification. Specifications of a composite abrasive material include hardness and elasticity; this presumes a careful choice of the type of abrasive powder and grain size, as well as the nature of the bond composite. The abrasive specification must be consistent with the composition and structure of the material being processed.[6]

- *Workpiece:* Properties to be specified include:

  - properties of volume: shape, size, composition, and structure of the material, as well as its physico-mechanical properties (elasticity, hardness, toughness, density, and thermal properties, i.e., conductivity and expansion).[7]

  - surface properties: these properties have a decisive influence on the friction process. The most important surface properties are roughness, composition, and microhardness.

- *Loose abrasive:* The choice of powder, abrasive spray, or abrasive fluid specification must also be consistent with the material being processed.

- *Process fluid:* The process fluid should be chosen,[8] to ensure the formation of mixed or boundary lubrication modes.[9] This involves the building of an extremely thin layer on the interfacial processing area. The properties of the layer are affected by fluid viscosity, oiliness, and roughness of the interacting surfaces. The fluid layer facilitates lubrication and cooling. The fluid should also contribute to wear-protection of the tool, as well as protecting the workpiece from corrosion; it may also have detergent power for cleaning the working surfaces and removing swarf.

The presence of additives in a fluid layer allows selective tribochemical reactions aimed at increasing the ability to undertake the machining process and improving the finished quality of the processed surface (Table 2.3).

2.2.3 Interrelationships Between System Elements

Interrelationships between system elements determine tribological processes in the contact area. The combined effects of these processes (Fig. 2.2) strongly influence the machining process. The tribological processes are briefly presented as follows:

- *Contact processes:*[9] Gives rise to apparent and real areas of contact and pressure, elastic and plastic deformations, and physico-chemical contact processes of absorption, adsorption, and chemisorption, as well as corrosion.

- *Friction processes:*[10] Characterized by transformation of mechanical energy into other types of energy, such as thermal, acoustic, electrical, and chemical energy. Friction processes bring about physico-chemical changes to the materials at the contact surface, as well as the spreading of these changes inside the volume of the workpiece.

Table 2.3. Properties of the Structural Elements of the Tribosystem

STRUCTURAL ELEMENT	KIND OF ELEMENT	PROPERTIES OF ELEMENT
Bonded abrasive tool	Shape Dimension Specification	Abrasive composite material: type of abrasive powder; grain sizes and concentration; composite grade and hardness; bond type.
Loose abrasive	Specification State of aggregation	Type of microabrasive dispersal state; grain size and carrier
Workpiece	Volume properties	Shape, dimensions and stock removal, composition and structure, and physico-mechanical properties.
	Surface properties	Roughness, composition and micro-hardness. Continuous, discontinuous.
Process fluid	Basic fluid additives	Air, water, mineral oil, synthetic oil, etc., Friction modifiers, E.P., detergent-dispersing, anticorrosion, antifoaming, antiwear, etc.
Materials from environment	Dust or fog Corrosive gases Accidental swarf	Corrosivity, chemical activity, abrasivity, concentration, etc.

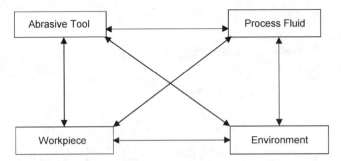

Figure 2.2. Interrelationships between the structural elements of a bonded abrasive tribosystem.

- *Tool wear processes:*[11] Includes wear of the abrasive tool. Wear processes modify the shape, dimensions, and structure of the active abrasive layer. With a correct tool specification, wear of the abrasive composite layer can lead to self-sharpening and dressing processes.[12]

- *Workpiece wear processes:* Abrasive processes acting on the workpiece material represent the essence of the machining process. These processes consist of the simultaneous action of three microprocesses:[13] chip formation, ploughing and sliding/rubbing.[14] The proportions and intensity of each of these three microprocesses describe the specific technological character of a particular abrasive process.

- *Lubrication processes:* The tool/workpiece couple is influenced by the machining conditions which may be dry or wet with a poor or plentiful supply of lubricant. The most realistic model is of boundary lubrication with multiple penetration points.[11] Lubrication influences various physico-chemical processes of contact (Table 2.4):

Eq. (2.4) $C = \{C_{(a,i;\,a,j)}\}$

The results of the tribological processes are reflected in the workpiece dimensional precision and surface quality of the finished part. During the abrasive process, the particles resulting from the abrasive tool wear, and the microchips from the machining process must be removed as swarf. Swarf represents a loss output.

Table 2.4. Interrelations of Structural Elements

GROUP OF PROCESSES	TYPE OF PROCESSES	DETAILS	DOMAIN
Contact	Contact area	Apparent and real areas	Mechanics
	Contact pressure	Apparent and real pressures	
	Deformations	Elastic and plastic deformation	Physics
	Absorption	Physico-chemical contact processes	Physico-chemical
	Adsorption		
	Chemisorption		Chemistry
	Chemical reaction	Chemical processes	
	Tribocorrosion	Tribochemical process	Tribo-chemistry
Friction	Abrasive grain/work material	Tribological abrasive processes with different effects, d.i. tribothermical, acoustical, electrical and chemical	Tribology Abrasive processing
	Bond/work material		
	Process fluid/couple tool-workpiece		
Wear	Abrasive wear (prevailing type)	Microprocesses of the abrasive wear type (abrasive processing) i.e., chip formation and ploughing (cutting) and sliding (rubbing or burnishing)	Tribology Abrasive processing
	Adhesive wear		
	Corrosive wear Surface fatigue wear		

Table 2.4. *(Cont'd.)*

GROUP OF PROCESSES	TYPE OF PROCESSES	DETAILS	DOMAIN
Lubrication	Wet process Dry process	The couple tool/workpiece is working in boundary lubrication with multiple penetration model; various kinds of tribochemical reactions	Tribology

The efficiency of an abrasive process is defined by material volume removed (V_w), removal rate (Q_w), abrasive tool wear volume (V_s), normal and tangential forces (F_n and F_t), stiffness of the tool/workpiece contact (K_m), process energy (e), and maximum surface roughness (R_t, R_a, or R_z) (Table 2.5).

The proper execution of a process may require careful consideration of the process thermal conditions,[15] in particular, grinding temperature (θ). The maximum temperature is of the greatest importance for the structure of the workpiece material and the abrasive tool composite layer. An important aspect of controlling thermal conditions is the efficient supply of process fluid during the machining process.

In an effort to unify the terminology, definitions, and symbols of abrasive machining processes, the International Institution for Production Engineering Research (CIRP) has made specific recommendations.[16] In this text, equivalent tribological parameters are also given where appropriate. Tribological parameters were established by the Organization for Economic Cooperation and Development (OECD).[17]

2.2.4 The Total Structural System

The total system (S) representing the structure of different abrasive processing models includes all three components of the tribosystem structure. The relationship has the general form of the following equation.

Table 2.5. Technological Parameters for Abrasive Processing

SYMBOL	TECHNOLOGICAL PARAMETERS	UNIT
d_s	Abrasive tool diameter	mm
b_s	Abrasive tool width	mm
V_a	Total abrasive tool active layer	mm^3
V_s	Tool volume wear	mm^3
V_w	Total removed volume from the workpiece	mm^3
Q_w	Removal rate	mm^3/s
F_n	Normal force	N
F_t	Tangential force	N
e	Processing energy	J
θ	Grinding temperature	°C
K_m	Tool/workpiece contact stiffness tool/workpiece	$N/\mu m$
t_s	Actual processing time	s
R_t	Maximum surface roughness	μm
C_s	Abrasive tool cost (unit price)	c.u./pcs.
C_m	Machine and labor costs (per unit time)	c.u./h

Eq. (2.5) $S = \{A, B, C\}$

Up to now, a total system relationship has been established only for particular cases and for a limited number of parameters. The establishment of general relationships that reflect the structure of an abrasive processing system remains an aim of much research. The definition of relationships for abrasive processes has the potential to explain the effect of various parameters on the process.[3] For example, in succeeding chapters we seek to establish a relationship for friction coefficient:

Eq. (2.6) $\mu = f \{X; S\}$

Another example is material removal rate from a workpiece. Removal rate has the same definition as volume wear rate. There is a simple relationship for removal rate:

Eq. (2.7) $Q_w = f' \{X; S\}$

In the present work, we use the established terminology of CIRP for these two tribological dimensions. The term grinding force ratio (μ) (Table 2.6)[6][14] will replace the term friction coefficient. The two parameters are identical from an analytical and dimensional point of view. The volume wear rate of the workpiece element is replaced in the terminology by removal rate (Q_w), both rates being obtained by relating the removed volume to the processing time.

2.3 THE THREE TRIBOSYSTEMS IN ABRASIVE MACHINING

Abrasive processes fall into one of three groups. These are:

- *Bonded:* Abrasive processes with two main elements.
- *Loose:* Abrasive processes with three elements.
- *Dressing:* For abrasive tools.

Other differences between processing technologies based on the various forms of the abrasive tool and workpiece shape, as well as on the kinematics provided by the machine-tool, are irrelevant in classifying the tribosystem. The three tribosystem groups are summarized below.

2.3.1 Bonded Abrasive Processes

An abrasive tool is used for operations such as grinding, cut-off, honing, and sometimes polishing. The main constituent is an abrasive layer with a composite structure. The abrasive layer is held in a tool holder or attached to a backing. The composite includes the grit within a matrix. The layer is made up of hard abrasive grit and bond material to retain the grit

Table 2.6. Basic Parameters for Process Optimization

SYMBOL	BASIC PARAMETER	UNITS	RELATIONSHIP NO.
Q'_s	Specific wear rate	$mm^3/s \cdot mm$	(2.8)
Q'_w	Specific removal rate	$mm^3/s \cdot mm$	(2.9)
$Q'_{w,n}$	Specific heat value	J/mm^2	(2.10)
G	Grinding ratio	—	(2.11)
μ	Grinding force ratio	—	(2.12)
q	Speed ratio	—	(2.13)
e_s	Specific energy	J/mm^3	(2.14)
h_e	Equivalent chip thickness	μm	(2.15)
F'_t	Specific tangential force	N/mm	(2.16)
P'_c	Specific grinding power	W/mm^3	(2.17)
$R_{a,f}$	Mean roughness of finished work	μm	measured
T	Abrasive tool life	h	(2.18)
c_s	Specific processing costs	$c.u./mm^3$	(2.19)

within a tridimensional structure. The grit and the bond have very different physical-mechanical properties. The hardnesses are about one order of magnitude different. The main contact between the tool surface and the workpiece takes place on the hard and sharp edges of the grit. The hardness of the grit is sufficient to plastically deform the workpiece material,[18] as required in ploughing and cutting the workpiece surface. The deformations of the abrasive particles are negligible in comparison with deflections of the bond.[19] The elastic properties of the tool itself are favorable for finishing and super-finishing actions, and are achieved by an appropriate choice of the type, hardness, and structure of the bond. The technology of manufacturing abrasive composites and their application in the processing of different materials represent the result of extensive studies on the process of abrasive wear.[20] The emergence of synthetic composites can be justified by the very high values of the abrasive-wear intensities achieved, in comparison with the other three basic wear mechanisms.[11] In abrasive machining, the aim is to achieve a wear process on the workpiece while minimizing wear on the tool.[17] Irrespective of whether abrasive wear is considered to be a benefit or a disadvantage, the lessons from tribology can be usefully applied to develop and utilize abrasive tools.

In the analysis of a tribosystem for bonded abrasive processing (Fig. 2.3), it is important not to overlook the process fluid, which has the functions of lubricating, cooling, and cleaning.

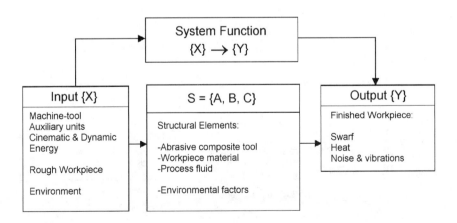

Figure 2.3. Tribosystem for bonded abrasive processing.

The process fluid is fed by means of a fluid delivery system. Separation equipment is required to continuously remove swarf from the fluid. An inadequate fluid delivery and extraction system can be very damaging to the workpiece quality and to the environment. Damage can be caused by the introduction of dust, fog, and corrosive agents into the atmosphere. Damage can be caused to the machine and to the health of persons working nearby.

2.3.2 Loose Abrasive Processes

Processes such as lapping or polishing employ loose abrasive technology and belong to a family of fine-finishing processes. The tribological couple consists of a workpiece and a supporting element, i.e., a lap or a polishing pad, with a low or medium hardness to pressurize the loose abrasive grains against the workpiece. The supporting abrasive element may be supplied by means of a fluid suspension to transport the grits. The fluid suspension also lubricates and cools the machining process. In fine-finishing operations, swarf must be removed by washing debris from the processed surface. The fluid used for the processing and washing are often evacuated from the system, together with the swarf, without recycling (Fig. 2.4).

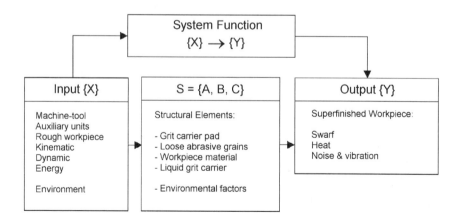

Figure 2.4. Tribosystem for loose-abrasive processing.

2.3.3 Dressing Processes

Dressing technology can be differentiated from other types of couples.[21] The workpiece is an abrasive tool with a friable and abrasive composite structure. Dressing is a process which includes *truing* or *profiling* the abrasive tool. The abrasive tool is made to conform to the shape and texture required for the subsequent grinding process. This tool shape is transferred later or simultaneously to the workpieces. The dressing tool may consist of a steel element and a hard or extra hard active *cutting* part, with a hardness value close to, but higher than, that of the processed piece (i.e., a grinding wheel). The contact pressures of the dresser on the wheel exceed the Hertzian level.[9] Consequently, the removal process is carried out through crushing and cutting mechanisms which act upon the grit and the bond structure of the abrasive layer. At the same time, an abrasive wear process is carried out on the hard insert of the dressing tool. In most cases, the dressing process takes place in the presence of the process fluid in order to reduce tool wear and, only secondarily, to improve the dressing action on the grinding wheel. As far as the functions of the cooling circuit and the environment are concerned, these conditions are the same as those in the grinding process. Grinding may be preceded by the dressing operation or, less usually, may be carried out at the same time, where it is important to continuously restore the cutting topography of the wheel (Fig. 2.5).[22]

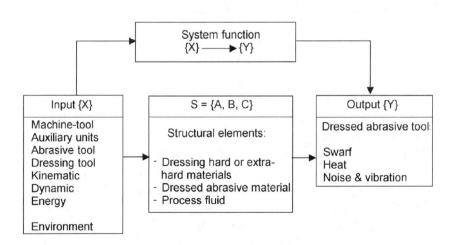

Figure 2.5. Tribosystem for dressing processing.

2.3.4 Basic Parameters of the Tribosystem Structure

Basic parameters of the tribosystem are employed for investigation and interpretation of the internal processes, which take place within the black box.

Parameters established experimentally during a working process are often specific to a particular operation, machine, or abrasive tool. Modeling, simulation, and optimization of abrasive machining are made possible by experimental testing, and then comparing results obtained with reference data, or with reference to basic physical principles. Basic parameters for this activity have been established at an international level.[16][25] Widely accepted parameters make it possible to draw conclusions concerning process efficiency. Because conclusions of an economic nature are often decisive in industry, relationships for working out costs of an abrasive operation are briefly presented (Table 2.7). However, basic parameters are of more concern, such as specific removal rate (Q'_w) and the grinding ratio (G), which are important for evaluating costs in relation to quality requirements.[23]

2.4 MODELING TRIBOSYSTEMS OF ABRASIVE PROCESSES

The systematic analysis of the ensemble of abrasive processing elements is necessary when the properties of the system result from interactions of its constituent elements rather than from direct summing of their primary functions. The modeling and simulating of the constituent processes will necessarily carry out the systematic analysis. Abrasive processing systems are characterized by the fact that a great number of external and internal elements take part. The simultaneous control of these elements during the development of the process is practically impossible. The various subprocesses have an unsteady, transitory character. The volume and surface properties of the elements, as well as their interrelationships, are subject to continuous changes rendering the modeling and simulation of the physical-chemical process complex. All this may lead to poor reproducibility of experimental results.

Table 2.7. Some Relationships used for Abrasive Process Optimization

Specific wear rate	$Q'_s = V_s/b_s \cdot t_s$	(2.8)
Specific removal rate	$Q'_w = V_w/b_s \cdot t_s$	(2.9)
Grinding ratio	$G = V_w/V_s$	(2.10)
Grinding force ratio	$\mu = F'_t/F'_n$	(2.11)
Speed ratio	$q = v_s/v_w$	(2.12)
Specific energy	$e_s = e/V_w$	(2.13)
Equivalent chip thickness	$h_e = Q'_w/v_s$	(2.14)
Specific tangential force	$F'_t = F_t / b_s$	(2.15)
Specific grinding power	$P'_c = v_c \cdot F'_t$	(2.16)
Abrasive tool life	$T = V_a \cdot G/Q'_w$	(2.17)
Specific processing cost	$c_s = [C_s/V_a \cdot G] + [C_m/Q'_w \cdot b_s]$	(2.18)

Tribological developments can have a great influence on the development of machining processes. When there is little time or money, modeling of a tribosystem may be achieved empirically[24] by using experimental results obtained within the laboratory for application to the industrial situation. A detailed study for CIRP[25] pointed out that most models in the field of abrasive processing are empirical. Consequently, there is a restricted range of applicability. Some lines of research are concerned, for example, with developing a new generation of abrasive tools or with elaborating on the processing technology for new materials. In such cases, special attention needs to be paid to the tribological principles that apply in machining, in order to establish new physical models of the processes (Fig. 2.6). Within a research laboratory, special instruments will be employed to measure appropriate parameters.[26][27] At the semi-industrial level, comparatively small machine tools with medium power, good precision, and high automation levels may be used.[28] The machines used must be appropriate from the kinematic viewpoint. Additional equipment may be required for investigation of the initial surface and structure of the materials employed, as well as the changes undergone during the machining process.

Taking these possibilities into account, some partial physical models can be designed, which influence the following:

- Abrasive tool surface topography.

- Uncut chip thickness.

- Size of wear particles.

- Magnitude of the cutting forces.

- Energy requirements of the process.

- Integrity and microgeometry of the couple element surfaces, etc.

The process of modeling and optimization in an industrial context will possibly consist of the following stages:

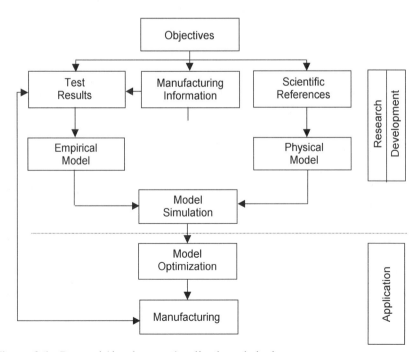

Figure 2.6. Research/development/application chain in many processes.

Models. A model is applied, consisting of an abstract representation of a process. The model is intended to establish a link between cause and effect, connecting the inputs and outputs of the tribosystem. The model is a representation of the real process aimed at foreseeing the effects of changes in the properties of the tribosystem on the results of the machining process.

Simulations. Simulations may be employed to predict a relationship between inputs and outputs. Simulation is an imitation of the process for short periods of time. It is obvious that modeling and simulation operations will be carried out only for processes having an almost steady character.

Long-term Tests. Finally, process optimization requires long-term tests, using values of set-up parameters, previously indicated by modeling or simulation. The conclusions from work carried out within a cycle of research-development-applications, as well as from the models established for a process, will prove valuable for calculating the basic parameters in Table 2.6, using the mathematical relationships in Table 2.7, and comparing these results to values previously determined as reference standards.

The number of tests should be carefully established beforehand, using a scientific method of planning the experimental work, e.g., Taguchi-Type,[30] in order to minimize testing costs. On the other hand, when there is enough technological information stored in the computer, as well as a program adequate to the investigated operations, optimization can be carried out more quickly with the help of calculation models.

2.4.1 The Influence of the Workmaterials in Modeling

In accordance with elasto-plastic properties and behavior determined by scratching the surface of the workpiece,[31] materials can be divided into two main categories: materials with either ductile or brittle structures. The two categories cannot be strictly separated, due to friction. The ductility of processed materials changes with a rise in temperature. Since processing temperatures can be very high, temperature becomes an important parameter and is discussed at length in Ch. 6, "Thermal Design of Processes."

Friction represents a large proportion of the energy used in an abrasive process[6] and makes a decisive contribution to the values of *flash temperature*.[32] In this way, friction influences the properties of the finished workpiece material. The wear properties of the processed material, which also depend on the abrasive tool and the process fluid type, directly influence machinability.[33] Therefore, we can directly apply the science of tribology to establish physical models in the field of abrasive processing. The numerous mathematical relationships found in tribological references can be used to characterize behavior of various industrial materials in the wear process.[11][34] A proportionality can be established between the specific removal rate of the workpiece material (Q'_w) and important physico-mechanical properties of the type,

Eq. (2.8) $Q'_w \approx C \cdot F'^{c1}_n \cdot E^{c2}/(Hv^{c3} \cdot K_{IC}^{c4})$

where C is the coefficient, c_1 through c_4 are the exponents, F'_n is the specific normal force, E is the modulus of elasticity, Hv is the Vickers hardness, and K_{IC} is the fracture toughness.

For the tribological characterization of the abrasive tool material, the property of specific resistance to wear is used, which is the inverse of the specific wear rate:[11]

Eq. (2.9) $R'_u = 1/Q'_s$

In place of the basic parameter of the grinding ratio we may utilize its inverse, the wear ratio:[16]

Eq. (2.10) $\Phi = 1/G$

In order to characterize the grit material, in addition to the empirical property of friability,[35] the *brittleness index*[36] can be calculated using the relation

Eq. (2.11) $B = [K_{IC} / Hv]^n$

where the exponent *n* equals 1, ... 2.

A model of an abrasive machining process must take into account the existence of two specific mechanisms, of which one will be dominant. These mechanisms are:

- *Brittle processing (fracture mode):* Characterized by the generation of structural cracks, lower residual stresses, a low value of grinding energy, and short chips.

- *Ductile processing (plastic flow):* Characterized by the shearing and sliding of the material layers, including changes of the processed surface composition and structure (for example, generation of an amorphous structure), tensile or compressive stresses in the sublayers, high values of specific energy, as well as long and curled forms of resulting chips.

In recent years, manufacturers and researchers have shown great interest in extending the ductile processing field for grinding and fine-finishing operations to very brittle materials. Considerable success has been achieved in avoiding the generation of surface and structural faults,[37] which can compromise the integrity and strength of machined parts.

2.4.2 The Influence of the Shape and Size of the Contact Surface

Modeling the contact between tool and workpiece for different abrasive machining processes starts with classification of the friction pairs.[9][38] According to this classification, the pairs (couples) are divided into four categories (Table 2.8). Contact modeling implies taking into account the form and structure of the tool. It is necessary to consider the abrasive grains size and concentration of the abrasive grains to represent a surface microgeometry comparable with the real tool in contact with the workpiece. Other characteristics of the system structure, i.e., the tool bond, the material of the test workpiece, and the process fluid, will be kept identical to the real pair.

2.4.3 The Influence and Measurement of Cutting Forces

The forces applied to the friction pair, or those forces that result from machining, may be measured and recorded continuously. The normal force, i.e., the load (F_n), may be recalculated as a real pressure, using the values of the grit surface density, as well as the nominal contact surface, both for the abrasive tool and for the test workpiece. The real contact pressure can be estimated using the test set-up (tribometer); the real contact pressure must be close to the same value as that of the effective processing machine, although the two values of normal forces may be different. The nature of application of the normal force in the test equipment must be identical to that of the modeled machine-tool (Table 2.1c).

Table 2.8. Modeling the Grinding Couple Contact Type

GRINDING COUPLE CONTACT TYPE	ABRASIVE MACHINING (EXAMPLES)	THEORETICAL AND EXPERIMENTAL MODELING OF GRINDING COUPLE
Point contact	Natural diamond single grain dressing	Hard pin with round or tapered end on rotating cylinder; Hard pin with round or taper end on rotating or translating flats.
Line contact	Cylindrical grinding; Centerless grinding; Surface grinding	Rotating abrasive cylinder on parallel rotating cylinder or flat surface.
Flat contact area	Form surface grinding; Tool and cutter grinding; Double disk grinding; Lapping; Polishing	Flat end pin on abrasive rotating flat; Flat end pin on abrasive translating flat; Abrasive translating flat-on-flat.
Curved contact area	Surface deep grinding; Creep-feed grinding; Honing; Polishing	Abrasive rotating cylinder on flat; Abrasive curved end pin on rotating cylinder.

The values of the friction force measured on the test machine will be considered as representative of the tangential (cutting) force values (F_t), allowing calculation of the basic parameter (μ) in Table 2.6. By watching the changes in force ratio, information can be obtained concerning the effect of wear and dressing on the real tool contact surface. Wear can result from the following processes:

- Blunting of sharp ends and edges of the abrasive grains.
- Grit breakage and pulling out.
- Transfer of the swarf to the active surface of the tool.
- Evidence of self-sharpening and spontaneous dressing effects.
- Carrying out dressing operations and its results.

Because of the small size of the friction couple, a tribometer provides an opportunity to perform some quick and easy tests and measurements of the couple wear. It lets investigators test the active and processed surfaces, the initial running-in of the couple for conformity of the tool and test piece surfaces, filtration of the process fluid, and necessary changes of fluid-type, etc.

Another important mechanical parameter is the peripheral speed of the abrasive tool (v_c), which will be maintained close to the real values in a machining operation. The drive motor should provide continuous variation of the rotational speed, as well as a slip-free transmission system. By attaching a microprocessor to the tribometer the variation of basic parameters can be monitored, which may be automatically calculated and screened.

It should be noted that many tribometers fail to give satisfactory results due to the lack of running accuracy of the machine and the intrusion of dynamic forces. It is important that the dynamic conditions are similar to those of the real machine. This problem becomes increasingly critical at high machining speeds.

The tribometer may be an ideal means for studying the effects on the abrasive machining processes of some auxiliary units, such as LASER-generation, HF-currents, electric or magnetic fields, ultrasounds, and active chemical agents, etc.

2.5 CONCLUSIONS

The following conclusions can be drawn regarding the analysis and testing of tribosystems of abrasive machining processes:

- The structure of the abrasive machining process is that of a tribosystem.

- The two simultaneously developing abrasive processes, i.e., the abrasive machining process and the abrasive wear process, have been studied for a long time in various domains of the technical sciences, but they are closely interconnected.[39]

- The modeling of abrasive machining processes should comply with tribological concepts. In experimental research, the friction force, i.e., the tangential cutting force, should be continuously monitored.[40][41]

REFERENCES

1. Ropohl, G., Systemtechnik – Grundlagen und Anwendung, Hanser Vlg., Munchen (In German) (1975)

2. Faurre, P., and Depeyrot, M., Elements of System Theory, Amsterdam, Holland (1977)

3. Czichos, H., Tribology: A Systems Approach to the Science and Technology of Friction, Lubrication, and Wear, Elsevier Sci. Publ. Co., Amsterdam (1978)

4. Czichos, H., and Habik, K. H., Tribologie Handbuch: Reibung und Verschleiss. Prueftechnik, Werkstoffe und Konstruktionselemente, Vieweg Vlg., Wiesbaden, (In German) (1992)

5. Zum Gahr, K. H., Microstructure and Wear of Materials, Elsevier, Amsterdam (1987)

6. Ott, H. W., Grundlagen der Schleiftechnik, Schleiftechnik Co., Pfaeffikon-ZH, (In German) (1993)

7. Toenshoff, H. K., Peddinghaus, J., and Wobker, H. G., Tribologische Verhaeltnisse zwischen Schleifscheibe und Werkstueck, *8th Int. Colloquium Tribology 2000,* Esslingen, Germany (In German) (1992)

8. Bowden, F. P., and Tabor, D., The Friction and Lubrication of Solids, Clarendon Press, Oxford (1964)

9. Pavelescu, D., Mushat, M., and Tudor, V., Tribology, Editura Dedactica & Pedagogica, Bucharest (In Romanian) (1977)

10. Nakayama, K., and Takagi, J., et al., Sharpness Evaluation of Grinding Wheel Face by Coefficient of Friction, *Proc. 4th Int. Conf. Prod. Eng.*, Tokyo (1980)

11. Rabinowicz, E., *Friction and Wear of Materials*, J. Wiley & Sons, NY (1994)

12. Marinescu, I. D., and Dimitrov, B., et al., Tribological Aspects in Utilization of Superabrasives in Metal Cutting, *Proc. 4th Europ. Tribology Congr.*, Ecully, France (1985)

13. Salmon, S. C., *Modern Grinding Process Technology*, McGraw-Hill, NY (1992)

14. Blaendel, K. L., Taylor, J. S., and Piscotty, M. A., *Summary Session Precision Grinding of Brittle Materials, Spring Topical Meeting of A. S. P. E.*, Annapolis, MD (1996)

15. Snoeys, R., Maris, M., and Peters, J., Thermally Induced Damage in Grinding, *Annals of the CIRP*, 27/1:206–216 (1978)

16. XXX, C.I.R.P.—Unified Terminology, Part III, Cutting, Grinding and Electromachining, *Edition CIRP*, Paris (1986)

17. X X X, Glossary of Terms and Definitions in the Field of Friction, Wear and Lubrication—Tribology OECD (1969)

18. Nakayama, K., Elastic Deformation of Contact Zone in Grinding, *Bull. Jpn. Soc. Precision Eng.*, 5(4):93–98 (1972)

19. Kato, K., Micromechanisms of Wear—Wear Modes, *Wear*, 153:277–295 (1992)

20. Childs, H. C., Fine Friction Cutting: A Useful Wear Process, *Tribology Int.*, 16(2):67–83 (1983)

21. Minke, E., Verschleissmodell zur Auslegung des Abrichtprozesses mit geometrisch-bestimmten Diamantschneiden, VDI-Z., 132(2):56–61 (In German) (1990)

22. Malkin, S., *Grinding Technology; Theory and Applications of Machining with Abrasives*, J. Wiley & Sons, NY (1989)

23. Marinescu, I. D., and Dimitrov, B., et al., Some Aspects Concerning Wear and Tool Life of Diamond Wheels, *Annals of the CIRP*, 23(1):251–254 (1983)

24. Profos, P., Modelbildung und ihre Bedeutung in der Regelungstechnik, VDI-Berichte 276:5–12 (In German) (1977)

25. Toenshoff, H. K., Peters, J., Inasaki, I., and Paul, T., Modeling and Simulation of Grinding Processes, *Annals of the CIRP,* 41(2):677–688 (1992)

26. Eyre, T. S., Development of Tribo-Test Methods, *Proc. 5th Int. Congr. Trib.,* 5:142–149, Espoo, Finland (1989)

27. Habig, K. H., Principles of Wear Testing, *Proc. Jpn. Int. Trib. Conf.,* II:1213–1218 (1990)

28. Patterson, H. B., The Role of Abrasive Wear in Precision Grinding, *Proc. Int. Grinding Conf.,* Philadelphia, PA (1986)

29. Brinksmeier, E., Inasaki, I., Toenshoff, H. K., and Peddinghaus, J., *CIRP Cooperative Work in Grinding,* pp. 1–18, IFW-Berlin, Germany (1993)

30. Steinborn, S., and Hartelt, M., Taguchi-Methoden – das japanische Geheimnis? Qualitaets-Z., 1:29–33 (in German) (1994)

31. Lamy, B., Effect of Brittleness Index and Sliding Speed on Morphology of Surface Scratching in Abrasive or Erosive Process, *Tribology Int.,* 17(1):35–38 (1984)

32. Bloch, H., The Flash Temperature Concept, *Wear,* 6(6):483–484 (1963)

33. Salje, E., Mushardt, H., and Damlos, H., Verschleisskenngroessen und Bedeutung zur Beschreibung und Bewertung von Schleifprozessen, Jahrbuch: Schleifen, Hohnen, Laeppen und Polieren, 50. Ausgabe, pp. 110–117 (In German) (1981)

34. Hutchings, I. M., Tribology: Friction and Wear of Engineering Materials, Ed. Arnold Publ., London (1992)

35. Belling, N. G., and Bialy, L., The Friatester – 10 years later, de Beers Ind. Diamond Information, p. L 33

36. Novikov, N. N., Dub, S. N., and Malnev, V. I., Microhardness and Fracture Toughness of Cubic Boron Nitride Single Crystals, *Sverhtverdye Materially,* 5(5):16–20 (In Russian) (1983)

37. Bifano, T. G., Dow, T. A., and Scattergood, R. O., Ductile-Regime Grinding of Brittle Materials, *Proc. Int. Congr. for Ultraprecision Technol.,* Aachen, Germany (1988)

38. Dimitrov, B., and Aelenej, M., Tribological Aspects of Testing and Optimization of Grinding Wheels, *Proc. Int. Symp.,* Tools, Gabrovo, (Bulgaria), (In Russian) (1982)

39. Zum Gahr, K. H., and Mewes, D., Werkstoffabtrag durch Furchungsverschleiss, *Metall.,* 37(12):1212–1217 (In German) (1983)

40. Flom, D. G., Tribology Modeling, *Proc. Workshop, Proc. Deformation Models to Predict Wear Behavior,* Columbia University, NY (1986)

41. Santner, E., Rechnergestuetzte Prueftechnik in der Tribologie Materialpruefung, 2(1/2):18–24 (1990)

3

Kinematic Models of Abrasive Contacts

3.1 INTRODUCTION

A *kinematic analysis* defines the removal rate and the penetration of the abrasive grains into the workpiece. Commencing with the speeds and motions of the tool and workpiece, an analysis proceeds to the physical consequences of the distribution of the cutting edges.

This chapter provides measures of the size, time of contact, and scale of the minute interactions which take place in the abrasive process. Penetration depth, the length of contact between an abrasive grain and a workpiece, the implications of grain size and spacing, grain speed and grain trajectory for grain forces, grain wear, and workpiece roughness will be considered. Grain shape and the effect of wheel dressing will also be covered. The scale of the grain-workpiece interactions form the basis for further study of all aspects of grinding behavior and of the tribology of the process.

Figure 3.1 shows an abrasive grain taking a circular path through the surface of a workpiece. The grain must be harder than the workpiece if it is to be an effective cutting tool. It is, therefore, assumed that the grain retains its shape and size during an interaction. Clearly, the material in the path of the grain must be displaced by the abrasive grain.

It does not matter whether the displaced material is simply ploughed to the sides of the groove or whether the displaced material is sheared out of the surface in some form of chip. The material in the path of the grain must go somewhere else. Therefore, the material displaced is referred to as the 'uncut chip' even though a chip may or may not be formed in a particular interaction. Examples of deformed chips are illustrated in Ch. 6, "Thermal Design of Processes."

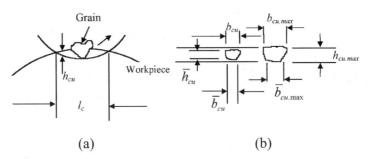

(a)   (b)

Figure 3.1. *(a)* The path swept by a hard grain. *(b)* The max and mean cross-sectional dimensions of the swept path. The material before being displaced is known as the "uncut chip" or sometimes as the "undeformed chip."

In Fig. 3.1, the uncut chip has a thickness, h_{cu}, which varies from 0 to $h_{cu.max}$. Similarly, the width of the chip, b_{cu}, varies from 0 to $b_{cu.max}$. The mean length of the uncut chips is assumed to be equal to the length of the contact zone, l_c. The mean volume of the uncut chip is given by the product of the mean chip thickness, the mean width, and the mean chip length.

Eq. (3.1) $$\overline{V}_{cu} = \overline{h}_{cu} \cdot \overline{b}_{cu} \cdot l_c$$

This is the mean volume that, when multiplied by the number of grain interactions, must conform with the volume of material removed from the workpiece. The total volume of the chips removed is equal to the volume of material machined from the surface. Even though some uncut chips do not lead to actual chips, other uncut chips lead to actual chips which are larger than the mean.

The mean size of the uncut chips must be consistent with the average size of the material removed by each grain. This is a volume constraint and is used later to define the mean uncut chip thickness.

Some examples of abrasive processes are reviewed in the following sections, and the basic process variables are defined in terms of the the machine control variables. The size and shape of the material removed by the grains and implications of process variability are considered. For a free abrasive process, the grain distribution is undefined, discussion commences with examples of fixed abrasive machining where it is possible to make reasonable assumptions concerning the grain sizes and locations in the abrasive tool.

3.1.1 Machine Control Variables

In fixed abrasive machining, the machine control variables are the tool or wheelspeed, v_s; the workspeed, v_w; the feedrate, v_f; and the depth of cut, a_e. The precise definition of these terms depends on the type of machine and the particular type of operation. Some examples are illustrated in Fig. 3.2.

3.1.2 Workpiece Material Removal

In surface grinding, the total volume of material removed from the workpiece in one pass is illustrated in Fig. 3.3.

Eq. (3.2) $$\overline{V}_w = b_w \cdot a_e \cdot L_w$$

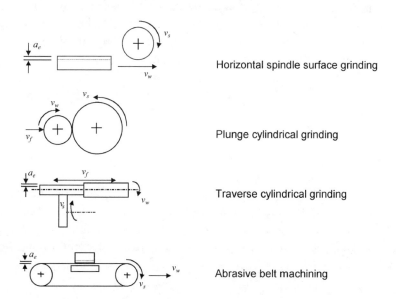

Horizontal spindle surface grinding

Plunge cylindrical grinding

Traverse cylindrical grinding

Abrasive belt machining

Figure 3.2. Examples of machine control variables in abrasive machining processes.

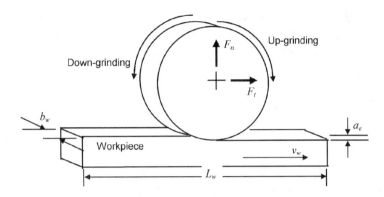

Figure 3.3. The difference between up-grinding and down-grinding.

The volume material removal rate from the workpiece is, therefore:

Eq. (3.3) $\qquad Q_w = b_w \cdot a_e \cdot v_w$

This figure also illustrates the difference between *up-grinding* and *down-grinding*. Down-grinding is where the surface speed of the wheel is in the same direction as the surface speed of the workpiece. Up-grinding is where the directions are opposite.

In surface grinding, the grinding wheel is fed downwards a set distance, a, known as the set depth of cut. Due to deflections of the various components of the machine-wheel-workpiece system, the real depth of material removed is less than the set depth of cut.

The real or "effective" depth of cut, a_e, is the actual depth removed and is equal to the set depth of cut less the deflections and the wheel wear during the pass, plus the thermal expansion. In general,

Eq. (3.4) $\qquad a_e = a - \delta - a_{sw} + a_t$

where a_e is the real depth of cut, a is the set depth of cut, δ is the system deflection, a_{sw} is the depth of wheel wear during the pass, and a_t is the thermal expansion of the wheel and workpiece. The real depth of cut can be accurately determined by measuring the workpiece before and after grinding.

3.1.3 Volume Tool Wear

The total volume of material removed from the grinding wheel corresponding to the wear depth, a_{sw}, is

Eq. (3.5) $V_s = b_w \cdot a_{sw} \cdot \pi \cdot d_s$

where d_s is the diameter of the grinding wheel. The wear depth on the grinding wheel can be measured by using part of the wheel surface for grinding so that the wear of the part of the wheel in contact forms a step. Subsequent to the grinding operation, a razor blade can be plunged into the grinding wheel to replicate the step onto the edge of the blade. The step on the blade can then be accurately measured.

Wear of the tool has a number of implications for the process which are almost always deleterious to control. Wear can lead to one or more of the following problems:

- An error in the depth of cut.
- An error in the finished size of the workpiece.
- Bluntness of the tool and increased forces leading to oversized workpieces.
- Resharpening of the tool and reduced forces leading to undersize workpieces.
- An increase or reduction in surface roughness leading to increased variability.
- Loss of form.
- The need for redressing leading to further loss of the wheel surface and dressing tool wear.
- A change in the number of active cutting edges on the tool.

This chapter establishes the kinematic removal conditions which relate to the variability of the number and shape of the cutting edges.

3.1.4 Grinding Ratio

The grinding ratio, G, is a measure of how much material is removed from the workpiece per unit volume of wheel wear.

Eq. (3.6) $$G = \frac{V_w}{V_s}$$

A value of a grinding ratio of $G = 1$ is very low and implies that the abrasive is not hard enough for the task. A grinding ratio of $G = 1,000$ is high and may sometimes correspond to a wheel that is too hard for the task. For example, if the abrasive tool wears by a slow glazing of the grains, the machining forces will rise and may lead to various problems such as poor size-holding, vibrations, and poor surface texture.

The subject of G-ratio is discussed further in the chapter on abrasive materials. Here, it is sufficient to note the importance of G-ratio as a measure of relative wear rates of the workpiece and tool.

3.2 BASIC ANALYSIS OF SURFACE GRINDING

3.2.1 Micromilling Analogy

The micromilling analogy of grinding was first presented independently in publications by Alden[1] and Guest[2] dating from 1914. Micromilling analysis is the starting point for most kinematic analyses.

Many variations have been presented on this theme and used as a basis for the interpretation of grinding behavior. A review undertaken by Tönshoff covers a range of typical parameters and predicted grinding behavior.[3] Tönshoff concludes that the use of kinematic models allows grinding conditions to be reproducible in different applications. The implications of kinematic conditions of material removal will quickly become clear as some examples are discussed.

Both Alden and Guest make the point that if the grain depth of cut is too large, the grinding wheel will wear rapidly. If it is too small, the wheel will glaze. The grain depth of cut should, therefore, lie within a range which allows the abrasive tool to cut efficiently.

The micromilling analysis is an analogy to the milling process, although unlike the milling cutter, the locations of the cutting edges in an abrasive tool are distributed in a random manner. The shapes of the cutting edges also vary randomly.

Abrasive machining is a stochastic process, in which the spacing of the grains, the depth of the grains below the surface of the tool, and the shape of cutting edges all vary randomly.

Considering the whole system, the abrasive process, which consists of a large number of random events, may have a consistency of average values for the machining parameters which is better than a supposedly non-random machining process, such as milling where the tool is subject to an unpredictable wear pattern. Randomness of grain spacing should not necessarily be seen as a disadvantage or as a major cause of process variability in size, average roughness, or form as long as the range of variations is not too great. Randomness is, of course, a source of variability in the behavior of individual grains and in the microsurface produced. The micromilling analogy can be considered as an attempt to characterize average process behavior and the nature of microvariations.

The micromilling analogy is represented in Fig. 3.4 where the real depth of cut is exaggerated for clarity. The spacing of the cutting edges is assumed to be uniform and equal to L. The feed of the workpiece per cutting edge is s.

The period from the time one grain engages the workpiece to the time the next grain engages is

Eq. (3.7) $$t_{gL} = \frac{L}{v_s}$$

The time for the workpiece to move the distance, s, is given by

Eq. (3.8) $$s = v_w \cdot t_{gL} = \frac{v_w}{v_s} \cdot L$$

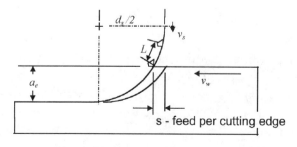

Figure 3.4. The micromilling analogy of grinding.

The feed per cutting edge, s, is illustrated in more detail in Fig. 3.5 and is used to define the maximum thickness of the uncut chip, $h_{cu.max}$. The uncut chip thickness is a measure of the depth of grain penetration into the workpiece.

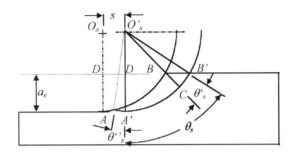

Figure 3.5. The geometry of the uncut chip.

3.2.2 Geometric Contact Length

The length of the uncut chip is equal to the length of contact between the wheel and the workpiece. The geometric contact length in Fig. 3.5 is defined as the arc length AB = A'B'.

For surface grinding, the geometric contact length is

Eq. (3.9) $$l_g = \frac{d_s}{2} \cdot \theta_s$$

However, l_g is usually evaluated from the chord length AB rather than the arc length, so that

Eq. (3.10) $$l_g = \sqrt{DB^2 + AD^2} = \sqrt{\left(d_s^2 \sin^2 \theta_s\right)/4 + a_e^2}$$

These two expressions give the same result if θ_s is small, and DB can be found from the principle of intersecting chords.

Eq. (3.11) $$DB = \sqrt{(d_s - a_e) \cdot a_e}$$

If a_e/d_s is small,

Eq. (3.12) $l_g = \sqrt{a_e \cdot d_s}$

More generally, for surface and cylindrical grinding processes, the following form can be used.

Eq. (3.13) $l_g = \sqrt{a_e \cdot d_e}$

where the effective wheel diameter is given by summing the curvatures of the wheel and workpiece

Eq. (3.14) $\dfrac{1}{d_e} = \dfrac{1}{d_s} \pm \dfrac{1}{d_w}$

The effective diameter is used because, as will be shown later, the contact length depends on the relative curvature of the wheel and the surface being machined. The effective diameter is an inverse measure of relative curvature.

Equations (3.12) and (3.13) are the usual expressions employed, although, in the next chapter it will be shown that the real contact length can be much larger than the geometric contact length.

Contact length has particular significance for the concentration of energy and forces in the contact zone and for rubbing wear of the abrasive grains..

If $a_e = 0.1 d_s$, which is a very large depth of cut for most operations, the error in the simplified expression for l_g in Eq. (3.12) is about 1%. The expression is a very good approximation even when the depth of cut is large.

It was assumed that the arc of contact was circular. This too, is a very good approximation in conventional grinding operations, if the grinding wheel remains circular. Figure 3.5 shows that if the grainspeed, v_s, is very much higher than the workspeed, v_w, a cutting edge coming into contact with the workpiece at point B follows a path which is very close to being circular and exits at point A. In conventional grinding operations, the grainspeed is approximately two orders of magnitude greater than the workspeed, that is $v_s/v_w \approx 100$. Clearly, the deviation from circularity is very small.

3.2.3 Kinematic Contact Length

At high workspeeds, the contact length as derived by geometry is increased. When the grinding wheel center, O_s, has moved forward a distance, s to O'_s in Fig. 3.5, a succeeding cutting edge comes into contact with the workpiece at B′ sweeping out the path B'A'.

The cutting edge exits from the workpiece at a point halfway between A and A′. In other words, the cutting edge is in contact for an extra angular rotation (θ'_s) so that the total arc of contact on the wheel is $\theta_s + \theta'_s$. The relative horizontal speed between the wheel and the workpiece is $v_s \pm v_w$ where the plus sign is for up-grinding and the minus sign is for down-grinding.

The extra distance to be travelled by the grain is $s/2$. The grain contact time is, therefore,

Eq. (3.15) $$t_{gc} = \left(l_g + \frac{s}{2} \right) \cdot \frac{1}{v_s}$$

The kinematic contact length is given by the distance traveled at the relative speed, $v_s \pm v_w$, during the grain contact time, t_{gc}.

Eq. (3.16) $$t_{gc} = \left(l_g + \frac{s}{2} \right) \cdot \frac{1}{v_s}$$

The contact length is increased for up-grinding and reduced for down-grinding. At higher workspeeds, the uncut chip shape is shorter and fatter for down-grinding than for up-grinding. Also, the shock loading on the grains is greater. This implies a greater tendency for grain fracture in down-grinding and less tendency to dull, if the speed ratio is sufficiently large.

3.2.4 Grain Penetration Depth or Uncut Chip Thickness

In Fig. 3.5, the maximum uncut chip thickness is BC. From the triangle BCB′

Eq. (3.17) $h_{cu.max} = s.\sin(\theta_s - \theta'_s)$

From the triangle, D′O′$_s$B

Eq. (3.18) $\sin(\theta_s - \theta'_s) = \dfrac{D'B}{O'_sB} = \dfrac{D'B'-s}{d_s/2 - h_{cu.max}}$

Since

$$D'B' = \sqrt{a_e d_e - a_e^2}$$

we can write

Eq. (3.19) $\sin(\theta_s - \theta'_s) = \dfrac{\sqrt{a_e d_s - a_e^2} - s}{d_s/2 - h_{cu.max}}$

The term $h_{cu.max}$ is always very small compared to the term $d_s/2$ in the denominator and may be ignored. The maximum uncut chip thickness from Eq. (3.17) is, therefore,

Eq. (3.20) $h_{cu.max} = 2s\sqrt{\dfrac{a_e}{d_s} - \dfrac{a_e^2}{d_s^2} - \dfrac{2s^2}{d_s}}$

This can be arranged in a simpler form, noting that $a_e^2 \ll a_e d_s$ and that, if v_w is much smaller than v_s, the last term $2s^2/d_s$ is relatively small. Neglecting this term, the usual expression is obtained for uncut chip thickness in surface grinding.

Eq. (3.21) $h_{cu.max} = 2s\sqrt{\dfrac{a_e}{d_s}}$

As previously shown, the more general form for surface and cylindrical processes is obtained by replacing d_s by d_e so that

Eq. (3.22) $$h_{cu.max} = 2s\sqrt{\frac{a_e}{d_s}} = 2L \cdot \frac{v_w}{v_s} \cdot \sqrt{\frac{a_e}{d_e}}$$

From the geometry of Fig. 3.5, this approximation corresponds to

Eq. (3.23) $$h_{cu.max} = s \cdot \sin\theta_s$$

where

Eq. (3.24) $$\sin\theta_s = 2\sqrt{\frac{a_e}{d_s}}$$

For values of θ between 0 and θ_s, the chip thickness is, therefore, given by

Eq. (3.25) $$h_{cu} = s \cdot \sin\theta = s \cdot \theta$$

This shows that the maximum penetration of the grains into the workpiece is the product of the feed per cutting edge and the angle subtended by the arc of contact. Increasing either of these two parameters increases the stress exerted on the abrasive grain, increasing the tendency to fracture or break away from the bond.

The thickness of the uncut chip can also be viewed as the cutting depth of the grain. The thickness increases with contact length, grain spacing, and speed ratio. It reduces with increasing wheel diameter.

Equation (3.25) also shows that the chip thickness increases linearly along the chip until it reaches a maximum. The thickness then reduces almost instantly, as illustrated in Fig. 3.6.

A typical maximum thickness of the uncut chip and the error involved in the approximation in Eq. (3.21) can be demonstrated with a typical example.

Wheel diameter: d_s: 100 mm

Grain spacing: L: 4 mm

Workspeed: v_w: 0.3 m/s

Wheelspeed: v_s: 30 m/s

Depth of cut: a_e: 0.003 mm

From Eq. (3.8), the feed per cutting edge, $s = 0.04$ mm,

$$\frac{2s^2}{d_s} = 3.2 \times 10^{-5} \text{ mm}$$

$$2s\sqrt{\frac{a_e}{d_s}} = 4.38 \times 10^{-4} \text{ mm}$$

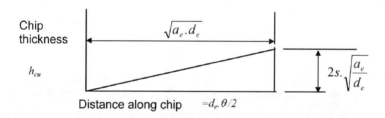

Figure 3.6. The thickness of the uncut chip increases linearly along its length.

The maximum thickness of the uncut chip is, therefore,

$$h_{cu.\max} = (4.38 - 0.32) \times 10^{-4} \text{ mm} = 0.4 \text{ microns}$$

The error in Eq. (3.21) is approximately 10%, which is acceptable for most purposes. A typical value of undeformed chip length can be obtained from the above example. From Eq. (3.12)

$$l_g = 0.55 \text{ mm}$$

To compare with a human hair, the uncut chip would have to be approximately 200 times thicker and have a length of 110 mm.

3.2.5 Uncut Chip Aspect Ratio

The previous expressions for uncut chip thickness and chip length are illustrated in Fig. 3.6, where the chip shape is shown to increase linearly along its length. This result is used in Ch. 6 to explain why the energy distribution in the contact zone is approximately triangular.

The length to thickness ratio of the uncut chip is given by Eqs. (3.12) and (3.21)

Eq. (3.26)
$$r_{cu} = \frac{l_g}{h_{cu.max}} = \frac{d_s}{2s} = \frac{v_s d_s}{2v_w L}$$

Generalizing, for surface and cylindrical grinding operations, we can replace d_s by d_e so that

Eq. (3.27)
$$r_{cu} = \frac{l_g}{h_{cu.max}} = \frac{d_e}{2s} = \frac{v_s d_e}{2v_w L}$$

In this example, the aspect ratio, r_{cu} equals 1,250. This demonstrates that the groove ploughed by an abrasive grain is very long compared to its depth. The grain travels a path which is almost parallel to the surface generating the characteristic long straight furrows, typical of ground surfaces.

3.3 CYLINDRICAL GRINDING CONTACTS

The analysis for external and internal cylindrical contacts has to take into account the diameter of the workpiece and the consequent effect on the shape of the uncut chip. The curvature of the workpiece has significant implications for the rate of wear of the abrasive grains and for the surface roughness of the workpiece so that different grades of wheel are required for the different curvatures experienced in internal and external grinding.

3.3.1 External Cylindrical Plunge Grinding

The micromilling analogy for the plunge external cylindrical grinding process is illustrated in Fig. 3.7. In plunge grinding, the grinding wheel is moved towards the workpiece at a plunge feedrate (v_f) .

The feedrate (v_f) is normally much smaller than the workspeed (v_w) while the workspeed is normally much smaller than the wheelspeed (v_s). The result is that the grinding wheel removes a layer of material in a spiral from the workpiece. However, the depth of cut is usually so small that the spiral can only be detected by sudden interruption of the grinding process.

3.3.2 Removal Rate in Cylindrical Plunge Grinding

The workpiece, the grinding wheel, and the machine structure all deflect as contact is established. Therefore, the real depth of cut, a_e, is smaller than the calculated depth of cut based on the feed rate. As described previously for surface grinding, the real depth of cut is affected by deflections, wheel wear, and thermal deflections.

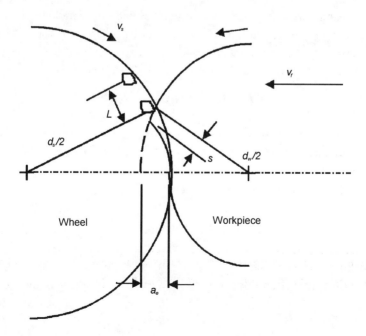

Figure 3.7. Micromilling analogy applied to plunge external cylindrical grinding.

After a period of infeeding at a constant rate, the deflections tend to stabilize and become constant until the infeed rate is changed or stopped. The removal rate from the workpiece, as for surface grinding, is given by

Eq. (3.28) $Q_w = b_w \cdot a_e \cdot v_w$

The real depth of cut is, therefore,

Eq. (3.29) $a_e = \dfrac{Q_w}{b_w \cdot v_w}$

If the wear-rate of the wheel is negligible, we can write

Eq. (3.30) $a_e = \dfrac{v_f}{n_w} = \pi d_w \dfrac{v_f}{v_w}$

where n_w is the rotational speed of the workpiece.

This leads to the usual expression for removal rate in cylindrical grinding which is sufficiently accurate for most purposes.

Eq. (3.31) $Q_w = b_w \cdot \pi \cdot d_w \cdot v_f$

3.3.3 Effect of G-ratio on Removal Rate

When grinding very hard materials, the wear-rate of the wheel often becomes very rapid. The removal rate from the workpiece and the calculation of the depth of cut must take the wear-rate into account. The removal rate from the workpiece becomes

Eq. (3.32) $Q_w = b_w \cdot \pi \cdot d_w \cdot (v_f - v_{f.\text{wear}})$

where $v_{f.\text{wear}}$ is the rate of reduction of the radius of the wheel due to wear.

The wear rate of the grinding wheel is,

Eq. (3.33) $Q_s = b_w \cdot \pi \cdot d_s \cdot v_f$

The value of $v_{f.\text{wear}}$ can be related to the G-ratio since

Eq. (3.34) $$G = \frac{Q_w}{Q_s} = \frac{b_w \cdot \pi \cdot d_w \cdot \left(v_f - v_{f.\text{wear}}\right)}{b_w \cdot \pi \cdot d_s \cdot v_{f.\text{wear}}}$$

Rearranging terms,

Eq. (3.35) $$v_{f.\text{wear}} = \frac{1}{1 + \dfrac{d_s}{d_w} \cdot G} \cdot v_f$$

Removal rate in cylindrical grinding, can now be expressed in terms of G-ratio by substituting this expression in Eq. (3.32).

Eq. (3.36) $$Q_w = \frac{b_w \cdot \pi \cdot d_w}{1 + \dfrac{1}{G}\dfrac{d_w}{d_s}} \cdot v_f$$

In an extreme case, where $G = 1$ and $d_w/d_s = 2$, the removal rate is reduced by two-thirds compared with the value from Eq. (3.31).

3.3.4 Effect of G-ratio on Depth of Cut

The real depth of cut is reduced by wheel-wear by the same factor as the removal rate. Applying the reduction factor in Eq. (3.36) to Eq. (3.30) gives the following equation:

Eq. (3.37) $$a_e = \frac{\pi \cdot d_w}{1 + \dfrac{1}{G}\dfrac{d_w}{d_s}} \cdot \frac{v_f}{v_w}$$

3.3.5 Geometric Contact Length in Cylindrical Grinding

The uncut chip shape in cylindrical grinding is illustrated in Fig. 3.8. From the principle of intersecting chords,

Eq. (3.38) $\qquad l_g \approx B'D' = \sqrt{d_s A'D'} = \sqrt{d_w ED'}$

Hence,

Eq. (3.39) $\qquad ED' = A'D' \cdot \dfrac{d_s}{d_w}$

From Fig. 3.8, the depth of cut is given by

$$EA = ED' + A'D'$$

so that

Eq. (3.40) $\qquad a_e = ED' + A'D' = \dfrac{l_g^2}{d_w} + \dfrac{l_g^2}{d_s}$

or

Eq. (3.41) $\qquad l_g = \sqrt{a_e \cdot d_e}$

where

Eq. (3.42) $\qquad \dfrac{1}{d_e} = \dfrac{1}{d_s} + \dfrac{1}{d_w} = \dfrac{d_s + d_w}{d_s \cdot d_w}$

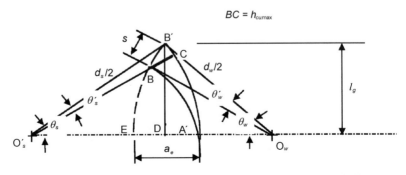

Figure 3.8. Geometry of the uncut chip for plunge external cylindrical grinding.

This derivation is the justification for the substitution of d_e for d_s in Eq. (3.12), mentioned previously in relation to surface grinding. The same argument can be shown to apply to kinematic contact length for cylindrical grinding and to uncut chip thickness.

3.3.6 Uncut Chip Thickness in Cylindrical Grinding

The maximum uncut chip thickness, $h_{cu.max}$, is given in Fig. 3.8 by the side BC of the right-angled triangle $BB'C$.

Eq. (3.43) $h_{cu.max} = s \cdot \sin(\angle BB'C)$

It can be seen that the angle $BB'C$ is equal to the sum of the angles $BB'D$ and $DB'C$. From right-angled triangles, we may write the angle may be written $BB'D' = \theta_w - \theta_w'$. Similarly, the angle $D'B'C = \theta_s - \theta_s'$ so that

Eq. (3.44) $h_{cu\,max} = s \cdot \sin[(\theta_s - \theta_s') + (\theta_w - \theta_w')]$

Making the usual approximations as justified previously for surface grinding, the following substitutions can be made

Eq. (3.45) $\theta_s \approx \sin\theta_s = \dfrac{2l_g}{d_s}$

Eq. (3.46) $\theta_w \approx \sin\theta_w = \dfrac{2l_g}{d_w}$

From arc lengths

Eq. (3.47) $\theta_s' = \dfrac{2s}{d_s} \cdot \cos[(\theta_s + \theta_w) - (\theta_s' + \theta_w')] \approx \dfrac{2s}{d_s}$

Eq. (3.48) $\theta_w' \approx \dfrac{2s}{d_w}$

leading to

Eq. (3.49) $h_{cu.\max} = 2s \cdot \dfrac{l_g}{d_e} - \dfrac{2s^2}{d_e}$

Substituting for l_g from Eq. (3.41) and ignoring the last term

Eq. (3.50) $h_{cu.\max} = 2s \cdot \sqrt{\dfrac{a_e}{d_e}} = 2L \cdot \dfrac{v_w}{v_s} \sqrt{\dfrac{a_e}{d_e}}$

as found previously for surface grinding.

For the same diameter wheel, the effective diameter, d_e, is smaller in external cylindrical grinding than in surface grinding. These equations demonstrate that the contact length is, therefore, reduced and the uncut chip thickness is increased in external cylindrical grinding.

The effect on the wheel is to reduce sliding wear and increase fracture wear, so that from this consideration, a harder wheel would be required. However, it should be noted that the reciprocating horizontal surface grinding process involves a shock loading on the wheel at the commencement of contact of each pass. Interrupted cuts greatly increase the fracture wear on a grinding wheel. A harder wheel is required to limit the damage caused by these shock loadings.

3.3.7 Basic Parameters for Internal Cylindrical Grinding

The internal grinding process is illustrated in Fig. 3.9. By reasoning similar to the previous cases, the geometric contact length is obtained

Eq. (3.51) $l_g = \sqrt{a_e \cdot d_e}$

and for the maximum undeformed chip thickness

Eq. (3.52) $h_{cu.\max} = 2s \cdot \sqrt{\dfrac{a_e}{d_e}} = 2 \cdot \dfrac{v_w}{v_s} \cdot L \cdot \sqrt{\dfrac{a_e}{d_e}}$

where

Eq. (3.53) $$\frac{1}{d_e} = \frac{1}{d_s} - \frac{1}{d_w}$$

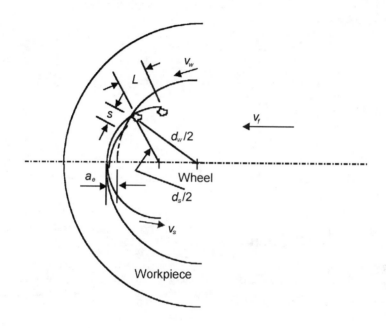

Figure 3.9. Internal cylindrical grinding. The chips are longer and thinner due to high conformity of curvature.

In internal grinding, d_e is much larger than d_s. Even though a grinding wheel for internal grinding is usually much smaller than for external grinding, the effective wheel diameter is much larger due to the conformal contact.

A consequence of the large effective diameter is that the contact length is very long, and the maximum uncut chip thickness is substantially reduced. This results in increased sliding wear of the abrasive grains and reduced fracture wear, so that a softer grade of wheel is required to prevent glazing.

3.3.8   Uncut Chip Aspect Ratio in Cylindrical Grinding

For either external grinding or internal grinding, the aspect ratio for length divided by maximum thickness in terms of equivalent diameter can be expressed as previously.

Eq. (3.54) $$r_{cu} = \frac{d_e}{2s} = \frac{v_s d_e}{2 v_w L}$$

where

Eq. (3.55) $$\frac{1}{d_e} = \frac{1}{d_s} \pm \frac{1}{d_w}$$

The plus sign is for external grinding and the minus sign for internal grinding.

3.3.9   Angle Grinding of Cylindrical Parts

In angle grinding of cylindrical parts, the grinding wheel axis is inclined at a wheel angle, β, to the workpiece axis. The wheel angle is commonly set at 30° in angle grinding, whereas in the more usual cylindrical grinding set-up, the wheel angle is set at 0°.

An example of shoulder and cylinder grinding is illustrated in Fig. 3.10 where the feed direction may be different from the wheel angle. The feed angle may be set at 0° to grind a cylindrical surface, at 90° to grind a shoulder, or at 45° to grind both at the same time. Other angles may be programmed on a CNC grinding machine depending on the stock removal required and the surface integrity requirements.

A consequence of the wheel angle is to reduce the curvature of the wheel when viewed parallel to the workpiece surface. This increases the effective wheel diameter. The adjustment is made by dividing d_s by $\cos\beta$.

Figure 3.10 compares angle grinding with a straight grinding set-up to achieve the same operation. Comparing the two methods, angle grinding greatly reduces the contact length on the shoulder.

In the straight grinding set-up, the contact extends across the width of the contact face which leads to problems of wheel rubbing and glazing. Even a small wheel angle of 1° greatly reduces the contact length and eases the problems of rubbing wear on the face of the wheel.

Another problem with straight shoulder grinding the face AB is that the material removal is concentrated at the outer edge of the wheel, a distance, a_e, along the surface BC while the surface AB is only rubbing. This causes rapid wear of the outermost grains. In practice, the operator always puts a small radius on the edge. The material removal is then concentrated on the radius which tends to increase in size as the grains wear. With a sharp corner, the contact area responsible for material removal on the face AB is $a_e \cdot l_g$.

This problem is avoided in angle grinding, since the material removal is spread across the line AB and the contact length is comparable with contact lengths in cylindrical grinding.

The relationships for depth of cut and contact length in Fig. 3.10 can be applied to any form grinding situation, by judicious application of the relationships at any point on the profile for particular values of a and b.

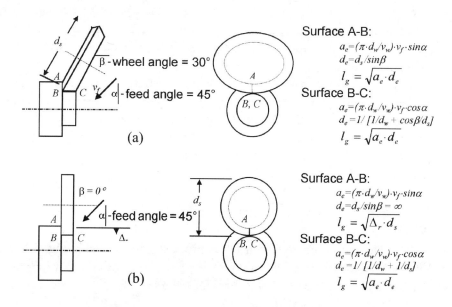

Surface A-B:
$$a_e = (\pi \cdot d_w/v_w) \cdot v_f \cdot \sin\alpha$$
$$d_e = d_s/\sin\beta$$
$$l_g = \sqrt{a_e \cdot d_e}$$
Surface B-C:
$$a_e = (\pi \cdot d_w/v_w) \cdot v_f \cdot \cos\alpha$$
$$d_e = 1/\left[1/d_w + \cos\beta/d_s\right]$$
$$l_g = \sqrt{a_e \cdot d_e}$$

Surface A-B:
$$a_e = (\pi \cdot d_w/v_w) \cdot v_f \cdot \sin\alpha$$
$$d_e = d_s/\sin\beta = \infty$$
$$l_g = \sqrt{\Delta_r \cdot d_s}$$
Surface B-C:
$$a_e = (\pi \cdot d_w/v_w) \cdot v_f \cdot \cos\alpha$$
$$d_e = 1/\left[1/d_w + 1/d_s\right]$$
$$l_g = \sqrt{a_e \cdot d_e}$$

Figure 3.10. Shoulder and cylinder grinding: *(a)* Angle grinding; *(b)* Straight grinding.

3.3.10 Comparisons of Surface, Internal, and External Grinding

Some comparisons of chip shape and effective diameter are given in Table 3.1 for surface grinding, external grinding, and internal grinding.

For the same size wheel, the effective diameter in a typical example is six times larger in internal grinding than in external grinding. The maximum chip thickness is halved and the contact length is doubled, with implications for wear performance as previously explained.

Table 3.1. Comparison of Chip Shapes for Surface Grinding, External Grinding, and Internal Grinding

PARAMETER	UNITS	SURFACE	EXTERNAL	INTERNAL
s	mm	0.04	0.04	0.04
d_s	mm	100	100	100
d_w	mm	—	50	200
a_e	mm	0.003	0.003	0.003
d_e	mm	100	33.3	200
l_g	mm	0.548	0.316	0.775
$h_{cu.max}$	µm	0.438	0.759	0.310
r_{cu}	—	1250	416	2500

3.3.11 Centerless Grinding

The plunge centerless grinding process is illustrated in Fig. 3.11. Material is removed from the workpiece by advancing the control wheel at a constant infeed rate, v_f. In some machines, the same result is achieved by advancing the grinding wheelhead.

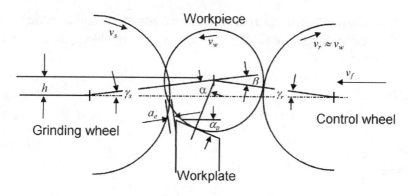

Figure 3.11. The geometry of centerless grinding.

In some countries, the control wheel is known as the regulating wheel. In either case, the name is given because the surface speed of the workpiece is controlled or regulated by the surface speed of the control wheel.

The workpiece is supported by a workplate, or workblade, and by the control wheel. The workplate is normally angled at 30° or 15° to the horizontal. The height of the workplate is adjusted to bring the center height (h) of the workpiece above the plane of the grinding wheel and control wheel centers.

By appropriate selection of the value of h, the angle β shown in Fig. 3.11 may be adjusted. The angles α and β define the support points on the workpiece. If these angles are zero, it is impossible to remove odd order lobing and 3 or 5 lobe shapes will be a frequent occurrence.

From Fig. 3.11,

Eq. (3.56) $\alpha = \alpha_p - \gamma_s$

where α_p is the angle of the workplate to the horizontal and γ_s is the angle of a line through the wheel contact and the horizontal.

The angle β is given by

Eq. (3.57) $\beta = \gamma_s + \gamma_r$

where γ_r defines the contact point on the control wheel.

For most purposes, to achieve an effective rounding mechanism and for stability, the angle β is set between 6° and 8°.[4] The workplate angle, α_p, is normally set to either 30° or 15°.

The rate of material removal under steady-state conditions depends on the infeed, v_f. The effective feedrate, v_{fe}, as modified by the setup geometry is

Eq. (3.58) $$v_{fe} = \frac{v_f \cdot \cos \gamma_s}{1 - K_1 + K_2}$$

where

Eq. (3.59) $$K_1 = \frac{\sin \beta}{\sin(\alpha + \beta)}$$

Eq. (3.60) $$K_2 = \frac{\sin \alpha}{\sin(\alpha + \beta)}$$

For small values of γ_s and β, $K_1 = 0$ and $K_2 = 1$ so that

Eq. (3.61) $$v_{fe} = \frac{v_f}{2}$$

Equations (3.30) and (3.31) must, therefore, be modified for centerless grinding leading to the depth of cut

Eq. (3.62) $$a_e = \frac{1}{2} \cdot \pi \cdot d_w \cdot \frac{v_f}{v_w}$$

and to the removal rate

Eq. (3.63) $$Q_w = \frac{1}{2} \cdot b_w \cdot \pi \cdot d_w \cdot v_f$$

3.4 IMPLICATIONS OF THE STOCHASTIC NATURE OF GRINDING

So far, it has been implicit that the uncut chip is representative of the various volumes of material displaced by the grains. Alternative measures of chip size give an indication of some effects of different shapes of uncut chip and factors leading to variability in the process.

Also, there are important implications of the variable distribution of cutting edges on the quality of the grinding process and the rate of wear of the grinding wheel or other abrasive tool. Most importantly, there should be consistency between the mean uncut chip and the total volume removed.

3.4.1 Mean Uncut Chip Thickness

The mean chip volume must be consistent with the total volume of material removed in grinding (see Sec. 3.1). This leads to a simple method for arriving at the mean uncut chip thickness.

The volume of material removed by the average grain is illustrated in Fig. 3.12. The average volume of material removed by an average grain is given by the product of the feed per cutting edge, the real depth of cut, and the mean chip width.

Eq. (3.64) $\qquad \overline{V}_{cu} = s \cdot a_e \cdot \overline{b}_{cu}$

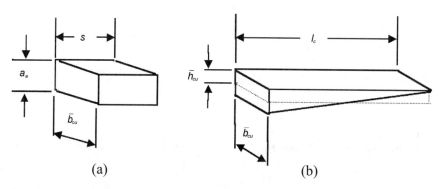

Figure 3.12. Volume removed by an average grain in a single interaction: *(a)* based on material removal rate; *(b)* based on mean uncut chip dimensions.

Previously, in Eq. (3.1), the mean chip volume was defined in terms of the mean chip thickness. For consistency with the volume given by Eq. (3.64), the mean chip thickness is, therefore,

Eq. (3.65) $\bar{h}_{cu} = s \cdot \dfrac{a_e}{l_c}$

Substituting

$$l_c = \sqrt{a_e \cdot d_e}$$

Eq. (3.66) $\bar{h}_{cu} = s \cdot \sqrt{\dfrac{a_e}{d_e}} = \dfrac{v_w}{v_s} \cdot L \cdot \sqrt{\dfrac{a_e}{d_e}}$

In other words, the mean chip thickness is half the maximum chip thickness as previously defined in Eqs. (3.22), (3.50), and (3.52).

3.4.2 Cutting Edge Density

The mean chip thickness is proportional to the grain spacing, L. The distribution of grain spacings is important for the chip thickness, hence for the forces that will be experienced by the grains.

The grain spacing can be considered in terms of the cutting edge density, defined as the number of active grains per unit area (C) of the cutting surface of the abrasive tool.

If the average spacing of the grains in the direction of motion is L and the spacing in the lateral direction is B, then

Eq. (3.67) $C = \dfrac{1}{L \cdot B}$

Employing the expression for s in Eq. (3.64), the mean uncut chip volume is

Eq. (3.68) $\bar{V}_{cu} = a_e \cdot \dfrac{v_w}{v_s} \cdot L \cdot \bar{b}_{cu}$

However, the lateral grain spacing is B. For consistency of the chip volume with the number of grain interactions per unit area of the workpiece,,

Eq. (3.69) $$\overline{V}_{cu} = a_e \cdot \frac{v_w}{v_s} \cdot L \cdot B$$

In other words, the material removal rate dictates that the penetration of the grain must increase until the mean chip width, $\overline{b}_{cu}$, is equal to the average spacing between grains, so that

Eq. (3.70) $$\overline{b}_{cu} = B$$

This is more easily visualized if the successive grains are imagined to be offset from one to the next, as in Figs. 3.13 and 3.14.

The mean chip volume is, therefore, given by

Eq. (3.71) $$\overline{V}_{cu} = \frac{a_e}{C} \frac{v_w}{v_s} \quad ...\text{surface and cylindrical grinding}$$

Eq. (3.72) $$\overline{V}_{cu} = \frac{\pi \cdot d_w}{C} \frac{v_f}{v_s} \quad ...\text{cylindrical grinding}$$

Eq. (3.73) $$\overline{V}_{cu} = \frac{1}{2} \cdot \frac{\pi \cdot d_w}{C} \frac{v_f}{v_s} \quad ...\text{centerless grinding}$$

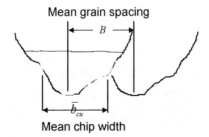

Mean grain spacing

Mean chip width

Figure 3.13. Mean grain spacing depends on the grain penetration and is equal to the mean chip width.

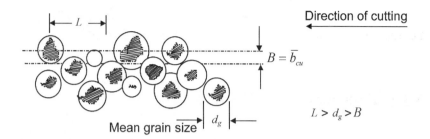

Figure 3.14. Section at the mean grain penetration depth below the surface of the abrasive showing the mean grain spacing with a random three-dimensional distribution.

In practice, the cutting edge density varies with depth below the surface of the abrasive tool. This is a consequence of the fact that not all cutting edges lie at the same height as each other. This means that C increases with depth of cut.

In Fig. 3.14, the shaded areas represent the areas which come into contact at a particular depth below the surface. With a very light depth of cut, these areas will be smaller compared to the case of a heavy depth of cut when these areas will be larger. As the shaded areas in contact increase, more cutting edges appear on any line parallel to the direction of motion. The result is that L reduces while B increases with increasing depth of cut. Assuming the grain is conical in shape or triangular in section is one way to take account of this phenomenon. As the grain depth increases, the chip width automatically increases. The triangular grain assumption discussed below predicts that the uncut chip thickness increases more slowly with depth of cut than with the more usual assumption of a constant value of L.

The variation of cutting edge density obviously has an effect on the mean chip volume, as well as on other chip parameters.

3.4.3 Mean Uncut Chip Cross-sectional Area

Dividing the mean chip volume from Eq. (3.69) by the chip length leads to the mean cross-sectional area of the uncut chips, A_{cu}. This is an alternative parameter to uncut chip thickness which also has a strong correlation with the mean tangential force on a grit. Dividing by geometric contact length, it is found that

Eq. (3.74) $\qquad A_{cu} = \dfrac{1}{2} \dfrac{v_w}{v_s} \cdot \sqrt{\dfrac{a_e}{d_e}}$

3.4.4 Equivalent Chip Thickness

Equivalent chip thickness is a parameter widely used, because of its simplicity, as a measure of the tool penetration depth. In experimental work, grain spacing is difficult to determine precisely and is a variable depending on depth of cut and wheel wear. Equivalent chip thickness is, therefore, commonly used as a basic variable in experimental trials. Equivalent chip thickness is defined as

Eq. (3.75) $\qquad h_{eq} = a_e \cdot \dfrac{v_w}{v_s}$

Comparison with the previous expressions for uncut chip volume shows that both parameters are proportional to equivalent chip thickness, although h_{eq} takes no account of grain spacing in the tool surface. It is evident, therefore, that grinding behavior correlates with a combination of h_{eq} and cutting edge density, C.

The average uncut chip thickness can be expressed in terms of the equivalent chip thickness.

Eq. (3.76) $\qquad \bar{h}_{cu} = h_{eq} \cdot \dfrac{L}{l_c}$

The two terms are of the same order of magnitude. It should be remembered, however, that the value of h_{eq} is the same for a very coarse grit wheel or for a very fine grit wheel for particular values of depth of cut and speed ratio. It follows that similar values of h_{eq} for two different wheel specifications will produce different values of surface roughness and different wear behavior of the grinding wheels.

Fortunately, it is a simple matter to make an approximate conversion from h_{eq} to $\bar{h}_{cu}$ using Eq. (3.76), if the grain spacing is estimated.

3.4.5 Grain Spacing

Manufacturers do not give figures for grain spacing in an abrasive tool due to the variability and uncertainty of the spacing. The spacing depends not only on the grain size, but on the proportion of bond material and the openness of the structure. A further complication is that the spacing can be artificially varied to some extent by the dressing procedures employed. The wear process in grinding can also vary the spacing of the cutting edges as the grains are fractured from the surface of the wheel.

A number of authors have employed direct methods to measure the grain density as described in detail by Shaw.[5] For example, Backer, et al.,[7] determined C by rolling a dressed wheel over a soot covered glass sheet and counting the number of contacts per unit area.

Other authors have used stylus or optical methods to determine the grain density from the peak count (P_c) at a particular depth below the wheel surface. None of these methods is straightforward and it is found that C increases depending on the depth sampled below the surface of the wheel.

Manufacturers give average values of grain size although the situation is further complicated by a common practice of employing a range of grain sizes and of mixing different grain sizes.

For an order of magnitude estimation of grain size, it is sometimes assumed that the average grain spacing is given by the average grain size, d_g. Based on this assumption which implies very close packing,

Eq. (3.77) $L \cdot B = d_g^2$

Thus if $d_g = 0.2$ mm, an order of magnitude estimate gives $L = 0.2$ mm, $B = 0.2$ mm and $C = 1/LB = 25$ grains per mm^2.

From values measured by Shaw,[5] the above estimate of C is more than twice the value obtained for a typical 80 grit white alumina wheel. Therefore, a better estimate might be obtained by assuming

Eq. (3.78) $L \cdot B = (1.5 \, d_g)^2$

This makes a sensible allowance for the bond volume and leads to

Eq. (3.79) $C = \dfrac{1}{2.25 d_g^2}$

It should be noted that this is a rough estimate and that the grain density actually varies with several parameters as previously discussed.

It was shown by Brough, et al.,[6] that the number of grains actively involved in cutting are distributed at depths below the surface of the tool. The variation of the number of grains with depth was found to be described by the Poisson distribution.

Different expressions in the literature for uncut chip thickness arise from different assumptions concerning grain spacing and shape. Some problems can be resolved if it is recalled that mean chip volume depends only on the mean number of grains contributing to the volume removal rate. As pointed out in Sec. 3.24, compatibility between removal rate and mean chip volume requires that as the mean grain density, C, increases with depth of cut, the mean uncut chip width, $\bar{b}_{cu}$, varies until it is equal to the mean grain spacing, B, as in Eq. (3.70).

The implications are further explored through the triangular grain model. However, a little thought leads to the conclusion that $L > d_g > B$ as illustrated in Fig. 3.14.

3.4.6 Effect of Grain Shape

Backer, Marshall, and Shaw[7] assumed that C = a constant and b_{cu} = $r \cdot h_{cu}$. This assumption is based on a triangular cross-section of the grains as illustrated in Fig. 3.15a.

The width of a chip, b_{cu}, is assumed to increase with the thickness, h_{cu}, since the width of the groove cut by a grain increases with its depth. This approach starts from the mean uncut chip thickness as given from Eq. (3.1) and (3.71) leading to

Eq. (3.80)
$$\bar{h}_{cu} = \frac{a_e \cdot v_w}{C \cdot \bar{b}_{cu} \cdot v_s \cdot l_c}$$

For a triangular cross-section, where the mean width is half the base width and for a constant chip width to thickness ratio, r,

Eq. (3.81)
$$r = \frac{b_{cu}}{h_{cu}} = 2\frac{\bar{b}'_{cu}}{h_{cu}} = 2 \cdot \frac{\bar{b}_{cu}}{\bar{h}_{cu}}$$

since $b_{cu} = 2 \cdot \bar{b}'_{cu}$ and the average width at any section is half the maximum width at that section. Substituting for $\bar{b}_{cu}$ in Eq. (3.80) leads to

Eq. (3.82) $$h_{cu} = \sqrt{\frac{2 \cdot a_e \cdot v_w}{C \cdot r \cdot v_s \cdot l_c}} = \sqrt{\frac{2 \cdot v_w}{v_s \cdot C \cdot r}} \sqrt{\frac{a_e}{d_e}}$$

Backer, et al., measured values of groove width and depth to obtain the ratio, r.

   This expression for the mean uncut chip thickness may be contrasted with the previous expression given by Eq. (3.66). It appears that there is a different functional relationship between the mean values and the machine control variables. In the earlier analysis, the mean values were made to conform to a rectangular cross-section of the material removed as shown in Fig. 3.12, whereas the triangular approach is shown in Fig. 3.15.

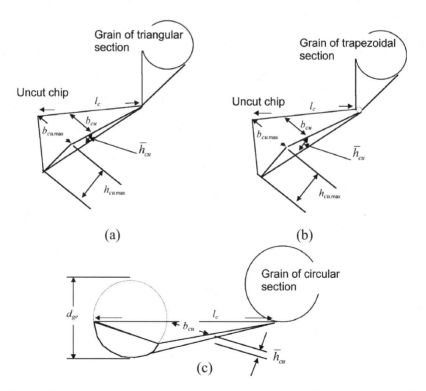

Figure 3.15. Grain shape assumptions: *(a)* the triangular grain; *(b)* the worn trapezoidal grain; *(c)* the circular grain.

The mean chip thickness is the value at the average cross-sectional area of the chip. The cross-sectional area is proportional to h_{cu}^2 so that the average along the length is related to the maximum uncut chip thickness for a triangular chip by

$$h_{cu.\max} = \sqrt{3} \cdot \overline{h}_{cu}$$

and

$$b_{cu.\max} = 2 \cdot \sqrt{3} \cdot \overline{b}_{cu}$$

These values may be compared with $h_{cu.\max} = 2 \cdot \overline{h}_{cu}$ and $b_{cu.\max} = \overline{b}_{cu}$ previously. The maximum chip thickness is not hugely different for the two cases, although the variation of chip width is much greater for the triangular assumption.

A conclusion from this analysis is that the maximum values of uncut chip thickness and chip width are actually a function of grain shape. The real grain shapes are not exactly rectangular or triangular. Actual shapes reflect the various initial shapes of the grains and the wear process experienced by the grains. With attritious wear, the grains develop small flats on the rubbing surface. This modifies the shape to a trapezoid as proposed by Malkin.[10]

It is probably of little importance in the analysis of the process what form is assumed for the chip shape, if the main concern is to obtain comparative measurements between kinematic conditions presented by different wheels and varying wheel conditions. It is more important to use the same model consistently when making comparisons.

Chen and Rowe showed that a wide range of grinding behavior could be reproduced using a computer simulation of the dressed and worn grains.[11][12] The shape of each cutting edge after dressing and wear was characterized and then approximated by a circle of equivalent diameter, d_{ge}, as illustrated in Fig. 3.10c.

In general, the chip thickness can be written in terms of h_{eq} as in Eq. (3.76) inserting an exponent and a factor to allow for grain shape.

Eq. (3.83) $$\overline{h}_{cu} = k_c \left(\frac{h_{eq}}{l_c} \right)^n$$ where $0 < n < 1$

$k_c = L$ for a rectangular grain cross section and

$$k_c = \sqrt{\frac{2}{Cr}}$$

for a triangular grain. For a spherical grain

$$k_c = \left[\frac{1}{C} \cdot \sqrt{\frac{3}{16 \cdot d_{ge}}} \right]^{\frac{2}{3}}$$

The exponents are $n = 1$ for a rectangular cutting edge, $n = 0.5$ for a triangular cutting edge, and $n = 2/3$ for a spherical cutting edge.

It is tempting to infer that the values for a rectangular cross section are more appropriate for blunt grains and the values for a triangular cross section are more appropriate for sharp grains. There will always be uncertainty over chip shape and the values may be expected to vary in the range indicated.

The general form of Eq. (3.83) is a justification for an empirical approach where parameters which are a function of uncut chip thickness are expressed in terms of a factor and an exponent. This is a frequently used approach for presenting experimental grinding results for forces, surface roughness, energy, grinding ratio, and so on.

3.4.7 Effect of Increasing Grain Density

To illustrate the effect of increasing grain density with mean chip thickness, Eq. (3.76) can be written in terms of a variable spacing, L, as shown in Fig. 3.16. According to this simplified assumption

Eq. (3.84) $L = L_o - k_l h_{cu}$

where L_o is the mean grain spacing at the outermost layer and k_l is the rate of reduction of grain spacing with mean chip thickness. Equation (3.76) then becomes

Eq. (3.85)
$$\bar{h}_{cu} = \frac{\dfrac{h_{eq}}{l_c} \cdot L_o}{1 + \dfrac{h_{eq}}{l_c} \cdot k_l} = \frac{\bar{h}_{cu.o}}{1 + \bar{h}_{cu.o} \cdot \dfrac{k_l}{L_o}}$$

where $\bar{h}_{cu.o}$ is the value of $\bar{h}_{cu}$ for $L = L$.

The effect of the reduced grain spacing can be seen in an example. If the mean grain spacing reduces from 4 mm to 3 mm for $\bar{h}_{cu.o}$ increasing from 0 to 0.2 μm based on L_o, the actual mean chip thickness is shown in Fig. 3.17 for a value $k_l = 5{,}000$.

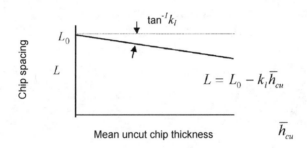

Figure 3.16. The mean grain spacing reduces with mean chip thickness.

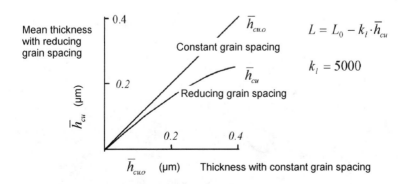

Fig. 3.17. Effect of reducing grain spacing with increasing cutting depth.

For this example, the mean chip thickness is reduced by approximately 30% compared to the constant spacing assumption. The effect is significant, but second order compared with the direct effect of depth of cut according to Eq. (3.67). A consequence of the increasing grain density is to reduce the impact of increasing depth of cut. The workpiece roughness becomes larger, but not so much as with constant grain spacing. The grains suffer increased stress, but not so much as with constant grain density.

3.4.8 Effects on Grain Wear

The discussion so far serves to demonstrate the importance of grain density on the surface for mean chip thickness which, in turn, has implications for wheel wear and grinding performance.

From a tribological viewpoint, the depth of cut is increased, the mean chip thickness is increased, which increases the probability of grain fractures and the rate of grain wear. It also increases the rate of bond fractures.

These types of wear are illustrated in Fig. 3.18. When the forces on the grain are small, the grain wears slowly due to attritious wear. This type of wear depends less on the hardness of the workpiece material than on the grain contact temperatures.[7]

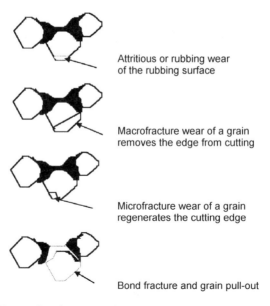

Attritious or rubbing wear of the rubbing surface

Macrofracture wear of a grain removes the edge from cutting

Microfracture wear of a grain regenerates the cutting edge

Bond fracture and grain pull-out

Figure 3.18. Types of grain wear.

The forces on the grain tend to be concentrated at the cutting edge, while the bond has a much larger area of contact with the grain. This allows the grains to be firmly held in the bond posts. The grains may be fractured either in the dressing process or in the grinding process. Fractures of the grain may remove part of the grain or remove the whole grain from the abrasive process.

In general, forces which cause fracture of the grain need to be larger than forces which cause attritious wear. Some polycrystalline grains are engineered to continually wear by microfracture which tends to produce a continual resharpening action with a low overall rate of wear. Very large forces on the grain can cause the bond posts to fracture and the whole grain to be pulled out of the surface.

It becomes apparent that grain spacing affects the nature of the grain forces and grain wear. It is also apparent that grain wear affects grain spacing as active cutting edges are removed from the process.

3.4.9 Irregular Grain Spacing

In practice, the grains are randomly distributed as in Fig. 3.19. This means that the spacing of some pairs of grains is greater than L, and for other pairs it is less. In general,

Eq. (3.86) $L_i = L \pm dL$

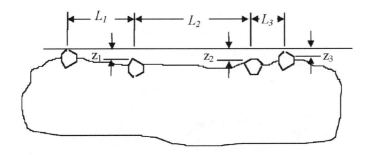

Figure 3.19. Irregular spacing and depths of grains actively cutting along a line of cut.

In this expression, the average spacing is L. If the maximum variation in grain spacing is $0.5L$,

$$L_i = L(1 \pm 0.5)$$

From Eqs. (3.52) and (3.66), the mean uncut chip thickness and the maximum uncut chip thickness also vary by $\pm 50\%$, as illustrated in Fig. 3.20.

Eq. (3.87) $\bar{h}_{cu.i} = (1 \pm 0.5)\bar{h}_{cu}$

The consequences of irregular grain spacing are two-fold:

1. Some grains experience forces corresponding to uncut chip thickness three times greater than other grains. This increased stress occurs for the grain coming into contact with the workpiece after a large space. The grain is much more likely to fracture or pull out, so that the spacing will tend to increase with time. Therefore, irregularity of grain spacing will increase with time leading to increasingly variable grinding performance.

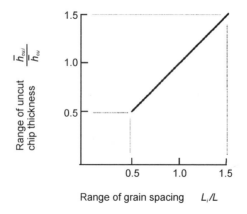

Figure 3.20. Effect of irregular grain spacing on chip thickness.

2. The variations of grain spacing lead to variations in the depth of the grooves cut in the workpiece surface. This has the consequence of increased irregularity of the workpiece surface roughness and also increased average roughness. The deepest grooves will be some 50% greater than the average groove.

3.4.10 Irregular Grain Depth

A similar effect on grinding behavior occurs due to the effect of irregular grain depths. The grain depths, Z_i, can be written in terms of the mean grain depth below the outermost surface.

Eq. (3.88) $Z_i = \bar{Z} \pm dZ$

The mean grain depth, $\bar{Z}$, obviously depends on the mean uncut chip thickness, $\bar{h}_{cu}$, and the variations in grain depth will be distributed about this mean.

Assuming that the grain depths lie within the range, $dZ/\bar{Z} = \pm 0.5$, the corresponding values of uncut chip thickness will be

Eq. (3.89) $\bar{h}_{cu.i} = \bar{h}_{cu} \pm dZ$

The mean value of Z must be equal to the value of $h_{cu.max} = 2 \cdot \bar{h}_{cu}$, so that

Eq. (3.90) $\bar{h}_{cu.i} = \bar{h}_{cu} \pm \bar{h}_{cu}$

In other words, the mean uncut thickness of individual chips lies in the range between zero and twice the mean for all chips, as illustrated in Fig. 3.21.

The outermost grains are much more likely to fracture and pull out. The outermost grains also suffer more attritious wear due to the longer time rubbing against the workpiece. Grains lying slightly below the surface suffer comparatively little wear.

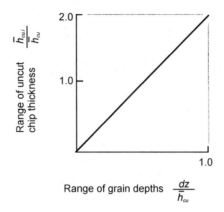

Figure 3.21. Effect of irregular grain depths on chip thickness.

3.4.11 Factors Affecting Workpiece Roughness

The most important factor affecting workpiece roughness is the grit size of the grains in the abrasive. A simplified explanation of this effect can be found from the kinematic roughness illustrated in Fig. 3.22.

A practical implication of this analysis is that the outermost grains suffer much larger forces and as each grain cuts a groove in the workpiece, a series of scallops is created according to the micromilling analogy. The theoretical peak-to-valley roughness, R_t, along the lay, that is, in the direction of the motion of the grains, is given by the height of the scallops. According to the principle of intersecting chords,

Eq. (3.91) $$R_t = \frac{1}{d_e} \cdot \left(\frac{s}{2}\right)^2 = \frac{1}{d_e} \cdot \left(\frac{L\, v_w}{2\, v_s}\right)^2$$

Thus, the primary determinant of variations in surface roughness along the lay is grain spacing on the wheel, which is closely related to grain size, workspeed, wheelspeed, and equivalent diameter.

In practice, Eq. (3.91) cannot be used to predict surface roughness with accuracy because other physical effects also play an important role, including ploughing and adhesion. Equation (3.91) suggests that depth of cut does not directly affect roughness and, therefore, that depth of cut can only affect roughness indirectly. For example, an increase in depth of cut causes grains to fracture or to pull out, thereby increasing the spacing, L, hence R_t.

Another viewpoint can be obtained for roughness across the lay. From the micromilling analogy, the grains are assumed to be spaced regularly. Assuming that the grains are approximately triangular in the cross section, the roughness can be related to the average chip thickness as in Fig. 3.23.

For a triangular cross-section, the average height is $1/\sqrt{3}$ times the maximum height. In other words,

Eq. (3.92)
$$R_t = \sqrt{3} \cdot h_{cu} = \sqrt{3} \cdot \frac{L}{l_c} \cdot h_{eq} = \sqrt{3} \cdot L \cdot \frac{v_w}{v_s} \cdot \sqrt{\frac{a_e}{d_e}}$$

It is, therefore, expected, and also experienced in practice, that workpiece roughness produced using a particular wheel and workpiece combination depends primarily on the speed ratio and secondarily on the depth of cut according to $\sqrt{a_e}$. This conclusion is confirmed by results from Salije.[3] Some typical values are shown in Fig. 3.24.

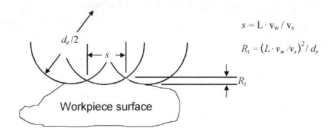

$s = L \cdot v_w / v_s$

$R_t = (L \cdot v_w / v_s)^2 / d_e$

Figure 3.22. Theoretical peak-to-valley roughness based on kinematics.

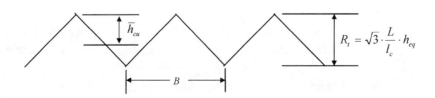

$$R_t = \sqrt{3} \cdot \frac{L}{l_c} \cdot h_{eq}$$

Figure 3.23. Theoretical peak-to-valley roughness based on the triangular grain shape.

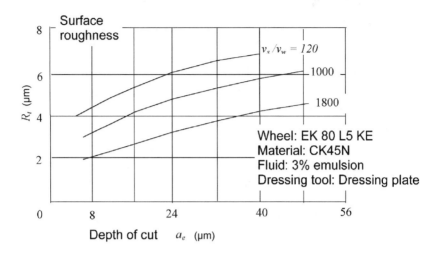

Figure 3.24. Typical values of surface roughness with increasing depth of cut and speed ratio.

The final roughness value of workpiece roughness at the end of a grinding operation depends on the final depth of cut. With spark-out, the depth of cut reduces and the roughness value is reduced.

The roughness predicted from Eq. (3.92) underestimates the real values due to ploughing. In ploughing, the material is displaced sideways and upwards by the grain as illustrated in Fig. 3.25.

In practice, many factors can affect and increase or reduce the roughness. It is impossible to predict roughness with any accuracy from purely analytical models. Some of these factors include:

1. Irregular grain spacing.

2. Irregular grain depths.

3. Effect of wheel dressing on grain density.

4. Effect of wheel wear on grain density.

5. Effect of wheel loading.

6. Effect of ploughing.

7. Deflection of the grains which reduces roughness.

8. Effect of spark-out which reduces roughness.

9. Adhesion between the workpiece and the grains which increases roughness.

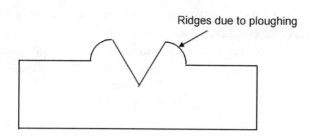

Figure 3.25. Surface roughness is increased due to ploughing.

Factors 1 to 4 are due to variability of grain spacing. Factors 5, 6, and 9 depend, at least partly, on temperature rise in the grinding contact zone.

A cooperative work by CIRP[9] showed that for a particular wheel and workpiece combination, roughness could be expressed empirically using a relationship of the same nature as Eq. (3.83).

Eq. (3.93) $\qquad R_a = R_1 \cdot h_{eq}^r$

where the constants relate to the particular wheel-workpiece combination and r is less than 1. The value of r is reduced if spark-out is employed. Typical values of r after spark-out lie between 0.15 and 0.6.

3.5 EFFECT OF DRESSING

The shape and distribution of the cutting edges on the surface of the wheel can be modified by the *dressing* process. Dressing is a process where the shape and topography of a grinding wheel are created. The wheel is machined with a dressing tool as illustrated in Fig. 3.26 for a single-point dressing tool.

The diamond tool is fed past the wheel at a feedrate corresponding to a feed per wheel revolution, f_d. The tool removes a layer of the wheel equal to the dressing depth, a_d.

The dressing depth is larger than the peak-to-valley roughness, R_t, corresponding to the dressing tool motion. The dressing feed, f_d, must be small enough to ensure that all the grains are dressed. This is ensured by making f_d/b_d less than unity, where b_d is the width of the dressing tool engaged with the wheel.

With coarse dressing, large values of a_d and f_d are employed so that the dressing helix is more pronounced and the density of the cutting edges on the wheel surface is reduced. With fine dressing the reverse is true, the density of the cutting edges is increased.

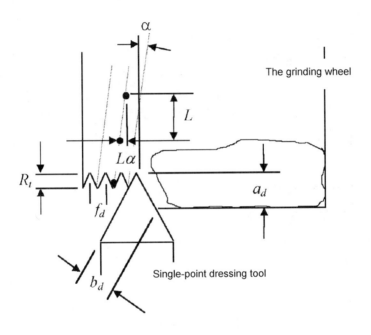

Figure 3.26. The single point dressing process.

It might be assumed that the lateral grain spacing is $\bar{b}_{cu} = f_d$. However, immediately after dressing, the active grains lie along the helix formed by the previous pass of the dressing tool. Therefore, the grains are offset from each other on average by a distance of $L\alpha$ as illustrated in Fig. 3.26.

In the grinding which takes place just after dressing, the helix pattern tends to disappear as the grinding wheel wears. The spacing then tends to revert towards the spacing based on the basic structure and composition of the wheel.

In fine dressing operations, both a_d and f_d are small. The dressing operation is often completed with one or more passes where $a_d = 0$. Fine dressing affects the wheel in an opposite sense to coarse dressing. The grain spacing is reduced with fine dressing and the grain density increased. The surface roughness of the workpiece is small immediately after dressing, but increases with time as more material is removed in the grinding process.

Fine dressing tends to produce a blunt wheel, but also produces cracks in the grains. The result is high initial power and grinding forces and an unstable process. Within a short time, particles of the grains fracture or pull out of the surface and the power and forces reduce.

In precision grinding, fine dressing conditions typically correspond to $a_d = 5$ mm and $f_d = 0.05$ mm/rev. Coarse dressing conditions typically correspond to $a_d = 25$ mm and $f_d = 0.25$ mm/rev. These values may be adjusted upwards or downwards depending on the grit size of the wheel and the width of the dressing tool.

3.6 SUMMARY OF PRINCIPAL KINEMATIC PARAMETERS

SYMBOL	EXPRESSION	MEANING
C	1/LB	Mean cutting edges per unit area.
L	$1/P_c(x)$	Mean grain spacing along motion.
B	$1/P_c(z)$	Mean grain spacing in lateral direction.
l_g	$\sqrt{a_e d_e}$	Geometric contact length.
$\bar{h}_{cu}$	$L \cdot \dfrac{v_w}{v_s} \cdot \sqrt{\dfrac{a_e}{d_e}}$	Mean thickness of the undeformed chips.
$\bar{h}_{cu}$	$\sqrt{\dfrac{2}{C.r} \cdot \dfrac{v_w}{v_s} \cdot \sqrt{\dfrac{a_e}{d_e}}}$	Value for triangular chip.

(Cont'd.)

SYMBOL	EXPRESSION	MEANING
r	b_{cu}/h_{cu}	Width/thickness ratio of chip.
r_{cu}	$\dfrac{d_e}{2L} \cdot \dfrac{v_w}{v_s}$	Length/thickness ratio of chip.
$h_{cu.max}$	$2\bar{h}_{cu}$ $\sqrt{3}.\bar{h}_{cu}$	Max. thickness of undeformed chips for *(i)* a rectangular chip and *(ii)* a triangular chip.
h_{eq}	$a_e.v_w/v_s$	Equivalent chip thickness.
$\bar{b}_{cu}$	B	Mean width of the undeformed chips.
$\bar{V}_{cu}$	h_{eq}/C	Mean volume of undeformed chips.
Q_w	$b.a_e.v_w$	Material removal rate from workpiece.
G	V_w/V_s	Grinding wear ratio.
d_e	$d_s d_w/(d_w \pm d_s)$	Equivalent wheel diameter (- for internal).
t_s	l_c/v_s	Grain contact time per interaction.
R_a	$k.\bar{h}_{cu} = R_1.h_{eq}^r$	Peak-to-valley roughness.

REFERENCES

1. Alden, G. I., Operation of Grinding Wheels in Machine Grinding, *Trans ASME*, 36:451–460 (1914)

2. Guest, J. J., Grinding Machinery, Edward Arnold, London (1915)

3. Tönshoff, H. K., Peters, J., Inasaki, I., and Paul, T., Modeling and Simulation of Grinding Processes, *Annals of the CIRP*, 41(2):677–688 (1992)

4. Rowe, W. B., Miyashita, M., and Koenig, W., Centerless Grinding Research and Its Application in Advanced Manufacturing Technology, *Annals of the CIRP*, Keynote paper 38(2):617–625 (1989)

5. Shaw, M. C., Principles of Abrasive Processing, *Oxford Science Publ.* (1996)

6. Brough, D., Bell, W. F., and Rowe, W. B., A Re-examination of the Uncut Models of Grinding and Its Practical Implications, *Proc. 24th Int. Machine Tool Design and Research Conf.*, Manchester, Pergamon Press, p. 261 (1983)

7. Backer, W. R., Marshall, E. R., and Shaw, M. C., The Size Effect in Metal-Cutting, *Trans ASME*, 74:61–72 (1952)

8. Reichenbach, G. S., Mayer, J. E., Kalpacioglu, S., and Shaw, M. C., The Role of Chip Thickness in Grinding, *Trans. ASME*, pp. 847–859 (May 1956)

9. Snoeys, R., and Peters, J., The Significance of Chip Thickness in Grinding, *Annals of the CIRP*, 23(2):227–236 (1974)

10. Malkin, S., Grinding Technology, Ellis Horwood, UK (1989)

11. Chen, X., and Rowe, W. B., Analysis and Simulation of the Grinding Process, Parts I–III, *Int. J. Machine Tools and Manuf.*, 36(8):871–906 (1996)

12. Chen, X., and Rowe, W. B., Analysis and Simulation of the Grinding Process, Part IV, *Int. J. Machine Tools and Manuf.*, 38(1–2):41–49 (1997)

4

Contact Mechanics

In this chapter, we first consider the size of the contact area between a grinding wheel and a workpiece, taking into account the contact stresses between the grains and the workpiece. Consideration is then given to mainly elastic effects of the contact stresses between the grains and the workpiece. Plastic effects are considered in greater detail in Ch. 5, "Forces, Friction, and Energy."

4.1 CONTACT AREA

The size of the contact region for abrasive machining processes has, until recent years, been largely disregarded. This is mainly because the importance of the contact area has not been understood. Estimates of contact area have varied by up to 600%.

The importance of contact length was supported with empirical evidence by Saljé.[1] The contact length is of considerable tribological importance since contact length governs the following:

- The length of the sliding contact of the abrasive grains, l_c.

- The intensity of the energy input to the workpiece, $q = E/b_w \cdot l_c$.

- The thickness of the uncut chips.

- The kinematic roughness of the workpiece.
- The time of contact of the abrasive grains.
- The number of abrasive grains in contact.
- The wear of the grains.

The overall or apparent area of contact, A_c, in Fig. 4.1 is given by

Eq. (4.1) $A_c = b_w \cdot l_c$

Early workers simply assumed that l_c was equal to the geometric contact length:

Eq. (4.2) $l_g = \sqrt{a_e \cdot d_e}$

This has since been shown to be an oversimplification in the majority of conventional grinding processes.

It is clear that the wear behavior of the abrasive tool and the temperature of the workpiece are affected by the length of the contact zone. Contact length affects many other aspects of performance including the surface roughness, the cutting forces, and the ability of the abrasive to cut the workpiece efficiently. It is, therefore, appropriate to examine contact behavior in more detail.

It was shown by Brown,[2] Verkerk,[3] and others including Gu[4] and Zhou,[5] that the length of the contact region, l_c, is substantially larger than the geometric contact length, l_g.

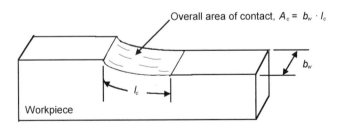

Figure 4.1. The overall or "apparent" area of abrasive contact.

The main reason for the increased area of contact is the deflection of the grinding wheel grains due to the effect of the normal force. This effect is strong for vitrified and resin bonded wheels where the elastic modulus of the porous abrasive grain-bond mixture is much lower than the elastic modulus of most workpiece materials. The deflection of individual grains is also increased due to high stresses at the discrete points of contact.

4.2 CONTACT LENGTH

4.2.1 Due To Deflection

Figure 4.2 shows a wheel of effective diameter (d_e) pressed into a plane surface distance (δ) by a normal force (F_n).

From the geometry of a circle and an intersecting chord, it is clear that the contact length is closely approximated by

Eq. (4.3) $\qquad l_f = 2 \cdot \sqrt{\delta \cdot d_e}$

Equation (4.3) is only accurate if the value of δ is small compared to d_e, which is obviously true for grinding wheels.

It is unimportant whether the deflection, δ, occurs by flattening of the wheel or by penetration of the wheel into the workpiece, since l_f is the chord length of the intersection between the two undeformed surfaces.

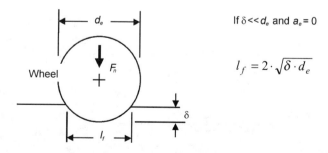

Figure 4.2. Contact length due to a normal force, F_n.

4.2.2 Due To Depth of Cut

Figure 4.3 illustrates the effect of depth of cut. As previously established in Ch. 3, "Kinematic Models of Abrasive Contacts," the geometric contact length, l_g, due to the depth of cut, a_e, is

Eq. (4.4) $l_g = \sqrt{a_e \cdot d_e}$

In the previous section, we took into account the deflection of the wheel and workpiece; here, it is assumed that the wheel and the workpiece remain rigidly in shape, so that no contact takes place after the material has been cut from the workpiece.

The roughness of both the workpiece surface and the grinding wheel surface causes variations in the grain depth of cut as discussed in the previous chapter. Some authors have proposed that roughness can be considered as variations in the total depth of cut.[6]

The geometric effect of roughness on depth of cut and, hence, on geometric contact length is illustrated in Fig. 4.4.

If the size of the workpiece before and after grinding is considered to be measured through the average center line of the roughness, the result of the roughness can either add or subtract from the geometric contact length. In Fig. 4.4, the roughness is illustrated by a wavy line. In practice, the height of the waviness is usually small in comparison with the pitch.

It is not totally clear in the figure whether the effect of the roughness adds or subtracts from the contact length since this depends on the position of the wave on the surface. In practice, both effects occur, which increase the spread of the distribution of contact lengths of individual grains.

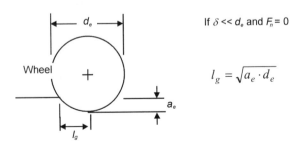

Figure 4.3. Contact length due to real depth of cut, a_e.

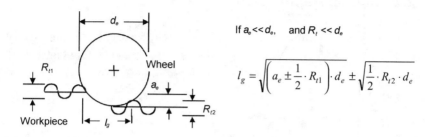

Figure 4.4. Geometric contact length taking into account the geometric effect of roughness.

For rough surfaces, it is proposed that the range of the variations in contact length due to surface roughness can be described by

Eq. (4.5)
$$l_g = \sqrt{\left(a_e \pm \frac{1}{2} \cdot R_{t1}\right) \cdot d_e} \pm \sqrt{\frac{1}{2} \cdot R_{t2} \cdot d_e}$$

If the roughness always added to the effect of the depth of cut, this would have a substantial effect on measured values of contact length. Even so, it can be seen that contacts will occur outside the expected range based on the mean depth of cut, and the maximum values of contact length will be increased.

4.2.3 Combined Deflection and Depth of Cut

Approximate Method. This method, although rather inaccurate, is included because it appears in the literature. It provides a simple way to visualize the effects of combined deflection and depth of cut as illustrated in Fig. 4.5. Here, it is assumed that the wheel is flattened by the deflection, δ, so that to achieve a real depth of cut, a_e, it is necessary to feed the wheel a distance, $a_e + \delta$, towards the workpiece. This leads to an over-estimate of the contact length, l_c.

Eq. (4.6)
$$l_c = \sqrt{(a_e + \delta) \cdot d_e} + \sqrt{\delta \cdot d_e}$$

It follows from Eqs. (4.3) and (4.4) that

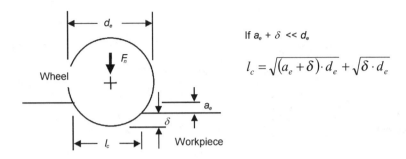

Figure 4.5. Over-estimate of contact length due to normal force and depth of cut.

Eq. (4.7) $l_c = \sqrt{l_g^2 + \dfrac{1}{4} \cdot l_f^2} + \sqrt{\dfrac{1}{4} l_f^2}$

Squaring both sides and evaluating terms for $l_g = l_f$ leads to

Eq. (4.8) $l_c^2 = l_g^2 + 1.62 l_f^2$ if $l_g = l_f$

Accurate Method. An accurate approach requires that we solve for the effects of l_g and l_f simultaneously as illustrated in Fig. 4.6. Here, the deflection, δ, is based on the effective wheel diameter, d_e, pressed against the curved surface of the workpiece having a contact diameter, d_{cu}.

The effective contact diameter for the two bodies is d_{ef}. Using these definitions, the results are consistent with the results of established methods in contact mechanics.

According to the definition given in the previous chapter for effective wheel diameter in grinding,

Eq. (4.9) $\dfrac{1}{d_e} = \dfrac{1}{d_s} + \dfrac{1}{d_w}$

The effective diameter, d_{ef}, for the purposes of calculating the deflection must be based on the combined effect of the effective wheel diameter, d_e, and the unloaded workpiece contact diameter, d_{cu}.

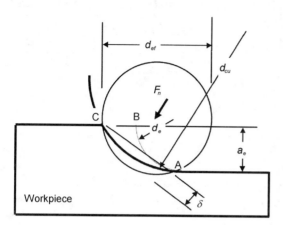

Figure 4.6. The arc of contact, diameter d_{ef}, combining effect of depth of cut, a_e, and deflection to normal force, F_n.

Eq. (4.10) $$\frac{1}{d_{ef}} = \frac{1}{d_e} - \frac{1}{d_{cu}}$$

The minus sign is necessary because the two contact curves are conformal. The diameters are illustrated in Fig. 4.6. The arc of contact in the absence of a normal force and in the absence of deflections is given by AB and

Eq. (4.11) $$l_g^2 = a_e \cdot d_e$$

Under the combined effect of the normal force and depth of cut, the arc of contact is given by AC and

Eq. (4.12) $$l_c^2 = a_e \cdot d_{cu}$$

The arc of contact is also the contact length, l_c, under the normal force, so that

Eq. (4.13) $$l_c^2 = 4 \cdot \delta \cdot d_{ef}$$

The expression for contact length can be evaluated in a more useful form. From Eqs. (4.11), (4.12), and (4.13)

$$d_e = \frac{l_g^2}{a_e}$$

$$d_{cu} = \frac{l_c^2}{a_e}$$

$$d_{ef} = \frac{l_c^2}{4\delta}$$

Substituting for these diameters in Eq. (4.10):

Eq. (4.14) $$\frac{4\delta}{l_c^2} = \frac{a_e}{l_g^2} - \frac{a_e}{l_c^2}$$

Rearranging terms:

Eq. (4.15) $$l_c^2 = l_g^2 + \frac{4\delta}{a_e} \cdot l_g^2$$

From the definitions of l_f and l_g given in Eqs. (4.3) and (4.4):

Eq. (4.16) $$\frac{4\delta}{a_e} = \frac{l_f^2}{l_g^2}$$

so that the accurate method provides the elegant and convenient result:

Eq. (4.17) $$l_c^2 = l_g^2 + l_f^2$$

It should be noted that we have employed d_e in Eq. (4.3) instead of d_{ef}. In other words, Eq. (4.17) is valid when l_f is defined in terms of d_e. This is a convenient form because we can calculate the terms l_g and l_f quite independently and find the combined length using Eq. (4.17). This is consistent with the theory proposed by Qi, Rowe, and Mills.[7]

In very shallow cut grinding, if the normal force is sufficiently large, $l_g \approx 0$. In other words,

Eq. (4.18) $l_c = l_f$ when $\delta \gg a_e$

In deep grinding, the deflections are small compared with the depth of cut, so that

Eq. (4.19) $l_c = l_g$ when $a_e \gg \delta$

4.3   SMOOTH BODY ANALYSIS

The relationships for elastic deflections between smooth bodies in contact is based on the analysis by Hertz in 1882 for spheres in contact.

For the smooth body analysis, it is assumed that the grinding wheel and the workpiece are two parallel cylinders pressed into contact. The surfaces are assumed to be perfectly smooth at the macro level and also at the micro level where asperities are normally observed. The result of this assumption is that 100% contact is achieved in the contact region. The situation is illustrated in Fig. 4.7.

When two cylinders are loaded against each other, the stress is concentrated around the contact region and the deformation takes place mainly in this area.

It is assumed that the concentration of stress in the contact region depends on the conformity of the surfaces. The conformity is defined as the relative curvature of the two surfaces, where curvature is the inverse of radius of curvature.

Based on this assumption, the contact is represented as in Fig. 4.2. The relative radius of curvature is $d_e/2$ and explained in Eq. (4.20):

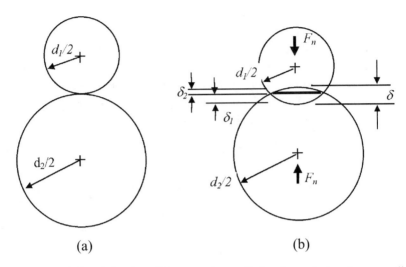

(a)   (b)

Figure 4.7. Relative deflection, δ, between two cylinders in contact under an applied normal force. *(a)* Unloaded and *(b)* loaded.

Eq. (4.20) $$\frac{1}{d_{ef}} = \frac{1}{d_1} \pm \frac{1}{d_2}$$

The plus sign applies for external diameters. If the cylinder, d_1, is pressed against an internal diameter, d_2, the negative sign applies.

The pressure distribution based on Hertz is parabolic and may, therefore, be described as illustrated in Fig. 4.8 by

Eq. (4.21) $$p = p_{max} \left[1 - \left(\frac{2x}{l_{fs}} \right)^2 \right]^{\frac{1}{2}}$$

where

$$x \leq \pm \frac{l_{fs}}{2}$$

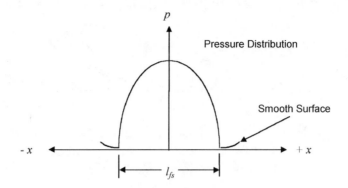

Figure 4.8. The parabolic pressure distribution for two perfectly smooth aligned cylinders pressed in contact based on Hertz.

The normal displacement, $\delta_1(x)$, of the surface of cylinder 1 depends on the elastic properties including Young's modulus, E, and Poisson's ratio, v. For consistency with the pressure distribution, we must make $\delta_1(x) = 0$, at $x = \pm\, l_{fs}/2$. Ignoring end effects and following the approach described by Williams,[8] we can express the deformation in the contact zone as

Eq. (4.22)
$$\delta_1(x) = \frac{1-v^2}{E_1} \cdot \frac{2\cdot p_{max}}{l_{fs}} \cdot \left[\left(\frac{l_{fs}}{2}\right)^2 - x^2 \right]$$

A similar expression applies for the second cylinder. Adding the two deflections, the total deflection, $\delta(x)$, may be represented as

Eq. (4.23)
$$\delta(x) = \frac{2\cdot p_{max}}{l_{fs} \cdot E^*} \cdot \left[\left(\frac{l_{fs}}{2}\right)^2 - x^2 \right]$$

where E^* represents the elastic properties of the two surfaces.

Eq. (4.24)
$$\frac{1}{E^*} = \frac{1-v_1^2}{E_1} + \frac{1-v_2^2}{E_2}$$

From Eq. (4.23) when $x = 0$ and $\delta(x) = \delta$, the maximum deflection is related to the maximum pressure by

Eq. (4.25) $p_{max} = \dfrac{2 \cdot E^*}{l_{fs}} \cdot \delta$

From Eq. (4.3), we note that the deflection, δ, is related to the contact length, l_{fs}, and the effective diameter of the two surfaces when undeformed. Substituting for δ in Eq. (4.25) gives

Eq. (4.26) $p_{max} = \dfrac{l_{fs} E^*}{2 \cdot d_{ef}}$

Integrating the pressure distribution in Eq. (4.22) to obtain the specific normal force and substituting from Eq. (4.26) for the maximum pressure yields

Eq. (4.27) $F_n' = \dfrac{F_n}{b_w} = \dfrac{\pi \cdot l_{fs}^2 \cdot E^*}{8 \cdot d_{ef}}$

The contact length for smooth bodies is, therefore, given by

Eq. (4.28) $l_c^2 = l_{fs}^2 = \dfrac{8 \cdot F_n' \cdot d_{ef}}{\pi \cdot E^*}$

When $a_e = 0$, the contact length for a specific applied force, F_n', is

Eq. (4.29) $l_f^2 = \dfrac{8 \cdot F_n' \cdot d_e}{\pi \cdot E^*}$

4.4 ROUGH SURFACE ANALYSIS

Equations (4.28) and (4.29) apply for smooth bodies. In practice, apparently smooth bodies viewed microscopically are far from smooth. Steel surfaces have to be pressed together very hard, with substantial plastic flow, before 100% contact can be achieved. This phenomenon, due to asperity contact, results in greatly reduced stiffness across a contact between two surfaces at light loadings.

There is now ample experimental evidence to show that abrasive machining cannot be accurately analyzed assuming smooth body contact.[9]–[11] A grinding wheel is designed to contact a workpiece on widely spaced cutting edges as illustrated in Fig. 4.9.

Each cutting edge deforms the workpiece plastically as it moves through the cutting zone at high speed. The stress at the grain contact points exceeds the yield stress applicable for the contact temperature of the workpiece material and may be assumed to reach a limiting plastic stress.

Figure 4.10 illustrates a modified pressure distribution, taking into account the roughness of real surfaces and a limiting value of stress for a plastic contact. The normal force is supported at a number of very small contact points. The pressure at each of these small contact points is limited by the plastic stress of the workpiece material. Penetration, or flattening, occurs until enough contact points have been established to support the normal force.

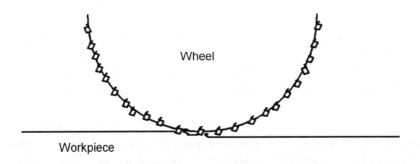

Figure 4.9. A grinding wheel is designed to contact a workpiece on widely spaced cutting edges.

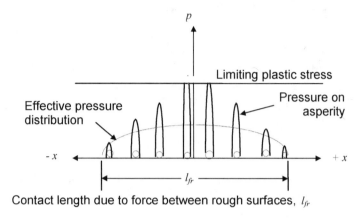

Figure 4.10. Pressures for rough surfaces in contact. The contact length is increased because the real area of contact is much smaller than the apparent area of contact.

Of course not all the contact points are plastic; some are plastic but many more are elastic. The result is that the effective pressure distribution remains approximately parabolic, tending to become more gaussian in shape as the roughness increases. More importantly, the contact length for rough surfaces (l_{fr}) has to increase substantially beyond the contact length for smooth surfaces (l_{fs}) for the same value of normal force.

The effective pressure distribution in Fig. 4.10 is shown by the dotted line. The effective pressure distribution is assumed to follow a Hertz pressure distribution according to Eq. (4.21) although the value of $p_{\max}$ is much lower than for smooth contact.

It is clear that the real area of asperity contact is much smaller than the overall apparent area of contact. This is consistent with the widely known explanation for a constant coefficient of friction between hard surfaces which is based on the finding that the real area of asperity contact increases proportionally to the normal force. This results in a traction force which is proportional to the normal force but independent of the apparent area of contact.

The equivalent Hertzian contact diameter (d_{ef}) between the un-loaded contact diameter (d_{cu}) of the workpiece cut surface and the equivalent wheel diameter (d_e) is defined according to Figs. 4.6 and 4.11.

Following the method of Greenwood,[12] we can define a roughness factor, R_r, which is the ratio of the rough and smooth contact lengths:

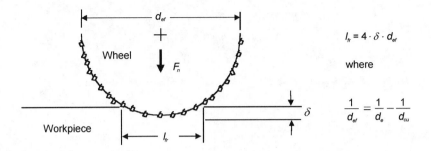

Figure 4.11. The equivalent Hertzian contact diameter, d_{ef}, when a normal force is applied between a rough grinding wheel of effective diameter, d_e, and a workpiece unloaded cut diameter, d_{cu}. (See also, Fig. 4.6.)

Eq. (4.30)

$$R_r = \frac{l_{fr}}{l_{fs}}$$

In experiments, the value of the roughness factor, R_r, was found to increase with roughness. The roughness was described by a roughness parameter (α), and the ratio of the roughness (σ_s) to the bulk deflection (δ).[13]

Eq. (4.31)

$$\alpha = \frac{\sigma_s}{\delta}$$

The sum of the variances of the asperity heights of the two contact surfaces is σ_s^2. For a particular case where $\sigma_s = 0.5$, the contact length ratio $R_r = 1.3$.

In grinding, it is difficult to describe or define σ_s for the grinding wheel. However, it is clear from measurements that R_r is usually an order of magnitude larger than the case described above. The roughness of a grinding wheel is clearly very significant.

Employing the definition of R_r given in Eq. (4.30), we can modify and generalize the expression given in Eq. (4.29) to apply to grinding:

Eq. (4.32)

$$l_{fr}^2 = \frac{8 \cdot R_r^2 \cdot F_n' \cdot d_{ef}}{\pi \cdot E *}$$

Multiplying both sides by d_e/d_{ef} we obtain

Eq. (4.33) $$l_f^2 = \frac{8 \cdot R_r^2 \cdot F_n' \cdot d_e}{\pi \cdot E^*} = \frac{d_e}{d_{ef}} \cdot l_{fr}^2 = \frac{d_e}{d_{ef}} \cdot l_c^2$$

Substituting for l_f in Eq. (4.17),

Eq. (4.34) $$l_c^2 = \frac{8 \cdot R_r^2 \cdot F_n' \cdot d_e}{\pi \cdot E^*} + a_e \cdot d_e$$

Equation (4.34) can be used if the normal force is known. An approximate method can be used based on the grinding power, which is often more readily available. While the approximate method is less accurate, any method which makes a reasonable estimate of the real contact length will be more accurate than simply ignoring the effect of deflections.

The contact length can be expressed in terms of power using an estimated value of the force ratio, $\mu = F_t/F_n$. This approach is used for on-line process monitoring for the purpose of avoiding thermal damage. In terms of grinding power, the contact length is

Eq. (4.35) $$l_c^2 = \frac{8 \cdot R_r^2 \cdot P \cdot d_e}{\pi \cdot \mu \cdot E^* \cdot v_s \cdot b_w} + a_e \cdot d_e$$

4.5 EXPERIMENTAL MEASUREMENTS OF ROUGHNESS FACTOR (R_r)

For a particular grinding wheel, it might be expected that the roughness factor, R_r, would be a constant. However, wheelwear, different workpiece materials, and the use of grinding fluid were all found to cause variations. The contact mechanics approach was substantially investigated by Qi in 1995.[11] Qi found that the new approach gave greatly improved correlation between measurement and prediction. The following sections summarize some of the important results achieved by Qi.

4.5.1 Comparison With Measurements by Verkerk

In 1975, Verkerk[3] was one of the first researchers to investigate contact length in grinding. The first attempts to test the new contact mechanics approach were, therefore, based on a comparison with the results obtained by Verkerk for flat surface grinding, as illustrated in Fig. 4.12. Since the contact length depends on a number of discrete contacts between grains and workpiece, there is some difficulty in defining what is meant by "measured contact length."

Some workers have interpreted the contact length as the distance between the first measured contact between a point on the workpiece and a grain and the last measured contact between the same point on the workpiece and a grain. However, vibrations between the workpiece and the wheel can artificially increase the measurement by this method. It is not always clear what approach has been employed by different researchers. Based on our measurements, it appears that Verkerk may not have included the sparse events which occur outside the main contact region and, therefore, achieved a conservative estimate of contact length. This seems to be a reasonable approach since these sparse events have little effect on the process.

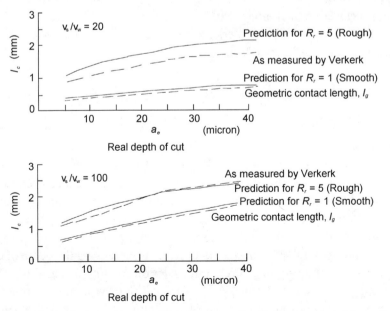

Figure 4.12. Predictions of contact length compared with the experimental results of Verkerk.

In Fig. 4.12, the predicted values are based on values of $R_r = 1$ and $R_r = 5$. These values were chosen because $R_r = 1$ corresponds to the smooth body analysis and $R_r = 5$ gave reasonable correlation for this set of experimental conditions. Results are also given for contact length based purely on l_g. It can be seen that the values based on smooth body mechanics predicted a contact length only slightly greater than l_g.

For the rough surface mechanics, the conditions employed were as follows:

$$E_w \text{ for steel: } 213 \text{ kN/mm}^2$$

$$E_s \text{ for vitrified alumina: } 49.6 \text{ kN/mm}^2$$

$$n_w \text{ for steel: } 0.29$$

$$n_s \text{ for vitrified alumina: } 0.22$$

$$d_s: 500 \text{ mm}$$

$$d_w: 90 \text{ mm}$$

$$v_s: 30 \text{ m/s}$$

$$v_w: \text{ various}$$

$$a_e: \text{ various}$$

It can be seen from Verkerk's results that measured values of contact length, l_c, were 1.5 to 3.5 times the geometric contact length, l_g. The rough surfaces contact model with $R_r = 5$ gave reasonable agreement with the experimental results of Verkerk, whereas the smooth contact model made very little difference from l_g.

4.5.2 Effect of Depth of Cut

Qi measured contact length by inserting an insulated electrical contact sensor in the workpiece. A voltage was applied to the sensor and the workpiece was connected to earth. When the grinding wheel passes over the sensor, contact is established with the workpiece. The contact length is estimated from the duration of the signal, and the extent of the contact is estimated from the magnitude of the signal. A typical contact signal is discussed in Sec. 4.5.4.

It is generally found that the effect of increasing the depth of cut is to increase the grinding force, almost in direct proportion, Fig. 4.13. Therefore, it is expected from Eq. (4.34) that l_f and l_g will increase roughly in the same proportion. In other words, the ratio, $R_L = l_c/l_g$ is expected to remain approximately constant with varying depth of cut.

Eq. (4.36) $\qquad R_L = \dfrac{l_c}{l_g}$

Therefore, it is expected that contact length will increase approximately with the square root of depth of cut. This expectation is confirmed by many results of which Figs. 4.12 and 4.14 are typical examples.

Grinding wheels: 19A60L7, 170 mm, 30 m/s

: B91ABN200, 174 mm, 30 m/s

Workpiece: AISI 1055, 0.1 m/s

Fluid: None

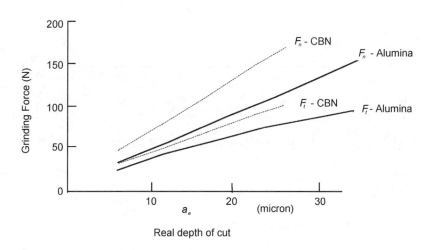

Figure 4.13. Grinding forces and depth of cut.

Variations of R_L from a constant value occur at smaller depths of cut and result from the variation of direct proportionality between force and depth of cut. It is seen that R_L tends to increase as depth of cut reduces towards zero due to the proportionately higher ratio of normal force and depth of cut.

Wheels: Alumina (A), 19A60L7V 170 mm

: CBN (B), B91ABN200 170 mm

Workpiece: AISI 1055

Process: Horizontal surface grinding

Workspeeds: 0.1 m/s and 0.3 m/s

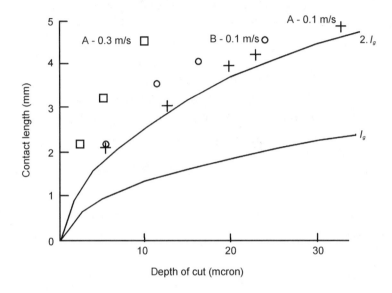

Figure 4.14. Measured contact lengths compared with geometric contact lengths in surface grinding.

4.5.3 Effect of Workspeed

In Fig. 4.14, it is seen that the contact length is greatly increased for the same depth of cut when workspeed is increased from 0.1 to 0.3 m/s. This result is inexplicable unless the effect of grinding force is taken into consideration. For the same depth of cut, geometric contact length is unchanged since

$$l_g = \sqrt{a_e \cdot d_e}$$

However, if we take the variation of normal force into account, according to Eq. (4.34), the explanation becomes clear. The grinding forces are greatly increased when workspeed is changed from 0.1 to 0.3 m/s. This is due to the higher material removal rate, which is increased by 300%, and the corresponding increase in the normal force is illustrated in Fig. 4.15.

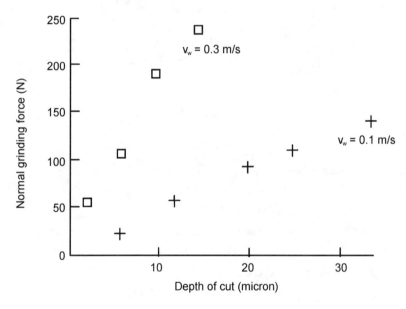

Figure 4.15. Effect of workspeed on normal force in surface grinding.

4.5.4 Evaluation of Roughness Factor, R_r, and Contact Length, l_c

Figure 4.16 shows an example of a smoothed contact signal used to evaluate R_r and l_c.

Wheels: Alumina (A), 19A60L7V, 170 mm
Workpiece: AISI 1055, v_w = 0.2 m/s
Process: Horizontal surface grinding, v_s = 30 m/s
Fluid: Synthetic oil, 2% in water emulsion

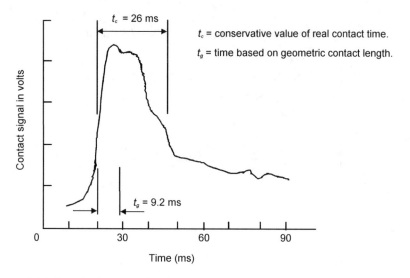

Figure 4.16. Smoothed contact signal.

The following example illustrates the calculation and evaluation of values of R_r.

Process: Surface grinding
Grinding wheel: 19A60L7V, d_s = 170 mm
Wheel properties: E_s = 49.6 kN/mm², v_s = 0.22
Workpiece: AISI 1055, v_w = 0.2 m/s, b_w = 15 mm
Workpiece properties: E_w = 213 kN/mm², v_w = 0.29
Depth of cut: a_e = 0.020 mm
Measured normal force: 225 N

The equivalent diameter of the wheel is given by

$$\frac{1}{d_e} = \frac{1}{170} + \frac{1}{\infty}$$

so that $d_e = 170$ mm. Geometric contact length:

$$l_g = \sqrt{a_e \cdot d_e} = 1.84 \text{ mm}$$

Geometric contact time of a point on the workpiece:

$$t_g = \frac{l_g}{v_w} = 9.22 \text{ ms}$$

Real contact time of a point on the workpiece estimated conservatively from Fig. 4.16:

$$t_c = 26 \text{ ms}$$

Real contact length:

$$l_c = t_c \cdot v_w = 5.2 \text{ mm}$$

Contact length due to the force:

$$l_f^2 = l_c^2 - l_g^2$$

$$l_f = 4.86 \text{ mm}$$

The roughness factor from Eq. (4.33):

Eq. (4.37) $R_r^2 = \dfrac{l_f^2 \cdot \pi \cdot E^*}{8 \cdot F_n' \cdot d_e}$

where

$$E^* = 42.57 \text{ kN/mm}^2$$

and

$$F_n' = 225/15 = 15 \text{ N/mm}$$

Evaluating the roughness factor from Eq. (4.37):

$$R_r = 12.5$$

This practical example, taken from a real grinding result, serves to demonstrate that real contact lengths in conventional grinding are typically 1.5 to 3.5 times greater than geometric contact lengths.

Roughness factors, R_r, averaged over a large number of results, typically range from 5 to 15. An average value for dry grinding of about 9 and for wet grinding of about 14 is expected for grinding common engineering steels. The value of R_r remains reasonably constant for a particular wheel-workpiece material combination, althougth the value of R_r is reduced as a wheel loses its sharpness due to wear or wheel loading. This might be expected from the reduced roughness of the wheel.

4.6 ELASTIC STRESSES DUE TO ABRASION

The forces at the contact points between the abrasive grains and the machined surface give rise to severe plastic deformations. This must be so, otherwise there would be no material removal. We will examine these plastic deformations in the Ch. 5, "Forces, Friction, and Energy." Before we do, we can obtain some important insights into the process from a consideration of the elastic stresses at a point of contact. For a more detailed consideration, the reader is referred to Suh[14] and Lucca.[16]

Point Load With Friction. Fleming and Suh[15] showed that a quick approximation to the sub-surface stresses may be found from the elastic solution for a point load, considering a line load in plane strain as illustrated in Fig. 4.17.

In polar coordinates, the stresses at a radius of r and angle of θ from the normal through the point of contact are given by

Eq. (4.38) $\qquad \sigma_{rr} = \dfrac{-2 \cdot F}{\pi \cdot r} \cdot \cos(\alpha + \theta)$

$$\sigma_{r\theta} = 0$$

$$\sigma_{\theta\theta} = 0$$

In other words, there is a compressive stress that is infinite at the point of contact and which reduces with increasing radius, r. In practice, an infinite stress is impossible due to material yield which gives rise to a finite size of the contact region. The material becomes plastic wherever the stress reaches the yield stress.

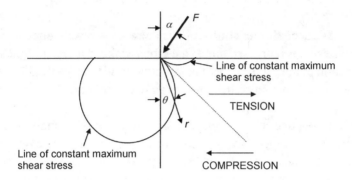

Figure 4.17. Effect of a point load (plane strain) inclined to a surface at an angle, α.

At a radius of r_d with $\alpha + \theta = 0$, the maximum shear stress becomes equal to the yield shear stress. Employing the Tresca yield criterion, yield occurs within a boundary:

Eq. (4.39)
$$K = \frac{F}{\pi \cdot r_d}$$

where K is the yield shear stress and r_d corresponds to the diameter of the circles of constant maximum shear stress.

In Fig. 4.17, there are two circles of maximum shear stress corresponding to the yield shear stress. The circle on the left is a region of compressive stress and the circle on the right is a region of tensile stress. Within the circles, the material is plastic.

It can be seen that with a cutting force moving from right to left, the material is subject to compression followed by tension. If the stresses exceed the yield stress, plastic strain occurs which has not been considered so far. When the load is removed, residual stresses arise in the surface of opposite sense to the loaded stress. After many passes of asperities across the surface, the residual stresses, due to the mechanical process, tend to be compressive.[17] However, the process is cyclic and contributes to crack propogation in crack-sensitive materials.

The friction coefficient is given by

Eq. (4.40) $$\mu = \frac{F \sin \alpha}{F \cos \alpha} = \tan \alpha$$

As the coefficient of friction is increased, the tendency to produce tensile stresses is increased due to the effect of the increased angle, α. This explanation of the occurrence of compressive and tensile stresses is termed a mechanical effect since the stresses arise due to a mechanically applied force.

In practice, tensile residual stresses are more usually caused by high grinding temperatures. The surface material in the region of high temperature yields due to the stresses caused by thermal expansion and due to the effect of the reduced yield stress at the elevated temperature. Elevated temperatures in the grinding contact zone are followed by rapid cooling due to the low bulk temperature of the workpiece. As the material contracts, it locks in tensile stresses in the surface. Thermal effects are discussed in Ch. 6, "Thermal Design of Processes."

In a practical abrasive contact of approximately spherical shape, the applied stresses at the surface according to Hertz are distributed with a semi-elliptical distribution as illustrated in Fig. 4.18. It can be shown that the maximum tensile stress without friction is caused by the radial stress component at the circular boundary of the abrasive contact:[14]

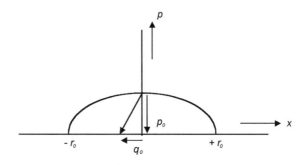

Figure 4.18. An elliptically distributed load as appropriate for a spherical or cylindrical contact of width, $2 \cdot r_0$.

Eq. (4.41)
$$\left(\sigma_{rr}\right)_{\max} = \frac{1-2\cdot v}{3} \cdot p_o$$

With friction, the maximum tensile stress moves outside the contact zone. This may initiate failure due to exceeding the maximum tensile stress, particularly for glassy materials. The stresses may be related to the normal and tangential grain forces:

Eq. (4.42)
$$p_o = \frac{3\cdot f_n}{2\cdot \pi \cdot r_o^2}$$

Eq. (4.43)
$$q_o = \frac{3\cdot f_t}{2\cdot \pi \cdot r_o^2}$$

where

Eq. (4.44)
$$r_o = \left(\frac{3\cdot f_t}{4} \cdot \frac{E^*}{d_{eg}}\right)^{1/3}$$

The effective diameter of the grain is d_{eg}.

It is found that the maximum shear stress occurs below the surface and nearer to the leading edge of the applied load. This is the position where yield initiates for ductile materials.

4.7 SUMMARY OF CONTACT STRESS IMPLICATIONS

The interactions of individual grains with the workpiece causes repeated compressive and tensile loading of the workpiece. The tensile stresses can cause crack propogation in brittle crack-sensitive materials. This effect is most pronounced with large values of maximum uncut chip thickness and high values of force ratio, μ.

The general tendency of the mechanical interactions of the grains with the workpiece is to cause residual compressive stresses. This effect is enhanced by low values of force ratio, μ. However, this effect will be reversed if the maximum temperature is sufficient to cause residual tensile stresses.

Elastic deflections of the grinding wheel and workpiece greatly increase the contact length of the grinding zone. This reduces maximum temperature for a particular grinding force, reduces surface roughness, reduces fracture wear of the grains, and increases the dulling wear of the grains.

The contact length is further increased with the use of grinding fluid which reduces the maximum temperature experienced.

There is some evidence that elastic deflections in the contact region can help to stabilize the grinding process against self-excited vibrations,[18]–[20] although it is important to maximize dynamic stiffnesses to avoid large vibration amplitudes and irregular wear of the grinding wheel with consequent poor workpiece quality.

REFERENCES

1. Saljé, E., and Möhlen, H., Fundamental Dependencies Upon Contact Lengths and Results in Grinding, *Annals of the CIRP*, 35(1):249–253 (1986)

2. Brown, R. H., Sato, K., and Shaw, M. C., Local Elastic Deflections in Grinding, *Annals of the CIRP*, 19(1):105–113 (1971)

3. Verkerk, J., The Real Contact Length in Cylindrical Plunge Grinding, *Annals of the CIRP*, 24(1):259–264 (1975)

4. Gu, D. Y., and Wager, J. G., New Evidence on The Contact Zone in Grinding, *Annals of the CIRP*, 37(1):335–338 (1988)

5. Zhou, Z. X., and Van Luttervelt, C. A., The Real Contact Length Between Grinding Wheel and Workpiece: A New Concept and A New Measuring Method, *Annals of the CIRP*, 41(1):387 (1992)

6. Vansevenant, E., A Sub-surface Integrity Model in Grinding, Dr. Ing thesis, Catholic University of Leuven (1987)

7. Qi, H. S., Rowe, W. B., and Mills, B., Contact Length in Grinding, *Proc. Inst. Mech. Engineers*, 211(J):67–85 (1997)

8. Williams, J. A., *Engineering Tribology*, Oxford Science Publ. (1994)

9. Rowe, W. B., Morgan, M. N., Qi, H. S., and Zheng, H. W., The Effect of Deformation on the Contact Area in Grinding, *Annals of the CIRP*, 42(1):409–412 (1993)

10. Rowe, W. B., Morgan, M. N., and Black, S. C., Validation of Thermal Properties in Grinding, *Annals of the CIRP*, 47(1):275–278 (1998)

11. Qi, H. S., A Contact Length Model for Grinding Wheel-Workpiece Contact, Ph.D. thesis, Liverpool, John Moores University (1995)

12. Greenwood, J. A., and Tripp, J. H., The Elastic Contact of Rough Spheres, *J. Applied Mechanics*, pp. 153–159 (Mar. 1967)

13. Johnson, K. L., *Contact Mechanics*, Cambridge University Press (1989)

14. Suh, N. P., *Tribophysics*, Prentice Hall (1986)

15. Fleming, J. R., and Suh, N. P., Mechanics of Crack Propogation in Delamination Wear, *Wear*, 44:39–56 (1977)

16. Lucca, D. A., A Sliding Indentation Model of the Tool-Workpiece Interface in Ultra-Precision Machining, *Trans ASME*, PD- 61:17–22 (1994)

17. Jahanmir, S., and Suh, N. P., Mechanics of Sub-surface Void Nucleation in Delamination Wear, *Wear*, 44:17–38 (1977)

18. Rowe, W. B., and Barash, M. M., Koenigsberger, F., Some Roundness Characteristics of Centerless Grinding, *Int. J. of Machine Tool Design & Research*, 5:203–215 (1965)

19. Sexton, J. S., Howes, T. D., and Stone, B. J., The Use of Increased Wheel Flexibility to Improve Chatter Performance in Grinding, *Proc. Inst. Mech. Engineers*, pp. 196, 291 (1982)

20. Rowe, W. B., Spraggett, S., and Gill, R., Improvements in Centerless Grinding Machine Design, *Annals of the CIRP*, 36:207–210 (1987)

5

Forces, Friction, and Energy

5.1 INTRODUCTION

Large grinding machines consume energy of the same order of magnitude as a family car traveling at 100 km/hr. Little wonder, then, that the grinding process can remove material at a high rate which may be limited only by the efficiency of the operating conditions.

The dissipation of this energy, typically 50 kW, equates directly to the production of heat. The process forces, although significant, are not as high as in chip removal processes such as milling, due to the much higher speeds of the abrasive. In comparison with a car, the lower speed range of the abrasive is typically 100 km/hr while speeds of 400 km/hr or even higher are sometimes employed.

In other abrasive machining processes such as honing, lapping, and polishing, the speeds of the abrasive are much lower, which limits the usable power in the process and, hence, the material removal rate.

The rate at which an abrasive machining operation can be carried out depends on the power which can be applied to the process and the efficiency of material removal. In other words, to remove material quickly, the following are needed:

- A high abrasive speed to carry energy to the machining point.

- A sharp abrasive tool and effective lubrication to minimize the forces and energy required.

- A large area of machining contact to maximize the surface being machined at any time.

When using a sharp abrasive tool, the forces are low, the job is completed quickly, little power is required, and the quality of the work produced will be very good for a correctly specified process.

In contrast, when using a blunt tool, the forces are high, the power required is high, the job takes a long time to complete, and various quality problems may be experienced. High vibration levels, noise, poor size and shape holding, thermally damaged workpieces, and even cracking or failure of the workpieces are typical of such problems.

The study of forces, friction, and energy is, therefore, as much about improving quality as about increasing output and reducing process costs.

5.2 FORCES AND POWER

In abrasive machining, a force acts between the abrasive tool and the workpiece. This force can be split into three components. Each of the three components has a distinct and different effect on process performance. The three components are:

- The tangential force, F_t.

- The normal force, F_n.

- The axial or side force, F_a.

The three components are illustrated for a grinding wheel in Fig. 5.1.

The tangential force, F_t, acts tangentially to the abrasive tool surface and also to the surface velocity of the abrasive tool. Due to the high speed of the wheel, the tangential force in grinding is mainly responsible for power dissipation, P, where:

Eq. (5.1) $P = F_t \cdot v_s$

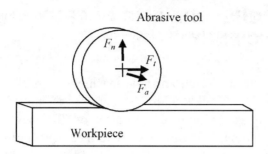

Figure 5.1. The three component forces in abrasive machining. The example is for horizontal axis surface grinding.

The normal force, F_n, is perpendicular to the abrasive tool surface. The normal force is usually much larger than the tangential force and acts directly to reduce the depth of cut. Therefore, F_n is mainly responsible for deflections, δ, of the workpiece, of the abrasive tool, and of the machine structure normal to the workpiece surface.

For a linear system, the deflection is related to the normal force by the total stiffness, k_m, of the machine-workpiece-tool system.

Eq. (5.2) $$\delta = \frac{F_n}{k_m}$$

Some machine elements may not be linear, although most designers try to achieve a linear system with high stiffness under all operating conditions.

The normal force, F_n, depends on the bluntness of the abrasive grains on the surface of the abrasive tool, and also depends on the hardness of the workpiece surface.

A side force, F_a, arises when there is a sideways movement of the abrasive tool.

The total machining force, F, is the vector sum of the component forces:

Eq. (5.3) $$F = \sqrt{F_t^2 + F_n^2 + F_a^2}$$

5.3 FORCES: SPECIFIC ENERGY AND EFFICIENCY

In energy terms, an abrasive machining process is efficient if material is removed quickly with low energy consumption. The specific energy is widely used as an inverse measure of efficiency.

In machining, specific energy is defined as the ratio of machining power to removal rate:

$$Specific\ energy = \frac{machining\ power}{rate\ of\ material\ removal}$$

Eq. (5.4)

$$e_c = \frac{P}{Q_w} = \frac{F_t \cdot v_s}{b_w \cdot a_e \cdot v_w}$$

An efficient process gives rise to a low specific energy and an inefficient process requires a high specific energy.

Since power and removal rate are both time related, the specific energy can also be defined as the ratio of the energy, E, to remove a volume, V_w, of the workpiece material. In other words, the specific energy is the energy per unit volume of removal:

Eq. (5.5) $e_c = \dfrac{E}{V_w}$

In units of the Système Internationale (SI), specific energy is usually expressed in joules per cubic millimeter and is obtained by dividing the power in watts by the removal rate in cubic millimeters per second. Care should be taken, however, when writing values into equations presented in this and subsequent chapters, to use a consistent set of units. In equations, SI units would normally require the specific energy to be converted into the less convenient form of joules per cubic meter.

When grinding ferrous materials, a specific energy of 10 J/mm^3 is very low, and the process is very efficient. This may be contrasted with large chip removal processes, such as turning and milling, where the specific energy may be an order of magnitude lower.

In fine abrasive machining processes, including creep-feed grinding, fine grinding, honing, lapping, and polishing, the specific energy may be higher than 100 J/mm^3. Clearly, such processes are relatively inefficient in energy terms. High specific energy values are more likely to be tolerated for finishing processes where high quality standards are demanded in terms of low surface roughness and small dimensional tolerances while machining hard or very hard surfaces.

However, despite the advantages of abrasive machining for ease of production of high quality parts, efficiency is important. In the first case above, an abrasive machine operating at maximum power can remove material 10 times faster than in the second case. The tangential force is the same in both cases, but the removal rate is dramatically different.

A high removal rate abrasive machining operation will normally be expected to lie at the low end of the specific energy range, although the specific energy is also dependent on the difficulty of machining the particular material. Some examples for different materials will illustrate this point.

The following examples illustrate typical values of forces and specific energy in grinding. The examples also illustrate the effect of a process fluid and of sharpness of the abrasive grains.

5.4 EXAMPLES OF MATERIALS AND THEIR GRINDING CONDITIONS

5.4.1 Gray Cast Iron

Figure 5.2 shows the normal and tangential forces measured when surface grinding a gray cast iron with a medium size 60 mesh grit alumina wheel using a water-based emulsion as the fluid. The measured forces increased slightly less than proportionately with depth of cut. The average force ratio is

$$\mu = F_t/F_n = 0.37$$

With a depth of cut $a_e = 20\,\mu m$, workspeed, $v_w = 0.3$ m/s, the tangential force, $F_t = 120$ N, and a wheelspeed, $v_s = 30$ m/s. The width of grinding contact is

b_w = 15 mm. The power and removal rate are given by $P = F_t \cdot v_s = 120$ × 30 = 3600 W, and $Q_w = b_w \cdot a_e \cdot v_w = 15 \times 10^{-3} \times 20 \times 10^{-6} \times 0.3 = 90$ mm³/s. The specific energy is

$$e_c = \frac{P}{Q_w} = \frac{3600}{90} = 40 \text{ J/mm}^3$$

reducing the workspeed to v_w = 0.1 m/s, Q_w = 30 mm³/s, and e_c = 70 J/mm³. It is seen that specific energy tends to be reduced at higher removal rates.

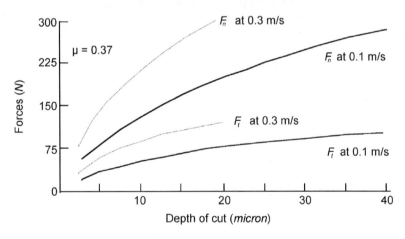

Process: Surface grinding
Wheel: 19A60L7V, 170 mm diameter
Workpiece: Gray cast iron, 15 mm wide
Fluid: 2% synthetic emulsion
Speeds: v_s = 30 m/s, v_w = 0.1 m/s and 0.3 m/s

Figure 5.2. Effects of workspeed and depth of cut on the grinding forces in horizontal axis surface grinding.

5.4.2 Medium Carbon Steel, AISI 1055

Figure 5.3 shows the forces measured when grinding a general purpose medium carbon steel with a fine 200 mesh grit cubic boron nitride (CBN) grinding wheel. In this example, results are given for both wet and dry grinding, and the forces increased proportionately with depth of cut, although there appears to be a small threshold force corresponding to the force required to initiate chip removal.

With a workspeed of $v_w = 0.1$ m/s and a depth of cut of $a_e = 20$ μm as described above, the removal rate is $Q_w = 30$ mm³/s, and the specific energies are determined.

$$\text{Dry grinding: } e_c = \frac{76 \times 30}{20 \times 0.1 \times 15} = 76 \text{ J/mm}^3$$

$$\text{Wet grinding: } e_c = \frac{47 \times 30}{20 \times 0.1 \times 15} = 47 \text{ J/mm}^3$$

Process: Surface grinding
Wheel: B91ABN200, 174 mm diameter
Workpiece: AISI 1055 steel, 15 mm wide
Fluid: None or 2% synthetic emulsion
Speeds: $v_s = 30$ m/s, $v_w = 0.1$ m/s

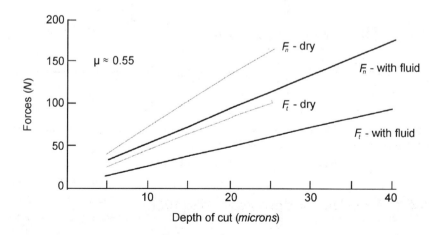

Figure 5.3. Effect of workspeed and depth of cut on the grinding forces in horizontal axis surface grinding with a sharp CBN wheel.

We find that dry grinding medium carbon steel with a fine grain CBN wheel, in this case, yielded a specific energy which is almost the same as wet grinding gray cast iron with a larger grain alumina wheel, in spite of the fact that the two materials were of the same hardness, $H_v = 210$–216 kg/mm^2.

Dry grinding the medium carbon steel with the 60 mesh grit alumina wheel, under otherwise identical conditions as used for the CBN wheel, yielded a value of specific energy, $e_c = 50$ J/mm^3. This value is lower than the value of $e_c = 76$ J/mm^3 using the 200 mesh grit CBN wheel. This tends to indicate a benefit from using a larger grain size for higher removal rates.

The value of $e_c = 50$ J/mm^3 using the alumina wheel is also lower than the specific energy of $e_c = 70$ J/mm^3 when using the same wheel to grind gray cast iron. In spite of the fact that the two workpiece materials were of a similar hardness, it is clear that the gray cast iron is more difficult to grind with this wheel. The gray cast iron showed a greater tendency to adhere to the grains of the grinding wheel which effectively reduces the sharpness of the grains. This means that a softer grade of wheel, which would have a stronger self-sharpening action under the influence of the forces, would have been better for the gray cast iron.

In Fig. 5.3, the benefits of lubrication are immediately obvious. The forces are reduced by almost 40%. The force ratio for this example was $\mu = F_t/F_n = 0.55$, a value which is higher than the value of $\mu = 0.37$ for gray cast iron. A high force ratio is indicative of a more effective cutting condition while a low value indicates a greater tendency towards a rubbing action.

5.4.3 C1023 Nickel-based Alloy

This material is a temperature-resistant alloy used in the aerospace industry for turbine blades. Because the material is temperature resistant, it retains its hardness at higher temperatures. This has the consequence of increasing the specific energy required in machining. Another consequence is a rapid rate of wheel wear as illustrated in Fig. 5.4.

The "fir tree-shaped" roots of turbine blades are produced by creep-feed grinding, a process that involves grinding the grooves to full depth in a single pass. The very large depth of cut and the associated long arc of contact results in high contact temperatures between the abrasive grains and the workpiece, leading to rapid dulling of the grains.

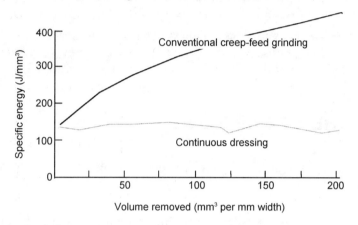

Process: Horizontal creep-feed grinding
Wheel: WA60/80FP2V
Workpiece: C1023 nickel-based alloy
Fluid: 4 mm
Speeds: v_s = 30 m/s, v_w = 0.38 μm/s

Figure 5.4. Specific energy with and without continuous dressing in creep-feed grinding *(based on Andrew, Howes, and Pearce, 1985)*.[26]

As the grains become dull, the specific energy increases rapidly from 150 J/mm^3 to more than 400 J/mm^3, as shown in Fig. 5.4. This demonstrates, very powerfully, the overriding importance of grain sharpness.

To overcome the problem of grain dulling, a process of continuous dressing was introduced. In this process, a diamond roll dresser is fed radially into the grinding wheel throughout the grinding pass. The dresser feedrate per grinding wheel revolution in Fig. 5.4 was f_d = 0.32 μm/rev. At the same time, a compensating feedrate was applied between the grinding wheel and workpiece to maintain a constant depth of cut. In this way, the sharpness of the grains was maintained constant, as evidenced by the constant values of specific energy in Fig. 5.4.

In summary, to reduce specific energy we should aim for the following:

- Large uncut chip thickness.
- Sharp abrasive grains.
- A well-lubricated process.

Adverse factors which increase specific energy are a lack of the above conditions and the following:

- High, hot hardness of the workpiece material.
- Chemical affinity between the workpiece and the grains.
- Insufficient chip removal space in the surface of the abrasive tool.
- A wheel with a hard bond which retains grits when they are blunt.

5.5 THE SIZE EFFECT

As already implied by the previous examples, large depths of cut and high removal rates tend to be more efficient in energy terms than small depths of cut.[2]

Figure 5.5 is typical of results which demonstrate a size effect.[3] By a size effect, we mean that the specific energy is reduced by an increase in depth of cut, or to be more precise, by depth of grain penetration. In the figure, specific energy is plotted against h_{eq} where $h_{eq} = a_e \cdot v_w / v_s$. As we have seen previously, for a particular wheel, h_{eq}, is both a measure of uncut chip thickness and a measure of removal rate. For a particular grinding wheel and wheelspeed, it is apparent that specific energy reduces as either depth of cut or removal rate increases.

The specific energy should remain constant if F_t increases in direct proportion with h_{eq}. We can see this from Eq. (5.4).

Eq. (5.6) $$e_c = \frac{F_t}{b_w \cdot h_{eq}}$$

However, in Fig. 5.5, the specific energy reduces as h_{eq} increase. This means that for a size effect, the forces must increase less than proportionately with h_{eq}. Where the forces increase in direct proportion, no size effect will be evident.

In Fig. 5.2, the forces increase less than proportionately with depth of cut, so that a size effect is evident. However, in Fig. 5.3, the forces increase almost in direct proportion with the depth of cut, so that the size effect is negligible.

Process: Cylindrical grinding
Wheel: A465-K5-V30W, d_s = 417 mm
Workpiece: Cast steel, HRC 60-62, d_w = 18.8 mm
Fluid: 6% synthetic emulsion
Speeds: v_s = 30 m/s

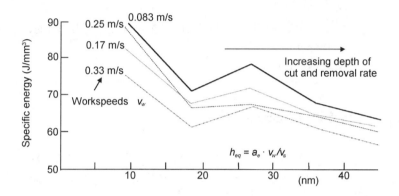

Figure 5.5. Effect of depth of cut and removal rate on specific energy.

A study of the size effect is enlightening because it leads to important conclusions about factors controlling efficiency, wear, and surface roughness.

In general, the specific force, that is, force per unit width of contact, tends to follow a relationship of the type:

Eq. (5.7) $$F_t' = F_1 \cdot h_{eq}^f$$

Values of f lie typically between 0.7 and 1.0 for a particular workpiece material and grinding wheel specification. From Eq. (5.6), it follows that the specific energy follows a relationship:

Eq. (5.8) $$e_c = F_1 \cdot h_{eq}^{f-1}$$

Equations (5.7) and (5.8) are useful ways of characterizing the cutting efficiency, hence, the size effect for a particular workpiece material, grinding wheel, and kinematic conditions. More generally, Chen and

Inasaki [1] related specific energy to an average cross-sectional area of the uncut chip and demonstrated a strong size effect for grinding ceramics.

There are several possible contributions to the size effect. Explanations are mainly based on the following:

- The number of chips/unit volume removed—the sliced bread analogy (Sec. 5.5.1).
- Differences between cutting, ploughing, and rubbing.
- A threshold force for cutting.
- Grain sharpness.

5.5.1 The Sliced Bread Analogy

When the depth of cut is increased, the mean chip volume of the uncut chip is increased as indicated by Eq. (3.69). When the depth of cut is reduced, the mean uncut chip volume is reduced. In other words, a volume of material removed is divided into a larger number of pieces which are smaller in volume. Therefore, it would be surprising if this did not consume more energy.

Considering the analogy of slicing a loaf of bread, if we cut the loaf into twice as many slices, we would expect to consume twice as much energy. This analogy is illustrated in Fig. 5.6. It would be expected that doubling the thickness of the slices should halve the work required to slice up the loaf.

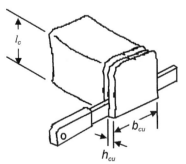

Volume per chip: $v_{cu} = h_{cu} \cdot b_{cu} \cdot l_c$

Cut surface area: $s_{cu} = 2 \cdot b_{cu} \cdot l_c$

Cut area/unit volume: $\dfrac{S_{cu}}{V_{cu}} = \dfrac{2}{h_{cu}}$

No. of chips/unit volume: $n_{cu} = \dfrac{1}{K}$

Figure 5.6. The sliced bread analogy illustrating the importance of the surface area sheared.

This analogy should not be taken too literally for abrasive machining because, if we double the number of chips per unit volume, the surface area of the chips created per unit volume is not exactly doubled as it was with the loaf. In fact, if the surface area were halved for each chip, we would expect the specific energy (that is, the energy per unit volume) to remain unchanged.

From the sliced bread analogy, the mean surface area per unit volume is

Eq. (5.9) $$\frac{S_{cu}}{V_{cu}} = \frac{2}{\overline{h}_{cu}}$$

It follows from Eq. (3.82), for the triangular chip:

Eq. (5.10) $$\frac{S_{cu}}{V_{cu}} = \sqrt{2 \cdot \frac{v_s}{v_w} \cdot C \cdot r \sqrt{\frac{d_e}{a_e}}}$$

where C is the the number of active cutting edges per unit area of the abrasive and r is the mean width-to-depth ratio of the uncut chip cross-section.

If the depth of cut is halved according to Eq. (5.10), the surface area created is increased by 19% and specific energy is increased accordingly. The exponent for the rate of increase in e_c can be found from Eq. (5.8), by writing $1.19 = 0.5^{f-1}$ so that $f = 0.75$.

The sliced bread analogy can be seen to offer an explanation of the size effect due to depth of cut. The sensitivity of the size effect to variations in C, the number of cutting edges per unit area of the wheel surface can be explained. From Eq. (5.10), specific energy to increase if C increases and the specific energy should reduce if C is reduced.

Therefore, the effect of changes in C, which take place with changes in depth of cut should be taken into account. The size effect is substantially changed if C varies strongly with depth of cut.

The effect of C helps to explain the sometimes confusing consequences of changing workspeed.

Increasing the workspeed would be expected to reduce the specific energy according to Eq. (5.10). In practice, the specific energy may either increase or reduce with changing workspeed. Therefore, secondary changes which may affect the specific energy, such as the variation in C must be examined.

The sensitivity to variations in the number of cutting edges, C, may be considered from a study of the effect of workspeed, v_w, in cylindrical grinding. The mean volume of the uncut chips appears to be unaffected by v_w according to Eq. (3.72). However, increasing v_w increases the mean uncut chip thickness as indicated by Eqs. (3.62) and (3.66). This has the effect of increasing the mean grain penetration and, under gentle grinding conditions, increases C.

As v_w is increased in cylindrical grinding, the mean uncut chip volume is, therefore, reduced according to Eq. (3.72). Under these conditions, it follows that increasing the workspeed will tend to increase specific energy due to an increase in C. This is the opposite of the direct effect of the workspeed on specific energy described above.

The result of a change of workspeed can be calculated, taking into account the effect on C. The calculation of the workspeed effect from Eq. (5.10) depends on three variables: C, a_e, and v_w. As an order of magnitude calculation, a strong positive relationship between C and the mean grain penetration, that is, the mean uncut chip thickness can be assumed. Assuming

$$C \propto \bar{h}_{ch}$$

and

$$a_e \propto \frac{1}{v_w}$$

for constant removal rate, it is found from Eqs. (5.9) and (3.82) that:

$$\frac{\bar{S}_{cu}}{\bar{V}_{cu}} \propto v_w^{1/6}$$

The strong relationship for C suggests a weak dependency of specific energy on workspeed, with specific energy increasing with workspeed.

For C, independent of mean uncut chip thickness, Eq. (5.10) shows, taking into account the effect of workspeed on depth of cut:

$$\frac{\bar{S}_{cu}}{\bar{V}_{cu}} \propto v_w^{-1/4}$$

The weak relationship for C suggests that specific energy should reduce as workspeed increases. In practice, examples are found of cases where specific energy may either increase or reduce with workspeed.

The next chapter shows that reducing workspeed can increase workpiece temperature so that, at low workspeeds, there may be problems related to thermal damage of the surface of the workpiece.

Initially, some softening of the material reduces the specific energy. However, experience shows that a sufficient increase in workpiece temperature not only softens the material, but increases the binding of deformed material to the wheel surface. This adhered material reduces cutting efficiency and causes surges in power. Typically, the power drops and then surges with large increases in power following a cyclic pattern.

As removal rates are increased, it may be necessary to increase workspeed to prevent the onset of thermal damage. However, at high workspeeds, the probability of chatter vibration increases.

The simple conclusion from experimental evidence is that reducing mean uncut chip volume tends to increase specific energy. However, the above discussion suggests that mean uncut chip thickness is more important than chip volume where variation in the density of active cutting edges is also taken into account. These simple conclusions are invaluable in trying to understand the implications of changing process parameters.

5.5.2 Cutting, Ploughing, and Rubbing

Hahn[4] recognized three stages of material deformation as a grain interacts with a workpiece: rubbing, ploughing, and cutting. These are illustrated schematically in Fig. 5.7 for an abrasive grain deforming a workpiece.

In the rubbing mode of deformation, material removal is neglible although friction is apparent. Rubbing is typical of a polishing operation or finish grinding with a smooth wheel after a long period without application of a depth of cut with a well-lubricated surface. Under these conditons, it is possible for a measurable normal force to be applied and to exist for many hours with a negligible rate of material removal.

In rubbing, the force on each grain is too small to cause large penetration of the workpiece. Elastic deformation and some plastic deformation takes place at the peaks of asperities evidenced by polishing of the workpiece and a slow process of smoothing.

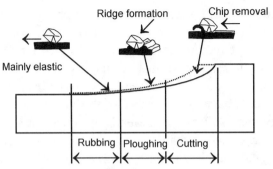

Figure 5.7. Rubbing, ploughing, and cutting regimes of deformation in abrasive machining.

Ploughing occurs when the penetration of the grains is increased. In the ploughing stage, scratch marks become evident and ridges are formed at the sides of the scratches. The scratch marks are evidence of significant penetration, but the rate of material removal remains negligible.

5.5.3 Threshold Force for Cutting

As the penetration of the grains is further increased, material removal rapidly increases and chips are produced. Hahn observed that there is a threshold force required before significant material removal takes place.

Clearly, below the threshold force, the specific energy approaches infinity. This forms the basis of the threshold force explanation of the size effect. As the cutting depth is increased, the proportion of rubbing and ploughing energy is reduced compared with the cutting energy, and the specific energy drops.

5.5.4 Grain Sharpness

Shaw[5] suggested that an abrasive grain might be idealized as a sphere, as shown in Fig. 5.8. The force applied to a grain was assumed to be of a similar nature to the force applied in a Brinell or Meyer hardness test.

The deformation process is constrained by an elastic-plastic boundary shown by the dotted circle. As the sphere moves in a horizontal direction, plastically deformed material is forced upwards. If the penetration depth, t, is sufficient, a chip is formed by being sheared from the workpiece surface.

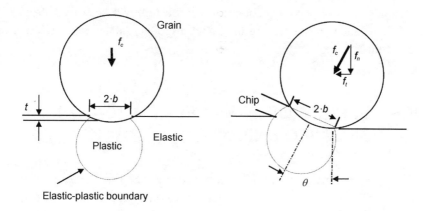

Figure 5.8. An abrasive grain modeled as a sphere.

Shaw approximated the specific energy as

$$\text{Eq. (5.11)} \qquad e_c = \frac{3 \cdot \pi}{4} \cdot \frac{H}{\beta} \cdot \frac{C'}{3} \cdot \left(2 + \mu \sqrt{\frac{d_g}{t}} \right)$$

where H is the hardness of the workpiece, β is an upward flow ratio, C' is a constraint coefficient, μ is a coefficient of friction, d_g is the diameter of the grain, and t is the grain penetration depth.

Essentially, the model adds the energies for deformation within the material and the energy for surface shearing. The model demonstrates that specific energy depends on the hardness of the material being cut, H, the significance of ploughing, β, the coefficient of friction at the surface, μ, the grain size, d_g, and the depth of penetration of the grain, t.

As the penetration depth, t, is reduced, the specific energy is increased. The ratio d_g/t is interesting because this term can be used to interpret the effect of grain sharpness.

A large value of d_g/t corresponds to a small value of θ in Fig. 5.8. This can be interpreted as a blunt grain with a large negative rake angle. For efficient cutting we need a large value of θ is needed, which corresponds to a small grain with a large penetration depth, t. A small value of d_g/t corresponds to a small negative rake angle which is more efficient.

The size effect based on the spherical grain model takes into account the change in their effective sharpness when grains penetrate to a larger depth relative to their size. While large grain abrasive tools have the advantage that there is more space for material removal, specific energy will only be reduced if the mean grain penetration, $\bar{h}_{cu}$, is increased relative to the grain size.

5.6 EFFECT OF WEAR FLAT AREA ON SPECIFIC ENERGY

Kannapan and Malkin[6] proposed that the specific energy could be split into three components, these being the component for chip formation, e_{ch}, for ploughing, e_p, and for sliding or rubbing, e_s. The three components correspond to the three mechanisms proposed by Hahn.

Eq. (5.12) $e_c = e_{ch} + e_p + e_s$

It was proposed that the sliding energy is proportional to the grain wear flat area. The partitioning of the energy into the three components was performed using the guidelines in the following sections.

5.6.1 Chip Formation Energy

The chip formation energy can be estimated by assuming that the energy absorbed by the chips is limited by the rapid reduction in shear stress as the melting temperature is approached.

Grinding and most abrasive processes concentrate a large amount of energy into a small volume of material in a short time duration. Calculations and measurements also show that the temperatures at a grain contact exceed the softening temperature of the material and approach the melting temperature. As a good approximation, the chip formation energy can be estimated as

Eq. (5.13) $e_{ch} = \rho \cdot C \cdot \theta_{mp}$

where ρ is the density of the material, C is the specific heat capacity, and θ_{mp} is the melting point temperature of the material. This gives a value approximately equal to $e_{ch} = 6$ J/mm^3 for ferrous materials.

Malkin[7] gives the difference in enthalpy between ambient temperature and the liquid state as 10.5 J/mm^3 for iron and steels. The material does not actually melt except in extreme cases, so this value may be a slight over-estimate. However, examinations of grinding swarf often show a dendritic structure characteristic of a casting process. This does suggest a temperature in the region of the melting temperature.

5.6.2 Sliding Energy

The sliding energy, e_s, is defined as that component of the energy which is proportional to the actual rubbing area of the grain wear flats. The grain wear flat area is sometimes expressed as a percentage, A, of the wheel surface area. The experimental technique to determine A is tedious, so few researchers attempt to carry out this measurement.

Relating sliding energy to wear flat area:

Eq. (5.14) $e_s = \mu_s \cdot F_{ns} \cdot v_s$

where μ_s is the coefficient of sliding friction and F_{ns} is the part of the normal force proportional to the area of the wear flats. The term $\mu_s \cdot F_{ns}$ is defined as that portion of the tangential force required to overcome sliding friction, F_{ts}, so that

Eq. (5.15) $e_s = F_{ts} \cdot v_s$

The normal and tangential forces are each split into three components according to the three values of energy.

Eq. (5.16) $F_t = F_{t \cdot ch} + F_{t \cdot p} + F_{t \cdot s}$

Eq. (5.17) $F_n = F_{n \cdot ch} + F_{n \cdot p} + F_{n \cdot s}$

Typical measurements of F_t and F_n for different values of A are illustrated in Fig. 5.9. Typically, for sharp grains, the grain wear flat area is less than 4% of the total surface of the wheel, and the contribution to the forces is relatively small.

A sharp transition in the slope for the forces shown for the two steels, AISI 52100 and SAE 1018, corresponds to the onset of burn at higher values of grain wear flat area. Sometimes, when a material undergoes severe thermal damage in grinding, the volume of deformed material increases. The increase in deformed material corresponds to an increase in the redundant energy input to the process. (Energy which deforms the material, but does not directly contribute to material removal is known as *redundant energy*.)

Values of F_{ts} can be read from results such as those shown in Fig. 5.9 so that the specific energy for sliding, e_s, can be determined by carrying out grinding force and grain wear flat measurements.

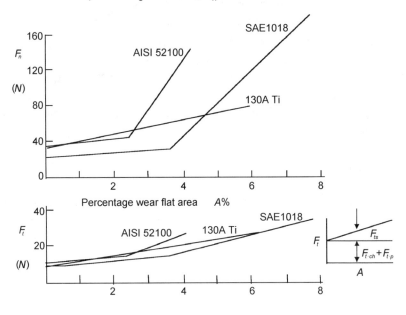

Process: Surface grinding

Wheel: 32A46, d_s = 200 mm

Workpiece width: b_w = 6.4 mm

Depth of cut: 20 microns

Speeds: v_s = 30 m/s, v_w = 0.077 m/s

Figure 5.9. Effect of wear flat area on normal and tangential forces *(based on Malkin)*.[7]

5.6.3 The Ploughing Energy

After deducting from the total specific energy, e_c, the chip formation energy, e_{ch}, and the sliding energy, e_s, the remainder is the ploughing energy, e_p.

According to Malkin,[7] the ploughing energy reduces with increasing removal rate, whereas the chip formation energy remains constant as illustrated in Fig. 5.10. According to this analysis, the size effect is associated with the ploughing energy. As the depth of cut increases, the ploughing energy increases proportionately slower than the removal rate. Therefore, the specific energy is reduced. Less clearly, it can be deduced that the size effect is also associated with the sliding energy when there are wear flats on the grains.

Process: Surface grinding
Wheel: Various, d_s = 200 mm
Workpiece: AISI 1095 HR, b_w = 6.4 mm
Depth of cut: 12.7–50.8 microns
Speeds: v_s = 30 m/s, v_w = 0.075–0.305 m/s

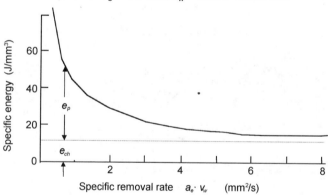

Figure 5.10. Specific energy for ploughing and chip formation *(based on Malkin)*.[7]

Whereas, previously, grain sharpness was interpreted in terms of an effective rake angle of the grains, the grain sharpness is now modified based on wear flat area. Both effects increase the specific energy with blunt grains, and both effects contribute to the size effect.

5.7 WEAR AND DRESSING CONDITIONS

The foregoing analysis places great emphasis on grain sharpness. Accordingly, the forces are expected to increase as the grains become blunt. However, grains sometimes fracture and pull out, in which case, forces reduce with tool wear. This effect may be particularly evident in grinding just after fine dressing and before the wheel has had sufficient time to stabilize as illustrated in Fig. 5.11.

Process: Cylindrical grinding

Wheel: A465-K5-V30W, d_s = 390 mm

Workpiece width: Cast steel, 60–62 HRC, d_w = 17 mm

Speeds: v_s = 33 m/s, v_w = 0.25 m/s

Dressing: a_d (mm), f_d (mm/rev)

coarse:	0.025	0.25
medium:	0.015	0.15
fine:	0.005	0.05

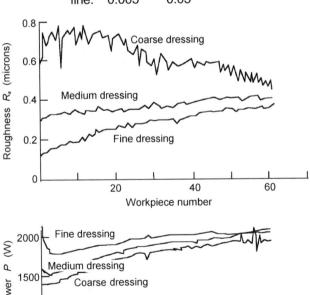

Figure 5.11. Effect of wear on power and roughness.

After dressing, the grains are more susceptible to fracture which reduces the grinding power. The surface of the wheel stabilizes and forces tend to steadily increase as the grains become blunt. The wear behavior is strongly dependent on the dressing conditions as described by Chen.[8]

Fine dressing conditions correspond to a small dressing depth, a_d, and small feed per revolution of the wheel, f_d. In precision grinding, the feed per revolution should always be sufficient to ensure several contacts between the dressing tool and the active grains of the wheel. An overlap ratio, U_d, can be defined as the result of dividing the contact width, b_d, of the dressing tool by the feed per wheel revolution, f_d:

Eq. (5.18)
$$U_d = \frac{b_d}{f_d}$$

The overlap ratio, U_d, should normally be greater than 2 and less than 10 to achieve satisfactory performance from a grinding wheel.

With fine dressing conditions, the number of active grains on the wheel surface is greatest, the grinding power is high, and the workpiece roughness is low. As fragile grain particles break out of the wheel surface, the power reduces and workpiece roughness increases.

With coarse dressing, whole grains are broken out of the surface, and the number of active grains on the surface is reduced. The workpiece roughness is much greater than after fine dressing. The grinding power is reduced after coarse dressing due to the size effect which, following the previous discussion, can be interpreted as due to the greater effective sharpness of the grains and due to the increased grain penetration.

As the wheel wears, the power levels for different dressing conditions tend to converge towards the same value. However, the effects of dressing do not totally disappear. Dressing too finely, or too coarsely, adversely affects the redress life of the tool. The redress life is defined as the volume of material which may be removed before the performance of the abrasive tool deteriorates, making it necessary to redress the tool. The volume of material removed during the redress life is directly proportional to the redress life.

5.8   EFFECT OF DRESSING TOOL WEAR

The most basic dressing tool is a large single diamond retained in a metal bond, as illustrated in Fig. 5.12. A dressing tool, as supplied by a tool manufacturer, may be lapped to an approximately conical shape.

With use, this sharp profile of the diamond dressing tool is rapidly worn away. The diamond sharpness ratio, γ, may be defined for a particular dressing depth, a_d, and dressing width, b_d:

Eq. (5.19) $$\gamma = \frac{a_d}{b_d}$$

Results from Chen[8] showed that a wide dressing tool, that corresponds to a low sharpness ratio, has a similar effect to coarse dressing: in both, the grinding power is reduced.

Process: Cylindrical grinding
Wheel: A465-K5-V30W, d_s = 390 mm
Workpiece width: Cast steel, 60–62 HRC, d_w = 17 mm
Speeds: v_s = 33 m/s, v_w = 0.25 m/s
Dressing: a_d = 0.015(mm), f_d = 0.015 (mm/rev)

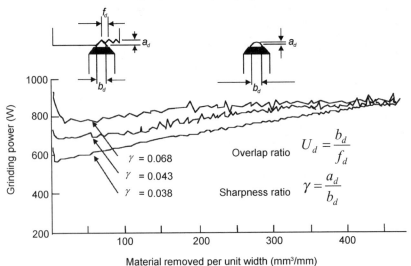

Figure 5.12. Effect of dressing tool wear on grinding power and grinding wheel wear.

A large sharpness ratio causes the subsequent grinding power to be greater than is the case after dressing with a low sharpness ratio.

Conversely, it seems that a sharp dressing tool leaves a higher number of active cutting edges on the surface than a blunt diamond, that is consistent with the higher initial grinding power. However, use of a sharp dressing tool also leaves cracks in the grains which lead to microfractures and rapid reduction in grinding power with further material removal in the grinding process.

Major problems can arise in industrial practice due to dressing tool wear. The process becomes variable whereas, ideally, a constant process is required. A constant process bestows an advantage that size, roughness, form, and roundness errors remain within closer tolerances than with a variable process. To overcome the problem of process variability, a variety of dressing tools are available with multiple diamond particles in the surface.

5.9 THE NATURE OF THE GRINDING FORCES

At any instant, there are a number of grains, n, in contact with the workpiece. The grinding force is the summation of the forces at any instant due to the various grains in contact as illustrated in Fig. 5.13.

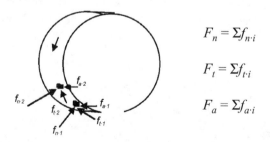

$$F_n = \Sigma f_{n \cdot i}$$

$$F_t = \Sigma f_{t \cdot i}$$

$$F_a = \Sigma f_{a \cdot i}$$

Figure 5.13. The forces on the wheel are the sum of the component forces on the grains.

Split into the three component directions, the force can be expressed as

Eq. (5.20) $\qquad F_n = \sum_{i=1}^{n} f_{n \cdot i}$

Eq. (5.21) $F_t = \sum_{i=1}^{n} f_{t \cdot i}$

Eq. (5.22) $F_a = \sum_{i=1}^{n} f_{a \cdot i}$

At any instant, the grinding force is a dynamically changing parameter which depends on the number of active grains in engagement at that moment. The force depends on the shape and depth of engagement of each moving active grain at each instant.

As explained in Ch. 3, "Kinematic Models of Abrasive Contacts," the active grains represent a random distribution of shapes and orientation with a random distribution of size and spacing. Normally, such situations are treated by attempting to determine how an "average" grain behaves, while taking into account the variations from the average, to predict how the whole process behaves.

In abrasive machining, there is a difficulty since it is not easy to predict how a single grain will behave, or how the combined action of a large number of grains will perform. The nature of the overall system behavior must be examined to try to explain it in terms of idealized single grains.

In orthogonal metal-cutting, there are three main zones of shearing in the workpiece. We can extend this concept to the action of an abrasive grain producing a chip as illustrated in Fig. 5.14.

The primary shear zone lies in the region ahead of the grain where the workpiece material changes direction as it is deformed and moves upward into the chip.

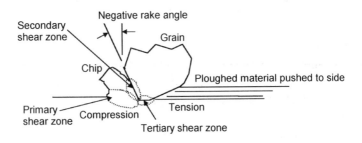

Figure 5.14. The concept of shear zones applied to an abrasive grain.

The secondary shear zone is within the chip where subsurface shear flow takes place as the chip moves upward against the surface of the grain. This shear zone is very thin and, as a consequence, the shear strain rates may be extremely high, $>10^7$, leading to high localized temperatures with implications for reduced yield stresses, increased tendency for material solution and diffusion, chemical interactions, and grain wear.

The tertiary shear zone is where the subsurface workpiece material shears under the grain. The tertiary shear zone can extend well below the surface as illustrated in Fig. 4.17.

All three zones are subject to compressive stresses. As the material moves behind the tool, there is a transition from compressive to tensile stresses.

Material deformation in abrasive machining takes place in three dimensions. Material is pushed upwards and sideways to form ridges on each side of the path of the grain. The tertiary shear zone, therefore, applies not only to the workpiece material under the grain, but also to the material at the sides of the grain.

The action of pushing the material upwards and sideways to form ridges is known as ploughing. The energy required is nonproductive since the ploughed material remains attached to the workpiece. In abrasive action, primary shear energy and secondary shear energy are generated both within the workpiece material remaining and within the chip material removed.

The effective rake angles are highly negative in abrasive machining which leads to a large compressive plastic zone ahead of and under the grain followed by a shallower tensile zone behind the grain, as illustrated in Fig. 4.17 for a point load. As explained in the previous chapter, this means that the mechanical effect of abrasive machining is to leave compressive residual stresses which are sometimes considered to be beneficial for bearing surfaces. However, thermal damage often leaves tensile residual stresses, which are almost always deleterious for bearing surfaces.

In most metal-cutting processes, the cutting force is dominated by the primary and secondary shear zones within the material removed and about to be removed as chips. In abrasive machining, due to ploughing, rubbing, and sliding, the process is usually dominated by shear within the surface material that will remain after the passage of a grain.

The overall force is the sum of all the shear forces required to make the material flow around the grains. The nature of the material flow depends on a number of factors including grain geometry, kinematics of the relative motion, the temperature and chemistry of the surface layers, and the presence of lubricant.

5.10 FORCE RATIO AND FRICTION COEFFICIENT

In abrasive machining, the ratio F_t/F_n is known as the force ratio. Due to its similarity to a coefficient of friction, the same symbol, μ, is used:

Eq. (5.23)
$$\mu = \frac{F_t}{F_n}$$

Typical values of μ in abrasive machining lie between 0.2 and 0.7. A value of 0.2 or lower corresponds to well-lubricated blunt abrasive grains in contact with the surface and very little penetration. High values correspond to sharp abrasive grains and deep penetration.

These values of force ratio may be contrasted with values for sharp positive-rake cutting tools where the force ratio is normally greater than 1.0 or even negative. A negative force ratio occurs where the positive rake angle is sufficiently large to produce a negative normal force. In abrasive machining, the effective rake angles are always negative.

The force ratio in abrasive machining is indicative of the sharpness of the grains. Further insight into the mechanics of the process may be gained by studying the force ratio and the effects of friction, with some simple examples from the theory of friction.

5.10.1 Blunt Asperity Contact (Adhesion Friction)

A blunt asperity contact is illustrated in Fig. 5.15. In this case, rubbing or, in other words, shear of the softer material, occurs. The depth of penetration by the hard asperity is very small.

Assuming the shear stress, τ, is equal to the shear strength, k, of the softer workpiece material as in Fig. 5.15a,

Eq. (5.24) $f_t = A_r \cdot \tau = A_r \cdot k$

where $\tau = k$ is the shear flow stress, and A_r is the real area of contact between the asperity of the abrasive grain and the workpiece.

The real area of contact under the asperity is given by the hardness, H, of the softer material, and the normal force, f_n:

Eq. (5.25) $\qquad A_r = \dfrac{f_n}{H}$

Approximating that,[9] and discussing in more detail later, $H = 6 \cdot k$ substituting in Eq. (5.25) and from Eq. (5.24), the coefficient of friction for adhesive conditions, μ_{ad}, is

Eq. (5.26) $\qquad \mu_{ad} = \dfrac{f_t}{f_n} = \dfrac{k}{H} \approx \dfrac{1}{6} = 0.17$

In fact, lower values are obtainable under lubricated conditions, but much higher values occur under either dry or wet conditions.

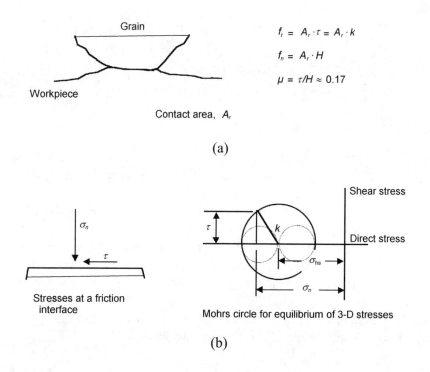

(a)

(b)

Figure 5.15. Adhesive friction based on the assumption of shear of the softer material. *(a)* The simple adhesion theory of sticking friction for a grain rubbing on a workpiece, and *(b)* for plastic shear of the softer material: $(\sigma_n - \sigma_{hs})^2 + \tau^2 = k^2$.

The above approach based on Bowden and Tabor[10] was later modified by Tabor[11] to take into account the junction growth at the asperity contact due to material deformation and adhesion. It was also assumed that the shear stress, τ, at the junction could be lower than the shear flow stress of the material, k. Defining an interface friction ratio, $f = \tau/k$, the modified relationship was

Eq. (5.27) $$\mu = \frac{1}{\delta^{1/2} \cdot \left(f^{-2} - 1\right)^{1/2}}$$

where δ is an empirical factor which Tabor took to be 9. With a value of the interface friction ratio, f, of 0.95, the coefficient of friction μ is 1.01. This model is very sensitive to the value of f. With $f = 0.9$, $\mu = 0.69$ and with $f = 1$, $\mu = \infty$.

This illustrates the crucial importance in sliding friction of very thin layers of low shear strength at the junction interface, which offsets the effect of junction growth.

An approximate explanation of Eq. (5.27) can be gained by considering the stresses at the contact from the Mohrs circle, Fig. 5.15b. Under plastic flow conditions at the junction, the stresses are related according to

Eq. (5.28) $(\sigma_n - \sigma_{hs})^2 + \tau^2 = k^2$

where σ_n is the normal stress at the interface, τ is the shear stress at the interface, σ_{hs} is the hydrostatic stress in a three-dimensional stress system, and k is the shear flow stress of the bulk material.

The hydrostatic stress, σ_{hs}, depends on the total stress system, but for the case where $f \to 0$, the direct stresses become compressive. As an order of magnitude, the compressive hydrostatic stress is $\sigma_{hs} \to 2 \cdot k$, and the normal stress is $\sigma_n \to 3 \cdot k$. If, on the other hand, in the special case where a tensile horizontal stress is applied to the workpiece equal to the normal stress on the surface, the two principal stresses in the plane under consideration are equal and opposite in the Mohrs circle. This means that $\sigma_{hs} = 0$ which corresponds to the application of pure shear, so that

Eq. (5.29) $\sigma_n^2 + \tau^2 = k^2$

and

Eq. (5.30) $\mu = \dfrac{1}{\sqrt{f^{-2}-1}}$

where $\mu = \tau/\sigma_n$ and $f = \tau/k$. Equation (5.30) bears some similarity to Eq. (5.27) and accounts for interface friction, but not for the junction growth phenomenon. It shows that under plastic flow conditions, reducing the interface friction greatly reduces the coefficient of friction for sliding motion between the two bodies. However, a larger interface pressure is required to maintain plastic flow with low interface friction.

As a note of caution, Eq. (5.29) represents a special case and underestimates the pressure required to achieve indentation due to the compressive hydrostatic stress. The solution of the stresses will be considered in more detail when we consider slip-line field and other upper bound methods which take into account the pattern of strains, and the associated energy required to achieve material flow.

With no sliding motion, as in the case of indentation with a flat wedge, it can be shown by upper bound methods[18] that the pressure required is $\sigma_n = p \approx 5.7k$, in which case, $\sigma_n^2 \approx 32 \cdot k^2$.

Therefore, as an approximate approach, let us consider a modification to Eq. (5.27) to allow for junction growth and redundant work:

Eq. (5.31) $\sigma_n^2 + a_1\tau^2 = a_2 k^2$

As pointed out by Williams,[12] a_1 is not necessarily equal to a_2. In this case, the coefficient of friction is given by

Eq. (5.32) $\mu = \dfrac{1}{\sqrt{a_1} \cdot \sqrt{\dfrac{a_2}{a_1} \cdot f^{-2} - 1}}$

If $a_1 = 32$, $a_1/a_2 = 0.9$, and $f = 1$, the coefficient of friction is reduced from infinity to $\mu = 0.53$.

5.10.2 Sharp Asperity Contact (Abrasive Cone Friction)

In this example based on Rabinowicz,[13] an abrasive grain is represented as a cone in Fig. 5.16.

As in the previous approach, using the definition of hardness:

Eq. (5.33) $f_t = r_b^2 \cdot \tan\alpha \cdot H$

Eq. (5.34) $f_n = \pi \cdot r_b^2 \cdot H$

where r_b is the maximum radius of the contact between the cone and the workpiece. From these equations, the force ratio gives the coefficient of friction, μ_{ab}, for abrasion as

Eq. (5.35) $\mu_{ab} = \dfrac{f_t}{f_n} = \dfrac{\tan\alpha}{\pi}$

For a sharp grain, the angle, α, will be larger than for a blunt grain. Assuming $\alpha = 45°$, $\mu_{ab} = 0.32$ for $\alpha = 45°$.

This value of force ratio is more representative of the force ratios measured in typical abrasive machining operations. It is closer to reality for efficient material removal and illustrates that the normal force is primarily responsible for indentation of the grains into the workpiece, whereas the tangential force is responsible for material removal and overcoming friction.

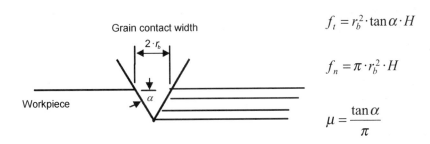

Figure 5.16. A simple abrasive theory of friction.

On further examination, the abrasive friction model is inaccurate because, in reality, contact only takes place between the leading and lower face of the grains and the workpiece. The above example can be modified assuming a normal stress, σ_n, on the leading face of the grain and a shear stress, τ. This yields

Eq. (5.36) $$f_t = r_b^2 \cdot \sigma_n \cdot \left(\tan\alpha + \frac{\tau}{\sigma_n} \cdot \sec\alpha \right)$$

Eq. (5.37) $$f_n = \frac{\pi \cdot r_b^2}{2} \cdot \sigma_n$$

Eq. (5.38) $$\mu_{ab} = \frac{2}{\pi} \cdot \left(\tan\alpha + \frac{\tau}{\sigma_n} \cdot \sec\alpha \right)$$

For the case where $\alpha = 45°$ and $\tau/\sigma_n = 0.17$:

$$\mu_{ab} \approx 0.64 + 0.15 = 0.79$$

The abrasive model of friction is more appropriate for cutting with sharp grains, whereas the adhesive model is more appropriate for a polishing operation with blunt grains.

Suh[9] postulates an asperity deformation model of friction which appears to be an appropriate model when two identical materials slide against each other under near-static conditions. Deformation friction would, therefore, appear to be a less appropriate model for friction in abrasive machining with sharp grains.

Some authors add together the coefficients of friction for the three mechanisms on the assumption that all three mechanisms take place together and the friction forces are additive. According to this proposal:

Eq. (5.39) $\mu = \mu_{ad} + \mu_{ab} + \mu_{def}$

However, this proposal is questionable. If the friction forces are broken down into smaller parts which act at different asperity contacts according to the different types of deformation, then the normal force and the tangential force should be broken down likewise. It would seem more reasonable to write

$$\mu = k_1 \cdot \mu_{ad} + k_2 \cdot \mu_{ab} + k_3 \cdot \mu_{def}$$

where $k_1, k_2,$ and k_3 represent the probabilities of each type of junction. This would be more in keeping with the approach to prediction of forces employed by Badger and Torrance.[14] The constraints on the values of k are

$$0 < k_i < 1$$

5.10.3 Combined Cone and Sphere Model

In practice, an abrasive grain does not have a sharp point as required by the cone model. A typical grain may be flat or have some rounding at the tip.

The analysis follows lines similar to the case given previously for the cone. The combined cone and sphere model gives a better correlation with practical experience that μ often reduces as grain penetration is reduced, as illustrated in Fig. 5.17.

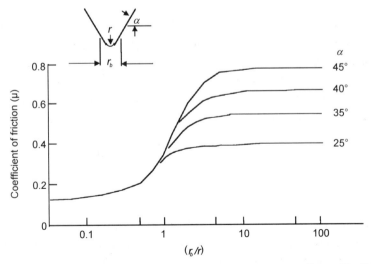

Figure 5.17. Combined cone and sphere model of abrasive friction according to Sin, Saka, and Suh.[27]

At small depths of penetration, when $r_b \leq 2 \cdot r \cdot \sin \theta$, only the spherical tip is in contact and the abrasive coefficient of friction is

Eq. (5.40)
$$\mu_{ab} = \frac{2}{\pi} \cdot \left\{ \left(\frac{r_b}{2 \cdot r} \right)^{-2} \cdot \sin^{-1}\left(\frac{r_b}{2 \cdot r} \right) - \left[\left(\frac{r_b}{2 \cdot r} \right)^{-2} - 1 \right]^{1/2} \right\}$$

At greater depths of penetration, when $r_b \geq 2 \cdot r \cdot \sin\alpha$

Eq. (5.41)
$$\mu_{ab} = \frac{2 \cdot \tan\alpha}{\pi} \left[1 - \frac{4 \cdot (\tan\alpha - \alpha)}{\tan\alpha} \cdot \left(\frac{r_b}{r} \right)^{-2} \right]$$

At small depths of penetration, the friction coefficient, due to abrasion, is smaller than the adhesive friction model. As the penetration is increased, the abrasive friction increases up to approximately 0.8 for $\alpha = 45°$. This is of the same order as for the sharp cone model.

5.11 ADHESIVE AND ABRASIVE WHEEL WEAR

The subject of adhesive wear is complex since wear is strongly dependent on the chemistry of the interacting materials, the lubricant, and the atmospheric environment, as well as the mode of material deformation. However, for a particular material pair operating in a constant environment, it is possible to relate the rate of adhesive wear to the normal load, the real area of contact, and the sliding distance as demonstrated by Archard.[15]

The volume, V, of material removed from the asperities in adhesive wear is given by:

Eq. (5.42) $V = K \cdot A_r \cdot L$

where the real area of contact is given by the normal force divided by the hardness so that $A_r = F_n /H$, the sliding distance is L, and K is the adhesive wear coefficient sometimes known as the Archard Constant. Archard's Law of adhesive wear is, therefore:

Eq. (5.43) $$V = K \cdot \frac{F_n}{H} \cdot L$$

Because of the sensitivity of the Archard Constant to the chemistry of the interactions, the value can vary over an enormous range from 10^{-2} to 10^{-8} as reported by Black.[16]

Archard's simple, but very useful, relationship describes the slow progressive wear of the abrasive grains under gentle grinding conditions with a hard wheel. Unfortunately, Archard's Law does not apply to the fracture wear of the abrasive grains described in Sec. 5.7. For the abrasive wear of the workpiece, it is more direct to use the simple kinematic relationships given in Ch. 3, "Kinematic Models of Abrasive Contacts," based on depth of cut and workspeed. However, Eq. (5.43) does reasonably apply for the period of steady wear of a grinding wheel between redressing intervals as illustrated in Fig. 5.18. From Eq. (5.42), the radial wheel wear, w_r, is given by:

Eq. (5.44) $$w_r = \frac{V}{A_r} = K \cdot L = K \cdot l_e \cdot N_s \cdot t$$

where K is the wear constant, L is the length of grain sliding contact, l_e is the real contact length, N_s is the rotational wheelspeed, and t is the total grinding contact time.

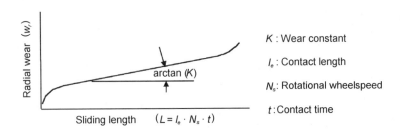

Figure 5.18. The rate of radial wear from an abrasive tool is proportional to the total distance for which the grains are in contact.

The above relationship for wear of an abrasive tool is confirmed by re-interpretation of data provided by Verkerk.[17] It is important to note that the contact length in Eq. (5.44) is the real contact length, l_e, and not the geometric contact length, l_g. This may be demonstrated by increasing the workspeed at a constant depth of cut. This increases the normal force and increases the real contact length, resulting in increased radial wheel wear. If geometric contact length is employed, it would appear that the radial wheel wear is unchanged. The advantage of using real contact length rather than geometric contact length in this case is that an explanation can be found for the increased wheel wear. In general, grinding performance is often better explained using real contact length.

It can be seen from Fig. 5.18 that the Archard's Wear Law does not apply at the beginning or at the end of the wear period. In the early wear period after dressing, small fractures in the grains and in the bond bridges lead to fracture wear. At the end of the wear period when the grains are blunt, fracture wear again predominates, as discussed in Sec. 5.7. Archard's Wear Law may also be assumed to apply to the dressing tool.

Although the above relationships have been derived on the basis of adhesive wear, a similar analysis applies to abrasive wear. For example, taking the abrasive cone model illustrated in Fig. 5.15, the volume of material removed in abrasion is

Eq. (5.45) $V = L \cdot r_b^2 \cdot \tan\alpha$

and, as before,

Eq. (5.46) $f_n = \dfrac{\pi \cdot r_b^2}{2} \cdot \sigma_n$

Eliminating r_b and substituting $\sigma_n = H/3$:

Eq. (5.47) $V = \dfrac{6 \cdot \tan\alpha}{\pi} \cdot \dfrac{f_n}{H} \cdot L$

or for a number of asperities:

Eq. (5.48) $V = K_{ab} \cdot \dfrac{f_n}{H} \cdot L$

where K_{ab} is a constant appropriate for the abrasive wear conditions, dependent on the sharpness of the grains.

It is hardly surprising that in many cases it is found that wear is proportional to normal force and sliding distance divided by hardness of the softer material. Archard's Law forms the basis for wear analysis of both fixed and loose abrasives.

5.12 SLIP-LINE FIELD SOLUTIONS

In principle, a valid slip-line field solution can be used to indicate the forces and energy involved in abrasive machining. In practice, slip-line fields are difficult to validate and are limited to plane-strain conditions. In spite of these difficulties, slip-line field solutions are applied to plane-strain and axi-symmetric situations and provide valuable insights into the nature of material flow processes and the effect of friction at the grain-workpiece interface.

As an introduction to solutions for abrasive grains, first let us consider indentation with a spherical tool, as illustrated in Fig. 5.19.

In Fig. 5.19, the material is shown in its undeformed state. Starting from this geometry, further penetration of the spherical tool will cause shear along the slip lines causing the material to flow sideways and upwards towards the free surface.

The slip lines are lines of maximum shear stress within a plastically deforming volume. The slip lines intersect with a free surface at 45° and with each other at 90°

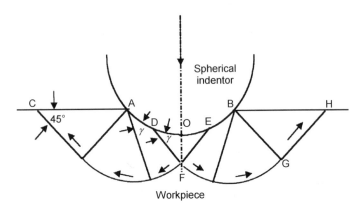

Figure 5.19. Slip-line field solution for spherical indentation with friction angle γ after Prandtl/Tomlenov.[28]

The slip lines meet the surface of the tool at the friction angle, γ. The friction angle, γ, is determined by the coefficient of friction, μ, on the spherical surface where sliding takes place. If the normal stress on the surface is σ_n, the shear stress on the surface, τ, will be the tangential frictional stress $\mu \cdot \sigma_n$. Therefore, for Coulomb friction, by reference to Mohrs Stress Circle, the angle, γ, between the shear stress on the surface of the tool and the plane of maximum shear can be found.

Eq. (5.49)
$$\gamma = \frac{1}{2} \cdot \cos^{-1} \cdot \frac{\mu \cdot \sigma_n}{k} = \frac{1}{2} \cdot \cos^{-1} f$$

where k is the shear flow stress of the material.

In regions where the friction and normal stress are sufficiently high, the material yields below the interface and the frictional stress is $\mu \cdot \sigma_n = k$. In this case, that is known as *sticking friction*, one slip line meets the surface tangentially and the complementary slip line meets the surface at 90°, so that $\gamma = 0$ for sticking friction.

The Prandtl-Tomlenov solution illustrated in Fig. 5.19 predicts a dead zone, DEF, which moves with the punch without deformation. Experiments by Rowe[18] confirmed the validity of this type of solution where the friction is high for strain hardening materials. For low friction materials and constant yield stress, the dead zone tends to disappear.

The effect of friction can be verified by making the construction illustrated in Fig. 5.19. With $\gamma = 45°$, the size of the slip-line field solution is reduced and the dead zone disappears. As a consequence, the indentation force is reduced and the energy required to penetrate the surface is reduced. In other words, friction at the sliding interface increases the redundant work required to deform the material and the depth of the deformation process.

Lortz,[19] based on the work of Tomlenov, developed a simplified plane-strain model to represent abrasion with a spherical grain. This solution is illustrated in Fig. 5.20 and models the situation where the spherical grain is moving sideways and takes into account the deformed shape of the material. While this solution is more complex than the previous example, the construction follows the principles previously outlined.

In Fig. 5.20, shearing takes place below the surface of the workpiece material left behind. The subsurface depth, w_m, of the deformed material remaining on the surface after the passage of the grain depends on the friction angle, γ.

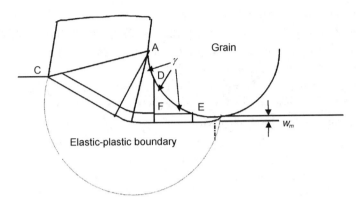

Figure 5.20. Proposed slip-line field for cutting with a spherical abrasive grain.[19]

In spite of the fact that friction plays such an important role in abrasive processes, it is only in recent years that the effect of friction on the mechanics of the process have started to be properly understood.

A major contribution to understanding was made by Challen and Oxley,[20] who proposed three models of deformation to take into account the different behavior of materials with different levels of roughness and varying effectiveness of lubrication.

As in Sec. 5.10, the effectiveness of lubrication is denoted by the ratio $f = \tau/k = \mu \cdot \sigma_n/k$, which is the ratio of the shear stress, τ, at the asperity interface and the material flow stress, k.

The Challen and Oxley models are illustrated in Fig. 5.21. The merit of the Challen and Oxley models is that predicted behavior is consistent with experimental measurements of friction and wear behavior under lubricated and dry conditions for three distinct situations:

- Rubbing/sliding with negligible wear—the wave model.
- Rubbing/sliding with wear—the wave removal model.
- Chip formation model.

5.12.1 Wave Model of Rubbing

For a hard abrasive grain with a small asperity angle, α, as illustrated in Fig. 5.21, a plastic wave moves along the surface without material removal, as in the drawing and extrusion processes. A similar phenomenon can also occur in rolling element bearings, where a plastic wave precesses around the bearing ring.

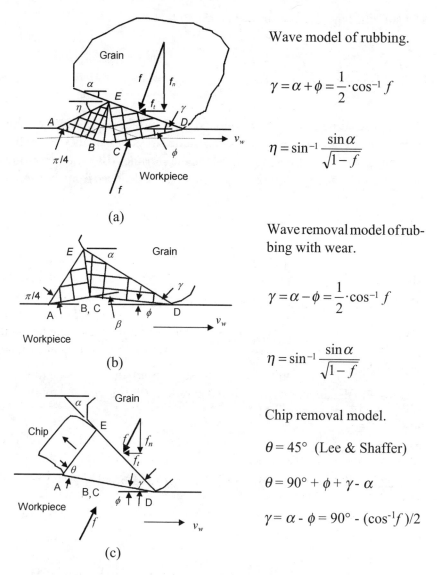

Wave model of rubbing.

$$\gamma = \alpha + \phi = \frac{1}{2}\cdot\cos^{-1} f$$

$$\eta = \sin^{-1}\frac{\sin\alpha}{\sqrt{1-f}}$$

(a)

Wave removal model of rubbing with wear.

$$\gamma = \alpha - \phi = \frac{1}{2}\cdot\cos^{-1} f$$

$$\eta = \sin^{-1}\frac{\sin\alpha}{\sqrt{1-f}}$$

(b)

Chip removal model.

$\theta = 45°$ (Lee & Shaffer)

$\theta = 90° + \phi + \gamma - \alpha$

$\gamma = \alpha - \phi = 90° - (\cos^{-1}f)/2$

(c)

Figure 5.21. Slip-line field models for increasing sharpness, α, after Challen and Oxley.[20]

The analysis is relatively straightforward. From Eq. (5.49), the friction angle at the asperity interface is

Eq. (5.50) $\gamma = \alpha + \phi = \dfrac{1}{2}\cdot\cos^{-1} f$

The slip lines meet a free surface at 45° so that

Eq. (5.51) $\eta = \sin^{-1} \dfrac{\sin \alpha}{\sqrt{1 - f}}$

The normal stress on the inclined grain surface is found by working from the free boundary where the hydrostatic stress σ_{hs} equals k. From the rotation, ψ, of the slip lines, moving along the slip line ABCD, and from the Hencky equations, $\sigma_n = \sigma_{hs} \pm 2 \cdot k \cdot \psi$, which can be deduced from the Mohrs circle,

Eq. (5.52) $\sigma_n = k \cdot \left(1 + \dfrac{\pi}{2} + 2 \cdot \phi - 2 \cdot \eta \right)$

The shear stress, τ equals $k \cdot f$ by definition, from Eq. (5.50)

Eq. (5.53) $\tau = k \cdot \cos 2(\alpha + \phi)$

The forces per unit width on the surface are found by resolving the stresses into the directions normal and parallel to the motion and multiplying by the length of contact ED.

Eq. (5.54)

$$f_t' = k \cdot \left\{ \left(1 + \dfrac{\pi}{2} + 2 \cdot \phi - 2 \cdot \eta \right) \cdot \sin \alpha + [\cos(2 \cdot \alpha + 2 \cdot \phi)] \cdot \cos \alpha \right\} \cdot \text{ED}$$

Eq. (5.55)

$$f_n' = k \cdot \left\{ \left(1 + \dfrac{\pi}{2} + 2 \cdot \phi - 2 \cdot \eta \right) \cdot \sin \alpha + [\cos(2 \cdot \alpha + 2 \cdot \phi)] \cdot \cos \alpha \right\} \cdot \text{ED}$$

The force ratio for abrasive machining under the "no wear" condition is then given by

Eq. (5.56) $\mu = \dfrac{f_t'}{f_n'} = \dfrac{A \sin \alpha + f \cos \alpha}{A \cos \alpha + f \sin \alpha}$

where:

$$A = 1 + \frac{\pi}{2} + 2 \cdot \phi - 2 \cdot \eta$$

We can now see clearly how the force ratio depends very strongly on two parameters: α, which is the inclination angle of the leading face of the asperity, and f, which is the ratio of the friction shear stress at the asperity sliding interface to the shear flow stress of the softer material. We can consider two extreme conditions:

> (i) *Low friction extreme: f = 0 and a small finite asperity angle, α.* At this extreme condition, which corresponds to low interface friction shear stress, the force ratio becomes $\mu_{f=0} = \tan\alpha$. The force ratio increases with roughness expressed by the angle, α. With $\alpha = 15°$, $\mu_{f=0} = 0.27$.

> (ii) *High friction extreme: f = 1 and $\alpha = 0$, corresponding to sticking friction at the interface and a very small asperity inclination.* Under these conditions, the friction coefficient becomes indeterminate, $\alpha_{f=0} = 1/A$. However, assuming $\eta = 45°$, $\mu_{\alpha=0} = 1$.

Both extreme conditions may be approached in practical abrasive machining. The low friction extreme, where $f \rightarrow 0$, may be experienced in "spark-out" where a blunt wheel is allowed to rub with negligible depth of cut. This condition is sometimes allowed in order to improve surface texture. With ample lubrication, a positive normal force is required and a small positive tangential force results; the material removal is negligible and the force ratio is low.

The high friction extreme also corresponds to a blunt wheel but, in this case, with very poor lubrication. This condition is more likely to be experienced with *wheel loading*. In dry grinding at low removal rates, the wheel tends to become blunt, the temperature in the contact zone rises, and the workpiece material softens and may adhere to the tips of the abrasive grains. Loaded wheels tend to give rise to higher grinding forces and temperatures. Under these conditions, loaded wheels can give rise to high force ratios due to adhesion between the loaded material in the wheel and the workpiece.

Where the wheel picks-up the workpiece material, the condition is similar to scuffing. Heating of the workpiece surface is accompanied by thermal expansion and outward surface movement towards the wheel in the locality where the high temperature occurs. This condition, when it happens, causes a surge in power as the wheel has to cope with an increase in material removal. The resulting surface texture is extremely poor.

5.12.2 Wave Removal Model of Wear

The high friction extreme described above is problematic to define in the form of a steady-state slip-line field when $\eta \geq 45°$. This is because as ϕ in Fig. 5.21c becomes negative, continuity of material flow cannot be maintained. However, Challen and Oxley[20] proposed that the wave builds up until it is removed by crack formation.

The wave removal model, illustrated in Fig. 5.21, is an attempt to deal with the unsteady transition between the nonwear wave model and a steady chip removal model. Material removal by rapid crack growth requires less energy than material deformation, so that the wave removal process is accompanied by a reduction in specific energy from the previous infinite value for the case of no wear.

5.12.3 Chip Removal Models of Abrasion

The simplest chip removal model adopted by Challen and Oxley was the early model due to Lee and Shaffer illustrated in Fig. 5.21. The model is oversimplified, in that no allowance is made for subsurface deformation in the region of D. The model is purely concerned with the minimum energy required to produce a chip.

It is assumed that a chip is produced by a shock line of velocity continuity along the line AD. After crossing the line AD, the material has an upwards chip velocity. The line AD is also a slip line. The direction of the line AD is given by the angle at which a slip line meets the friction face DE of the asperity. From the Mohrs Stress Circle, the direction is defined by:

Eq. (5.57) $\alpha - \phi = \dfrac{\pi}{2} - \dfrac{1}{2} \cdot \cos^{-1} f$

where α is the inclination angle of DE, ϕ is the angle of the slip line AD, and f is the friction ratio, $f = \tau/k$. The line AE is stress free and, therefore, the slip line AD is at 45° to AE and $\theta = 45°$.

Working from the free surface along AD, the hydrostatic stress, σ_{hs} = k, the shear flow stress so that the principal stress $\sigma_1 = 2 \cdot k$ and the orthogonal principal stress $\sigma_3 = 0$. The principal stresses define the direction of the resultant force acting on the line AD. When this force is scheduled into the tangential and normal forces, the force ratio is given by:

Eq. (5.58) $$\mu = \frac{f_t}{f_n} = \tan\left(\frac{\pi}{2} - \theta + \phi\right)$$

where the forces per unit width are:

Eq. (5.59) $$f_t' = 2 \cdot k \cdot t \cdot \frac{\sin\left(\alpha - \dfrac{\pi}{4} + \dfrac{1}{2} \cdot \cos^{-1} f\right)}{\cos\left(\alpha + \dfrac{1}{2} \cdot \cos^{-1} f\right)}$$

Eq. (5.60) $$f_n' = 2 \cdot k \cdot t \cdot \frac{\cos\left(\alpha - \dfrac{\pi}{4} + \dfrac{1}{2} \cdot \cos^{-1} f\right)}{\cos\left(\alpha + \dfrac{1}{2} \cdot \cos^{-1} f\right)}$$

With $\theta = 45°$ using the relationship in Eq. (5.53):

Eq. (5.61) $$\mu = \tan\left(\alpha - \frac{\pi}{4} + \frac{1}{2} \cdot \cos^{-1} f\right)$$

In slip-line field solutions, as with previous empirical indentation models, the three most important parameters that determine the forces are the material shear flow stress, k, the asperity inclination angle, α, and the ratio, f, of the sliding interface shear stress, τ, to the shear flow stress, k.

It is interesting to examine the case when the asperity angle is 45°, a value which might not seem unrealistic for a reasonably sharp grain. If the friction factor on the face is $f = 0.5$, the value $0.5\cos^{-1}f = \pi/6$ or 30°. In Eq. (5.53), the angle of the shear plane is $\phi = 45° - 60° = -15°$. In effect, this means the plastic zone must extend below the depth of penetration of the asperity and some of the material flows under the grain as in the wave model.

It becomes impossible to construct the slip-line field solution for chip removal with high values of the friction factor, f, and for low values of α. The limiting case is when the sliding friction on the asperity is zero, $f = 0$. In this case, $\phi = 0$ and the tangential force is infinite.

In practice, material can flow sideways around an asperity as well as upwards and downwards, so ploughing and chip removal can take place with forward angles of inclination smaller than 45°.

However, a physical insight gained from this simplified cutting model is that reducing the sliding friction, f, encourages chip removal rather than subsurface deformation. Also, reducing the sliding friction, f, increases the tangential force rather than the normal force. In other words, the force ratio, μ, increases with grain angle or sharpness and also with improved lubrication. Improved lubrication and increased grain sharpness lead to more effective material removal. Results according to Challen and Oxley are illustrated in Fig. 5.22.

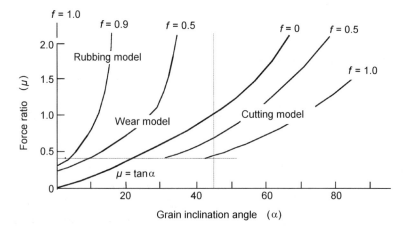

Figure 5.22. Effect of grain inclination angle and friction ratio on force ratio based on Challen and Oxley.[20]

Comparing the wave model and the cutting model, improved lubrication reduces the tangential force, hence, the force ratio in rubbing friction, but increases the tangential force and increases the force ratio under cutting conditions with negative rake grits. In both cases, improved lubrication is a beneficial result. As pointed out by Challen and Oxley, the force ratio and energy expended is reduced if the process shifts from a rubbing mode to a cutting mode for any particular values of α and f. Where the solutions overlap, the minimum energy principle dictates that cutting takes priority over rubbing as the more appropriate solution.

Remembering that not all grains in contact with the workpiece are involved in chip removal; any abrasive machining operation must be a combination of rubbing and cutting abrasion. Even those grains which are actively involved in chip removal are involved in rubbing contact for part of the contact time. Even under cutting conditions, part of the grain surface will be in rubbing contact. Taking all this into account, the models tend to support the frequent observation that better lubrication reduces the machining forces and the specific energy since rubbing and ploughing predominate in most cases.

5.13 THREE-DIMENSIONAL PYRAMID MODEL OF GRINDING

Williams[12] points out that chip removal can be promoted at much smaller values of forward angles of inclination if the asperity allows oblique cutting. This might be expected, considering the shape of the centuries old plough used in agriculture. That is to say, if the material is encouraged to flow sideways and upwards around the cutting edge, the tool will appear sharper, whereas an asperity that requires a chip to flow directly upwards acts as if it is blunt.

This situation is illustrated in Fig. 5.23, where a pyramid-shaped abrasive grain is characterized by two angles, the forward angle of inclination α, otherwise known as the attack angle, and the dihedral angle, $2 \cdot \phi$, at the base of the pyramid of the grain. With a small value of ϕ, the grain behaves more like a knife, as illustrated in the sliced bread analogy. With a large value of $\phi = 90°$, the grain behaves more like the relatively blunt two-dimensional case previously described. Sharpness in this model is reflected in a large value of α and a small value of ϕ. The figure illustrates the mode of deformation which includes a moving wave or prow which builds up ahead of the grain, a ridge thrown up and remaining at the side of the grain, and an obliquely-formed chip. This model, therefore, encompasses three important aspects of abrasive action.

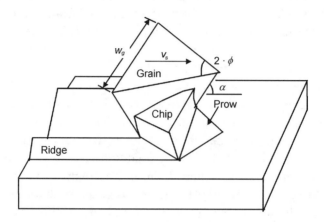

Figure 5.23. Pyramid model of abrasion by Williams and Xie.[21]

Williams and Xie[21] developed this model for a single grain and applied a Gaussian distribution to the heights of the abrasive peaks, successfully deriving equations for the forces and coefficients of friction for previously published experimental results of abrasion.

Badger and Torrance[22] developed a simulation of the grinding process based on the Williams and Xie model. The topography of the surface of the grinding wheel was based on circumferential profilometry using a stylus instrument. The surface of the grinding wheel was characterized by the number of grain asperities per unit area, C, and the mean asperity attack angle, α.

For the cutting mode of deformation, where material is removed on either side and deformation occurs from the creation of a ridge on either side of the pyramid and shearing of material beneath the surface, the coefficient of friction and the specific wear were fitted by the empirical expressions:

Eq. (5.62)
$$K \approx 0.003 \frac{\tan^3 \alpha}{f \cdot k \cdot l^{0.5}} \sqrt{\frac{H_b}{H_s}}$$

Eq. (5.63)
$$\mu \approx \sqrt{\frac{2}{\pi}} \cdot \frac{\tan\alpha}{l^{0.5}} \left[1 - f \cdot \left(1 + \frac{\pi}{4 \cdot \tan^2\alpha}\right)^{0.5}\right]$$

where, according to Badger and Torrance, the specific wear can be expressed in the more usual terminology for abrasive machining as

$$K = \frac{h_{eq}}{F_n}$$

and

$$F_t = \mu \cdot F_n$$

As previously discussed in Sec. 5.12, the friction factor for the grain-workpiece interface is

$$f = \frac{\tau}{k}$$

The grain spacing is expressed in dimensionless form as

$$l = \frac{L}{0.5 \cdot w_g}$$

where L is the grain spacing and w_g is the pyramid base width.

The bulk hardness of the workpiece is H_b and the surface hardness of the workpiece is H_s. The model allows for the surface hardness to be greater than the bulk hardness. For values of $H_s/H_b = 1$, the transition from ploughing to cutting was found to occur at an attack angle of $\alpha = 6°$ and $\alpha = 12°$ for $H_s/H_b = 1.25$. These values are very much lower than the values found from the previous two-dimensional models.

Badger and Torrance found good correlation between theory and experimental results assuming interfacial friction values of $f = 0.1$ for neat oil as the grinding fluid and $f = 0.4$ using a water-based emulsion.

5.14 LIMIT CHARTS

A major difficulty for the application engineer is to know how to set the abrasive machining conditions, such as workspeed and depth of cut, for a particular grinding wheel and material combination, workpiece diameter, and grinding width. The process of selecting machining conditions is made simpler if the engineer has experience or approximate knowledge of the specific energy for the particular type of application. From the specific energy, it is a simple matter to select an appropriate power level using Eq. (5.4)

In practice, there may be other constraints on selection of machining conditions which impose an operating envelope. Process optimization requires that machining conditions are selected within the envelope imposed by process constraints. Charts which describe the process constraints are known as limit charts.

Examples of limit charts derived for plunge-feed centerless grinding are illustrated in Figs. 5.24 and 5.25. Removal rate is proportional to infeed rate, v_f, in plunge centerless grinding. For the particular conditions, the optimum workspeed is approximately given by $v_w = 0.25$ m/s when grinding with medium carbon steel, AISI 1055, or with gray cast iron. At this workspeed, the infeed rate could be increased up to the maximum power available from the machine without offending the roughness, burn, or chatter constraints.

Doubling the wheelspeed from 30 m/s to 60 m/s allowed the machine to deliver more power to the process. The higher removal rate improved the cutting efficiency of the process due to the size effect, so that doubling the wheelspeed tripled the removal rate.

Machining the gray cast iron with a resin-bonded C48BBT carbide wheel was very efficient in energy terms, allowing removal rates nearly twice as great as achieved when grinding the medium carbon steel with a WA60MVRC vitrified alumina wheel. For the cast iron at $v_w = 0.3$ m/s, the power was 32 kW. The removal rate at an infeed rate of $v_f = 0.58$ mm/s, a workpiece diameter $d_w = 40.5$ mm and a grinding width $b_w = 65$ mm was

$$Q_w = (\pi/2) \cdot d_w \cdot b_w \cdot v_f = 2400 \text{ mm}^3/\text{s}$$

The specific energy for these conditions was, therefore,

$$e_c = \frac{32,000}{2,400} = 13.3 \text{ J/mm}^3$$

Process: Centerless grinding
Wheel: WA60MVRC, d_s = 500 mm
Workpiece: AISI 1055 steel, d_w = 50 mm, b_w = 65 mm
Wheelspeeds: v_s = 30 and 60 m/s

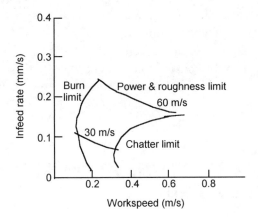

Figure 5.24. Limit chart for centerless grinding AISI 1055 steel with various speed combinations.

Process: Centerless grinding
Wheel: C48BBT, d_s = 500 mm
Workpiece: Gray cast iron, d_w = 40.5 mm, b_w = 65 mm
Wheelspeeds: v_s = 30 and 60 m/s

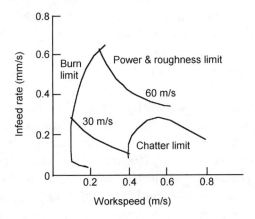

Figure 5.25. A limit chart for centerless grinding gray cast iron with various speed combinations.

Studies of limit charts reveal substantial benefits from optimizing the process for minimum specific energy by maximizing removal rates.

Adaptive control of feed rate is a way of ensuring that the maximum feedrate is employed subject to imposed constraints on power or maximum temperature.[23]-[25] The use of intelligent computer techniques also makes it possible to improve size-holding capability, as well as provide an operator with advice on parameter selection.

Some caution needs to be exercised so that the grinding wheel is not allowed to glaze thereby causing forces to increase. In these situations, the feedrate conditions are reduced. To overcome this problem, the wheel must be redressed or more appropriate machining conditions chosen.

5.15 PROCESS OPTIMIZATION AND WHEELSPEED

For at least three decades, there has been a trend towards higher wheelspeeds. Whereas previously a wheelspeed of 25–30 m/s was the norm, wheelspeeds of 45–60 m/s are now commonplace, and there are reports of speeds in excess of 140 m/s for some applications.

The application of increased wheelspeeds does not come without increased costs. The machine must be stiffer and more vibration resistant to cope with the higher wheelspeeds. Also, the grinding wheels must be stronger to resist increased bursting stresses, and the grinding fluid must be applied more effectively to penetrate the faster airstream around the grinding wheel. Higher power must be delivered to the machine to make effective use of the increased speeds and overcome the increased losses in the bearings and the increased losses due to windage of other rotating elements. The stronger guards must be fitted to safeguard against wheel bursting.

However, as demonstrated in an example in Sec. 5.14, doubling the wheelspeed allowed the removal rates to be tripled. It is tempting to assume that higher wheelspeeds will automatically increase efficiency. This is not actually true. Simply increasing wheelspeed, without changing other grinding conditions, will usually reduce efficiency. This can be seen from kinematic considerations shown in Eq. (3.71). Increasing v_s reduces the average uncut chip volume and increases specific energy.

Improvement in grinding efficiency can only be achieved by increasing removal rate and wheelspeed together.

There is a straightforward way to determine whether increasing wheelspeed increases or reduces efficiency. To do this, the specific energy at the lower speed is measured under grinding conditions which appear favorable. The wheelspeed and workspeed are then increased in proportion, maintaining a constant depth of cut. For cylindrical plunge grinding, this means increasing the feedrate in the same proportion.

The result is that the kinematic conditions are similar at both the high speed and at the low speed. The uncut chip thickness and the geometric contact length are unchanged. The power is measured and the specific energy at the high speed calculated.

The change in specific energy that results is the speed effect. Of course, vibration levels as well as the fluid application must be carefully controlled to ensure that these factors do not allow the grinding conditions to deteriorate.

If the specific energy is increased, the speed effect has made the process less efficient. If the specific energy is reduced, the process is more efficient. Either way, the removal rate will be increased and the power requirement increased. However, if the maximum power is constrained, for example, by the capacity of the machine, a greater removal rate will be achieved at the maximum power under the conditions where the specific energy is a minimum.

An example of the speed effect is shown in Fig. 5.26 for four different values of v_f/v_w, corresponding to four different values of uncut chip thickness and chip length. Each value of v_f/v_w corresponds to a constant value of mean uncut chip thickness. The speed ratio was maintained constant at $v_s/v_w = 140$ throughout the tests.

Each diagram represents a constant kinematic condition for a particular value of mean uncut chip thickness. The variation in specific energy is, therefore, purely a consequence of the effect of speed on the process. At higher speed, there is less time for conduction and the process is closer to being conducted under adiabatic conditions. Thermal effects on process efficiency are discussed in Ch. 6, "Thermal Design of Processes."

The removal rate increases in each diagram in proportion to wheelspeed, v_s. Moving from the front diagram to the rear, removal rate increases with the square of mean uncut chip thickness. At any particular speed, the maximum removal rate is 2.5 times the minimum removal rate. It is seen that with the mean uncut chip thickness equal to 0.8 μm, the specific energy of $e_c = 25$ J/mm^3 is minimum at a wheelspeed of $v_s = 45$ m/s.

Process: Centerless grinding

Wheel: WA60MVRC, d_s = 500 mm

Workpiece: AISI 1055 steel, d_w = 50 mm, b_w = 65 mm

Wheelspeeds: v_s = 30 and 60 m/s

Speed ratio: v_s/v_w = 140

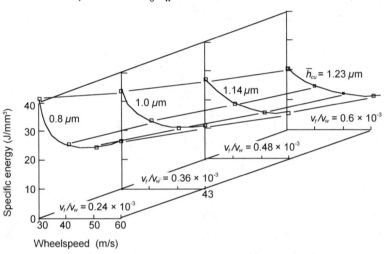

Figure 5.26. Experiment to determine optimum feed and wheelspeed combination for minimum specific energy at a constant speed ratio of v_s/v_w = 140. Repeating the experiments at a speed ratio of v_s/v_w = 200 further reduced the values of specific energy.

When the mean uncut chip thickness was increased to 1.23 μm, the specific energy was less sensitive to wheelspeed. The minimum specific energy was reduced to e_c = 15 J/mm³ and remained almost constant between v_s = 40 m/s and v_s = 60 m/s.

The maximum specific energy was e_c = 42 J/mm³ and occurred at the lowest wheelspeed of v_s = 30 m/s with the lowest removal rate. The maximum specific energy was 2.7 times the minimum value which has major implications for the temperature in the grinding zone, as well as for the maximum achievable removal rate. Substantial cost benefits can be obtained by optimizing the process for high speeds and removal rates.

REFERENCES

1. Chen, C., Jung, Y., and Inasaki, I., Surface, Cylindrical and Internal Grinding of Advanced Ceramics, Grinding Fundamentals and Applications, *Trans ASME*, 39:201–211 (1989)

2. Backer, W. R., Marshall, E. R., and Shaw, M. C., The Size Effect in Metal-cutting, *Trans ASME*, 74:61–72 (1952)

3. Rowe, W. B., and Chen, X., Characterization of the Size Effect in Grinding and The Sliced Bread Analogy, *Int. J. Production Research*, 35(3):887–899 (1997)

4. Hahn, R. S., On The Mechanics of The Grinding Process Under Plunge Cut Conditions, Trans ASME, *J Engineering Industry*, pp. 72–80 (1966)

5. Shaw, M. C., A New Theory of Grinding, *Proc. Int. Conf. Science Industry, Monash University, Australia*, pp. 1–16 (1971)

6. Kannapan, S., and Malkin, S., Effects of Grain Size and Operating Parameters on the Mechanics of Grinding, Trans ASME, *J. Engineering Industry*, 94:833–842 (1972)

7. Malkin, S., *Grinding Technology*, Ellis Horwood, Publ. (1989)

8. Chen, X, Strategy for Selection of Grinding Wheel Dressing Conditions, Ph.D. Thesis, Liverpool, UK, John Moores University (1995)

9. Suh, N. P., *Tribophysics*, Prentice Hall (1986)

10. Bowden, F. P., and Tabor, D., The Area of Contact Between Stationary and Moving Surfaces, *Proc. Royal Society*, A169:391–413 (1939)

11. Tabor, D., Junction Growth in Metallic Friction, *Proc. Royal Society*, A251:378–393 (1959)

12. Williams, J. A., *Engineering Tribology*, Oxford Science Publ. (1994)

13. Rabinowicz, E., *Friction and Wear of Materials*, John Wiley & Sons (1965)

14. Badger, J. A., and Torrance, A. A., A Computer Program to Predict Grinding Forces from Wheel Surface Profiles Using Slip-line Fields, *Proc. Int. Seminar Improving Machine Tool Performance, Jul. 1, 6–8, 1998, San Sebastien* (1998)

15. Archard, J. F., Contact and Rubbing of Flat Surfaces, *J. Appl. Physics*, 24:981–988 (1953)

16. Black, A. J., Kopalinsky, E. M., and Oxley, P. L. B., Asperity Deformation Models for Explaining Metallic Sliding Friction and Wear, *Proc. Inst. Mechanical Engineers*, Part C, 207:335–353 (1993)

17. Verkerk, J., Characterization of Wheel Wear in Plunge Grinding, *Annals of the CIRP*, 26(1):127 (1977)

18. Rowe, G. W., *Elements of Metal Working Theory*, Edward Arnold (1979)

19. Lortz, W., A Model of the Cutting Mechanism in Grinding, *Wear*, 53:115–128 (1979)

20. Challen, J. M., and Oxley, P. L. B., An Explanation of the Different Regimes of Friction and Wear Using Asperity Deformation Models, *Wear*, 53:229–243 (1979)

21. Williams, J. A., and Xie, Y., The Generation of Wear Surfaces by the Interaction of Parallel Grooves, *Wear*, 155:363–379 (1992)

22. Badger, J. A., and Torrance, A. A., Comparison of Two Models to Predict Grinding Forces From Wheel Surface Topography, private communication (1999)

23. Rowe, W. B., Allanson, D. R., Pettit, J. A., Moruzzi, J. L., and Kelly, S., Intelligent CNC for Grinding, *Proc. Inst. Mechanical Engineers*, 205:233–239 (1991)

24. Rowe, W. B., Thomas, D. A., Moruzzi, J. L., and Allanson, D. R., Intelligent CNC Grinding, *Proc. Inst. Electrical Engineers, Manufacturing Engineer*, 238–241 (1993)

25. Rowe, W. B., Li, Y., Inasaki, I., and Malkin, S., Applications of Artificial Intelligence in Grinding, Keynote Paper, *Annals of the CIRP*, 43(2):521–531 (1994)

26. Andrew, C., Howes, T. D., and Pearce, T. R. A., *Creep-Feed Grinding*, Holt, Rinehart, and Winston (1985)

27. Sin, H. C., Saka, N., Suh, N. P., Abrasive Wear Mechanisms and the Grit Size Effect, *Wear*, 55:163–190 (1979)

28. Tomlenov, A. D., Eindringen eines abgerundeten Stempels in ein metall unter Vorhandsein von Reibung, Vestn. Mashinostr., 40:56–58 (1960)

6

Thermal Design of
Processes

6.1 INTRODUCTION

Well-designed abrasive machining processes usually enhance workpiece surface quality, producing low roughness, compressive or neutral residual stresses, and improved fatigue life. This is particularly true for lapping, polishing, and honing, but also for low stress grinding.

Conversely, abusive machining can lead to a range of forms of surface damage. Generally, it is found that the higher the machining temperature, the greater the damage caused to the surface.

Thermal effects are usually deleterious to surface integrity as examples below demonstrate. The main changes can be summarized as one or more of the following:

- Diffusion leading to grain growth, precipitation, and softening.
- Phase transformations leading to re-hardening.
- Thermal effects leading to expansion, contraction, possible cracking, and tensile residual stresses.
- Chemical reactions leading to increased oxidation.

In this chapter, we consider surface damage and go on to explore relationships between process energy, selection of grinding conditions, and aspects of process design to achieve high productivity and surface integrity.

6.2 EXAMPLES OF SURFACE DAMAGE

Some examples of grinding damage to materials, gray cast iron and AISI 52100 1%C 1% Cr bearing steel, are illustrated below. Both materials are commonly ground. The bearing steel is usually hardened and tempered before grinding to produce a hard wear-resistant surface.

6.2.1 Discoloration

Oxidation colors are the most obvious and often the first indications observed relating to thermal damage. Colors ranging from light straw to dark blue are produced at temperatures from 450°C and upwards in conventional grinding processes. Light straw colors are produced at lower temperatures, and dark blue colors are produced at higher temperatures. The degree of oxidation is time-dependent so that discoloration occurs at lower temperatures with very low grinding speeds.

Figure 6.1 shows the striping produced on a gray cast iron cylindrical workpiece. In this case, the stripes are dark blue and indicate seriously abusive grinding conditions employed in a centerless grinding operation. The problem was eliminated by increasing the workspeed without reducing the material removal rate. The striping is also indicative of vibration and temperature surging, which are often experienced with high temperature burning. In this example, power surging and vibration caused chatter patterns on the workpiece.

6.2.2 Softening

Grinding temperatures which may not be high enough to cause severe thermal damage, as in the previous case, may still be sufficiently high enough to reduce the hardness of a bearing surface and undermine the wear-resistant properties of the surface.

In Fig. 6.2, a hardened and tempered bearing steel has been further tempered by the grinding process. The grinding process produced a light straw color indicative of a lightly burned surface. The specimen cross-section was carefully polished and etched using a 2% nital solution to allow the structure to be visible under a microscope at 400× magnification. The black area is a bakelite mounting medium adjacent to the workpiece surface. Near the surface, a darker structure is visible where carbon has come out of the solution. This change in the etched appearance is indicative of diffusion which has taken place during the grinding process. The lighter microstructure deeper under the surface is the unaffected tempered martensite produced in the original heat treatment.

Microhardness measurements taken across the cross-section, Fig. 6.3, show that tempering has softened the surface. If this was a bearing surface, the result would be reduced wear resistance due to the lower hardness.

Figure 6.1. Severe workpiece thermal damage in centerless grinding due to low workpiece speed. The dark blue stripes and chatter are a consequence of high grinding temperatures.

Figure 6.2. Microstructure at 400X magnification under the surface of a hardened and tempered bearing steel AISI 52100. The material near the surface has been tempered due to grinding at a temperature sufficient to cause light straw temper colors.

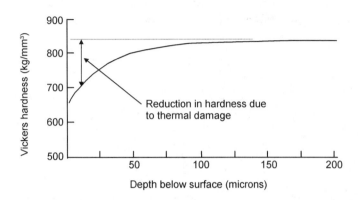

Figure 6.3. Microhardness variations under the workpiece surface caused by temper burn. AISI 52100 bearing steel ground with light temper colors.

A lack of temper colors after grinding is not a guarantee that the surface is undamaged because oxidation colors are easily removed by the spark-out part of the grinding cycle, leaving a damaged surface underneath. Typically, a surface is damaged to a depth of the order of 100 μm. This depth is likely to be too deep to be removed by the spark-out element of the cycle. A fine feed for a sufficient depth can overcome the problem, but often it is safer to avoid the problem in the first instance.

Softening is a diffusion process which is time dependent. The degree of softening, for a particular temperature, is greater at low workspeeds because there is more time for diffusion to take place. Of course, softening is only a problem with materials which can be hardened. Softening is not a problem with low carbon steels.

6.2.3 Re-hardening

Figure 6.4 shows a section of a hardened bearing steel which has been abusively ground leading to dark blue temper colors.

The cross-section shows a white martensitic layer. In this case, the surface has been heated above the austenitizing temperature indicated in Fig. 6.5.

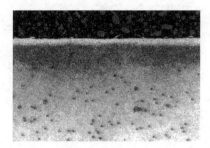

Figure 6.4. Microstructure of abusively ground AISI 52100 bearing steel at 100X magnification. The white band is a martensitic layer at the surface. The material under the white layer has been softened by tempering in the grinding process. The dark particles include carbides of manganese and chromium.

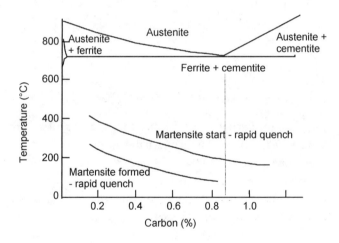

Figure 6.5. The iron-carbon diagram.

The transformation to martensite is rapid, taking place in microseconds, since no diffusion is required. The material cools rapidly after grinding due to the low temperature of the bulk material below the surface. Quenching leads to a thin layer of brittle martensite which is susceptible to cracking. At greater depths, the maximum temperature achieved is lower than at the surface. The cooling rate is also lower. This leads to a softened layer below the hardened layer as shown by the microhardness measurements in Fig. 6.6.

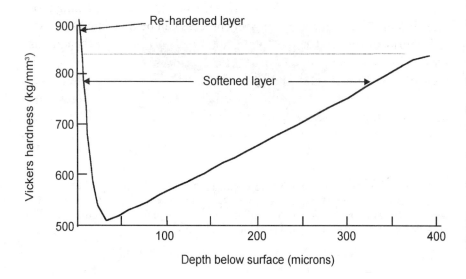

Figure 6.6. Microhardness variations under the workpiece surface caused by abusive grinding. Bearing steel AISI 52100 ground with blue temper colors leading to re-hardening burn.

Surprisingly perhaps, it is found that high temperatures are not always deleterious to surface integrity. In the region of the critical temperature, the transformation to austenite is accompanied by a 1–2% volume reduction which creates a tensile stress because volume reduction is opposed by the bulk material deeper below the surface. This is the opposite of the dominant general thermal expansion effect. On quenching, the reverse occurs and under the right conditions compressive stresses can be set up at the surface, as explained below. In this context, Brinksmeier and Brockhoff[1] proposed using grinding as a new hardening process. It was demonstrated that high temperatures could be used to achieve surface hardening without causing surface cracks.[33] To explain these results, it is necessary to take into account all of the expansions and contractions which take place at the surface and in the subsurface during the heating and cooling processes. Briefly, the explanation offered is that when martensite is formed where previously there was soft ferrite and pearlite, the resultant martensite occupies a 4% larger volume, which leads to a compressive stress at the surface.

There is relatively little experience using this new hardening process. It remains to be seen whether the process can be translated successfully into industrial production. It is also not completely clear what conditions are required for successful operation of the process. Very low workspeed, typically 0.008 m/s, is required both to obtain a substantial hardened layer and also to reduce the thermal gradients below the surface and allow time for the required austenitizing which is a diffusion process. Cooling then produces an acceptable form of martensite.

6.2.4 Cracks

Heating of the surface is accompanied by thermal expansion and contraction. As the wheel passes a point on the workpiece, the surface expands due to heating. After the wheel has passed, the surface rapidly cools and is quenched by the mass cooling from the subsurface. At this stage the material contracts.

If the thermal expansion phase is sufficient to cause plastic yield, the subsequent cooling contraction will lead to tensile residual stresses. In severe cases, this may lead to cracking (Fig. 6.7), where the crack extends downward from the surface.

Figure 6.7. A surface crack produced by abusive grinding; AISI 52100 bearing steel.

6.2.5 Spheroidal Swarf

Individual pieces of swarf are insufficient evidence of thermal damage to the workpiece since the temperature at the grain-chip interface is very localized. However, the nature of the swarf is likely to be indicative of the nature of the grain contact conditions.

Some swarf removed from a workpiece, as viewed under a microscope, is illustrated in Fig. 6.8a–c. Some of the chips are wire-like and slightly curled which is typical of swarf produced by sharper cutting edges at lower temperatures. There are also examples of spheroidal swarf which suggests higher temperatures. Closer examination shows that the surface of some of the spheroids clearly has the appearance of a material which has become molten or near molten. Malkin found spheroids that were hollow and suggested that the material structure was dendritic, meaning the material had become molten.[2] Shaw found that spheres are formed by small platelets bonding together.[32] Either way, the symmetry suggests a high temperature process. Platelets are usually considered to be evidence of high temperature rubbing contact with blunt asperities. Sometimes a partly formed sphere was found which had the appearance of platelets.

6.2.6 Tensile Residual Stresses

It has been mentioned that tensile stresses may be set up by thermal expansion and contraction as the wheel passes a point on the workpiece. First, the temperature rises rapidly and the surface layer expands. Another consequence of the elevated temperature is that the yield stress of the workpiece material in the region of heating falls.

The yield stress depends on the temperature and also on the previous heat treatment of the material. If the temperature is sufficient, the yield stress at the elevated temperature is exceeded. The result is that the surface is plastically deformed due to the thermal effect in a much larger region than the immediate area surrounding an individual grain resulting from the mechanical interaction.

After heating and being plastically deformed, the surface cools and contracts producing tensile residual stresses. The tensile residual stresses in the outermost surface are balanced by compressive stresses of a lower magnitude extending below the surface.

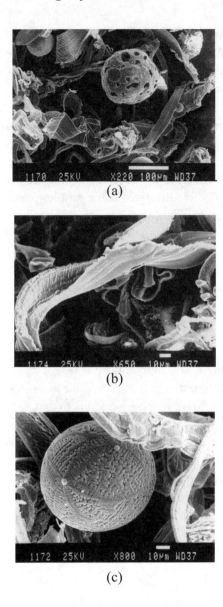

(a)

(b)

(c)

Figure 6.8. *(a)* Typical grinding swarf viewed under a microscope. Some of the swarf is straight or slightly curled which is typical of ductile oblique machining and some of the swarf particles are formed almost into hollow spheres which is typical of higher temperature interaction between blunt grains and the workpiece. There are also platelets of swarf. *(b)* A long thin chip. The underside is smooth having experienced ductile shearing. The upper surface shows the ripples typical of a ductile chip formed under compression. The chip is wide compared to its thickness. *(c)* A spheroid of AISI 52100 material which has become near molten. The structure at 800X magnification has an appearance rather like dendrites typical of the casting process. This form of swarf is a typical result initiated by a high temperature interaction during the removal process.

Because the yield stress of the material reduces with increasing temperature, the onset of tensile residual stresses tends to occur close to the softening temperature for the particular material.

Typical transition temperatures measured by McCormack, Rowe and Chen[3] are listed in Table 6.1. It should be noted, however, that the transition temperatures can vary substantially according to the heat treatment specification. For the AISI 1055, the transition can vary from 100–330°C.

For low-stress grinding, the grinding temperature should be kept below the transition value throughout the grinding cycle.

Machine parts which have to withstand high surface stresses should be protected from adverse stresses induced by machining. It is usually considered that tensile residual stresses are more damaging than compressive residual stresses although in some applications compressive residual stresses may also be considered undesirable.

6.2.7 Mechanically Induced Stresses

In most cases, low-temperature abrasive machining tends to enhance the physical properties of the surface due to mechanical effects, while high-temperature abrasive machining tends to damage the surface due to thermal effects.

It is generally considered that the primary mechanical effect of abrasive machining is due to the effect of indenting blunt abrasive grains into the ductile surface to create compressive stresses in the same way as sand-blasting or shot-peening a surface. This effect is usually beneficial. Measurements of residual stresses after low-temperature abrasive machining indicate that compressive residual stresses do result, in practice.

Table 6.1. Residual Stress Transition in Grinding

MATERIAL	TYPICAL TRANSITION TEMPERATURE (°C)
AISI 1055 (BS070 M55 Plain carbon steel)	330
AISI 52100 Bearing steel	400
M2 Tool steel	560

The tangential and normal movement of the grains, particularly if the grains present to the workpiece a bluff face with a large attack angle, can give rise to cracks which are particularly damaging to crack-sensitive workpiece materials such as hard ceramics.

6.3 THERMAL MODELING—KEY DEVELOPMENTS

Some companies use a power monitor in critical machining processes as a way of checking that the process is operating satisfactorily. If the power level exceeds the normal operating level, it is known that there is a risk of thermal damage. The operator stops the process and makes adjustments to restore operating efficiency. The adjustment may consist of redressing the grinding wheel, changing the grinding conditions, or even changing a dressing tool or grinding wheel.

The state of research has now advanced to the point where there is the potential for using the PC to monitor the power and compute the max temperature in many plunge grinding operations. This raises the prospect of continually monitoring a process and being able to guarantee surface integrity.

Thermal analysis has greatly clarified the importance of the various physical processes involved in abrasive machining leading to radical developments such as creep grinding and high-efficiency deep grinding (HEDG). Therefore, the following analysis is justified in terms of the benefits obtainable with improved process understanding.

Some findings in the development of thermal modeling are listed in Table 6.2. Many other authors have tackled particular applications of thermal modeling of abrasive machining processes, however, the papers listed below are key concepts required to deal with a range of situations.

Outwater and Shaw[4] modeled heat transfer to the workpiece based on a sliding heat source at the shear plane, whereas Hahn[5] reasoned from energy considerations that the principal heat generation is at the grain-workpiece rubbing surface. This is because the shear plane energy assumption cannot account for the much larger energy experienced in practice. The heat generation is more accurately described by considering the energy to be dissipated at the contact between the workpiece and the grain, neglecting the shear plane energy.

Table 6.2. Key Developments Listed By Year, First Author, and Reference

KEY DEVELOPMENTS	YEAR	FIRST AUTHOR	REF. #
Shear plane energy partitioning model	1952	Outwater	[4]
Energy partition between grain and workpiece	1962	Hahn	[5]
Real contact length $l_e > l_g$	1966	Makino	[6]
Fluid convection over l_e	1970	Des Ruisseaux	[7]
Limiting chip energy and sliding energy	1971	Malkin	[8]
Fluid boiling and limiting fluid energy	1975	Shafto	[9]
Triangular heat flux	1978	Snoeys	[10]
Wheel/work/fluid/chip partitioning	1980	Werner	[11]
Conical grain (1D) model	1989	Lavine	[12]
Transient contact model	1991	Rowe	[13]
Force/contact length model	1993	Rowe	[14]
Critical temperatures for tempering	1995	Rowe	[15]
Grain temperatures	1996	Ueda	[16]
Effective abrasive thermal properties	1996	Rowe	[17]

In practice, shear plane and wear flat energies are important. However, there is a limit to the shear zone energy which can be carried away by the chips, as pointed out by Malkin.[8]

It is practically impossible to predict workpiece temperatures with any accuracy purely from theory because actual energies depend greatly on the extent of fracture and subsurface redundant work. The starting point for temperature prediction is based, in most cases, on measured power levels from which specific energy levels can be determined for the particular abrasive machining conditions.

Many authors have applied over-simplifications in calculating machining temperature. For example, it would be tempting to assume that all the machining energy goes into the geometric contact area of the workpiece, ignoring all other heat sinks. This is satisfactory for some abrasive machining processes, such as lapping and honing, but greatly over-estimates the surface temperatures and is quite useless for any practical control of the conventional grinding process.

Useful estimates of temperature can be obtained by systematic application of the key principles listed above. Temperature estimates can provide better process understanding leading to process improvements. It is also possible that temperature estimation can provide a tool for integration into process monitoring and control systems.

In 1966, Makino[6] measured temperatures using a thermocouple and found that the actual length of the heat source was two to three times the geometric contact length. The reasons for this are discussed in Ch. 4, "Contact Mechanics." The assumption of a geometric contact length over-estimates the heat flux density in the contact zone and, as a result, predicts artificially high temperatures.

In 1993, Rowe, et al.,[14] found that the contact length could be predicted based on the combination of the geometric contact length and the elastic contact length due to the forces described in Ch. 4, "Contact Mechanics."

The effect of surface cooling by a grinding fluid was analyzed in 1970 by Des Ruisseaux and Zerkle.[7] It was found that significant convective cooling would not occur in the contact region for conventional shallow grinding at typical depths of cut and likely values of convection coefficients. Of course convective cooling is very important outside the contact region where most of the heat is extracted by the fluid. However, fluid cooling outside the contact region is ineffective if the temperature rise of the workpiece within the contact region is already sufficient to cause thermal damage.

The role of convective cooling within the contact area is now being re-evaluated. As thermal models become more refined, it appears that convective cooling in conventional grinding usually extracts less than 10 % of the energy within the contact zone.

In creep-feed grinding with large depths of cut, convective cooling inside the contact region is of much greater importance, as shown by Shafto in 1975.[9] In creep-feed grinding, convective cooling sometimes extracts more heat from the contact zone than all the other elements combined. However, Shafto found that fluid boiling limits the energy which may be extracted by fluid cooling.

Howes[18] in 1987 found that fluid boiling severely limits cooling and lubrication in shallow-cut grinding when the temperatures in the contact zone exceed the boiling temperature of the fluid. Therefore, we may conclude that the maximum energy which can be convected by the fluid is limited by the energy required to cause boiling of the fluid.

Malkin[8] realized from a simple calculation of the melting energy of the chips that the maximum energy which could be convected by the chips was limited. For ferrous materials, this energy is approximately 6 J/mm^3. Further consideration leads to the assumption that this forms a reasonable basis for the estimation of the energy convected by the chips.

Werner, et al.,[11] used finite element analysis to model heat flows to the workpiece, wheel, chips, and fluid in creep-feed grinding. This was one of the first attempts to model the effect of all four heat sinks simultaneously.

Rowe, et al.,[19] modeled the energy partitioning to all four heat sinks analytically. Upper and lower bound estimates of the energy to the workpiece provided confidence limits on the temperature calculations. In this early method used by Rowe, it was assumed that a grinding wheel could be modeled as a homogeneous mass. In later papers, it was shown that it was as simple but more accurate to model the wheel as a set of discrete contacts within the contact zone based on the two-dimensional model of heat transfer into a plane grain proposed by Hahn.[5]

Lavine[12] modeled heat flows into the workpiece, the fluid, and the grain using a conical grain model. The conical grain model, based on a one-dimensional approximation for two-dimensional heat dispersion, appears at first to give advantages over the plain grain model. However, it was later realized that the conical grain model only works for very small cone angles. For larger angles typical of abrasive grains, the plane grain model is more accurate and much simpler.

Rowe, et al.,[13] used the cone model with a more direct partitioning approach to demonstrate the effect of transient heating of the abrasive grains. It was shown that this reduces the energy partitioned to the workpiece, particularly with high conductivity grains.

Rowe, et al.,[17] demonstrated that the transient solution could also be applied to Hahns plane grain model.

Rowe, et al.,[15][20] investigated critical temperatures for onset of thermal damage for several ferrous materials. The critical temperature for the onset of temper colors was found to lie between 450°C and 500°C under most conditions.

During the course of the work by Morgan, et al.,[21] it became apparent that the accuracies of the thermal properties of the grains were crucial to the accuracy of solutions for cubic boron nitride (CBN) grinding. This is because CBN grains are much more conductive than conventional abrasives. Large discrepancies between published values produced unacceptable uncertainties. Careful experimental work showed that it was advisable to determine effective abrasive properties from temperature measurement experiments and correlation with the thermal model to be employed. This inverse technique also calibrates the effect of unknown factors in the system.

6.4 RATE OF HEAT GENERATION

The power required for the abrasive machining process is the power remaining after deducting the power required to overcome friction in the machine tool elements and to provide the pumping power for the hydraulics and process fluid delivery.

The process power is almost exactly equal to the rate of heat generation in the abrasive contact zone if we deduct the power required to accelerate the fluid at the contact with the abrasive tool. A negligible proportion of the energy inputs provides acceleration of the chips and some is locked into the deformed material.

Contact temperatures are primarily dependent on the average heat flux generated in the contact zone and the duration of heating. The distribution of the heat flux is of secondary importance.

The rate at which energy is generated is approximately proportional to the rate of material removal. We have already seen from Ch. 3, "Kinematic Models of Abrasive Contacts," that the removal rate varies almost linearly across the contact zone as in Fig. 3.6. Therefore, the heat flux is most intense at the leading edge of the contact zone as illustrated in Fig. 6.9.

Early assumptions of a uniform heat distribution are not supported by temperature measurements. It is more correct to assume a triangular heat flux, as in Fig. 6.10. Fortunately, the maximum potential error in the maximum temperature estimated by using the uniform heat flux distribution is limited to approximately 5% increase in the temperature.

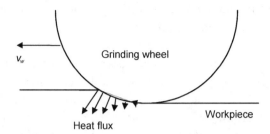

Figure 6.9. The heat flux generated in the grinding process is distributed on the 'inclined' contact surface and is concentrated towards the leading edge of the wheel in both upcut and downcut grinding. The inclined surface is approximately the chord shown as a dashed line.

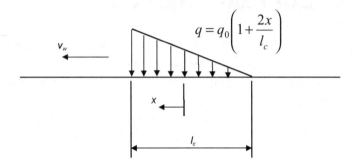

Figure 6.10. The assumption of a triangular heat flux generation in the contact zone.

The term heat flux is commonly used to describe the power density which is the same as the heat flowrate per unit area. The reader should be aware that heat flux tends to be defined in the literature in different ways depending on the requirements of the situation.

The average heat flux, q_0, in the contact zone is

Eq. (6.1) $$q_0 = \frac{P}{l_c \cdot b_w}$$

where l_c is the length of the contact zone and b_w is the width of the contact zone.

It can be seen that the flux which determines the maximum temperature of the workpiece depends on the power and also on the size of the contact zone.

The triangular heat flux distribution in Fig. 6.10 is defined at any point, x, along the contact length by

Eq. (6.2) $$q = q_0 \left(1 + \frac{2x}{l_c} \right)$$

Of course, we need to remind ourselves that the flux described by Fig. 6.10 is an averaging of very short bursts of very intensive energy at randomly distributed small areas, as shown in Fig. 6.11. Each small area represents the instantaneous location of an abrasive grain as it passes through the contact zone.

For grinding, typical durations of the heat inputs are summarized in Table 6.3. Each grain interaction moves past a point on the workpiece at wheelspeed so that a typical heat pulse is experienced for about 1 microsecond.

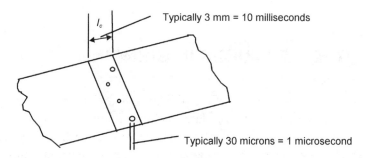

Figure 6.11. Contrasting typical durations of energy bursts as individual grains pass points on the workpiece and the overall duration of the passage of the contact zone. Consequently, the individual energy spikes are much more intense than the average energy across the contact zone.

Table 6.3. Typical Heating Duration

CONTACT	DURATION	TYPICAL VALUE
Grain at workpiece point	$2 \cdot r_0/v_s$	1 microsecond
Grain in contact zone	l_c/v_s	100 microseconds
Workpiece in contact zone	l_c/v_w	10,000 microseconds

A section of the workpiece will experience many such pulses over a much longer duration. The overall duration of the succession of pulses is 10,000 microseconds.

A grain typically moves across the whole contact length in 100 microseconds. Therefore, the grain experiences a heat pulse for a period approximately 100× longer than the workpiece. This allows the surface of the grain to reach a steady state temperature which has been estimated to be close to the melting temperature by Ueda, et al.[16]

The microsecond duration of the heat pulse due to an individual grain is too short to cause much more than a very localized plastic deformation along the path of contact and some localized oxidation. The effect of many such pulses averaged over a period, measured in milliseconds rather than microseconds, is sufficient to cause significant temperature rise at depths up to and often exceeding 0.1 mm.

In looking for the causes of surface thermal damage, we are more concerned with the variation of the average heat flux according to Eq. (6.2) than with the much more shallow thermal damage caused by individual grains.

6.5   TEMPERATURES IN GRINDING

Direct measurements of temperatures in abrasive machining processes are made difficult by the large number of short random discrete events taking place as individual grains pass through the contact zone. Ueda, et al.,[16] measured the temperature of abrasive grains just after the grains left contact with the workpiece. Measurement was made by detecting infrared radiation using a fiber-optic linked to a two-color pyrometer.

It was said that cooling is extremely rapid. The maximum temperature at the end of cutting diminishes to one quarter within 1 ms. It was estimated that the maximum temperature of the grain at the end of cutting is approximately equal to the melting temperature of the workpiece. This is in agreement with expectation because grinding energies greatly exceed the specific energy required to melt the material removed.

The workpiece material softens as it approaches the melting temperature and it is difficult to dissipate further energy in the material removed. The surplus energy is dissipated within the remaining workpiece material, the grains of the abrasive, and the fluid.

The results from Ueda are useful, as will be described later in this chapter, for analyzing the proportion of the energy conducted into the workpiece. However, the conclusion that the material removed reaches a temperature close to melting tells us little about the temperatures which give rise to thermal damage and tensile stresses. For this purpose, it is necessary to measure the maximum background surface temperature of the workpiece as it moves through the contact zone.

Nee and Tay[22] measured the surface temperatures using two insulated thermocouple electrodes housed in a split workpiece. The junction was formed by the grinding action. The thickness of the junction was 0.46 mm. Rowe, et al.,[15] reduced the size of the junction to less than 0.1 mm by using thin foil thermocouples. A thin thermocouple has greater discrimination of a local temperature and a faster response time. A further reduction in size to 0.05 mm was achieved by using the workpiece as one of the electrodes.

Under ideal conditions, contact between the electrodes is achieved throughout the passage through the contact zone under both wet and dry grinding conditions. There is a tendency towards open-circuits, particularly in wet grinding, which means that the procedure may have to be repeated several times until continuous contact is achieved. It is also important to ensure that the integrity of the insulation is maintained, particularly when grinding with water-based fluids.

An idealized temperature measurement is shown in Fig. 6.12, illustrating the difference between a 'spike' temperature, T_{wg}, experienced as an individual grain passes over the sensor and the background surface temperature, T_w.

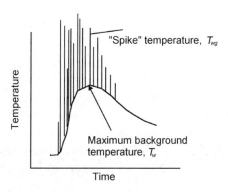

Figure 6.12. Interpretation of an idealized thermocouple measurement of workpiece surface temperature.

In practice, a thermocouple is incapable of responding quickly enough to reproduce the spike temperature accurately. The spike temperature is observed but attenuated. The thermocouple is, however, capable of responding to the background temperature without significant distortion. The duration of the background temperature pulse is measured in milliseconds which is sufficiently long to give rise to significant diffusion of heat to depths in excess of 0.1 mm.

Figure 6.13 shows the effect of depth of cut on maximum background temperature. The temperature increases approximately in proportion to the square root of the depth of cut. These results are almost identical when grinding an AISI 1055 steel with either a 60 grit alumina wheel or a 200 grit CBN wheel. The fine CBN wheel gives rise to higher forces and specific energy than the coarser alumina wheel but due to the high thermal conductivity of the CBN grains the temperatures are not increased.

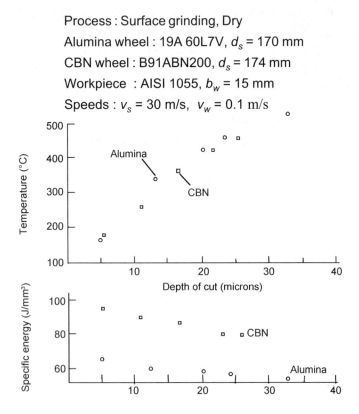

Figure 6.13. Effect of depth of cut on temperature, T_{max}, for a case where the specific energy with CBN is higher than with an alumina wheel.

An important point here is that temperatures are dependent on the specific energy of the process, on the thermal properties of the abrasive, and on the thermal properties of the workpiece. This point is convincingly demonstrated by contrasting the results in Fig. 6.13 with the results in Fig. 6.14. In Fig. 6.14, the workpiece material is M2 tool steel. In this example, the specific energy using a 200 grit CBN wheel is much lower than the specific energy using a 200 grit alumina wheel.

Corresponding to the sharper condition of the CBN wheel, there is an impressive reduction in the workpiece temperature. Many such experiments confirm the validity of these conclusions.

The use of a grinding fluid can be very important for the reduction of temperatures. Figure 6.15 shows an example where an alumina wheel is used to grind M2 tool steel. In this example, it can be seen that a substantial reduction is achieved using a 2% oil in water emulsion. Here, the operation is shallow cut grinding so that the contact area and the contact time are relatively small and there is little scope for convective cooling in the contact zone.

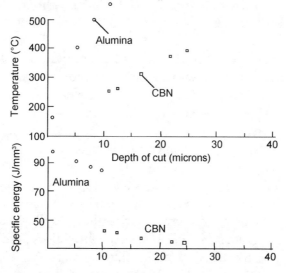

Process : Surface grinding, Wet
Alumina wheel : A200V, d_s = 174 mm
CBN wheel : B91ABN200V, d_s = 174 mm
Workpiece : M2 tool steel, b_w = 15 mm
Speeds : v_s = 30 m/s, v_w = 0.25 m/s

Figure 6.14. Effect of depth of cut on temperature, T_{max}, for a case where the specific energy with CBN is lower than with an alumina wheel.

Process : Surface grinding, Wet and Dry

Alumina wheel : A200V, d_s = 174 mm

Workpiece : M2 tool steel, b_w = 15 mm

Speeds : v_s = 30 m/s, v_w = 0.25 m/s

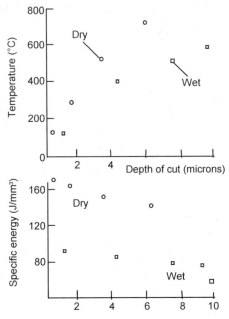

Figure 6.15. In shallow cut grinding, the main effect of the grinding fluid is to reduce the specific energy and hence the temperatures.

The main effect of fluid on the temperature is to lubricate the grinding process in shallow-cut grinding to provide convective cooling within the contact zone. The reduced specific energy is sufficient to explain the lower workpiece temperature in the above example.

A different conclusion is drawn for creep-feed grinding where the contact area and the contact time may be one or two orders of magnitude larger than in shallow-cut grinding. This is illustrated in Table 6.4.

Andrew, et al.,[23] showed that, in creep-feed grinding, convective cooling contributes a very substantial proportion of the heat dissipation. The success of a creep-feed grinding operation is strongly dependent on frequent or continuous dressing to maintain low values of specific energy and also on effective delivery of fluid into the contact zone. It was shown by Shafto[9] that fluid boiling is accompanied by a rapid increase in temperature.

Table 6.4. Comparison of Typical Values of Depth of Cut, Contact Length, and Contact Time in Shallow-cut and Creep-feed Grinding

	CREEP GRINDING	SHALLOW GRINDING
Depth of cut	7 mm	0.04 mm
Wheel diameter	200 mm	200 mm
Workspeed	0.005 m/s	0.2 m/s
Contact length	37.4 mm	2.8 mm
Contact time	7.48 s	0.014 s

At first sight, it would appear sensible to present temperature results for various process variables such as wheelspeed and workspeed. However, no sensible interpretation of such results can be obtained without reference to the accompanying values of specific energy. This should now be clear from the above examples. To illustrate this point, consider the case where grinding wheelspeed is increased. If the effect of the increased wheelspeed is to increase specific energy, the workpiece temperature will be increased. If the effect of the increased wheelspeed is to reduce specific energy, the workpiece temperature will be reduced.

For a more complete understanding of the effect of any process variable, it is necessary to examine more closely the heat transfer in the region of the abrasive contact.

6.6 HEAT CONDUCTION IN THE WORKPIECE

The heat flux, q_w, conducted into the workpiece is only a proportion of the total heat flux, q_t. Defining this proportion as R_w, the heat flux which enters the workpiece is

Eq. (6.3) $\qquad q_w = R_w \cdot q_t$

Since q_t varies as a function of position in the contact zone, as in Eq. (6.2), we can write more generally

Eq. (6.4) $q_w(x) = R_w \cdot q_t(x)$

Typically, the workpiece partition ratio, R_w, varies according to the type of abrasive, the workpiece material, the specific energy, the grinding fluid, and the contact length. Grinding with an alumina wheel in a dry shallow-cut operation, the value of R_w may be as high as 90%, whereas in well-lubricated creep-feed grinding the partition ratio may be less than 5%. The determination of R_w will be discussed in a later section of this chapter.

First, a method is needed to determine the heat conduction into the workpiece. For a general discourse on the theory of heat conduction, the work by Carslaw and Jaeger[24] is recommended.

The following general approach can be used to obtain a temperature solution for any shape of heat flux distribution across the contact length.

The band of heat in a practical grinding contact is usually considered as a series of moving line source elements. The solution for the moving line source can be obtained using Bessel functions which are available from tables or as functions in mathematical software. The line source temperatures must be summed over the length of the grinding contact. This summation process is expressed by the integrals in Eqs. (6.5) and (6.6) assuming a semi-infinite workpiece.

The band heat source is usually considered to lie in the flat plane where $z = 0$. For this case, the solution for a band heat source moving in the x-direction with flux, q, varying in strength with position, a, within the contact length is given by

Eq. (6.5a) $$T = \int_{-l_c/2}^{+l_c/2} \frac{q}{\pi \cdot k} \cdot e^{\frac{v(x-a)}{2 \cdot \alpha}} \cdot K_0 \left\{ \frac{v\left[(x-a)^2 + z^2\right]^{1/2}}{2 \cdot \alpha} \right\} da$$

where $K_0[u]$ is the modified Bessel function of the second kind of order zero for an argument of value u.

The integration variable, a, defines various positions of the line source elements within the heat band.

Eq. (6.5b) $-\dfrac{l_c}{2} \leq a \leq \dfrac{l_c}{2}$

For deep cuts, as experienced in creep-feed grinding, the moving heat should be expressed in terms of the angle, ϕ, between the line of motion and the plane of the band source. This situation is illustrated with values of temperature in Fig. 6.18. For the inclined band source, the equation to be solved is

Eq. (6.6)

$$T = \int_{-l_c/2}^{+l_c/2} \frac{q}{\pi \cdot k} \cdot e^{\frac{v(x - a\cos\phi)}{2 \cdot \alpha}} \cdot K_0 \left\{ \frac{v\left[(x - a\cos\phi)^2 + (z - a\sin\phi)^2 \right]^{1/2}}{2 \cdot \alpha} \right\} da$$

The thermal properties of the workpiece material are represented by k, ρ, c, and α, where k is the thermal conductivity, ρ is the density, c is the specific heat capacity, and

Eq. (6.7) $\qquad \alpha = \dfrac{k}{\rho \cdot c}$

is the diffusivity.

For special cases, a simpler approach can be employed as shown in later sections. However, Eqs. (6.5) and (6.6) give a more rigorous analysis of workpiece temperatures. Accurate solutions to Eq. (6.6) for a curved grinding arc are given by Rowe and Jin.[34]

6.7 FLUX DISTRIBUTION

Surface temperatures calculated from Eq. (6.5) are given in Fig. 6.16 for a uniform heat flux distribution and in Fig. 6.17 for a triangular flux. It can be seen that the maximum temperatures are not greatly different for the two flux distributions. The main difference is the position of the maximum temperatures relative to the position of the center of the contact zone which extends from

Eq. (6.8) $\qquad -1 \leq \dfrac{2 \cdot x}{l_c} \leq +1$

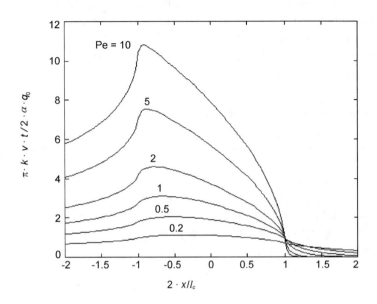

Figure 6.16. Temperature solutions for a uniform heat flux, $q(x) = q_0$ and width $-1 < 2 \cdot x/l_c < +1$.

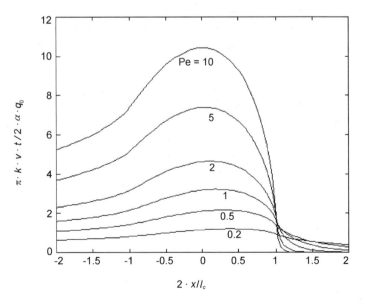

Figure 6.17. Temperature solutions for a triangular heat source solution, $q(x) = q_0(1+2 \cdot x/l_c)$ and width $-1 < 2 \cdot x/l_c < +1$.

For the uniform heat flux, the maximum temperature occurs at the trailing edge of the contact zone at high values of Peclet number, Pe, typical of grinding. For the triangular heat flux, the maximum temperature occurs at the midpoint of the band. This point has significance for the correct interpretation of measured temperature signals.

For a uniform flux, the contact length is equal to the distance from the initial contact to the maximum temperature position.

For a triangular heat flux, the contact length, l_c, is twice the distance from the initial contact to the maximum temperature position.

The contact lengths differ by a factor of two, which is a factor of great importance when determining the magnitude of the flux from Eq. (6.1) and affects the calculation of the energy entering the workpiece and future temperature calculations based on information from experiments.

It is not generally accurate to determine the average flux, q_0, using the geometric contact length, l_g. In Ch. 4, "Contact Mechanics," it was shown that the real contact length may be more than twice the l_g. Taking into account the real contact length and the real flux distribution yields, more accurate predictions of temperature also helps to explain the physical behavior experienced in practice.

The flux distribution reflects the contact pressure in the contact zone. According to the laws of cutting and friction with asperities of negative rake, we expect the heat generation to be approximately proportional to the cutting pressure. The flux distribution therefore represents the variation in contact pressure.

For very deep cuts, the flux is distributed on an inclined plane as illustrated in Fig. 6.9. The resulting temperatures are much lower than would be expected by modeling the heat as a distribution in the direction of motion. Therefore, it is more accurate to model the heat source as distributed along the contact length represented as an inclined plane or as an arc. This is shown in Fig. 6.18. In the figure, it appears that the band source width reduces with the angle of inclination. This is because the length of the band source projected onto the x-y plane reduces with inclination.

Not only are the temperatures along the contact surface reduced when comparing Fig. 6.18 with Fig. 6.17, but even more impressively, the temperatures in the plane of the finished surface are greatly reduced. For example, with the angle of inclination $\phi = 30°$, the maximum temperature on the finished surface is less than one fifth of the value for $\phi = 0°$.

On the plane of the finished workpiece surface, the maximum temperature is close to the end of the contact length where the wheel leaves contact with a position on the workpiece. However, if we examine the temperatures along the contact arc, the maximum temperature moves towards the start of contact with increasing angle. This is because the temperatures correspond more closely with the shape of the flux distribution with increasing angle.

Temperatures for a moving inclined triangular heat source.

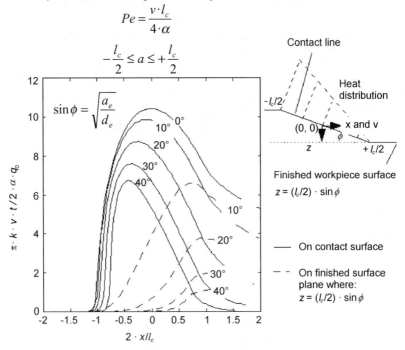

Figure 6.18. Temperatures (solid line) along the contact surface with the wheel and (dotted line) on the plane of the finished workpiece surface. The temperatures reduce with increasing angle, ϕ. The value of the Peclet Number used was $B = 10$.

Some examples of flux distribution for different pressure distributions are illustrated in Fig. 6.19.

The selection of the most appropriate distribution depends on the type of abrasive machining process. A uniform flux is most appropriate where the pressure is applied uniformly, as in polishing or possibly in honing.

A triangular flux is most appropriate where the production of chips is approximately in proportion to the uncut chip thickness, as in grinding. This assumption is supported by measurements of grinding temperature by Rowe, et al.[20]

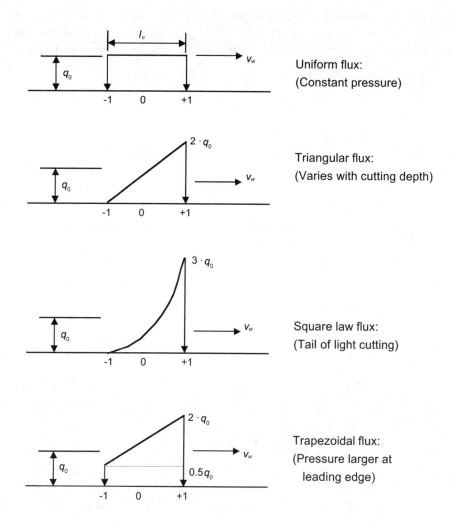

Figure 6.19. Dependence of heat flux distribution on the contact pressure.

The square law flux was proposed for situations where there is a transition from cutting to an extended zone of light rubbing, which might be encountered in light grinding with a well-lubricated blunt wheel.

A trapezoidal flux is most appropriate for an abrasive tool of restricted length where the cutting pressure is heavier at the leading edge. This is probably the most appropriate distribution for honing.

The average flux is given by Eq. (6.1). The flux distribution can be described by equations of the form

Eq. (6.9) $q_0 = $ constant ... uniform $(n = 0)$

Eq. (6.10) $q_1 = q_0 \left(1 + \dfrac{2 \cdot a}{l_c}\right)$ ... triangular $(n = 1)$

Eq. (6.11) $q_2 = \dfrac{3}{4} \cdot q_0 \left(1 + \dfrac{2 \cdot a}{l_c}\right)^2$ ... square law $(n = 2)$

Eq. (6.12) $q_{tr} = q_0 \left(1 + \dfrac{a}{l_c}\right)$ ... trapezoidal

More generally, for a center-referenced heat flux used in analysis, as illustrated in Fig. 6.19

Eq. (6.13) $q_n = q_0 \cdot (n+1) \cdot \left(\dfrac{1}{2} - \dfrac{a}{l_c}\right)^n$

6.8   PECLET NUMBER

Temperatures in each of Figs. 6.16 and 6.17 are expressed in terms of Peclet Number in order to compress a large range of situations into one chart. Peclet Number is a dimensionless parameter proportional to the sliding speed. It is also proportional to the contact length of the sliding heat source and inversely proportional to the thermal diffusivity of the material under the heat source.

For the background workpiece temperature, the relevant sliding speed is the workspeed, v_w. For this case

Eq. (6.14a) $Pe = \dfrac{v_w \cdot l_c}{4 \cdot \alpha}$

More generally, to cope with a range of tribological situations, Peclet Number is given as

Eq. (6.14b) $$Pe = \frac{v \cdot b_h}{2 \cdot \alpha}$$

where b_h is the half-width of the heat source, v is the sliding speed, and α is the thermal diffusivity as defined by Eq. (6.7).

In Fig. 6.18, the motion and the contact length no longer lie in the same plane, so the equivalent value to the Peclet Number is expressed as B defined in the same way as Pe in Eq. (6.14b).

In most cases, grinding conditions correspond to a high value of Peclet Number. For a typical cast iron, see Table 6.1.

$$v_w = 0.2 \text{ m/s}$$

$$l_c = 0.002 \text{ m}$$

$$a = 14.4 \times 10^{-6} \text{ m}^2/\text{s}$$

$$Pe = \frac{0.2 \times 0.002 \times 10^{-6}}{4 \times 14.4} = 6.94$$

Some typical values of thermal properties for common engineering materials are given in Table 6.5. Values for abrasive grains are given in the Appendix.

Table 6.5. Thermal Properties of Typical Engineering Materials

	CONDUCTIVITY (W/mK)	DENSITY (kg/m^3)	SPECIFIC HEAT (J/kgK)
Cast iron (260)	53.7	7,300	511
AISI 1055	42.6	7,840	477
M2	23.5	7,860	515
AISI 52100	34.3	7,815	506

6.9 TEMPERATURES IN THE CONTACT AREA

Temperatures may be read from charts such as Figs. 6.16 to 6.18 or calculated from Eqs. (6.5) and (6.6) if the energy entering the workpiece is known. For shallow-cut processes using alumina or silicon carbide abrasives, a crude approximation can be made that all the energy enters the workpiece. This will over-estimate the temperature rise. More accurate methods of energy partitioning will be discussed later in Sec. 6.13. For the time being, it is simply assumed that the energy entering the workpiece can be determined.

At high sliding speed (Pe > 5), surface temperatures may be estimated from the equation for a moving plane heat source. Within the contact zone, temperature builds up with time of contact.

6.9.1 A Uniform Heat Flux

For a uniform plane heat source on a semi-infinite body

Eq. (6.15) $$T = \frac{2 \cdot q_0}{\sqrt{\pi \cdot (k \cdot \rho \cdot c)_w}} \cdot t^{1/2}$$

As the abrasive tool passes the workpiece, the duration of heating can easily be shown to be

Eq. (6.16) $$t_c = \frac{l_c}{v_w} = \frac{4 \cdot \alpha \cdot Pe}{v_w^2}$$

The maximum temperature for high *Pe* values occur when $t = t_c$. From Eqs. (6.15) and (6.16), the maximum temperature is given by

Eq. (6.17) $$\frac{T \cdot \pi \cdot k \cdot v}{2 \cdot \alpha \cdot q} = 3.54 Pe^{1/2} \; ... \; (Pe > 10)$$

At low speeds, the maximum temperature occurs when $t = 0.5\, t_c$, so that

Eq. (6.18) $\qquad \dfrac{T \cdot \pi \cdot k \cdot v}{2 \cdot \alpha \cdot q} = 2.51 Pe^{1/2} \ ... \ (Pe < 0.2)$

For intermediate speeds, we can write an approximate expression which is useful for quick calculations

Eq. (6.19) $\qquad \dfrac{T \cdot \pi \cdot k \cdot v}{2 \cdot \alpha \cdot q} = \left(\sqrt{2\pi + Pe/2}\right) \cdot Pe^{1/2} \ ... \ (0.2 < Pe < 10)$

Since the maximum temperatures are not strongly dependent on flux distribution, Eqs. (6.17), (6.18), and (6.19) can be used for a range of situations. A factor close to one may be employed for the particular flux distribution, if required for greater accuracy. The next section shows that we need to multiply by approximately 0.95 to adjust these values for a triangular distribution.

6.9.2 Triangular Heat Flux

In some situations, it is required to calculate the temperatures throughout the contact zone. An example arises in creep-feed grinding where the long arc of contact gives rise to strong fluid convection from the workpiece surface. Convection of heat is proportional to temperature, so that it is necessary to evaluate temperatures throughout the region to determine the magnitude of the convection to the fluid.

For the steady moving triangular plane source with similar assumptions to the previous example, a solution can be obtained by integrating the temperature solution for an instantaneous line source with varying flux strength. The line source solution is

Eq. (6.20) $\qquad T = \dfrac{q_0}{\sqrt{\pi \cdot (k \cdot \rho \cdot c)_w}} \cdot t^{-1/2} \cdot e^{-z^2/4\alpha t}$

The flux can be expressed as a function of time from start, $t = 0$, to the end of the arc of cut, $t = t_c$.

Eq. (6.21) $\qquad q = 2 \cdot q_0 \cdot (1 - t/t_c)$

The moving band temperatures are then given by

Eq. (6.22) $$T = \int_0^t \frac{2 \cdot q_0}{\sqrt{\pi \cdot k \cdot \rho \cdot c}} \cdot t^{-1/2} \cdot \left(1 - \frac{t}{t_c}\right) \cdot e^{-z^2/4\alpha t}$$

Evaluating the integral at the surface

Eq. (6.23) $$T = \frac{4 \cdot q_0}{\sqrt{\pi \cdot k \cdot \rho \cdot c}} \cdot t^{1/2} \cdot \left(1 - \frac{2 \cdot t}{3 \cdot t_c}\right)$$

The maximum temperature is given when $t = 0.5t_c$ from which it is found that

Eq. (6.24) $$\frac{T \cdot \pi \cdot k \cdot v}{2 \cdot \alpha \cdot q_0} = 3.34 Pe^{1/2} \ \dots \ (Pe > 10)$$

6.9.3 Trapezoidal Heat Flux

Employing the same approach for the trapezoidal heat flux illustrated in Fig. 6.19, the surface temperatures are given by

Eq. (6.25) $$T = \frac{q_0}{\sqrt{\pi \cdot k \cdot \rho \cdot c}} \cdot t^{1/2} \cdot \left(3 - \frac{4 \cdot t}{3 \cdot t_c}\right)$$

The maximum temperature can be shown to occur at the position given by $t = 3/4t_c$, where

Eq. (6.26) $$\frac{T \cdot \pi \cdot k \cdot v}{2 \cdot \alpha \cdot q} = 2 \cdot \sqrt{\pi} \cdot Pe^{1/2} = 3.54 Pe^{1/2} \ \dots \ (Pe > 5)$$

This is the same as the value for uniform heat flux although the position is different.

6.9.4 Maximum Temperatures

It has been demonstrated that the maximum temperatures for the various flux distributions are not greatly different from each other. Therefore, the maximum temperatures can be estimated from Eqs. (6.17) to (6.19).

The expressions for the maximum temperatures can be shown in many different forms. One of the simplest forms for abrasive machining contacts is

Eq. (6.27)
$$T = C \cdot R_w \cdot \frac{q_0}{\beta} \cdot \sqrt{\frac{l_c}{v_w}}$$

where

$$\beta = \sqrt{(k \cdot \rho \cdot c)_w}$$

is the thermal property of the workpiece material, l_c is the contact length, v_w is the workspeed, q_0 is the heat flowrate (or power) per unit area, and R_w is the proportion of the power which enters the workpiece. We will come to variations of R_w in Sec. 6.10.

A constant, C, is determined from Eqs. (6.17) to (6.19). For a triangular flux distribution, the values of C are given in Table 6.6

Table 6.6. Temperature Factors For A Triangular Flux

Pe	C
>10	1.06
0.2 < Pe < 10	$\dfrac{0.95}{\pi} \cdot \sqrt{2.\pi + \dfrac{Pe}{2}}$
< 0.2	0.76

If l_g is used instead of l_c for the contact length, an error is introduced in shallow-cut grinding. For a grinding condition where the contact length $l_c \approx 2 \cdot l_g$, it is found that using l_g instead of l_c predicts a temperature which is over-estimated by 41%. This can be shown by writing Eq. (6.27) in terms of power.

Eq. (6.28) $$T = C \cdot \frac{R_w \cdot P}{\beta \cdot b_c \cdot l_c} \sqrt{\frac{l_c}{v_w}}$$

Substituting $0.5\, l_c$ for l_c we see that the result is proportional to the square root of l_c and the error is $1.41 - 1 = 41\%$.

An approximate correction for this case could be achieved for a uniform flux by changing $C = 1.13$ to $C = 0.8$ to compensate for the error. For the triangular flux we must reduce C from 1.06 to 0.75.

The determination of maximum temperature depends on the partition ratio, R_w. We must, therefore, give further consideration to the heat flows in abrasive machining.

6.10 HEAT FLOWS IN THE CONTACT AREA AND PARTITIONING

The total heat in the contact area flows out along four paths (Fig. 6.20). For convenience the total machining power is represented as the total heat flux according to Eq. (6.1). This is achieved by dividing the power by the contact area. We can, therefore, write

Eq. (6.29) $q_t = q_w + q_s + q_{ch} + q_f$

where q_w is the heat flux which enters the workpiece within the contact zone, q_s is the heat flux which enters the grinding wheel or abrasive tool, q_{ch} is the heat flux which is carried away by the chips, and q_f is the heat flux which is carried away by the fluid within the contact zone.

Partition ratios can be defined as the proportions of these fluxes to the total flux so that

Eq. (6.30) $1 = R_w + R_s + R_{ch} + R_f$

where $R_w = q_w / q_t$ and so on.

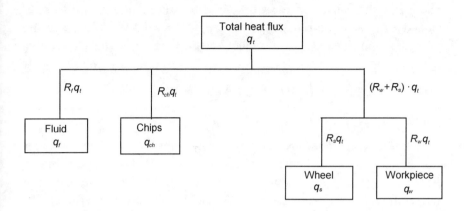

Figure 6.20. Heat generated in the contact area is carried away by the fluid, the chips, the abrasive, and the workpiece. Partition ratios express the proportions of the total heat carried away by each element.

The workpiece partition ratio, R_w, is required in Eq. (6.27) to calculate the maximum workpiece temperature.

From maximum temperature measurements, R_w can also be determined from Eq. (6.27). This is the basis on which measurements can be compared with predictions so that thermal models can be refined to achieve agreement over a wide range of operating conditions.

The problem of predicting temperatures must now be faced. We need to know R_w and to find this value we must determine R_s, R_{ch}, and R_f. This is not a trivial problem.

In the general case where R_s, R_{ch}, and R_f are all of significant magnitude, we must estimate the maximum temperature by considering each heat flow in turn. Some examples are given in the following section.

6.10.1 Heat Flow to the Chips

Assuming the chips reach a temperature close to but not greater than the melting temperature as previously discussed, the heat flux to the chips is

Eq. (6.31) $$q_{ch} = \rho \cdot c \cdot T_{mp} \cdot \left(\frac{a_e \cdot v_w}{l_c} \right)$$

where ρ is the density of the workpiece material, c is the specific heat capacity, T_{mp} is the melting temperature, a_e is the real depth of cut, v_w is the workpiece speed, and l_c is the contact length.

6.10.2 Heat Flow to the Process Fluid

The heat flow to the process fluid will depend on whether the contact zone temperatures remain below the boiling temperature or whether this temperature is substantially exceeded. Where fluid boiling is avoided, the heat convected by the fluid is proportional to the average surface temperature, T_{av}, the contact area, $b \cdot l_c$, and the convection coefficient, h_f. The heat flow rate to the fluid per unit area of the contact surface is the flux, q_f, and is, therefore,

Eq. (6.32) $q_f = h_f \cdot T_{av}$

In general, the average temperature in the contact area is approximately two-thirds the maximum temperature, so that for nonboiling situations

Eq. (6.33) $q_f = \dfrac{2}{3} \cdot h_f \cdot T_{max}$

It is usually more convenient to compute the maximum temperature and this can be used as a measure of the average temperature for the purpose of estimating the fluid convection effect.

In many cases, fluid boiling temperatures are exceeded. In shallow-cut grinding, convection plays a relatively small role in heat dissipation within the contact area, so that it is less important if the fluid boils.

In deep grinding, fluid convection predominates so boiling is more significant. If the boiling temperature is only just reached, there will still be substantial cooling due to convection in a substantial part of the contact area and also evaporative cooling with nucleate boiling. However, if the boiling temperature is substantially exceeded in creep-feed grinding, a numerical solution might be expected to allow a more accurate estimate of temperatures taking convection into account.

The supply of fluid in sufficient volume is, therefore, important even if boiling temperatures are exceeded not only for cooling but also for lubrication and to minimize wheel loading.

6.10.3 Heat Flow to the Workpiece and the Abrasive

The heat flow, q_{ws}, shared by the workpiece and the abrasive given by rearrangement of Eq. (6.29) is

Eq. (6.34) $\qquad q_{ws} = q_w + q_s = q_t - q_{ch} - q_f$

The problem now is to distinguish the proportions of q_{ws} shared by the abrasive and the workpiece. There are several methods of partitioning the heat flow between the abrasive and the workpiece. This problem will be discussed in Sec. 6.13 since approximate values quoted in the case studies which follow are readily available for either conventional or CBN abrasives. These models provide a workpiece-abrasive partition ratio, R_{ws}, where

Eq. (6.35) $\qquad R_{ws} = \dfrac{q_w}{q_w + q_s}$

For a particular workpiece-abrasive combination, R_{ws} remains reasonably constant.

6.10.4 Heat Flow to the Workpiece

From Eqs. (6.34) and (6.35) the heat flow to the workpiece is

Eq. (6.36) $\qquad q_w = R_{ws} \cdot (q_t - q_{ch} - q_f)$

6.10.5 Maximum Workpiece Temperature

From Eq. (6.27) and Eq. (6.36), we can now write the maximum workpiece temperature in terms of the workpiece-abrasive partition ratio.

Eq. (6.37) $\qquad T_{max} = R_{ws} \cdot \left(q_t - q_{ch} - q_f \right) \cdot \dfrac{C}{\beta} \sqrt{\dfrac{l_c}{v_w}}$

For the nonboiling case, we can substitute for q_{ch} and q_f from Eqs. (6.31) and (6.33) so that

Eq. (6.38)
$$T_{max} = \frac{q_t - \rho \cdot c \cdot T_{mp} \cdot \dfrac{a_e v_w}{l_c}}{\dfrac{\beta}{R_{ws} C} \cdot \sqrt{\dfrac{v_w}{l_c}} + \dfrac{2}{3} \cdot h_f}$$

With suitable, even approximate, values of R_{ws}, Eq. (6.38) can be used to estimate maximum temperature in an abrasive contact zone.

Equation (6.38) can be written concisely, if we express Eq. (6.27) in the form of a convection coefficient albeit for conduction into the workpiece.

Eq. (6.39)
$$T_{max} = \frac{3}{2} \cdot \frac{q_w}{h_w}$$

where

Eq. (6.40)
$$h_w = \frac{3}{2} \cdot \frac{\beta}{C} \cdot \sqrt{\frac{v_w}{l_c}}$$

Replacing the term for the flux to the chips, as q_{ch}, and replacing the workpiece conduction term by the convection term, h_w, leads to the concise expression

Eq. (6.41)
$$T_{max} = \frac{3}{2} \cdot \frac{q_t - q_{ch}}{h_w / R_{ws} + h_f}$$

The nature of the underlying assumptions are probably clearer from Eq. (6.41) while Eq. (6.38) shows the terms which have to be used for calculation.

6.11 CASE STUDIES ON PROCESS VARIATIONS AND PROCESS DESIGN

The following case studies demonstrate some developments in process design and show how process differences can be explained based on temperature effects.

6.11.1 Case Study 1: Shallow-cut Grinding–10 μm Cut

Calculations of the maximum temperature in the contact zone for the following grinding conditions are shown below.

d_e : 200 mm : equivalent diameter

e_c : 40 J/mm^3 : process specific energy

e_{ch} : 6 J/mm^3 : specific energy to chips

v_s : 30 m/s : wheelspeed

v_w : 0.3 m/s : workspeed

a_e : 10 μm : depth of cut

l_c : 3 mm : contact length

k_w : 34.3 W/mK : workpiece conductivity

ρ_w : 7,810 kg/m^3 : workpiece density

c_w : 506 J/kgK : workpiece specific heat

R_{ws} : 0.8 : work-wheel partition ratio

b_w : 15 mm : contact width

h_f : 10,000 Wm^{-2}K^{-1} : convection coefficient for fluid

The Peclet Number is given by,

$$Pe = \frac{v_w \cdot l_c}{4 \cdot \alpha} = 25.9$$

Therefore, we may use Eq. (6.27) with a value of $C = 1.07$. For this we need to evaluate the flux terms based on the removal rate and the process variables.

The removal rate per unit width of cut (specific removal rate) is given by $Q_w' = a_e \cdot v_w = 3 \text{ mm}^2/\text{s}$. The total process heat flux is given by,

$$q_t = \frac{e_c \cdot Q_w}{b_w \cdot l_c} = \frac{e_c \cdot Q_w'}{l_c} = 40 \text{ W/mm}^2$$

The energy to the chips represented as a heat flux is given by,

$$q_{ch} = \frac{e_{ch} \cdot Q_w}{b_w \cdot l_c} = \frac{e_{ch} \cdot Q_w'}{l_c} = 6 \text{ W/mm}^2$$

The thermal property of the workpiece material is given by,

$$\beta = \sqrt{k_w \cdot \rho_w \cdot c_w} = 11{,}640 \text{ J/m}^2\text{Ks}^{0.5}$$

The coefficient for conduction into the workpiece is given by,

$$h_w = \frac{3}{2} \cdot \frac{\beta}{C} \cdot \sqrt{\frac{v_w}{l_c}} = 163{,}200 \text{ W/m}^2\text{K}$$

The maximum temperature is given by,

$$T_{max} = \frac{3}{2} \cdot \frac{q_t - q_{ch}}{h_w / R_{ws} + h_f} = 238°\text{C}$$

A temperature of 238°C means that the workpiece is below the tempering and oxidation region. With water-based coolants, fluid boiling will be experienced but this is unlikely to be a problem in shallow-cut grinding. At higher temperatures where thermal damage is a problem, action should be taken to reduce the specific energy. This may be achieved by one of several methods such as redressing the wheel, using a coarser dressing feed, using a more open or softer wheel, and improving the coolant supply for more effective lubrication. Problems experienced may also be overcome by reducing the depth of cut to reduce the total heat flux, or even reducing the wheelspeed to increase the uncut chip thickness.

If the calculation had been carried out ignoring heat flows to the chips, the fluid, and wheel, the calculation is simplified but the maximum temperature is over-estimated by 54%.

$$T_{max} = \frac{3}{2} \cdot \frac{q_t}{h_w} = 368 \text{ °C}$$

compared with the more accurate result of 238°C.

Using the geometric contact length instead of the real contact length increases the over-estimation to approximately 118%.

Taking the chip energy into account in this example reduced the temperature predicted by 15%.

Taking fluid convection into account with an approximate convection factor reduced the predicted temperature by approximately 5%. Of course, in this example, water-based fluid would have boiled and part of the cooling would have come from evaporative cooling rather than conventional convective cooling.

More recent work by Rowe and Jin[34] demonstrated much higher values of fluid convection coefficients where boiling was avoided. It was found that values of h_f of 290,000 W/m²K for water are typical in grinding and 23,000 W/m²K with oil.

6.11.2 Case Study 2: Creep Grinding–1 mm Cut

The conditions are the same as in the previous case except that the depth of cut is increased and the workspeed is reduced to maintain the same removal rate.

a_e : 1 mm : depth of cut

l_c : 14 mm : contact length

v_w : 0.003 m/s : workspeed

The new results which may be compared with the previous results are

Pe : 1.21

C : 0.79

Q_w' : 3 mm²/s

q_t : 8.57 W/mm²

$$q_{ch} : 1.28 \text{ W/mm}^2$$

$$q_t\text{-}q_{ch} : 7.29 \text{ W/mm}^2$$

$$h_w : 0.0102 \text{ W/mm}^2\text{K}$$

$$h_f : 0.01 \text{ W/mm}^2\text{K}$$

$$T_{\max} = \frac{1.5 \times 7.29}{0.0102 / 0.8 + 0.01} = 480\ ^\circ\text{C}$$

Although the removal rate is unchanged, the heat flux is reduced to 21% of its previous value. This is due entirely to the change in the contact length and confirms the importance of correctly estimating contact length.

The factor, h_w, for conduction into the workpiece is reduced by 16 times due to the reduced workspeed and the increased contact length. As a result of the reduction in h_w, the conduction into the workpiece is only slightly larger than the convection to the fluid, since h_f is only slightly smaller than h_w.

The resultant temperature is more than doubled, which appears to suggest that reducing workspeed is not a good idea. Also, since the maximum temperature is well above the boiling temperature of a process fluid, it must be assumed that 'burn-out' will occur and the maximum temperature will be even higher. The grinding conditions, therefore, should be changed if possible to reduce the temperatures to below the fluid 'burn-out' temperature.

If it is assumed that due to boiling the heat extracted by the fluid is reduced to 40% of its previous value, that is assuming the fluid is effective up to 140°C introducing a factor of 140/(480 × 2/3), then the temperature can be estimated for $h_f = 4000 \text{ W/m}^2\text{K}$. This yields $T_{\max} = 643^\circ\text{C}$.

In practice, a hot spot will grow from the region of the maximum temperature because convection cooling no longer applies uniformly across the contact leading to an even higher peak temperature. It can also be visualized, depending on the direction of wheel rotation, that one side or the other will be starved of coolant once boiling has occurred. This will further extend the very hot region.

Since the operation is creep-feed grinding, we should consider the effect of the inclined heat source as expressed by Eq. (6.6). The contact angle is given approximately by

$$\sin \phi = \sqrt{\frac{a_e}{d_e}}$$

For this case, the angle $\phi = 4°$. Referring to Fig. 6.18, it is seen that the maximum temperature on the contact surface is reduced for this angle by less than 5%.

This suggests the surface may be severely burned. For higher temperatures, special measures need to be taken, such as high pressure wheel cleaning and oil lubrication to prevent wheel loading and serious loss of grinding efficiency. If at all possible, such high temperatures will be avoided. Where dry conditions apply, Lee, Zerkle, and Des Ruisseaux[25] proposed that $h_f = 500$ W/m²K approximately.

6.11.3 Case Study 3: Creep-feed Grinding–20 mm Cut

The high temperatures in the previous case are to be overcome by increasing the depth of cut to 20 mm and reducing the workspeed to 0.015 mm/s, maintaining the specific removal rate at 3 mm²/s.

a_e : 20 mm : depth of cut

v_w : 0.00015 m/s : workspeed

Accordingly,

l_c : 63.2 mm : contact length

q_t : 1.9 W/mm² : total flux

q_{ch} : 0.28 W/mm² : flux convected by chips

$q_t - q_{ch}$: 1.62 W/mm²

h_w : 0.00112 W/mm²K : factor for workpiece conduction

h_f : 0.01 W/mm²K : fluid convection factor

The maximum temperature is reduced to

$$T_{max} = \frac{1.5 \times 1.62}{0.0011/0.8 + 0.01} = 213 \text{ °C}$$

Whereas in Case Study 2, the move to creep-feed grinding increased temperature, the further increase in depth of cut and reduction in workspeed reduces temperature. The reduction in temperature is due to the increased contact length.

The predicted temperature is still above the boiling temperature for a water-based fluid and, therefore, the calculation which assumed $h_f = 10,000$ W/m^{2}K suggests that the predicted temperature is likely to be an under-estimate. On this basis, a further reduction in workspeed would be required to bring the temperature down and ensure a completely satisfactory process.

If we now take the angle of inclination into account according to Eq. (6.6) we should find that the maximum temperature is reduced. First, calculating the angle of inclination using the expression given in the previous example, we find that, $\phi = 17.5°$. Referring to Fig. 6.18, the maximum temperature for 17.5° is reduced by approximately 20%. This corresponds to increasing the workpiece convection factor, h_w, by 25%. This reduces the maximum temperature to 207°C which is a rather small reduction. The reduction in temperature is smaller than might be expected due to the predominance of fluid convection over workpiece conduction. The effect of the angle of inclination on temperature would have been much more substantial in dry grinding.

Figure 6.21 illustrates the effect of increasing depth of cut at a reduced removal rate, $Q_w' = 1$mm^2/s. The process can now be carried out with cool grinding, either at small depths of cut or at larger depths of cut in excess of 8 mm, as long as there is a plentiful supply of fluid to achieve the value of $h_f = 10,000$ W/m^{2}K.

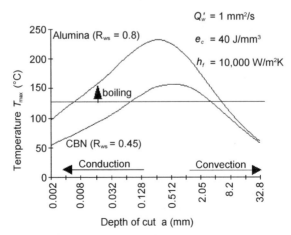

Figure 6.21. Temperatures in conventional and creep-feed grinding using alumina and CBN wheels. No allowance has been made for fluid boiling. The boiling temperature shown is an approximate value for oil in water emulsion.

In conventional grinding, successful grinding can be achieved at temperatures well above boiling since below the peak values in the mid-range conduction predominates over convection (shown in Fig. 6.21).

In creep-feed grinding, with depths of cut greater than 0.2 mm in Fig. 6.21, convection becomes increasingly important to achieve effective cooling in this region. However, at a depth of cut of 8 mm, it is possible to achieve very cool grinding, $T_{max} = 110°C$.

The temperature model demonstrates similar values to the results presented by Werner, et al.,[26] who found there was an intermediate range of depths of cut where thermal damage was likely.

Cubic boron nitride (CBN) abrasives reduce temperature substantially as shown in Fig. 6.21. The reduction is due to the high thermal conductivity of CBN. The value of thermal conductivity is reflected in the value of R_{ws}.

Specific energy is an inverse measure of the cutting efficiency of the grinding wheel and workpiece material combination for the particular process conditions. Figure 6.22 is a chart of predicted temperatures for various values of specific energy at a constant specific removal rate $Q_w' = 1$ mm^2/s, using an alumina wheel to machine AISI 52100.

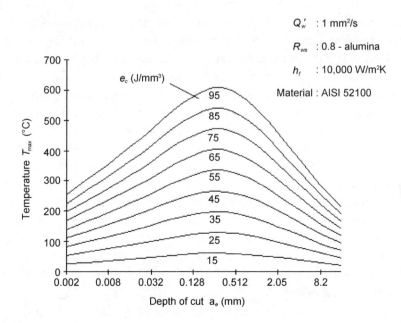

Figure 6.22. Effect of specific energy on maximum temperature.

6.11.4 Case Study 4: High Efficiency Deep Grinding (HEDG)

A development in recent years combines the advantages of deep grinding as demonstrated in the previous examples with the advantages of high workspeeds as employed in conventional grinding. Combining these two factors minimizes the heat conducted into the workpiece. High wheelspeeds, 80–200 m/s, are also employed to minimize the grinding forces.

The concept of high efficiency deep grinding (HEDG) was first proposed by Guhring.[27] The principles were largely developed by Werner and further described by Tawakoli,[28] and Rowe and Jin.[34]

To achieve HEDG requires large depths of cut as in creep-feed grinding, combined with a large power supply and drive motor to cope with the extremely high removal rates.

Above all, the process requires very low specific energy values to be achieved to limit temperature rise. Therefore, it helps if the materials machined at high wheelspeeds are of the type known as "easy-to-machine." These include free machining cast irons and steels. "Difficult-to-machine" materials must be machined at lower wheelspeeds.

The use of CBN abrasives also helps to reduce temperature rise. Cubic boron nitride (CBN) wheels of the electroplated or metal-bonded type are mainly used to withstand the bursting stresses at the high speeds involved and for chemical stability at high temperatures. High speed silicon carbide wheels can also be used but at lower speeds and have excellent temperature stability. Care must be taken to ensure appropriate guarding in case of a wheel burst. Alumina wheels have lower temperature stability.

The requirements for successful grinding are even more critical in HEDG than in other abrasive processes. In particular, several coolant nozzles may be required to ensure high velocity fluid penetrates the boundary layer around the wheel and is absorbed into the pores of the wheel.

Sometimes a pressurized shoe is fitted around a portion of the circumference and the sides of the wheel to ensure fluid penetration. In addition, high pressure jets are employed to clean away swarf from the wheel surface. This is because wheel loading quickly destroys the cutting efficiency of the wheel. Other aspects of machine design for HEDG are described by Tawakoli.[28]

The principle of HEDG is very interesting since a new physical principle is involved. Previously, it was argued that the chip temperature could not exceed the melting temperature of the material. This limits the energy which can be absorbed by the chips. Now we introduce a new physical limitation to the analysis. The background temperature of the workpiece contact surface cannot exceed the melting temperature of the material. A third physical limitation is that the temperature at the grain contact with the workpiece cannot exceed the melting temperature.

According to Eqs. (6.27) and (6.38), this places a limitation on the energy generated by the process. In other words, low specific energy can be achieved if either contact surface is allowed to approach the melting temperature, when shear forces are greatly reduced. The high wheelspeeds assist in maintaining the high contact temperatures and the rapid removal of the softened layer. High workspeeds are essential to prevent excessive conduction of heat down to the level of the finished workpiece surface.

Assuming the temperature of the steel is limited to about 1,500°C at the contact surface we can interpret the situation as illustrated in Fig. 6.23. If the specific energy in conventional grinding is 40 J/mm³, the contact temperature rapidly increases with depth of cut until the melting temperature is approached.

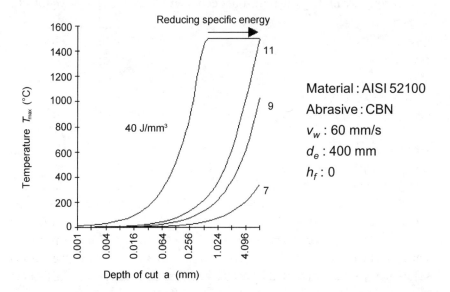

Figure 6.23. Possible explanation for high efficiency deep grinding (HEDG). Under HEDG conditions, contact temperatures are limited to the melting temperature of the workpiece material. Specific energy is reduced because the shear forces are reduced.

Under these conditions, the specific energy will fall to a fraction of its former level. In the example shown, the specific energy, at a depth of cut of 8.2 mm, converges towards the curve for 11 J/mm^3, almost a quarter of the value in conventional grinding.

Tawakoli quotes one example of a specific energy of 7.05 J/mm^3. This must be close to the minimum ultimately achievable for steels, since it is close to the melting energy of the material.

When HEDG conditions are achieved, the increases in specific removal rate are spectacular. Figure 6.24 shows the specific removal rates corresponding to Fig. 6.23. At the depth of cut of 8.2 mm, the specific removal rate is 492 mm^2/s compared with conventional values of the order of 3 mm^2/s, an increase of two orders of magnitude.

For lower specific removal rates, $Q_w' < 20$ mm^2/s and for the highest specific removal rates, $Q_w' > 70$ mm^2/s, Tawakoli recommends down-grinding to obtain lower forces and less wheel wear. For intermediate values, Tawakoli recommends up-grinding to reduce thermal damage. Up-grinding gives lower temperatures for the intermediate values because this is the condition where convective cooling is effective. Up-grinding may allow the fluid to act on more than half the arc of contact before boiling occurs. In down-grinding, the temperatures where the fluid enters the contact zone are higher and burn-out occurs within a shorter arc of contact.

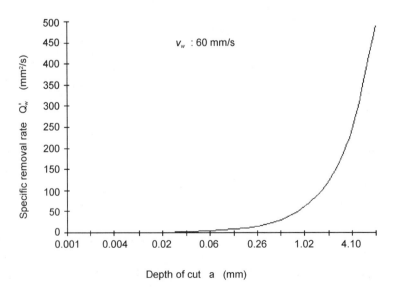

Figure 6.24. Specific removal rate against depth of cut.

For smaller depths of cut and lower removal rates, the conditions are similar to conventional grinding and the cooling is secondary to the importance of forces and wheel wear. For the highest removal rates in HEDG, convection again becomes secondary to the importance of reducing forces and wheel wear. Down-grinding is employed, which imparts higher impact forces on the abrasive grits and helps to maintain cutting efficiency.

The question obviously arises as to whether the very high temperatures at the surface will damage the workpiece surface which remains after the machining pass. To answer this question requires an analysis of the temperatures below the contact surface. This is the topic we will tackle next.

6.12 SUBSURFACE TEMPERATURES

The temperatures, with depth, z, below the surface, are obtained by integration of Eq. (6.22) leading to the following solution for $Pe > 10$ and $0 < t < t_c$.

Eq. (6.42)

$$T = \frac{4 \cdot q_w}{\sqrt{\pi \cdot k \cdot \rho \cdot c}} \cdot t^{1/2} \cdot \left(1 - \frac{2}{3} \cdot \frac{t}{t_c} - \frac{z^2}{6 \cdot \alpha \cdot t_c}\right) \cdot e^{-z^2/(4 \cdot \alpha \cdot t)}$$

$$- \frac{2 \cdot q_w \cdot z}{k} \cdot \left(1 - \frac{t}{t_c} - \frac{z^2}{6 \cdot \alpha \cdot t_c}\right) \cdot \left[1 - erf\left(\frac{z}{\sqrt{4 \cdot \alpha \cdot t}}\right)\right]$$

Equation (6.42) can be solved using a mathematical software package. The first term is the most significant term but both terms are required for accuracy.

Figure 6.25 shows values predicted for temperatures below the position of the maximum temperature for a depth of 0.032 mm. The depths below the surface are plotted exponentially to reveal the features of interest using a spreadsheet which provides smoothing, leading to the impression of slightly negative temperatures where the temperature drops back to zero. Negative temperatures should be ignored.

Slowly at first and then more rapidly, the temperature reduces below the contact temperature. The finished surface (corresponding to the position where the wheel exits from contact) lies at a depth of $a_e/4$ below the contact surface at the mid-position. It can be seen that at this position, where $z/a_e = 0.25$, the temperature is almost exactly equal to the contact temperature.

A check of the angle of inclination of the heat source shows that $\phi = 1°$. The inclined angle solution is not required for such a small angle.

Increasing the depth of cut, it is found that a point is reached where the temperature on the finished workpiece surface is greatly reduced. This is shown in Fig. 6.26 which demonstrates the condition sought in high efficiency deep grinding. Figure 6.26 is calculated from Eq. (6.42) with the values of q_w solved inversely from the contact surface temperatures.

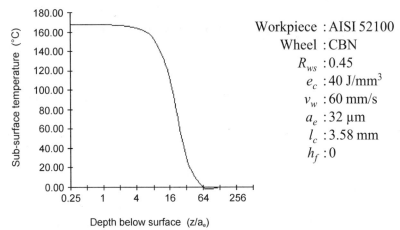

Workpiece : AISI 52100
Wheel : CBN
R_{ws} : 0.45
e_c : 40 J/mm^3
v_w : 60 mm/s
a_e : 32 μm
l_c : 3.58 mm
h_f : 0

Figure 6.25. Sub-surface temperatures under T_{max}.

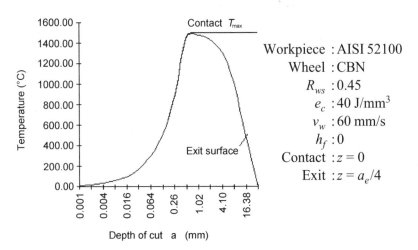

Workpiece : AISI 52100
Wheel : CBN
R_{ws} : 0.45
e_c : 40 J/mm^3
v_w : 60 mm/s
h_f : 0
Contact : $z = 0$
Exit : $z = a_e/4$

Figure 6.26. Surface temperatures in grinding with increasing depth of cut.

A check of the values of the angle of inclination shows that the solutions shown in Fig. 6.26 are valid for depths of cut up to 1 mm. Above this value we should strictly use Eq. (6.6) to obtain the solutions. However, the application of Eq. (6.6) only serves to demonstrate that the temperature under the midpoint of the heat source falls away even more rapidly than shown with very deep cuts as illustrated in Fig. 6.18.

When the depth of cut reaches 16 mm, the angle of $\phi = 16°$. At this angle, with high values of Peclet Number, the maximum temperature occurs near the exit from the arc of contact. Using Eq. (6.6) for the exit temperature shows a surface temperature at the exit of less than 4°C.

At very large depths of cut with high workspeeds, the high temperature material at the point of the maximum temperature is removed before heat can penetrate significantly down to the level of the exit workpiece surface. This means that if the wheel can keep removing the material under these near-molten conditions, as well as remove material in the region where the material is cooler, the finished workpiece surface may remain cold. Clearly, a more complete explanation of the physical behavior in the region justifies a more detailed investigation. However, this explanation is sufficient to indicate the potential of HEDG. A cool finished surface is made possible by the combined effects of high workspeed, a large depth of cut, a large angle of inclination of the heat source, and a low specific energy.

At low workspeeds where a substantial proportion of the heat penetrates below the exit surface, the maximum subsurface temperatures occur under the midpoint of the contact length and serious problems are more likely to be encountered. Bulk deformation of the subsurface and expansion of the subsurface as it is being machined lead to a very rough surface and unpredictable behavior. The process may be accompanied by power surging and loading of the wheel.

At high workspeeds, it should be possible to keep the subsurface cool and avoid some of the problems which ensue when the subsurface is allowed to become hot. It should be noted that high efficiency deep grinding is still not well understood. However, the foregoing thermal analysis provides a basis for further investigation of this exciting new process.

6.13 WORK-ABRASIVE PARTITION RATIO – R_{ws}

We have seen from the previous case studies on process design using thermal analysis that the sharing of the heat between the workpiece and the abrasive as represented by R_{ws} plays an important role.

It was shown that CBN abrasive which is more conductive than alumina, reduces the workpiece temperature. In order to determine R_{ws} we must look more closely at the heat flows in the contact zone, as illustrated in Fig. 6.27.

The estimation of the heat flow to the chips and fluid were previously considered. The purpose of this analysis is to determine how the heat flow $q_{ws} = q_w + q_s$ is shared between the workpiece and the abrasive.

Eq. (6.43) $q_{ws} = q_w + q_s = q_t - q_{ch} - q_f$

The work-abrasive partition ratio, R_{ws}, is then defined as

Eq. (6.44) $R_{ws} = \dfrac{q_w}{q_w + q_s}$

To determine R_{ws}, it is necessary to analyze the contact between the grains and the workpiece. Figure 6.28 illustrates the heat flows into a grain and into the workpiece.

The heat flux into the grains may be related to the grain contact temperature by the factor h_g, in the same way as for other heat fluxes.

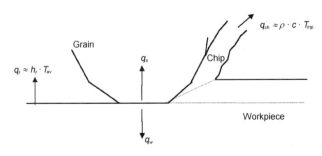

Figure 6.27. The power flux is shared between four heat sinks: the workpiece, the grain, the chip, and the fluid.

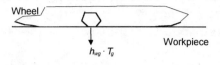

(a) Heat flow into the workpiece at a grain contact.

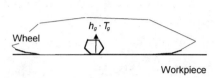

(b) Heat flow into the abrasive at a grain contact.

Figure 6.28. Heat transfer at the contact between a grain and the workpiece.

Eq. (6.45) $\qquad q_g = h_g \cdot T_g$

Correspondingly, the heat flux into the workpiece at the grain contacts may be expressed as

Eq. (6.46) $\qquad q_{wg} = h_{wg} \cdot T_g$

so that

Eq. (6.47) $\qquad q_{wg} + q_g = (h_{wg} + h_g) \cdot T_g$

The sharing of heat between a grain and the workpiece at each grain contact is the mechanism by which heat is shared between the abrasive and the workpiece over the whole contact area. By comparing Eq. (6.48) with Eq. (6.45), we can, therefore, write

Eq. (6.48) $\qquad R_{ws} = \dfrac{q_{wg}}{q_g + q_{wg}} = \dfrac{h_{wg}}{h_g + h_{wg}}$

We now need expressions for h_g and h_{wg} which are given in the following sections.

6.13.1 Analysis of h_{wg}

The grain moving over the workpiece represents a heat source sliding at wheelspeed, typically $v_s = 30$ m/s. The width of the heat source corresponds to the dimensions of the contact area of the grains. For sharp grains this has been estimated as 20–100 μm.

For a typical steel workpiece of thermal diffusivity 8.68×10^{-6} m²/s, the lowest Peclet Number is

Eq. (6.49) $Pe = \dfrac{v_s l}{2 \cdot \alpha} = 34.5$

Since this value is greater than 10, we may safely assume steady state conditions apply for conduction into the workpiece.

The solution for a heat source is relatively insensitive to shape. Rearranging Eq. (6.17) for a band source with the expression for Pe from Eq. (6.49) and since the average temperature is approximately two-thirds of T_{max},

Eq. (6.50) $h_{wg} = 0.94 \beta_w \cdot \sqrt{\dfrac{v_s}{r_0}}$ ... band source

For a circular contact, Archard[29] found that

Eq. (6.51) $h_{wg} = \dfrac{\beta_w}{0.974} \cdot \sqrt{\dfrac{v_s}{r_0}}$ ... circular source

Clearly, a factor of one is sufficiently accurate for our purpose where the grains are irregular.

6.13.2 Analysis of h_g

In order to use compatible solutions for heat flux into the grains and into the workpiece, the contact shape is assumed to be a circle. As demonstrated above, the solution is insensitive to whether the shape is actually an infinitely wide band or a circle. However, the same assumption should be employed for both cases.

The heat conduction into the grains is a transient two-dimensional problem. The classical solution is for a stationary heat flux applied to a circular semi-infinite grain. Although the grain is not always a semi-infinite body it has been shown that the heat transfer characteristics of the abrasive are dominated by the thermal properties of the grain and are largely unaffected by the thermal properties of the bond.[21]

The solution for conduction into a circular contact is given by Carslaw and Jaeger.[30]

Eq. (6.52)
$$h_g = \frac{k_g}{2 \cdot r_0 \tau} \cdot \left[\frac{1}{\sqrt{\pi}} - ierfc\left(\frac{1}{2 \cdot \tau} \right) \right]$$

where k_g is the conductivity of the abrasive grain, r_0 is the contact half-width, $ierfc()$ is the integral complementary error function available from tables and

Eq. (6.53)
$$\tau = \sqrt{\frac{\alpha_g \cdot x}{r_0^2 \cdot v_s}}$$

x is the distance traveled by the grain from the commencement of contact so that $t = x/v_s$ is the time of contact. When steady state is achieved

Eq. (6.54)
$$h_g = \frac{k_g}{r_0}$$

so that ignoring transient effects

Eq. (6.55)
$$R_{ws} = \left(1 + \frac{0.974 \cdot k_g}{\beta_w \cdot \sqrt{r_0 \cdot v_s}} \right)^{-1}$$

This solution was proposed as the partition ratio, R_w, by Hahn,[5] although we now know we must consider convection by the chips and fluid.

Equation (6.55) is adequate for most purposes to estimate R_{ws}. However, there is an error introduced by assuming steady state conditions. This error can be corrected by solving the transient case.

For the transient case, it is necessary to integrate the transient solution for partition ratio across the contact length, l_c.

Eq. (6.56) $$R_{ws} = \frac{1}{l_c} \int_0^{l_c} \left(1 + \frac{h_g}{h_{wg}}\right)^{-1} \cdot dx$$

Performing the integral the solution is found in the form

Eq. (6.57) $$R_{ws} = \left(1 + \frac{0.974 \cdot k_g}{\beta_w \cdot \sqrt{r_0} \cdot v_s} \cdot \frac{1}{F}\right)^{-1}$$

where the transient function, F, was determined[13][31] as

Eq. (6.58)
$$F = 1 + \frac{1}{6 \cdot \tau^2} + \frac{4 \cdot \tau}{3 \cdot \sqrt{\pi}} \left\{1 - \left[\exp\left(\frac{1}{4 \cdot \tau^2}\right)^{-1}\right] \cdot \left(1 + \frac{1}{4 \cdot \tau^2}\right)\right\}$$
$$- \left(1 + \frac{1}{6 \cdot \tau^2}\right) \cdot erf\left(\frac{1}{2 \cdot \tau}\right)$$

and where

Eq. (6.59) $$\tau = \left(\frac{\alpha_g \cdot l_c}{r_0^2 \cdot v_s}\right)^{1/2}$$

In the steady state $F = 1$.

A close approximation to Eq. (6.58), given by Black,[31] is

Eq. (6.60) $$F = 1 - e^{-\tau/1 \cdot 2}$$

Equation (6.55) shows that R_{ws} increases with the grain contact half-width, r_0, and with the wheelspeed, v_s. These effects are, to an extent, offset by the transient function, F.

Figure 6.29 gives some typical values for the bearing steel AISI 52100 ground with alumina and CBN abrasives. The values were calculated for a wheelspeed of 30 m/s.

The values of R_{ws} are sensitive to the effective thermal conductivity of the abrasive. Figure 6.29 is based on a value of 240 W/mK for CBN and a value of 35 W/mK for alumina abrasive.

Figure 6.29 reinforces the conclusion that the abrasive grains should be sharp to minimize the heat entering the workpiece. Fortunately, CBN grains which are much harder than alumina grains tend to retain their sharpness better.

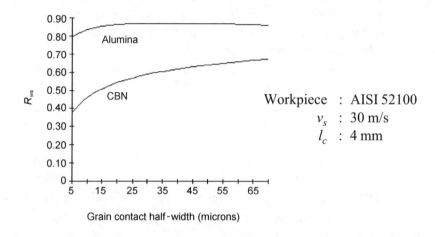

Grain contact half-width (microns)

Figure 6.29. Workpiece-abrasive partition ratio, R_{ws}.

APPENDIX

Table 6.7. Approximate Thermal Properties of Typical Abrasive Grains[21][32]

	CONDUCT-IVITY (W/mK)	DENSITY HEAT (kg/m³)	SPECIFIC HEAT (J/kgK)	$\beta = \sqrt{k \cdot \rho \cdot c}$ (J/m²sK)
Diamond	2000	3,520	511	60,000
Cubic boron nitride	240 (Pure – up to 1,300)	3,480	506	20,600 (48,000)
Silicon carbide	100	3,210	710	15,100
Aluminium oxide	35	3,980	765	10,300

REFERENCES

1. Brinksmeier, E, and Brockhoff, T., Utilization of Grinding Heat As A New Heat Treatment Process, *Annals of the CIRP*, 47(1):275–279 (1996)

2. Malkin, S., *Grinding Technology*, Ellis Horwood, Publ. (1989)

3. McCormack, D., Rowe, W. B., and Chen, X., Residual Stresses In Ground Components, Institute of Materials Seminar, Stratford (Nov. 1998)

4. Outwater, J. O., and Shaw, M. C., Surface Temperatures In Grinding, *Trans ASME*, 74:73–78 (1952)

5. Hahn, R. S., On The Nature Of The Grinding Process, *Proc. 3rd Machine Tool Design Research Conf.*, pp. 129–154 (1962)

6. Makino, Suto, and Fokushima, An Experimental Investigation Of The Grinding Process, *J. Mechanical Laboratory*, Jpn., 12(1):17 (1966)

7. Des Ruisseaux, N. R., and Zerkle, R. D., Temperatures In Semi-infinite and Cylindrical Bodies Subject to Moving Heat Sources and Surface Cooling, *J. Heat Transfer*, 92:456–464 (1970)

8. Malkin, S., and Cook, N. H., The Wear of Grinding Wheels, Part 2— Fracture Wear, ASME, *J. Eng. Industry*, pp. 1129–1133 (Nov. 1971)

9. Shafto, G. R., Creep-feed Grinding, PhD Thesis, U. Bristol (1975)

10. Snoeys, R., Maris, M., and Peters, J., Thermally Induced Damage in Grinding, *Annals of the CIRP*, 27(2):571–581 (1978)

11. Werner, P. G., Younis, M. A., and Schlingensiepen, R., Creep-feed—An Effective Method to Reduce Workpiece Surface Temperatures in High Efficiency Grinding Processes, *Proc. 8th N. American Metalworking Research Conf.*, SME, pp. 312–319 (1980)

12. Lavine, A. S., Thermal Aspects of Grinding: Heat Transfer to Workpiece Wheel and Fluid, Collected Papers in Heat Transfer, *ASME*, 123:267–274 (1989)

13. Rowe, W. B., Morgan, M. N., and Allanson, D. R., An Advance In the Thermal Modeling of Thermal Effects In Grinding, *Annals of the CIRP*, 40(1):339–342 (1991)

14. Rowe, W. B., Qi, H. S., Morgan, M. N., and Zhang, H. W., The Effect of Deformation in The Contact Area in Grinding, *Annals of the CIRP*, 42(1):409–412 (1993)

15. Rowe, W. B., Black, S. C. E., Mills, B., Qi, H. S., and Morgan, M. N., Experimental Investigation of Heat Transfer in Grinding, *Annals of the CIRP*, 44(1):329–332 (1995)

16. Ueda, T., Sato, M., and Nakayama, K., Cooling Characteristics of The Cutting Grains in Grinding, *Annals of the CIRP*, 45(1):293–298 (1996)

17. Rowe, W. B., Black, S. C. E., Mills, B., and Qi, H. S., Analysis of Grinding Temperatures by Energy Partitioning, *Proc. Inst. Mechanical Engineers*, 210:579–588 (1996)

18. Howes, T. D., Neailey, K., and Harrison, A. J. Fluid Film Boiling in Shallow-Cut Grinding, *Annals of the CIRP*, 36(1):223–226 (1987)

19. Rowe, W. B., Pettit, J. A., Boyle, A., and Moruzzi, J. L., Avoidance of Thermal Damage in Grinding and Prediction of the Damage Threshold, *Annals of the CIRP*, 37(1):327–330 (1988)

20. Rowe, W. B., Black, S. C. E., Mills, B., Morgan, M. N., and Qi, H. S., Grinding Temperatures and Energy Partitioning, *Proc. Royal Society*, Part A, 453:1083–1104 (1997)

21. Morgan, M. N., Rowe, W. B., Black, S. C. E., and Allanson, D. R., Effective Thermal Properties of Grinding Wheels and Grains, *Proc. Inst. Mechanical Engineers*, 212B:661–669 (1998)

22. Nee, A. Y. C., and Tay, O. A., On the Measurement of Surface Grinding Temperature, *Int. J. Machine Tool Design and Research*, 21(3):279 (1981)

23. Andrew, C., Howes, T., and Pearce, T., Creep-feed Grinding, Rinehart and Winston (1985)

24. Carslaw, H. S., and Jaeger, J. C., Conduction of Heat in Solids, Oxford Science Publ., Oxford University Press (1959)

25. Lee, D. G., Zerkle, R. D., and Des Ruisseaux, N. R., An Experimental Study of Thermal Aspects of Cylindrical Grinding, ASME Paper WA-71/ Prod - 4 (1971)

26. Werner, P. G., Younis, M. A., and Schlingensiepen, R., Creep-feed—An Effective Method to Reduce Workpiece Surface Temperatures in High Efficiency Grinding Processes, *Proc. 8th Metalworking Research Conf.*, SME, pp. 312–319 (1980)

27. Guhring, K., Hochleistungs-schleifen, Dissertation, RWTH, Aachen (1967)

28. Tawakoli, T., High efficiency Deep Grinding, VDI-Verlag and Mechanical Engineering Publ. (1993)

29. Archard, J. F., The Temperature of Rubbing Surfaces, *Wear*, 2:438–455 (1958)

30. Carslaw, H. S., Jaeger, J. C., Conduction of Heat in Solids, Clarendon Press, Oxford (1946)

31. Black, S. C. E., The Effect of Abrasive Properties on the Surface Integrity of Ground Ferrous Components, PhD Thesis, Liverpool John Moores University (1996)

32. Shaw, M. C., Principles of Abrasive Processing, Oxford Science Publ. (1996)

33. Brockhoff, T., Grind-hardening: A Comprehensive View, *Annals of the CIRP*, 48(1):255–260 (1999)

34. Rowe, W. B., and Jin, T., Temperatures in High Efficiency Deep Grinding (HEDG), *Annals of the CIRP*, 50(1):205–208 (2001)

7

Molecular Dynamics for Abrasive Process Simulation

7.1 INTRODUCTION

To meet the continuously rising demands for high accuracy and high quality products, abrasive processes like grinding and polishing are playing an important role as finishing processes in manufacturing. However, as abrasive processes are very complex, with a large number of characteristic parameters that influence each other, reproducibility of machining accuracy and quality are the most significant factors for the application of such processes. In order to fulfill the demands, set-up parameters are commonly determined by machining tests that are both time consuming and costly. Consequently, the process quality depends extensively on the experience of the operator.[1][2]

One way to reduce the dependence of the operation on a skilled operator is to establish a reliable process model, that is to store information and knowledge of that process, and to employ simulation technology. Modeling and simulation serve to analyze, to predict the results, and to broaden the understanding of a process on the basis of such a model. Focusing on machining-process modeling, many different types of models have been developed and employed, ranging from macroscopic to microscopic ones,[3]–[10] covering analytical, empirical, and physical approaches.

In principle, independent from the level of abstraction, all models have an abstract presentation as to link, cause, and effect of a real process in common. Simulation is the imitation of a process on the basis of a model.[3]

The characteristic features of abrasive processes are the small amount of material which is taken away by each grain and the localized grain/workpiece contact with small resulting forces.[1][2] The inherent disadvantage of the localized machining and the small irregularly shaped and moving grains is a limited or difficult access to the grain/workpiece contact area for measurements during machining. As in the finishing process or in precision machining, the actual material removal can be limited to the very surface of the workpiece, where only a few atoms or layers of atoms will be taken away. At this range, measurements of typical quantities of interest are even more difficult or impossible. The described inherent measurement problems and the lack of more detailed data limit the development of analytical and empirical models.

Common phenomenological cutting process models for isotropic and homogeneous materials are based on continuum mechanics and macroscopic integral process quantities. At the microscopic level, these models can no longer be applied because of the lack of detailed, local information and atomistic phenomena which determine the process. In order to address these problems, a modeling method with a strong physical foundation and detailed microscopic information is needed.[1][3][6][11]

In physics and physical chemistry, a method called molecular dynamics (MD) was developed.[12][13] Today, MD is a well-established physical modeling method, that is also used in materials science, with a comprehensive representation of material properties on the atomic level. An MD simulation appears to meet the necessary requirements for machining process simulation at a microscopic scale. In this chapter, the use of MD for the simulation of microscopic machining processes is demonstrated. A selection of relevant examples of MD simulations for abrasive processes are given in Sec. 7.4.

7.2 CONCEPT AND BASIC ELEMENTS OF MOLECULAR DYNAMICS

Molecular dynamics comprises macroscopic, irreversible thermodynamics and reversible micromechanics. The thermodynamic equations form a link between the micromechanical state, i.e., a set of atoms and molecules, and the macroscopic surroundings, i.e., the environment. The thermodynamic equations introduce the quantities system temperature and hydrostatic pressure into the model and allow the determination of energy changes involving heat transfer. In mechanics, it is common to consider energy changes caused by displacement and deformation.

By the term "mechanical state" of a microscopic system, a list of present coordinates (r) and velocities [$v(t)$] of the constituents is implied.[13] For the information about the state of the system to be useful, equations of motion capable of predicting the future must be available. As the governing equations of motion for a system of constant total energy, Newton's equations of motion can be used:

Eq. (7.1)   Newton's equations of motion:
$$d[v_i(t)]/dt = 1/m_i \, \Sigma_{i<j}[F_{ij}(R_{ij},\alpha,...)]$$

Eq. (7.2) $d[R_i(t)]/dt = v_i(t)$ with $i,j = 1$ to n

The resulting force on an atom, i, is expressed by integral over-all force contribution, F_{ij}. Numerically this is calculated as a sum over all forces acting on each atom, i [Eq. (7.1)]. Hence, two bodies at a close distance interact through this sum of force contributions in the equation of motion (see Fig. 7.1).

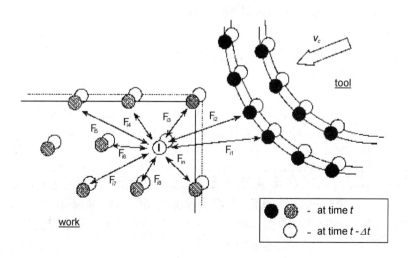

Figure 7.1. Calculation of forces.

The force contribution is calculated by employing partial derivatives of so-called potential functions, which describe the energetical relation between atoms with respect to the separating distance, bonding angle, and possibly, the bonding order. This is also where different materials can be considered as well.

To advance the atoms in space, the equation of motion has to be integrated with respect to time, first to obtain the new velocity, and second for the new position of each atom. Numerically, this operation is more efficiently carried out by approximation schemes, for instance, using finite difference operators and the so-called Verlet or Stoermer algorithm.[12][13]

Eq. (7.3)   Verlet algorithm (velocity form):
$$R_i(t + \Delta t) = R_i(t) + \Delta t \cdot v_i(t) + \tfrac{1}{2} \cdot 1/m_i \cdot \Delta t^2 \cdot F_i(t)$$

Eq. (7.4) $v_i(t + \Delta t) = v_i(t) + h/(2 \cdot m_i) \cdot [F_i(t + \Delta t) + F_i(t)]$
with $i = 1$ to n

With the present positions $[R_i(t)]$, velocities $[v_i(t)]$, and forces $[F_i(t)]$, the new positions and forces at $t + \Delta t$ can be calculated along with the new velocity.

Given the equations of motion, forces, and boundary conditions, i.e., knowing the current mechanical state, it is possible to simulate future behavior of a system. Mathematicians call this an initial value problem. A reasonable distribution of the initial velocities can be obtained from the Maxwell-Boltzmann distribution function.

7.2.1 Material Representation: The Potential Function

While the original molecular dynamics theory is well based within physics, empirical elements were introduced from the materials science area in order to match the results of experiments with the theoretical, and so far physical, model. The key to computational efficiency of atomic-level simulations lies in the description of the interactions between the atoms at the atomistic, instead of the electronic, level. This reduces the task of calculating the complex many-body problem of interacting electrons and nuclei, as in quantum mechanics, to the solution of an energetic relation involving, basically, only atomic coordinates.[13]

The central element of the MD-code is the force calculation. The force calculation is the most time-consuming part in an MD program and therefore determines the whole structure of the program. Efficient algorithms for the force calculation are important for large-size systems, i.e., for systems with a large number of atoms.[12]

Pair Potentials. A model of a material which can form liquid and solid condensed phases at low temperatures and high pressures was first described by van der Waals. Such condensed phases require both attractive and repulsive forces between atoms.[13] Since the simplest possible representation of many-body interactions is the sum of two-body interactions, the so-called pair potentials were the first descriptions of such type. A typical course of the functions is shown in Fig. 7.2.

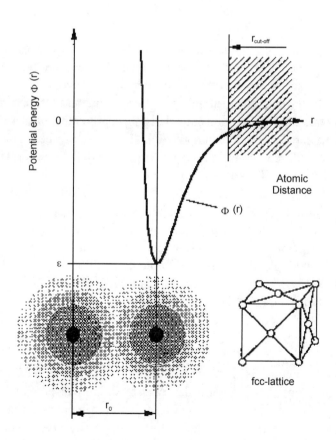

Figure 7.2. Course of a pair potential function.

The best-known pair-potential functions are the Lennard-Jones and the Morse potential, see Eqs. (7.5) and (7.6). The well-depth of the functions are given by the parameter, ε, and D for the minimum potential energy or sublimation energy, while σ and r_o are constants that define the position of the energy minimum:

Eq. (7.5)   <u>Lennard-Jones:</u>

$$V_{LJ}(r) = 4 \varepsilon \left[(\sigma/r)^{12} - (\sigma/r)^6 \right]$$

Eq. (7.6)   <u>Morse:</u>

$$V_M(r) = D \left[e^{-2\alpha(r-ro)} - 2e^{-\alpha(r-ro)} \right]$$

These parameters are derived from experimental data like lattice constant, thermodynamic properties, defect energies, and elastic moduli.

    The potentials describe chemically active materials as bonds that can be established or cut at the long-range part. They represent reasonable descriptions for two-body forces to the extent that they account for the repulsion due to overlapping electron clouds at close distance and for attraction at large distances due to dispersion effects. Generally, in solids, a shielding effect is expected to make interactions beyond the first few neighbors of limited physical interest. Hence, potential functions are commonly truncated at a certain cut-off distance, preferably with a smooth transition to zero, and result in so-called, short-range forces. The long-range Coulomb forces are usually beyond the reach of MD model sizes.[13]

    Many-body Potentials. Pair potentials only stabilize structures with equal next-neighbor distances like fcc and hcp structures, basalt planes, and triangular lattices. However, it is not possible to correctly describe all elastic constants of a crystalline metal using pair potentials. For a better representation of metals, many-body interactions need to be included into the function, for example, the well-known potentials following the embedded atom method (EAM).[14]–[16] In one of the introduced MD simulations, the Finnes-Sinclair type EAM potential by Ackland, et al., was employed.[16] Embedded atom method potentials have been developed and tested for complex problems such as fracture, surface reconstruction, impurities, and alloying problems in metallic systems.

    The structure of brittle or nonmetallic materials with covalent or ionic bonds cannot be satisfactorily described by simple pair-potentials. Ionic materials require special treatment because Coulomb interactions have poor convergence properties unless the periodic boundaries are implemented with care.[12] For the diamond lattice structure of covalently bonded semiconductors like silicon and germanium, it is necessary to treat the strong directional bonding explicitly by including terms that describe the interaction between three or more atoms considering bond angles and bond order. Many important semiconductor and opto-electronic materials show the diamond lattice structure or the similar cubic zinc-blend structure; this also applies to some ceramics.

One of the first potentials for silicon was developed by Stillinger and Weber[17] and a newer, improved one, by Tersoff,[18] which was employed in one of the later MD examples.

Boundary Conditions. Boundaries are an intrinsic, vital part of models. Thermodynamic properties are thought of as bulk properties. Thermodynamic properties are only considered to be reliable if the volume analyzed is large enough so that local surface effects and fluctuations have negligible effect on the system. The system size to be chosen needs to be made big enough to decrease the influences of boundaries.[13]

Besides the option of free surfaces, that would result in a cluster in free space, basically two types of boundaries are common in MD simulations: fixed and periodic boundaries. The simplest type, in terms of realization, is the fixed atom boundary which confines all freely propagating atoms inside or provides support for them at one or more sides. It is simply realized by taking away the dynamics of such atoms. The consequences of such infinitely hard boundaries for the simulation can be significant as no energy can pass through the boundary and phonons will be reflected at it. Hard boundaries used alone are poor representations of the surrounding environment/material.

Periodic boundary conditions (PBC) were introduced to avoid the hard boundary reflection and allow the study of bulk and bulk/interface structures without the strong boundary influence in small models.[4][5][12][13] It is assumed that the bulk of the material is made of many similar systems along the axis perpendicular to the periodic boundary plane, i.e., there are no surfaces along this axis. The system reacts as if there are identical systems at both sides of the PBC, exposed to the same conditions and changes. In effect, the system is connected to itself and atoms at one side interact with atoms on the other side or transfer through the PBC from one side to the other. Figure 7.3 shows a sketch of an indentation model (triangular indenter on top of a workpiece), where a one-axis PBC is considered perpendicular to the horizontal axis.

A consequence of periodic boundaries is that energy and phonons are not reflected, but travel through the system by means of PBC. One- or two-axis PBC can be employed where symmetry axes are available and the lattice structure allows an undisturbed bonding through the PBC planes. Additionally, a deformation compatibility across the PBC has to be fulfilled by an appropriate alignment of preferred slip systems relative to the PBC, in order to avoid artificial deformation patterns.

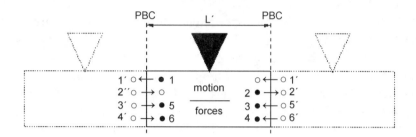

Figure 7.3. Motion through (atoms 1,2) and forces at (atoms 3–6) periodic boundaries (PBC).

7.3 REQUIREMENTS FOR MOLECULAR DYNAMICS SIMULATIONS OF ABRASIVE PROCESSES

The fundamental part of an abrasive process is the interaction of two bodies, where one is carrying out work upon the other one. Therefore, the MD model needs to include the area of contact of these two bodies and their surfaces on which they interact with each other. Hence, full three-axes PBC are not applicable since the surfaces have to be along one axis. Also, a relative motion between both bodies need to be applied to account for the cutting speed. The process requires that cutting, thrust, and tool forces are balanced or accommodated at the system boundaries in order to measure, for instance, tool forces, or to avoid unwanted translational and rotational motion by the tool or the workpiece.

By exerting work upon each other, energy is added to the bodies and thereby to the system, whereupon temperature would rise. Implementing thermostat areas along the boundaries allows the temperature of the system to be controlled by releasing energy to the not-modeled environment. The machining operation represents a massive deformation process. In order to reduce interaction with the boundaries and to allow for sufficient elastic deformation, a reasonably big-sized model has to be determined by tests or on the basis of experience. Experience shows that two-dimensional MD models have little meaning, unless the process in question is restricted to a plane and the all-important information can be accounted for in such a model. In most of the cases, a 3-D model or at least a semi- or quasi-2-D model is the better choice, since material properties and structures are significantly better represented.

In principle, it is desirable to limit the MD model to the closer area of the tool/work contact, including a sufficient distance to the boundaries, in order to allow for a total model size which is as close as possible to the real process. Following this concept, the system layout requires implementing the cutting speed as a motion of the workpiece relative to a fixed tool. For this purpose, it is best to embed the whole model into a frame system, through which the workpiece atoms propagate with cutting speed from one side to the other side. In this way, new atoms of undisturbed structure enter the system at one side, and others leave the system as part of the chip or in a deformed, new arrangement behind the tool, at the generated surface (see Fig. 7.4). This procedure allows steady-state conditions to be achieved with a relatively small number of atoms. Long simulations can be carried out as required to ensure steady-state conditions. The procedure also allows for change in orientation of the material entering the system. This makes it possible to change the orientation and simulate machining through grain boundaries within the progression of one simulation. Mechanical or thermodynamic properties and quantities with larger time constants can be determined.

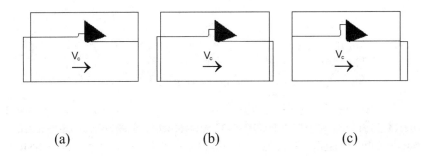

(a)   (b)   (c)

Figure 7.4. Simulation concept for constant cutting conditions on the basis of atomistic modeling.

The use of large, fast computer systems, like parallel-processor computer systems, is as desirable for MD as fast algorithms. Regarding fast algorithms, the basic idea of the introduced techniques is to reduce the calculation of forces and stresses to the necessary ones. In principle, the force calculation is at least an N^2 operation, even for simple pair potentials. Methods to speed-up the calculation, like the so-called bookkeeping technique, the next-neighbor-cell method,[4][12] or the use of tabulated force functions,[4][5][19] employ static or dynamic tables at different levels of the calculation procedure. Using these techniques, the program codes become significantly more complex, but change the dependence of the calculation time on the system size from an N^2 relation to an N^1 relation. Further details about these techniques can be found in the literature.[4][5][12][19]

7.4 APPLICATION EXAMPLES FOR MOLECULAR DYNAMICS SIMULATION OF ABRASIVE PROCESSES

In the following sections, results from four applications of molecular dynamics simulation will be discussed, each focusing on specific features of abrasive machining at the microscopic level or on analyzing atomistic models for quantities and data of interest.

First, results of an orthogonal cutting process simulation of a ductile, single-crystalline material will be presented.[10] Here the main interest was in obtaining stress and temperature distributions for micromachining processes. The second example demonstrates the capability of MD for investigating the pile-up formation in abrasive machining.[7][8] Machining polycrystalline materials is more common than single-crystalline ones. The influence of grain boundaries in polycrystalline materials on the microcutting process was investigated by the third example.[9] The fourth example deals with phenomena observed in a different class of material, the brittle semiconductor material silicon.[19]

7.4.1 Orthogonal Cutting of a Ductile, Single-crystalline Material

Figure 7.5 shows an orthogonal cutting model, which was employed to study the chip formation and the surface generation at a cutting edge of an abrasive on a grinding wheel. The model was designed to improve on some of the common short-comings of the MD cutting models, for instance, the sole use of pair potentials, infinitely hard and thermally nonconducting tool atoms. In order to correctly consider the crystal structure of the workpiece and its orientation, a 3-D model is necessary. By employing one-dimensional PBC along the *y* axis, it was possible to reduce the model to a quasi-2-D type with a small width and to correctly model the 3-D crystal structure of the copper workpiece as well.

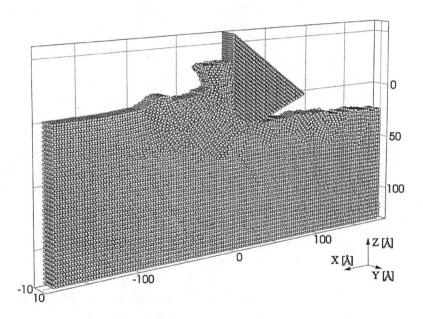

Figure 7.5. Dimensions of the 3D cutting model.

The model contained 11,000 tool atoms (dark, at the top center) and 71,000 work atoms. Atoms at the bottom and to the very left and right hand side of the model represent fixed boundaries, but were shifted with cutting speed to create the relative motion between work and tool. Atoms in the layers next to these hard boundary atoms had thermostat properties. Similar conditions were defined for the tool (fixed layers of atoms at the oblique, upper-right plane to hold and to position the tool, also to measure forces). Employing the before-mentioned next-neighbor-cell method, the all sur-rounding system frame (Fig. 7.4) was divided into small sub-cells. On their path through the cells, the workpiece atoms changed their status and properties from boundary atoms on entering the model, to thermostat and to freely propagating atoms away from the boundaries, and back. This procedure was controlled by the cells, i.e., the status of the cell where the atoms belonged at a certain time. To allow for a smooth entry of initially hard atoms into the thermostat area without causing any shock by a sudden compression upon release of initially fixed atoms, a so-called semipermeable boundary (SPB) was introduced. Here, initially fixed atoms gradually pass with cutting speed into the thermostat cell. Once they have crossed the SPB, they can freely move, but do not return into the boundary.

In order to investigate the heat conduction through the diamond tool, the tool atoms have to propagate freely. For this tool/workmaterial combination, no plastic and no significant elastic deformation in the tool was expected; the tool structure was not of special importance. Therefore, the interactions between carbon-carbon atoms were approximated by a Lennard-Jones pair potential and an fcc lattice structure with the same atomic density as diamond. Its surfaces were formed by preferred diamond/fcc cleavage planes, and its orientation was chosen arbitrarily. The interaction potential between tool and work atoms was derived from data found by Shimada, et al.[20] The work atom interactions are described by the Acklands EAM potential for copper.[16] The cutting speed was restricted to 100 m/s and 50 m/s. A lower speed was not practical to simulate due to computational limitations.

Figure 7.6 shows a so-called deformation graph in a 2-D projection. It shows areas of plastic deformation, dislocations, and large elastic deformations in the sub-surface region. The method is based on horizontal and vertical connections between initial-neighbor atoms. Deformations show as sharp equilateral folds in neighboring layer lines within the otherwise rectangular structure or by narrowing mesh spacing as in the case of strong elastic deformation. For large displacements between initial-neighbor atoms, the bond was considered to be broken and was not drawn anymore. In this way, highly deformed areas, like the chip and the newly generated surface, show few initial-neighbor lines.

Deep running dislocations, observed in 2-D molecular dynamics cutting simulations using still pair potentials,[4][5][7][21] could not be confirmed by employing this 3-D model and the better EAM potential. This model predicts intensive plastic deformation at the generated surface with a thickness of only a few atom layers. At the same strain, a 2-D model always predicts larger dislocations than its 3-D counterpart.

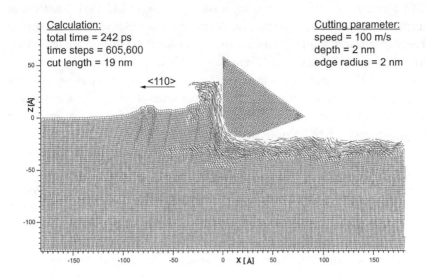

Figure 7.6. Deformation graph, view <110>.

In comparison to a smaller depth of cut (1 nm, $a/r = 0.5$), the generated surface has a bigger deformed layer ($a/r = 1$) and a higher dislocation density. Also, the surface appears rougher, although the surface level varies (only four layers of atoms). The stress field around the tool tip activates dislocations which run deeper into the bulk material than at a 1 nm depth of cut. The thickness of the deformation layer is nearly doubled.

The cutting process changes drastically when changing the ratio of depth of cut to the cutting edge radius from 0.5 to 1.0, also depending on the crystalline orientation of the work. At a ratio of $a/r = 1.0$, the tool begins to utilize more of its rake face for the chip formation. With an increase in depth of cut, the portion of twin dislocations in chip formation increases over dislocation slipping. Such twinned areas can be seen ahead of the tool and the chip. The lower energy requirement for twinning makes the chip removal process at a larger depth of cut more efficient and the cutting forces only increase proportionally.

Obtaining Process Information. So far, most of the MD results of cutting process simulations were presented as atomic, discontinuous sets of instantaneous data at individual atom sites, such as snap shot atom positions, relative displacement, and instantaneous atomic temperature, but no stress distribution. Besides the limited meaning of instantaneous atomic temperatures and stresses, looking at such large sets of 10,000, 100,000, or even millions of data is not practical from a point of view of efficient data analysis. Further, it makes an attempt to compare MD results with continuous mechanics results difficult, if not impossible. A more practical representation of MD results is necessary. Many properties of interest strongly fluctuate with each iteration step. Thus, when aiming at macroscopic thermodynamic properties, suitable time intervals for averaging these properties have to be identified.

Simulations showed that an average of about 1,000 time steps led to sufficiently stable mean properties, but still provide a certain time resolution in order to study some details of the process. Considering the basics of MD and the physical nature of these quantities, the results can now be represented in the form of gradual distributions as so-called contour plots, with a certain resolution in space as well as in time. The representation of stresses and temperature in terms of continuous distributions allows a direct comparison of continuous mechanics results and MD results and is, therefore, considered to be an important step for further use and spread of MD as an analysis tool in engineering.

In the following sections, the maximum shear stress and temperature distributions for the orthogonal cutting process are shown (Figs. 7.7 and 7.8). With the help of these distributions, it is possible to determine where energy is stored and to support the understanding of how it is dissipated. They also allow the determination of how deep the process influences the work and where new dislocations can occur, since areas of high shear stress are potential sources for formation or extension of dislocation.

Because of the complex shearing and chip formation processes, the shear stress distribution is complex as well. The ratio of $a/r = 1$ leads to the formation of a number of twin dislocations, which clearly show up in the deformation graph (Fig. 7.6), where all high stresses concentrate in this rather shallow depth, ahead of the tool in cutting direction. The shear stress concentration well ahead of the tool suggests the formation of a new effective shear plane with a small effective shear angle and rising chip cross-section.

Figure 7.8 shows the temperature distribution in the workpiece and in the tool during the cutting process. The hottest area is the chip area and the area of shearing. At the tip of the tool, the material is deformed at a high stress level, whereby a lot of heat is generated. The high temperature areas that extend from the chip under the tip of the tool are an important source

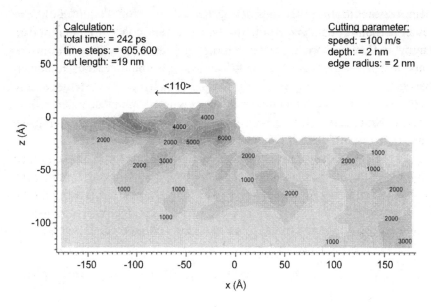

Figure 7.7. Maximum shear stress distribution(N/mm^2), view <110>.

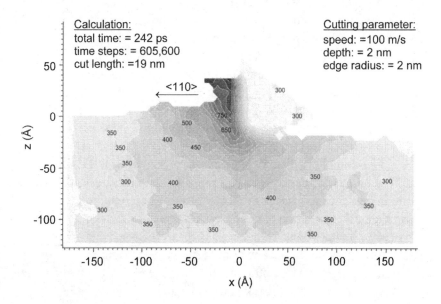

Figure 7.8. Temperature distribution, T in (K), view <110>.

of heat generation to the areas of shearing. Away from the areas of heat generation, the temperature distribution shows a concentric shape. Between tool and work a strong temperature gradient exists. The high conductivity of the diamond quickly passes heat to the thermostatic layer of the tool, where energy is withdrawn. This ability keeps the temperature in the tool nearly constant and heat accumulation is avoided.

Note that only the thermal conductivity through phonons was considered regarding the temperature distribution in this work. The conductivity by electrons was neglected even though it is one order of magnitude larger than that of the phonons. Hence, the temperature level as well as the local gradients would actually be lower than shown in this work. New algorithms were developed to describe thermal conductivity more accurately by considering both the electron and the phonon conductivity.[22]

7.4.2 A Model for Investigating the Pile-up Formation in Abrasive Machining

The following figures demonstrate the capability of molecular dynamics simulation to aid in the study of abrasive machining processes. The models were set up to investigate the pile-up phenomenon and the chip formation as a function of shape and orientation of abrasives. For these types of problems, full 3-D models are necessary, as shown in Figs. 7.9 and 7.10, due to the character of the phenomenon.

The Lennard-Jones potential function was employed for the copper workpiece interactions in this model. For the tool, the abrasive grain was diamond. The diamond grain was kept stiff, i.e., no interactions among carbon atoms and no heat transfers through the grain. For the process simulation, the cutting speed was set to a fairly high value of 100 m/s, as a concession to the expected calculation time.

The complex shape of abrasives can hamper the evaluation of specific geometry influences, as a super-position of influences of all active cutting edges and rake faces is taking place. The computer simulation allows a clearer distinction of geometry influences as an ideal tool design and is easier to achieve and to analyze. Therefore, extreme tool orientations for the rake face were chosen and two are shown here. Figure 7.9 shows the conditions for a grain with a straight, aligned rake face. Clearly the initial stage of chip formation at the rake face is visible and pile-up at its sides is still very small with this straight aligned grain.

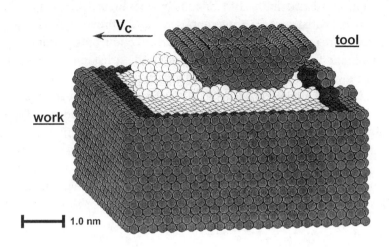

Figure 7.9. Grain with a straight aligned rake face in cutting direction.

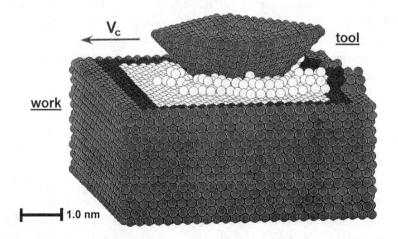

Figure 7.10. Rotated grain with oblique rake faces in cutting direction.

The second extreme tool orientation is achieved with the same grain model, but it is turned on the vertical axis by 45°. Figure 7.10 shows this configuration, with the same observation time as the previous model (Fig. 7.9). Here, there is no straight plane in cutting direction, but a leading edge with two grain oblique faces which causes a high thrust to the sides. The comparison with the previous figure makes the different mechanism evident. The straight aligned grain led to less intensive pile-up formation and more effective material removal than the grain with oblique rake faces. Hence, it is possible with MD to simulate the influence of grain shape and orientation on the efficiency of abrasive processes. On the basis of bigger, more suitable MD models, it will be possible to determine the energy dissipation by a direct analysis of elastic and plastic work and the microscopic mechanisms that determine the surface roughness perpendicular to the cutting direction.

7.4.3 Machining of Ductile, Polycrystalline Materials

In microcutting experiments on several materials, it has been observed that the burr height at the edges of machined grooves depends on the workpiece material and its crystalline structure.[21] When crossing grain boundaries in micromachining of polycrystalline copper, it was found that the burr height drastically increased from values of 0.3 mm to 1.0 mm and larger, accompanied by increased surface roughness. As many materials of interest have a polycrystalline rather than a single-crystalline structure, it is important to understand the mechanism of burr formation and the influence of grain boundaries on the surface integrity. The similarities between the conditions in microcutting and those at grains in abrasive machining make these observations important for abrasive processes as well.

The complexity of this microscopic phenomenon requires a 3-D model for its investigation. Figures 7.11 to 7.13 show different stages of the simulation and different views of the employed model. The model contained up to 15,000 atoms in the observation space, an all-confining simulation box or frame structure, including the atoms of a stiff diamond tool and the freely propagating copper workpiece.

In Fig. 7.11, a front view of the simulation box for constant cutting conditions is shown. Atoms with a specific crystalline orientation enter the model in front of the figure, and flow at cutting speed, toward the back. The chip formation at the rake face of the tool is slightly tilted to the right side, which causes the chip to flow not only up the rake face, but partially around the tool as well. On the left edge of the rake face some material piles up, forming the observed burr.

When atoms of different structures are added, a grain boundary forms. As the tool approaches the grain boundary, dislocations ahead of the tool suddenly extend to the grain boundary which changes the structure and density in this area and leads to some pile-up of material (see Fig. 7.12).

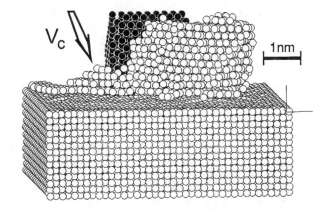

Figure 7.11. Front view of the model at constant cutting conditions.

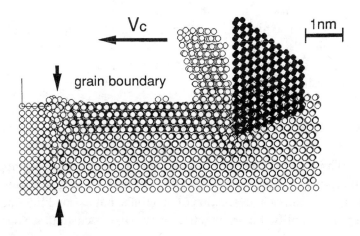

Figure 7.12. Cross-section of the model, parallel to cutting direction.

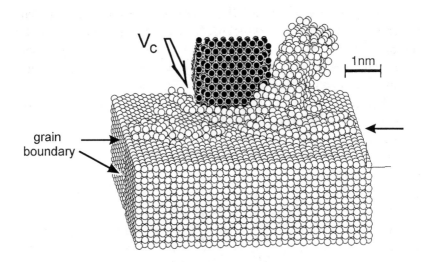

Figure 7.13. Front view of the model at constant cutting conditions.

As Fig. 7.13 shows, the observed dislocations do not run straight ahead of the rake face; they run along typical slipping planes of the {111} surface, to the left and right side. Dislocation formation and motion is a mechanism of stress relief. The related energy is partially used for the deformation work, is stored at the dislocation core, and is lost as friction which causes the local temperature to rise. The extension of the dislocations towards the grain boundary iss combined with a relaxation of the stress field at the tool, which is believed to be the cause for the interruption of the chip formation visible in Fig. 7.13. With further progress of the simulation, a new chip begins to form and more material is piled up at the boundary in front of the tool as well. The pile-up at the boundary increases the actual depth of cut and, coincidently, the burr height. When the tool passes the grain boundary, the burr height decreases in the new grain decisively, i.e., much smaller than shown in Fig. 7.11.

7.4.4 Indenting a Brittle Semiconductor Material

The indentation testing model was used to study the impact of machining on hard and brittle materials. Indentation testing is not only seen as a means for hardness evaluation of materials, but its conditions are also viewed as a simplified representation of abrasives penetrating a surface.

Chipping and cracking easily occur in machining of brittle materials. Microcracking is a critical issue because the cracks can reduce the mechanical strength of machined components considerably.[1][2] Because of its importance in the electronic industry and the availability of a potential function, silicon was chosen as the representative material.

Figures 7.14 and 7.15 show snap shots of indentation simulations on the (121) surface of silicon. The indentation depth was 1.5 nm in Fig. 7.14 and 0.75 nm in Fig. 7.15. The diamond structure of silicon required a 3-D model. However, in order to observe the formation of cracks more easily, a quasi-2-D model was also employed. It had only a shallow depth of four atomic layers, just enough to model the silicon structure correctly. Note that not all atoms in the observation plane had the same position in depth, and others were hidden at lower levels beneath the visible atoms. In both directions of the horizontal plane, perpendicular to the direction of indentation, periodic boundaries at the surface were considered and fixed boundaries at the bottom balance the indentation force.

The material in the direct vicinity of the indenter tip deforms plastically and elastically in the surrounding area. This elastic/plastic deformation by the indenter leads to a tensile load at the surface on the right side of the indenter, where a crack is then initiated at the transition from the plastically deformed to the elastically deformed area. While crack initiation has not been observed for indenters with a tip radius smaller than 2 nm, indenter tip radii equal to or larger than 2 nm have caused crack initiation. The dependency of crack initiation on the indenter tip radius is well known from cutting and grinding of brittle material.[1][2][11][19][23][24]

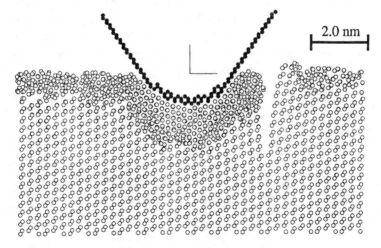

Figure 7.14. Crack pattern in indentation on silicon with a 2 nm indenter.

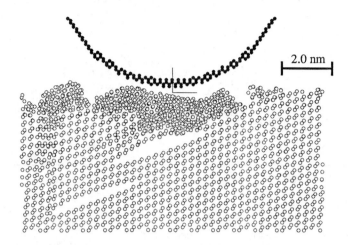

Figure 7.15. Crack pattern after small indentation on silicon with a 5 nm indenter.

Figure 7.15 shows a crack pattern for a larger indenter (5 nm), that was generated after taking the indentor off the workpiece surface. The indentation was reversed when a zone of plastic deformation was formed in the vicinity of the tip (depth 0.75 nm), but before any crack occurred. When the tip was taken off the specimen, residual stresses formed cracks, starting from the transition of the plastically deformed to the elastically deformed area under the tip, as well as from the core of the vertical dislocation on the left side of the model.

The large indenter caused an intensive elastic deformation in the model, whereupon the vertical dislocation was formed. When the indenter was taken off the specimen, the elastically deformed areas gradually recovered. Plastic deformation and dislocation formation were combined with stress relief and energy release. The formed dislocation did not recover much, and its ability to accommodate all expansion by elastic deformation was insufficient on the right side of the dislocation, whereupon a crack was initiated. The plastically deformed area under the indenter tip could not fully recover as its expansion coefficient is different from that of the single crystal structure. The stresses at the transition between the plastically deformed and undeformed structure caused crack initiation. Crack patterns, due to residual stresses in indentation on brittle materials, have been described in the literature.[1][2][23][24]

7.5 SUMMARY AND OUTLOOK

The capability of molecular dynamics (MD) simulation for abrasive machining processes has been demonstrated by a number of applications in this field. With its well-based physical concept, MD modeling provides a comprehensive representation of material properties, where elastic and plastic materials, as well as, hard and brittle materials are treated within one framework.

Molecular dynamics provides a unique insight into abrasive processes at the nanometer level. Once the atomic interaction is described, atomic contact, friction, fracture, and plastic deformation result from the interaction with neighbor atoms. With this powerful tool, the engineer is given a better insight into the essential elements of the processes during machining, such as localized material behavior. Molecular dynamics results can serve as verification or input data, i.e., as validation and support for analytical or extended continuous mechanics models. The specific patterns of the temperature distribution can help to derive suitable input data for the heat source in thermal-mechanical FEM models, for example. Micromechanical flow and failure criteria, like flow stress, crack propagation, and initiation, can be verified and improved by means of MD simulation. For nanometer level cutting, the MD results can either be taken directly or be used in input variable evaluation for micromachining models.

The main problem today lies in the lack of a comprehensive number of realistic interatomic potential functions for all materials, in particular for compounds and alloys. This is an active field of development, combining new computational techniques with novel experiments. The progress in the development of new potential functions is supported by activities in physics, physical chemistry, and materials science.

With the spreading of high-performance, parallel computing facilities, the main bottle-neck is shifting to efficient data processing and representation. The first example presented in this chapter provided the idea and some means of such improved processing and presentation methods for MD information.

At present, the direct combination of models and parallel application of different methods at specific length scales (quantum physics + atomistic + meso-scale + continuum models) is being investigated in physics and materials science. The idea of making the MD models as big as possible is aiming at another goal, that is bridging the gap between MD and FEM, i.e., continuous mechanics, respectively. First approaches to combine atomistic methods, such as molecular statics (MS) or MD with FEM models, were carried out several years ago.

On the basis of atomic interaction dynamics, MD provides the chance to study the friction conditions in the cutting process directly, as seen at the chip/tool contact. The introduction of chemical potentials, correctly accounting for activation energies and reaction dynamics, will significantly improve the quality of calculation and widen the field of application of MD simulation to chemical and abrasive wear. Fundamental tribology and friction investigations are already on the way and international conferences have established whole MD sessions. In recent years, many molecular modeling research groups have been formed in various fields and institutions around the world.

REFERENCES

1. Inasaki, I., Toenshoff, H. K., and Howes, T. D., Abrasive Machining in the Future, *Annals of the CIRP*, International Institution for Production Engineering Research, 42(2) (1993)

2. Inasaki, I., Grinding of Hard and Brittle Materials (Keynote paper) *Annals of the CIRP*, 36(2):463–471 (1987)

3. Toenshoff, H. K., Peters, J., Inasaki, I., and Paul, T., Modeling and Simulation of Grinding Processes, Keynote paper, *Annals of the CIRP*, 41(2):677–688 (1992)

4. Rentsch, R., Process Modeling by Means of Molecular Dynamics (MD), *2nd Int. Conf. Machining of Advanced Materials, Aachen, Germany*, 30.9:175–195, VDI—Berichte No. 1276 (Jan. 10, 1996)

5. Rentsch, R., Dissertation, *Molecular Dynamics Simulation for Nanometer Chip Removal Processes*, Keio U., Yokohama, Jpn., 150 pages (1995)

6. Inasaki, I., Application of Simulation Technologies for Grinding Operations, *2nd Int. Conf. Machining of Advanced Materials*, Aachen, Germany, 30.9:197–211, VDI—Berichte No. 1276 (Jan. 10, 1996)

7. Rentsch, R., and Inasaki, I., Investigation of Surface Integrity by Molecular Dynamics Simulation, *Annals of the CIRP*, CH, Hallwag Publ., Berne, 44:295–298 (Jan. 1995)

8. Rentsch, R., and Inasaki, I., Molecular Dynamics Simulation For Abrasive Processes, *Annals of the CIRP*, CH, Hallwag Publ., Berne, 43:327–330 (Jan. 1994)

9. Rentsch, R., and Inasaki, I., Simulation of Single Point Machining of Polycrystalline Metals by Molecular Dynamics (MD), International Progress in Precision Engineering (M. Bonis, et al., eds.) *Proc. 8ᵗʰ Int. Precision Engineering Seminar (IPES)*, Compiegne, F, Elsevier, pp. 347–350, (May 1995)

10. Rentsch, R., Influence of Crystal Orientation on the Nanometric Cutting Process, Serie Berichte aus Fertigungstechnik, Precision Engineering – Nanotechnology, euspen, (P. McKeown, et al., eds.) Bremen, D, Shaker Verlag, Aachen, 1:250–253 (1999)

11. Ikawa, N., Donaldson, R. R., Komanduri, R., Koenig, W., McKeown, P. A., Moriwaki, T., and Stowers, I. F., Ultraprecision Metal Cutting–The Past, the Present and the Future, *Annals of the CIRP*, 40(2):587–594 (1991)

12. Allen, M. P., and Tildesley, D. J., Computer Simulation of Liquids, Clarendon Press, Oxford (1987)

13. Hoover, W. G., Computational Statistical Mechanics, Studies in Modern Thermodynamics 11, Elsevier Science Publ., Amsterdam-Oxford-New York-Tokyo, 313 pages (1991)

14. Daw, M. S., and Baskes, M. I., Embedded-atom Method: Derivation and Application to Impurities, Surfaces, and Other Defects in Metals, Phys. Rev. B 29, 12:6443–6453 (1984)

15. Finnis, M. W., and Sinclair, J. E., A Simple Empirical N-body Potential for Transition Metals, *Philosophical Magazine, A.,* 50(1):45–55 (1984)

16. Ackland, G. J., Tichy, G., Vitek, V., and Finnis, M. W., Simple N-body Potentials for the Noble Metals and Nickel, *Philosophical Magazine, A.,* 56(6):735–756 (1987)

17. Stillinger, F. H., and Weber, A. W., Computer Simulation of Local Order in Condensed Phases of Silicon, Phys. Rev. B, 31(8):5262–5271 (1985)

18. Tersoff, J., Modeling Solid-State Chemistry: Inter-Atomic Potentials for Multicomponent Systems, Phys. Rev. B, 39(8):5566–5568 (1989) and 41(5):3248 (1990)

19. Rentsch, R., and Inasaki, I., Indentation Simulation on Brittle Materials by Molecular Dynamics; Modeling, Simulation, and Control Technologies for Mfg., *SPIE Proc.*, 2596:214–224, (R. Lumia, ed.) Philadelphia, PA. 22–26.10 (1995)

20. Shimada, S., Ikawa, N., Tanaka, H., Ohmori, G., and Uchikoshi, J., Feasibility Study on Ultimate Accuracy in Mirocutting using Molecular Dynamics Simulation, *Annals of the CIRP*, 42(1):91–94 (1993)

21. Rentsch, R., Inasaki, I., and Brinksmeier, E., Influence of Material Characteristics on the Micromachining Process, Materials Issues in Machining-III and The Physics of Machining Processes-III, (D. A. Stephenson, and R. Stevenson, eds.) pp. 65–86, TMS Publ., Cincinnati, OH (Oct. 1996)

22. Caro, A., Electron-Phonon Coupling; Molecular Dynamics Codes, Radiation Effects and Defects in Solids, 130–131:187–192 (1994)

23. Lawn, B. R., and Evans, A. G., A Model for Crack Initiation in Elastic/Plastic Indentation Fields, *J. Materials Sci.*, 12:2195–2199 (1977)

24. Lawn, B. R., and Wilshaw, R., Review, Indentation Fracture: Principles and Applications, *J. Materials Sci.*, 10:1049–1081 (1975)

8

Fluid Delivery

8.1 THE ROLE OF PROCESS FLUIDS

For most purposes, the importance of the process fluid cannot be overstated. The use of a process fluid may be essential for one or more reasons related to the various functions of the fluid. This includes dry fluids since compressed inert gases or air may sometimes be employed where a liquid fluid would not be acceptable. In most cases, the fluid will be either water based or an oil. A further possibility is solid lubrication which is more likely to be appropriate for slow speed processes. In special cases, abrasive machining processes may be undertaken dry, where it is not permitted to contaminate the workpiece, for example.

The process fluid, often referred to as coolant, has wider functionality than cooling. The main functions in shallow grinding are flushing, bulk cooling, and lubrication. In creep grinding, the additional function of process cooling is important. In general, the functions of the fluid include:

- Mechanical lubrication of the abrasive contacts.
- Chemo-physical lubrication of the abrasive contacts.
- Cooling in the contact area, particularly in creep grinding.
- Bulk cooling outside the contact area.

- *Flushing* or the transport of the debris away from the abrasive process.

- Transport of abrasive to a loose abrasive process.

- Entrapment of abrasive dust and metal process vapors.

The lubrication of the cutting action is not well understood. Cutting fluids lower the temperatures in shallow grinding mainly by reducing friction, hence the forces and wheel dulling. By reducing friction, power is reduced so that less heat is generated.

A plentiful channel of coolant on and around the workpiece achieves bulk cooling and flushing of the swarf. This is essential even if very little fluid enters the contact region between the grinding wheel and workpiece. However, lubrication depends on fluid actually entering the contact region between the grinding wheel and the workpiece. Lubrication does not necessarily require a large volume to achieve a significant reduction in wheel wear, but lubrication will be ineffective if no fluid enters the grinding zone at all. Even small quantities of fluid entering the contact zone can be beneficial to process efficiency, and there have been reported trends towards minimum quantity lubrication (MQL) as a means of reducing environmental and disposal costs.

Traditional delivery of the process fluid to slow-speed abrasive machining processes is illustrated in Fig. 8.1. A simple low-pressure feed is typically supplied from a pump with less than 0.1 MPa (1 bar) pressure head.

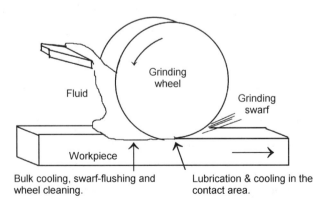

Figure 8.1. The role of a process fluid in grinding.

However, the requirements for modern high-speed abrasive processes need careful planning to obtain the benefits of optimal lubrication,
cooling, and wheel cleaning. Large flow-rates and high energy consumption
is the price sometimes paid to achieve a satisfactory fluid delivery. A single
flood nozzle as shown in Fig. 8.1 may not be sufficient to satisfy the various
functions expected of the process fluid. A medium flowrate nozzle with a
medium pressure head, for example, 4 bar, may be used to supply the contact
area. A 100 bar high-pressure nozzle may be used continuously or discon-
tinuously as a cleaning jet to force fluid at high velocity into the pores of the
wheel. The aim is to remove loose and weakened grits from the cutting
surface. The high pressure jet also removes loose swarf from the wheel
surface and thus slows down the rate of wheel loading.

Bulk cooling and swarf removal from the bed of the machine only
require low pressure, but with a moderately high flowrate. Therefore, it is
worth considering whether the various functions of bulk cooling, swarf
removal, wheel cleaning, contact zone lubrication, and process cooling may
best be served with multiple nozzles. Each nozzle can then be designed to
achieve a particular purpose in the most cost-effective way.

The impact of machining and process fluids on the working environment is a matter of increasing legislation which should also be given
consideration. These concerns range from potentially carcinogenic effects
of ingesting workpiece material particles or of ingesting process fluid, and
extend through to the irritant effects of metals, oils, and bacteria through
inhalation or direct contact with the skin. Increasingly, the answer to these
problems is seen as total machine enclosure with fume extraction systems.
The debate also ranges over the benefits of oil versus water-based fluids and
the desirability of minimum fluid application.

In examining the total life cycle costs of a process, consideration
should be given to the costs of fluid disposal and replacement and of
workpiece cleaning. A well-designed mineral oil system may be considered
more desirable than a water-based emulsion because of the virtually
infinite life of the fluid, the good lubricating properties, the corrosion
inhibiting properties, and the absence of bacterial contamination. However,
a water-based fluid has a higher thermal conductivity which is useful for
bulk cooling of the workpiece and, if the machine is not fully enclosed, is
usually the preferred solution. For high wheelspeeds, the machine should be
fully enclosed because substantial volumes of mist are produced that
require containment. If oil is used, thought needs to be given to fire hazards.

The proportion of air in the process environment must be kept below the level at which flash fires break out.

The thermal properties of a typical mineral oil and of water are compared with air in Table 8.1. The data shows that water has a greater propensity for cooling due to its higher thermal conductivity, the greater specific heat capacity, and the greater value of heat of evaporation, whereas oil is a better lubricant.

There are many papers dealing with fluid delivery including the examples listed at the end of the chapter. Reference 6 lists 160 relevant papers on the subject. This chapter, however, looks more particularly at the mechanics of fluid delivery for high speed grinding, starting with the problems caused by the air barrier.

8.2 OVERCOMING THE AIR BARRIER IN HIGH SPEED GRINDING

When grinding at wheelspeeds above 45 m/s, it becomes increasingly difficult to feed fluid to the grinding contact with a simple flood nozzle, as illustrated in Fig. 8.1, due to the turbulent air barrier.[1] The air barrier becomes stronger with increasing wheelspeed, and the methods of fluid delivery used in traditional workshop grinding are totally unacceptable.

The presence of an air barrier, even at low speeds, is illustrated in Fig. 8.2. Transparent plastic plates were provided at each side of the wheel and workpiece to allow the fluid passing through the contact area to be measured. Even with a gap of 0.08 mm between the wheel and the workpiece, no through flow occurs. Air entrained by the grinding wheel at a speed of 33.5 m/s, recirculates as it approaches the narrow gap and pushes the fluid away from the wheel.[1]

A slightly higher velocity of the fluid or a slightly larger gap between the wheel and workpiece is sufficient at low wheelspeed to cause fluid to overcome the recirculating air, and it then becomes entrained within the boundary layer and passes through the grinding contact area. This is shown in Fig. 8.3 for a gap of 1 mm between the wheel and the workpiece. Of course, in grinding there is no gap between the wheel and the workpiece, although the pores in the surface of the wheel are equivalent to a small gap. The effect of porosity is discussed next.

Table 8.1. Typical Values of Thermal Properties of Water, Mineral Oil, and Air at Atmospheric Temperature and Pressure (atp)

PROPERTY	SYMBOL	UNITS	MINERAL OIL	WATER	AIR
Density	ρ	kg/m^3	900	1,000	1.2
Specific heat capacity	c_p	kJ/kgK	1.9	4.2	1.0
Thermal conductivity	k	W/mK	130	600	26
Heat of evaporation	r	kJ/kg	210	2,260	0

Figure 8.2. The influence of the air barrier on fluid supply.

Figure 8.3. Fluid passes through the contact area at a gap of 1 mm between the wheel and workpiece and is then thrown tangentially outwards away from the wheel. A few smaller particles are carried around the wheel periphery within the air barrier.

A few small particles of fluid and swarf get carried around the periphery within the air barrier, but mostly the fluid exits tangentially outward within a very short distance from the contact. Centripetal acceleration is required to hold fluid or metal particles inside the air barrier. For large particles, the centripetal acceleration is too great to be provided by local inward acting pressures, and the fluid is thrown off in a tangential direction. In this respect, the wheel can be visualized as a rotating pump throwing the fluid off tangentially. It should be made clear that fluid is not thrown radially outwards by centrifugal force. This idea arises from a misunderstanding of D'Alembert's Principle. The outflow is tangential as evident in Fig. 8.3. The fluid emerges from the grinding contact at a speed much greater than the workspeed along tangential directions, not at all in the radial direction.

The example in Fig. 8.6 demonstrates a centripetal acceleration since the fluid clings to the periphery of the wheel. This is made possible by atmospheric pressure and hydrodynamic pressures acting radially inwards on the fluid. The tangential outflow results in a reduction in pressure close to the wheel surface. Hydrodynamic action creates high pressures close to the wheel surface in the region indicated in Fig. 8.4.

Entrained flow within a boundary layer close to a moving surface is termed Couette flow as illustrated in Fig. 8.4 for porous and nonporous wheels. The entrained velocity of the boundary layer approaches the surface velocity and is normally assumed to be equal to the surface velocity with zero slip.

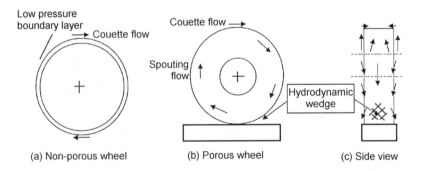

Figure 8.4. The nature of the air barrier. *(a)* A non-porous wheel: Couette entrained flow and a low pressure boundary layer; *(b)* a porous wheel: additional spouting flow through and at sides of the wheel; and *(c)* pressures increased in converging gap.

Laser doppler anemometry and use of computational fluid dynamics (CFD) have been used to confirm and further explore boundary layer behavior.[1] The air thrown off must be replaced. This occurs mainly by way of the flow of air from the sides of the wheel towards the mid-plane where the pressure is lowest. With a very wide nonporous wheel, the pressure gradient increases radially outwards from the wheel periphery near the mid-plane so that a Poiseuille flow takes place radially inwards from the atmosphere to help to counterbalance the flowrate of air thrown off tangentially.

Entrained Couette flow also takes place at the sides of the wheel and creates additional boundary layers in these areas. Air spirals outward, accelerated tangentially along the side faces of the wheel.

In a practical situation, the wheel is porous, so the flow of air or process fluid also takes place through the wheel moving outwards in a spiral and inwards towards the mid-plane of the wheel. The outward air flow through the wheel and along the sides of the wheel is sometimes termed *spouting*. The spouting flow is an additional contribution to the peripheral air barrier.

An example of air flow in the boundary layer close to the entry to the grinding contact from CFD is shown in Fig. 8.5. These air flows have been confirmed by measurement using laser doppler anemometry. This figure clearly shows the region of reversed flow where the air moving towards a converging volume recirculates back and away from the contact zone.

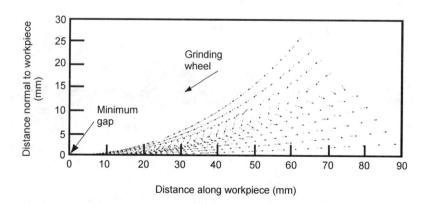

Figure 8.5. Directions of the air flow in the boundary layer predicted from CFD.

Process fluid delivered in the region of reverse flow is directly opposed by the air barrier. It is better to deliver the process fluid above the region of reverse flow where the boundary layer can assist the transport of the process fluid. The benefit of this approach is seen in Fig. 8.6 where the fluid is carried within the boundary layer into the contact zone.

Pressures measured in the air gap between the grinding wheel and the workpiece are shown in Fig. 8.7 for a wheel rotating at 33.5 m/s and 25 mm wide. The wheel diameter was 178 mm. The maximum pressure in the hydrodynamic wedge occurs at the mid-plane of the wheel, the pressures reduce toward the sides of the wheel. The hydrodynamic pressures can be used to assist fluid delivery. The nature of the air flows in the air barrier including the effects of the hydrodynamic wedge are shown in Fig. 8.4.

If the converging gap is filled with liquid instead of air, the hydrodynamic pressures are greatly increased, and these pressures can even help to force the process fluid into the pores of the wheel. An illustration of the hydrodynamic effect is given in the next section. However, before this can occur, liquid must replace the air in the boundary layer in the region of the converging gap.

Figure 8.6. Process fluid delivered above the region of reversed flow is transported into the grinding contact where the surface speed of the wheel is 27 m/s and of the process fluid is 3.5 m/s at 4.2 l/min. The jet was applied at 34° to the wheel tangent, which gave approximately optimal results.

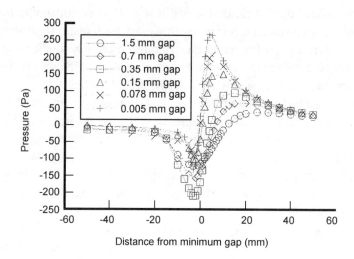

Figure 8.7. Hydrodynamic pressures in the air gap between a grinding wheel and a workpiece. As contact is approached, the pressures become maximum. Negative pressures occur in the diverging gap at the exit from the contact.

A jet of water-based fluid was applied at three different heights and angles to the wheel surface. At the lowest height with a near tangential jet, the recirculating air obstructed the jet. At an increased height corresponding to 90° around the wheel from the grinding contact position with a jet inclined at 15° to the wheel radius, the jet stream was entrained, but became separated from the wheel surface within 45° around the periphery. Varying the position of the jet around the periphery, the best result was achieved with a jet inclined at 56° to the wheel radius and located about 34° ahead of the contact area, as illustrated in Fig. 8.6. From measurements of the useful flowrate, which is the flowrate actually carried through the grinding contact, the best grinding results were achieved when the maximum useful flowrate was achieved.[1]

A countermeasure sometimes used to reduce the effects of the air barrier is to apply a scraper in the boundary layer ahead of the position of fluid delivery. Another countermeasure is to seal the sides of the wheel to reduce the volume of air entering the pores of the wheel.

Work by Trmal and Kaliszer[2] has shown the benefits of using scraper plates. Using a pitot tube to measure the velocity of the boundary layer, a decrease in air velocity occurred as the scraper plate was moved towards the wheel periphery. This is supported by Campbell[7] who investigated the hydrodynamic pressure at the wheel-workpiece interface caused by the passage of cutting fluid beneath the wheel. At a critical wheelspeed, this pressure approached zero, i.e., the boundary layer of air was preventing cutting fluid from passing beneath the grinding wheel. Introducing a scraper plate allowed the grinding wheelspeed to be increased by 20% before the hydrodynamic pressure again approached zero. In other words, the presence of substantial hydrodynamic pressures in the converging gap between the wheel and the workpiece is evidence that fluid is successfully being delivered. A larger peak hydrodynamic pressure indicates a larger delivery of fluid to the contact.

The effect of a scraper and side sealing on the boundary layer velocity, as measured by a pitot tube placed 20 mm from the contact zone, is shown in Fig. 8.8.

The use of a scraper plate to deflect the boundary layer reduces the effect of the air barrier, but does not completely overcome the problem even at these low speeds. The boundary layer quickly re-establishes itself within a short distance of the scraper plate, so it is important to place the scraper just upstream in the boundary layer and as close as possible to the nozzle.

The effect of sealing is also shown. About 90% of the area of the side faces of the wheel was sealed with silicone rubber. The use of the mask reduces the effect of the air barrier, and the use of masking together with a scraper plate just ahead of a nozzle both assist the delivery of fluid to the grinding contact.

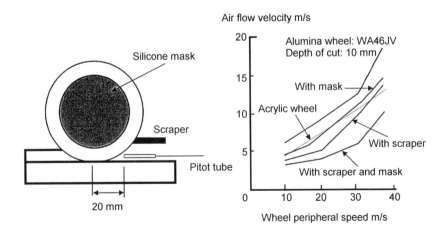

Figure 8.8. The effects of a scraper and of a mask to counter the air barrier.

8.3 NOZZLES FOR HIGH SPEED GRINDING

Delivery of cutting fluid via a nozzle in the form of a jet can have two benefits. First, the fluid can be delivered with a velocity great enough to penetrate the boundary layer of air, and second, if applied at a high enough velocity, it may be used to clean the wheel mechanically by removing loose metal from the pores. The purpose of the nozzle is to direct the flow to the optimal position to feed the grinding contact and to increase the velocity of the process fluid by restricting the cross-sectional area of the jet stream. The pressure energy of the fluid upstream of the nozzle is converted into kinetic energy and heat energy within the nozzle. A well-designed turbulent flow nozzle converts more of the pressure energy into kinetic energy than a laminar flow nozzle. A turbulent nozzle converts approximately half the pressure energy into kinetic energy.

A laminar flow nozzle has the advantage that the jet stream coherence and velocity will be maintained for a greater distance than a turbulent flow nozzle. Laminar flow can only be achieved with a Reynolds number less than 2,000. This implies low jet velocities compared with peripheral wheel velocity that can be seen in the following equation. The Reynolds number is given by

Eq. (8.1) $$\text{Re} = \frac{\rho \cdot v_{jet} \cdot d_{jet}}{\eta}$$

Thus, if it is required to achieve a velocity of 100 m/s with a 1 mm diameter nozzle using water as the process fluid, the Reynolds number is,

$$\text{Re} = \frac{1,000 \times 100 \times 0.001}{0.001} = 100,000$$

Clearly, the nozzle for this case must be designed using turbulent flow equations.

Research into the effects of a coherent jet as opposed to a jet of cutting fluid dispersed upon exit from the nozzle has been carried out by Webster, Cui, and Mindek.[8] Measuring the grinding temperature determined that where a coherent jet was maintained the grinding temperature was reduced, compared with a dispersed jet. To maintain a coherent jet, a round nozzle was used with concave converging nozzle walls prior to exit. Flow conditioners were used up-stream to reduce fluid turbulence. The use of the term coherent, in this context, does not necessarily mean that the jet is laminar. A large turbulent jet stream may remain reasonably uniform for quite long distances if the momentum of the stream is large compared with the surface friction from the atmosphere, and if the exit from the nozzle is smooth and free of interference. Any interference with the flow causes immediate break-up and dispersion of the jet.

Methods used to get fluid into the boundary layer as it enters the grinding contact area principally involve one or more of the following techniques used in combination:

- High velocity of the process fluid applied across the boundary layer and towards the contact area.

- High pressures applied to force process fluid into the pores of the wheel.

- Methods of disrupting the air boundary layer and seeking to substitute a process fluid layer.

- Use of the boundary layer to assist fluid delivery rather than oppose fluid delivery, i.e., delivery in the same direction as the wheel velocity.

In traditional workshop grinding, fluid is fed to the process through a simple flood nozzle attached to a flexible pipe or joint. The operator pushes the nozzle away when setting and measuring. In consequence, the nozzle is replaced in various positions and the flowrate may often be reduced by the operator to prevent splashing. The result is inconsistent grinding performance. Many grinding problems have arisen due to this practice and due to insufficient delivery of fluid to the correct location.

The flood nozzle, illustrated in Fig. 8.1, is likely to be quite ineffective for high speed grinding. To achieve high velocity at the nozzle exit, it is necessary to use a pump of adequate flowrate and power to ensure that the velocity can be achieved with a nozzle of the required diameter.

For high wheelspeeds, a variety of nozzles and fluid delivery systems have been applied. Nozzle designs mainly fall into one of two groups as illustrated in Fig. 8.9:

- Jet nozzles which introduce fluid at high velocity.

- Shoe nozzles which seek to replace the air layer with a process fluid.

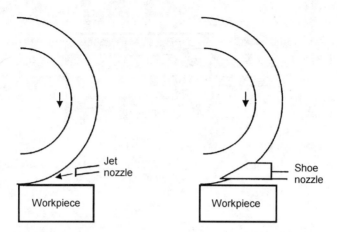

Figure 8.9. *(a)* A jet nozzle uses the velocity of the process fluid to create a liquid boundary layer which enters the contact area. *(b)* A shoe nozzle pressurizes the process fluid into the wheel surface to create a liquid boundary layer that enters the contact area.

It has also been proposed that feeding the fluid through the pores of the wheel overcomes the air barrier.[3][10] However, for wheels at very high speeds, the design of the wheel becomes more complex, and there are few applications of this technique. Also, the fluid must be carefully filtered, otherwise the wheel quickly becomes blocked with swarf.[13]

The action of jet nozzles and shoe nozzles may be enhanced by the application of sealing for the side walls of the wheel and by use of a scraper designed into the structure of the nozzle.

8.3.1 The Turbulent Orifice Nozzle

The turbulent nozzle is an orifice of short length to diameter ratio, ideally less than 0.25. The orifice nozzle is illustrated in Fig. 8.10.

The ideal nozzle has a smooth convergent section leading the flow into a smooth orifice section. A concave convergent section is preferred to a convex convergent section at the inlet to the orifice. The exit from the nozzle should be either a smooth flat face or a smooth chamfer to avoid interfering with the jet stream. The design equations for an orifice are as follows:

Eq. (8.2)
$$p_p = \frac{\rho \cdot v^2_{orifice}}{2 \cdot C_v^2}$$

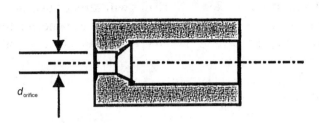

Figure 8.10. An orifice nozzle has a very short length-to-diameter ratio and produces turbulent flow with a vena contracta in the jet stream close to the exit.

Eq. (8.3) $Q_f = C_a \cdot \dfrac{\pi \cdot d_{orifice}^2}{4} \cdot v_{orifice} = C_d \cdot \dfrac{\pi \cdot d_{orifice}^2}{4} \cdot \sqrt{\dfrac{2 \cdot p_p}{\rho}}$

Eq. (8.4) $H_p = p_p \cdot Q_f$

where p_p is the pumped pressure available at the nozzle inlet, Q_f is the flowrate, H_p is the pumping power, ρ is the density of the fluid, $d_{orifice}$ is the orifice diameter and $v_{orifice}$ is the fluid velocity leaving the orifice.

The value of the velocity coefficient, C_v, is typically 0.95 to 0.98 for turbulent flow. The value of the area contraction coefficient, C_a, is typically 0.63. The orifice discharge coefficient, C_d, is the product of $C_a \cdot C_v$. The value of C_d varies with Reynolds number and typically ranges from 0.3 for Re = 10^2 to 0.56 for Re = 10^6.

Example 1. As an orifice design example, a jet velocity of 100 m/s with an orifice of 1 mm using a water-based emulsion having a viscosity of 0.003 Ns/m² must be produced.

Application of the design equations yields the following values:

Re = 33,000

The pumped pressure immediately upstream of the jet is p_p = 50 bar or 1450 lbf/in². The flowrate delivered is Q_f = 0.044 l/s or 0.58 gal/min. The pumping power required to force the fluid through the jet is H_p = 0.22 kW.

Calculations show that a fluid velocity of 100 m/s requires a high pump pressure. This pressure is further increased by the need to allow for losses in the pumping system and piping.

Example 2. A velocity of 10 m/s will suffice but a larger delivery using oil as the process fluid is required. The nozzle diameter is to be 10 mm using an oil having a viscosity of 0.03 Ns/m^2. Calculations find that Re = 3,000, p_p = 0.48 bar or 7 lbf/in^2, Q_f = 0.49 l/s or 6.1 gal/min, and H_p = 24 watts.

The pumping power required to force the fluid through the orifice is 24 W. Allowing for pump losses, pipe bends, and other flow disturbances, the power capacity of the pump needs to be increased by a factor of at least 300%.

It is clear from these two examples that the required nozzle pressure is strongly dependent on nozzle velocity and increases with $(v_{orifice})^2$. The nozzle pressures required for nozzle velocities below 10 m/s are quite low. The pump pressures must be increased depending on the flow disturbances and restrictions in the pipework.

8.3.2 The Transitional and Laminar Flow Capillary Nozzle

A capillary tube of large length (l) to diameter (d) ratio, as illustrated in Fig. 8.10, is designed to achieve laminar flow with low delivery velocities.

The design equations for flowrate and pumping pressure assumed for low and transitional values of Reynolds number are:

Eq. (8.5) $$Q_f = \frac{\pi \cdot d_{cap}^2}{4} \cdot v_{cap}$$

Eq. (8.6) $$p_p = \frac{32 \cdot \eta \cdot l_{cap} \cdot v_{cap}}{d_{cap}^2} + \frac{\rho \cdot v_{cap}^2}{2}$$

Example 3. A nozzle is to be designed for laminar flow. The velocity of the jet is to be 5 m/s, and the capillary is to be 10 mm diameter and 200 mm long. The process fluid is to be neat oil with a viscosity of 0.045 Ns/m^2. Calculations show that both terms in Eq. (8.6) are significant: Re = 967, Q_f = 0.39 l/s or 5.2 gal/min, p_p = 0.26 bar or 3.8 lbf/in^2, and H_p = 10 watts.

8.3.3 The Slot Nozzle

For wider grinding contacts, the supply of fluid must be dispersed across the width of the contact. This can be achieved using a slot nozzle. The design of a slot is illustrated in Fig. 8.11.

The slot nozzle's advantage is that laminar flow is more readily achieved with high flowrates due to the narrow slot thickness which can be employed. The slot nozzle is also easy to design and manufacture either for laminar or turbulent flow. A large chamber between the supply pipe and the slot is helpful to allow the flow to find a smooth path. A wedge-shaped chamber may be employed as illustrated in Fig. 8.6. The design equations for a slot nozzle are as follows:

Eq. (8.7) $$Q_f = v_{slot} \cdot h_{slot} \cdot w_{slot}$$

Eq. (8.8) $$p_p = \frac{12 \cdot \eta \cdot l_{slot} \cdot v_{slot}}{h_{slot}^2} + \frac{\rho \cdot v_{slot}^2}{2}$$

Eq. (8.9) $$Re = \frac{\rho \cdot v_{slot} \cdot h_{slot}}{\eta}$$

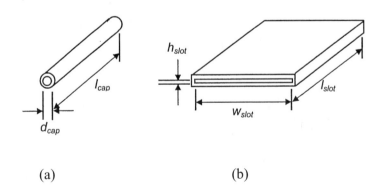

(a)   (b)

Figure 8.11. Capillary and slot nozzles designed to achieve laminar flow: *(a)* capillary nozzle; *(b)* slot nozzle.

The pumping power is calculated in the same way as for the orifice, Eq. (8.4). The length of the slot, l_{slot}, should be at least 20 times the slot thickness, h_{slot}, for laminar flow. The length of the slot should be as short as possible for turbulent flow. A smooth, concave converging section is appropriate for turbulent flow, as in the case of an orifice nozzle.

Example 4. A slot nozzle is designed to produce a jet velocity of 10 m/s across a wheel width of 80 mm. The slot thickness is to be 0.5 mm and the slot length is to be 10 mm. The viscosity of the process fluid is 0.003 Ns/m². The calculations, allowing for transitional flow, are Re = 167, Q_f = 0.4 l/s or 5.3 gal/min, p_p = 0.59 bar or 7.8 lbf/in², and H_p = 23.6 watts.

8.3.4 The Shoe Nozzle

The shoe nozzle is illustrated in Fig. 8.9. The sides of the nozzle fit around the sides of the wheel as a shoe fits around a foot. By making the shoe nozzle a reasonably close fit to the wheel, the air barrier is reduced in the region where the fluid is introduced.[13] The plates at the sides of the wheel disrupt the spiraling flow along the sides of the wheel that is a significant source of boundary layer air. The combination of a scraper inherent in the shoe design and the side plates allow fluid to be entrained with the wheel at much lower pressures than required with a jet nozzle. Also, the flowrate can be supplied much more economically and effectively through the grinding contact, if the flowrate is reduced to the minimum necessary to produce a liquid layer of useful flow approximately two to three millimeters thick.[11]

Allowance must be made for the extra energy to be supplied to the spindle motor which has to accelerate the fluid up to wheelspeed. Energy and momentum requirements are covered in the next section. Fluid penetration into the wheel using a shoe nozzle is discussed in Sec. 8.6 which deals with the particular problems of fluid delivery for creep grinding.

8.4 ENERGY AND MOMENTUM REQUIREMENTS OF THE PROCESS FLUID

The total power (H_t), required to supply the process fluid, is the sum of the pumping power (H_p) and the momentum power (H_m) exerted by the spindle to accelerate the fluid up to wheelspeed.

Eq. (8.10) $H_t = H_p + H_m$

The system pumping power requirement is given by:

Eq. (8.11) $H_p = p_s \cdot Q_f$

where p_s is the system-pumped pressure, assuming the flow in the supply lines takes place at low Reynolds numbers. At high Re, a term for the kinetic energy in the supply line, $\rho \cdot Q_f \cdot v_f^2/2$ should be added into Eq. (8.11).

The momentum power supplied by the wheel spindle can be estimated for a shoe nozzle or for a jet nozzle applied perpendicular to the wheel surface by assuming that the total flowrate supplied through the nozzle is accelerated up to the peripheral speed of the wheel. While undoubtedly, a significant proportion of the flowrate is not accelerated up to this speed, there will be additional losses and the estimate should be reasonable. Based on this assumption, the momentum power is the power required for the process fluid to achieve the kinetic energy corresponding to wheelspeed:

Eq. (8.12) $H_m = \dfrac{\rho \cdot Q_f \cdot v_s^2}{2}$

For a jet tilted at angle α normal to the wheel surface, the wheelspeed in Eq. (8.12) might be replaced by $v_s - v_{\text{orifice}} \cdot \sin\alpha$.

Example 5. A shoe nozzle is to be used to supply 0.4 l/s at a pressure of 0.5 bar of water-based fluid to a grinding wheel having a peripheral speed of 100 m/s. Assuming the system pumped pressure is 400% of the nozzle pressure, the power requirements associated with fluid delivery can be estimated:

$$H_p = 4 \times (0.5 \times 10^5) \times (0.4 \times 10^{-3}) = 80 \text{ watts}$$

$$H_m = 1/2 \times 1,000 \times (0.4 \times 10^{-3}) \times (100)^2 = 2,000 \text{ watts}$$

In this example, the momentum power to deliver the fluid which is supplied by the spindle motor may be very much greater than the pumping power requirement supplied by the coolant pump. However, as suggested by Klocke, et al.,[11] the flowrate to be accelerated can be reduced to the level that just fills the gap between the wheel and the workpiece. This aspect will be considered in the next section.

8.4.1 Velocity Requirement For A Jet Nozzle

Before proceeding to examine the useful flowrate, the velocity requirement for a jet nozzle must be considered. The following analysis is rather approximate for several reasons, but can be used to obtain an estimate of the right order which works quite well in practice. It is assumed that the jet is distributed across the width of the wheel, so that the result is independent of the wheel width. Based on practical research, the momentum of the jet stream must be greater than the momentum of the air barrier, if the jet is to overcome the air barrier and reach the surface of the wheel.

The rate of momentum per unit width of the wheel, of a jet of thickness, h_{jet}, is given by:

Eq. (8.13) $\qquad M_f' = \rho_f \cdot h_{jet} \cdot v_{jet}^2$

The rate of momentum per unit width of the wheel, of the air boundary layer of thickness, h_{air}, moving at the wheelspeed is given by:

Eq. (8.14) $\qquad M_{air}' = \rho_{air} \cdot h_{air} \cdot v_s^2$

Ideally, the width of the jet should be, at least, equal to the width of the wheel and the minimum velocity of the jet compared to the velocity of the wheel,

Eq. (8.15) $\qquad v_{jet} = v_s \cdot \sqrt{\dfrac{\rho_{air} \cdot h_{air}}{\rho_f \cdot h_{jet}}}$

Example 6. The minimum jet speed must be determined assuming a jet thickness of 1 mm and an air layer thickness of 5 mm, (the distance of the nozzle from the wheel). The density of air at 20°C and the atmospheric pressure is 1.2 kg/m³. The wheelspeed is 100 m/s.

$$v_{jet} = 100 \cdot \sqrt{\frac{1.2 \times 5}{1,000 \times 1}} = 25 \text{ m/s}$$

There has been insufficient research on the effect of nozzle angle to make a firm recommendation as to the effect on the velocity requirement. Although a jet aimed against the wheel velocity would appear to have the best chance of disrupting the air barrier, in practice the jet is usually aimed to merge smoothly with the wheel velocity. This strategy has the advantage that the jet can be located closer to the grinding contact. As mentioned previously however, if the jet is aimed tangentially in the direction of the wheel velocity, the jet is easily deflected outwards and away from the wheel surface.

8.5 USEFUL FLOWRATE THROUGH THE GRINDING CONTACT

The *useful flowrate* is defined as the flowrate that actually enters the grinding contact. Much of the flowrate supplied in grinding is usually diverted sideways and fails to enter the grinding contact. Engineer, et al.,[9] using a conventional flood nozzle, found useful flowrates which were 5–30% of the total flowrate depending on the nozzle position and wheel porosity. The excess flowrate is useful for bulk cooling, but it is a waste of energy to accelerate excessive quantities of fluid to wheelspeed if it is simply used for bulk cooling. The maximum percentage of the fluid aimed at the wheel is required to enter the grinding contact while fluid required for bulk cooling can be aimed at the workpiece.

A method is needed to determine the maximum useful flowrate, Q_{fu}, which can reasonably be achieved. This depends on the porosity of the wheel, ϕ_{pores}, and the mean depth of the pores, h_{pores}.

Eq. (8.16) $Q_{fu} = \cdot \phi_{pores} \cdot h_{pores} \cdot b_s \cdot v_s$

The mean thickness of the useful fluid layer, h_{fu}, can be estimated as

Eq. (8.17) $$h_{fu} = \frac{Q_{fu}}{b_s \cdot v_s} = \phi_{pores} \cdot h_{pores}$$

where b_s is the width of the grinding contact and v_s is the peripheral wheelspeed.

The pores on the surface of the wheel entrain and pump the process fluid through the grinding contact in a similar action to the flow through a rotary pump. When the pores near the surface of the wheel are full of fluid, it obviously becomes more difficult to increase the useful flowrate.

Engineer[9] found that the maximum useful flowrate was greater for wheels of large grain size having higher porosity near the surface of the wheel. The maximum useful flowrate was also increased by dressing a more open surface on the wheel for the same reason.

Example 7. If the mean depth of the pores on the surface of a grinding wheel is approximately 0.1 mm, and the porosity of the wheel is 30%, the maximum useful flowrate for a wheelspeed of 30 m/s with a wheel 19 mm wide can be estimated. From Eq. (8.16), the maximum useful flowrate is estimated as $Q_{fu} = 30\% \times (0.1 \times 10^{-3}) \times (19 \times 10^{-3}) \times 30 = 17.1 \times 10^{-6}$ m^3/s $\equiv 0.017$ l/s. The mean depth of the fluid film from Eq. (8.17) is 0.03 mm.

These order of magnitude calculations suggest that for this wheel running at this speed, the maximum useful flowrate will be approximately 0.017 l/s. If this flowrate can be made to enter the grinding contact, there will be little point in supplying extra fluid. However, depending on the type of nozzle and its location, it may be necessary to aim 2 to 20 times the maximum useful flowrate at the wheel.[9]

Engineer[9] using the conventional flood nozzle found that the *percentage utilization* of the flow, defined as the ratio of useful flow divided by total flow, tended to be a constant for a particular wheel and speed. For the flood nozzle as shown schematically in Fig. 8.12, two nozzle locations were tried. With $a_1 = 2$ mm and $a_2 = 0$ mm, the useful flowrate was more than doubled compared with $a_1 = 7$ mm and $a_2 = 5$ mm. The results shown in Fig. 8.12 are given for a more distant location. The proportionality between useful flowrate and total flowrate was demonstrated for useful flowrates up to 14 ml/s.

Klocke,[11] using a very low-pressure shoe nozzle, indicated a higher percentage utilization. It was suggested that the useful flowrate should increase with total flowrate until a maximum value of useful flowrate is achieved. For higher total flowrates, the useful flowrate is expected to remain approximately constant and the percentage utilization will be reduced.

Hydrodynamic pressures can help to pressurize fluid into the pores of the wheel and increase the useful flowrate. Of course, if the fluid can be pressurized into the sub-surface pores, it is possible to achieve greater useful flowrates and greater mean film thickness.[12] With wheels of a more open structure, the grains may more easily break out of the surface which further increases the effective porosity.

For hydrodynamic pressures to occur, it is essential that the converging gap is filled with fluid. The pressures can be calculated using standard techniques, and the pressures can be measured by placing a pressure transducer in a hole in the workpiece surface.[1][11] The magnitude of the maximum pressure provides a measure of whether the converging gap is full or not. With a smaller section of the converging gap completely filled, the maximum pressure will be smaller than when a large section is completely filled.

In an internal grinding operation with a more flexible spindle, hydrodynamic forces can be a source of problems in grinding. The hydrodynamic forces tend to separate the wheel and workpiece making it difficult to control the removal rate and the final size of the workpiece.

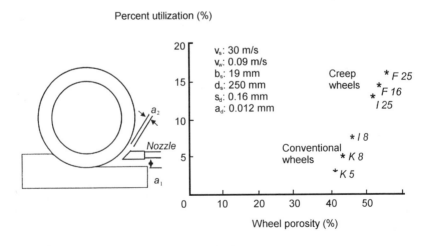

Figure 8.12. The useful flowrate as a percentage of total flowrate for a flood nozzle supplying fluid at up to approximately 85 ml/s.[9]

8.6 MECHANICS OF COOLING IN CREEP GRINDING

In shallow grinding, the fluid is often ineffective in providing significant convective cooling in the contact zone as demonstrated in Ch. 6, "Thermal Design of Processes." However, in low temperature creep grinding, fluid cooling in the contact zone is the predominant mechanism of heat transfer from the process.

A set-up used to detect the presence of a fluid film in the contact zone by temperature measurement is shown in Fig. 8.13. A brass workpiece was mounted in acrylic resin to conserve heat and heated on the undersurface. The grinding wheel was lightly touched on the workpiece along a blind groove ground by the same wheel. Initially the air barrier prevents the fluid from entering the contact zone. As the flowrate is increased, the fluid overcomes the air barrier and penetrates into the contact zone. The result is that the measured temperatures decrease. In this way, temperature measurement can be used to detect the commencement of useful flow through the grinding contact.

The results shown in Fig. 8.14 confirm that, as the wheelspeed is increased, it becomes more difficult for the fluid to enter the contact. A higher flowrate from the nozzle is required before the temperatures start to fall.

Figure 8.14 shows that, using the heater set-up, it is possible to determine the flowrate at which penetration is achieved. The momentum per unit width of the nozzle flow, M_f', was calculated from Eq. (8.14). Similarly, the momentum per unit wheel width of the air flow, M_{air}', was calculated from Eq. (8.15). The critical limit for the momentum of the jet flow was determined using the temperature measurement and also judged by an electrical resistance method. Making estimates of the air film thickness, the results confirmed the criterion for coolant penetration:

Eq. (8.18) $M_f' > M_{air}'$

An electrical resistance method was used to measure the effective fluid film thickness in the contact area. The calibration method is illustrated in Fig. 8.15. The fluid film forms an arm in an electrical resistance bridge circuit using an acrylic workpiece and an acrylic wheel. The resistance of fluid films of known thickness set up at low wheelspeeds are used to calibrate the output current. Thereafter, the measurement of the current in the bridge circuit is used to measure the effective film thickness with real grinding wheels and workpieces using measurement inserts in the workpieces.

The results demonstrate that the equivalent fluid film thickness is an important parameter in the determination of temperature in creep grinding as illustrated in Fig. 8.16.

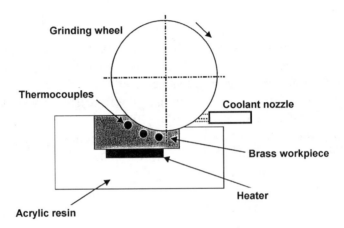

Figure 8.13. Detection of the penetration of coolant into the contact arc by temperature measurement.

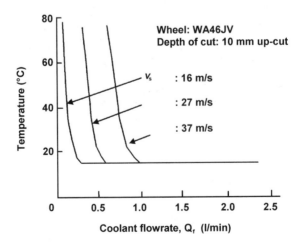

Figure 8.14. Detection of coolant penetration through temperature measurement. At higher wheel speeds, more flowrate is required from the nozzle to achieve penetration into the arc of grinding contact.

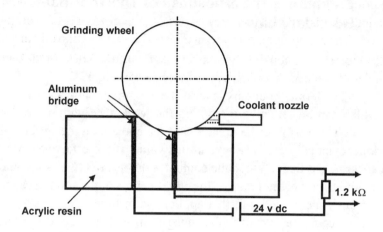

Figure 8.15. Measurement of equivalent coolant film thickness by the electrical resistance method.

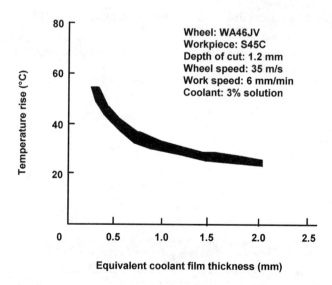

Figure 8.16. Relationship between the equivalent fluid thickness and the temperature rise in low-temperature creep grinding.

8.6.1 Estimating The Thickness of The Thermal Boundary Layer

By application of heat transfer theory, the thickness of the thermal boundary layer can be estimated, as illustrated in Fig. 8.17.

The thermal boundary layer is the maximum thickness of coolant, y_b, which can be effective in cooling the workpiece within the contact length, l_c. This assumes the thickness of the coolant on the surface of the workpiece actually extends beyond this value. If the effective fluid film thickness is less than y_b, then the coolant is not as effective as it could be. If the fluid film is greater than y_b, the effectiveness of the cooling cannot be significantly increased by further application of fluid.

Of course, a layer of coolant must be present throughout the long arc of contact which occurs in creep grinding.

Heat conduction in the fluid is based on Fourier's Law applied for the y direction.

Eq. (8.19) $$\alpha \frac{\partial^2 T}{\partial y^2} = \frac{\partial T}{\partial t}$$

where α is the thermal diffusivity of the coolant, and T is the temperature rise at position y, and time is t.

The solution of Fourier's equation for the given boundary conditions and sliding contact conditions is:

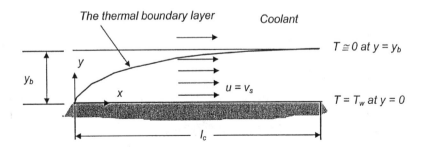

Figure 8.17. Simplified model of the thermal boundary layer of the coolant in the grinding zone.

Eq. (8.20)
$$\frac{T}{T_w} = 1 - \frac{\sqrt{\pi}}{2} \cdot erf\left(\frac{y}{2} \cdot \sqrt{\frac{v_s}{\alpha \cdot x}} \right)$$

where the error function

$$erf(a') = \frac{2}{\sqrt{\pi}} \cdot \int_0^{a'} e^{-a^2} da$$

An approximation of the error function can be found for small values of a' by series expansion and integration term by term. This series is rapidly convergent for values of $a' < 1$.

Eq. (8.21)
$$erf(a') \cong \frac{2}{\sqrt{\pi}}\left(a' - \frac{a'^3}{3} \right) \cong \frac{2}{3} \cdot a'$$

for $a' < 0.5$. Substituting the approximate solution in Eq. (8.21):

Eq. (8.22)
$$\frac{T}{T_w} = 1 - \frac{y}{2} \cdot \sqrt{\frac{v_s}{\alpha \cdot x}}$$

The thickness of the thermal boundary layer is where the temperature is reduced to zero, $T/T_w = 0$, so that:

Eq. (8.23)
$$y_b = 2 \cdot \sqrt{\frac{\alpha \cdot x}{v_s}}$$

The thickness of the thermal boundary layer increases with distance x and with thermal diffusivity. The thickness decreases with wheelspeed v_s.

Example 8. The minimum thickness of water-based fluid required to satisfy the thermal boundary layer for low temperature creep grinding with a contact length of 3 mm can be estimated. The wheelspeed is 100 m/s and the thermal diffusivity of the fluid is assumed to be $140 \times 10^{-6}\, m^2/s$ (see Table 8.1). The thickness of the thermal boundary layer is estimated to be:

$$y_b = 2 \times \sqrt{\frac{(140 \times 10^{-6}) \times (3 \times 10^{-3})}{100}} \cong 130\, \text{microns}$$

The diffusivity of water is many times greater than the diffusivity of steel, so that conduction into the coolant predominates over conduction into the workpiece, if the presence of the coolant can be ensured.

8.6.2 Application of Coolant in Creep Grinding

Important findings in the application of coolant in creep grinding have been achieved and described by Shafto,[14] Powell,[15] and Andrew, Howes, and Pearce[16] at the University of Bristol. These findings relate to the effect of coolant burn-out, the effect of the nip, the effect of slots in the wheel, and the extent of coolant penetration into the grinding wheel.

Coolant Burn-out. The coolant provides efficient cooling in creep grinding up to and including the temperature at which nucleate boiling takes place. Above this temperature, a vapor blanket greatly reduces heat transfer until radiation starts to become efficient, at temperatures approaching 1,000°C.[14]

Coolant burn-out is the term associated with the effect of temperatures rising above the boiling temperature for the process fluid. The consequence of coolant burn-out is likely to be a rapid rise in grinding temperatures and the incidence of thermal damage to the workpiece. Coolant burn-out is prevented by improving the efficiency of the abrasive process that reduces the grinding power, reducing the removal rate that reduces the grinding power, or improving the effectiveness of fluid delivery that overcomes an insufficiently useful flowrate. The critical values of heat flux (alternatively termed power flux) for the onset of burn typically varied between 5 and 25 W/mm^2 depending on the workspeed, depth of cut, and the angle of inclination which may be represented by the maximum normal infeed rate, (MNIR = $v_w \cdot \sin \theta$).

The Nip. As the grinding wheel commences engagement with the workpiece in up-cut grinding or disengages in down-cut grinding, there is a region in which there is no nip to constrain the fluid in proximity to the wheel. This situation is illustrated in Fig. 8.18. The nip is necessary to provide the converging wedge, thus generating the hydrodynamic pressures which help to induce fluid into the grinding contact.

Powell showed that heat flux before the burn occurred could be considerably increased by using a nip. The increase varied between 12% when using jet speeds in excess of 20 m/s compared to 50% at a jet speed of 9 m/s. The critical heat flux was highest at jet speeds exceeding 20 m/s and, for the particular example, critical values greater than 30 W/mm^2 were achieved.

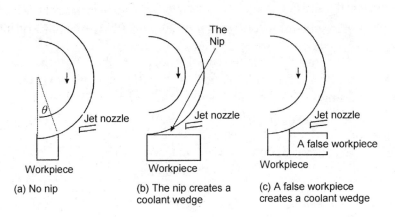

(a) No nip   (b) The nip creates a coolant wedge   (c) A false workpiece creates a coolant wedge

Figure 8.18. The converging wedge, known as a nip, can create hydrodynamic pressures to constrain the process fluid into the contact. In the absence of a nip, burn often occurs as the wheel enters or leaves contact with the workpiece. This is a consequence of the loss of fluid supply at the ends of the workpiece. A false workpiece can be used at either end of the workpiece as appropriate for up-cut or down-cut grinding to overcome the problem.

Coolant Penetration into the Wheel Using a Shoe Nozzle. Once fluid has been introduced into the contact, the assumption is that the fluid is constrained by the grinding wheel and remains in contact with the workpiece throughout the contact, unless burn-out conditions are achieved. On this basis, ensuring the required effective thickness of fluid film applied at entry to the contact would be sufficient. So far, the effective film thickness has been considered on the basis of the surface porosity of the wheel using Eqs. (8.17) and (8.18). The thickness of the useful fluid layer, h_{fu}, needs to be greater than the thickness of the thermal layer, y_b. In order to ensure this condition, pressure to force the fluid deeper into the pores of the wheel may be necessary. This can be achieved using a pressurized shoe nozzle.[15] The situation is illustrated in Fig. 8.19.

Within the arc subtended by the shoe, the coolant penetration depth h increases with angle θ. The maximum depth of penetration occurs at the end of the arc subtended by the shoe and is given by:[16]

Eq. (8.24) $$h_p = \left(\frac{9 \cdot p \cdot \theta^2}{4 \cdot K \cdot \phi_f^2 \cdot \omega^2} \right)^{\frac{1}{3}}$$

where p is the shoe pressure, θ is the angle subtended by the shoe, K is the permeability of the wheel, ϕ_f is the flow porosity of the wheel, and ω is the angular velocity of the wheel.

The permeability of the wheel is not generally known and needs to be measured for a particular wheel. Andrew[16] gave two examples for creep grinding wheels having a permeability of approximately 1.5×10^7 Ns^2/m^4 and flow porosity of 0.28 and 0.31. Measurements of fluid penetration were made by dye staining, and Eq. (8.24) gave a reliable prediction.

An analysis of outflow was also made.[15] The outflow gave rise to a reducing value of the fluid layer, h, with increasing angle, θ. The theory predicts a greater outflow by a factor of approximately two times than was actually measured. This suggests Eq. (8.25) would be safe for estimating the minimum fluid layer carried into the contact.

Eq. (8.25)
$$\frac{dh}{d\theta} = -\left(\frac{\rho \cdot d_s}{2 \cdot K \cdot \phi_f^2} \right)^{1/2}$$

where ρ is the fluid density and d_s is the wheel diameter.

Both shoe and jet nozzles provided satisfactory fluid delivery, however, the pressurized shoe nozzle has the advantage over a jet nozzle in that it can provide satisfactory delivery of fluid even in the absence of a nip. This appears to be because the pressurized fluid, forced deeper into the pores when using a shoe, does not rely on the hydrodynamic action of the nip for transport into the grinding contact.

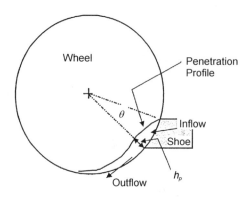

Figure 8.19. A pressurized shoe nozzle forces the process fluid to depth, h, into the pores of the wheel. After leaving the shoe, the fluid is carried around the wheel, and the outflow is thrown off tangentially.

Bulk Temperature of the Coolant. Coolant bulk temperature is clearly important, particularly when using a water-based process fluid. The boiling temperature of a typical water-based emulsion is between 120° and 140°C. If the bulk temperature is increased from 20° to 40°C, the maximum temperature rise is reduced by approximately 17–20%. This means the maximum removal rate must be correspondingly reduced to avoid the incidence of burnout.

8.7 SUMMARY OF CONCLUSIONS

The air barrier makes fluid delivery more difficult at high wheelspeeds.

By attention to nozzle design and location, hydrodynamic effects can be used to assist fluid delivery.

The fluid layer transported into the grinding contact needs to be at least as great as the thickness of the thermal boundary layer.

To increase the thickness of the useful fluid layer, it may be necessary to force the fluid into the subsurface pores of the wheel.

The process fluid has several functionalities. To achieve each of these requirements economically may require multiple nozzles, each designed for a different purpose.

Fluid delivery systems should be designed for optimal fluid delivery while minimizing the total power requirements.

REFERENCES

1. Ebbrell, S., Woolley, N. H., Tridimas, Y. D., Allanson, D. R., and Rowe, W. B., The Effects of Cutting Fluid Application Methods on the Grinding Process, *Int. J. Machine Tools and Manufacture*, 40:209–223 (2000)

2. Trmal, G., and Kaliszer, H., Delivery of Cutting Fluid in Grinding, *Proc. 16th Int. Machine Tool Design and Research Conf.*, UMIST, p. 25 (1975)

3. Pahlitzsch, G., Features and Effects of A Novel Cooling Method, *Microtecnic*, 8:4 (1953)

4. Tsunama, Y., Aoyama, T., Inasaki, I., and Yonetsu, S., Improvement of Grinding Fluid Application Creep Feed Grinding, State of Grinding Fluid in the Arc of Contact, *Trans. Jpn. Soc. Precision Engineers*, 48(11):112 (1982)

5. Tsunama, Y., Aoyama, T., Inasaki, I., and Yonetsu, S., Improvement of Grinding Fluid Application Creep Feed Grinding, Relation Between State of Grinding Fluid Application and Workpiece Temperature, *Trans. Jpn. Soc. Precision Engineers*, 50(7):7 (1982)

6. Brinksmeier, E., Heinzel, C., and Wittmann, M., Friction, Cooling, and Lubrication in Grinding, *Annals of the CIRP*, Keynote paper, 48(2):581–598 (1999)

7. Campbell, J. D., Optimized Coolant Application, *1ˢᵗ Int. Machining and Grinding Conf., MI, USA,* Soc. Mfg. Engineers (Sep. 12–14, 1955)

8. Webster, J. A., Cui, C., and Mindek, J., Grinding Fluid Application System Design, *Annals of the CIRP*, 4(1):333–338 (1995)

9. Engineer, F., Guo, C., and Malkin, S., Experimental Measurement of Fluid Flow Through the Grinding Zone, *Trans. ASME*, 114:61–66 (Feb. 1992)

10. Graham, W., and Whisto, M. G., Some Observations on Through the Wheel Coolant Application in Grinding, *Int. J. Machine Tool Design and Research*, 18:9 (1978)

11. Klocke, F., Baus, A., and Beck, T., Coolant Induced Forces in CBN High Speed Grinding with Shoe Nozzles, *Annals of the CIRP*, 49(1):241–244 (2000)

12. Guo, C., and Malkin, S., Analysis of Fluid Flow Through the Grinding Zone, *J. Engineeering for Ind.*, 114:427–434 (1992)

13. Fisher, R. C., Grinding Dry With Water, *Grinding and Finishing*, pp. 32–34 (Mar. 1965)

14. Shafto, G. R., *Creep-feed Grinding*, Ph.D. Thesis, University of Bristol (1975)

15. Powell, J. W., *The Application of Grinding Fluid in Creep-feed Grinding*, Ph.D. Thesis, University of Bristol (1979)

16. Andrew, C., Howes, T. D., and Pearce, T. R. A., *Creep-feed Grinding*, Holt, Rinehart, and Winston (1985)

9

Electrolytic In-process Dressing (ELID) Grinding and Polishing

9.1 INTRODUCTION

This chapter introduces and reviews abrasive processes assisted by electrolytic in-process dressing (ELID) technique. This *in situ* dressing method is used for metal-bond wheels and is relatively new. As illustrated by the examples given below, the introduction of this technique has been highly successful when fine grain wheels were efficiently used to obtain very low surface roughness, and when hard ceramics had to be machined using a very small grain cutting depth in order to avoid failure by cracking.

The basic system, principles, and characteristics of ELID abrasion mechanisms are introduced first. The success and wide application of ELID principles to ceramic grinding are explained. Fourteen applications of the ELID principle to modern abrasive processes are documented to illustrate the scope of application.

9.2 BASIC SYSTEM

Electrolytic in-process dressing (ELID) grinding was first proposed by the Japanese researcher Hitoshi Ohmori back in 1990.[1] The most important feature is that no special machine is required. Power sources from conventional electro-discharge or electro-chemical machines can be used for ELID, as well as ordinary grinding machines.

The basic arrangement of the ELID system for surface grinding is shown in Fig. 9.1. The essential elements of the ELID system are a metal-bonded grinding wheel, a power source, and a high pH electrolytic coolant. The metal-bonded wheel is connected to the positive terminal of a power supply with a smooth brush contact, while the fixed electrode is connected to the negative pole. The electrode is made out of copper and must cover at least one sixth of the wheel's active surface and a width that is two millimeters wider than the wheel rim thickness. The gap between the wheel and the active surface of the electrode is 0.1–0.3 millimeters and can be adjusted by mechanical means.

The grinding wheel is dressed as a consequence of the electrolysis phenomenon that occurs between the wheel and the electrode, when direct current (dc) is passed through a suitable grinding fluid that acts as an electrolyte.

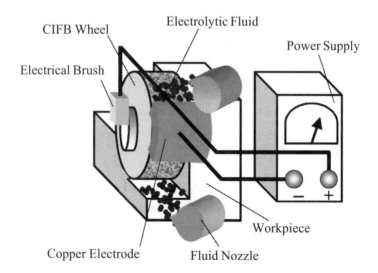

Figure 9.1. Basic arrangement for ELID grinding.

9.3 BASIC PRINCIPLES

Electrolytic in-process dressing (ELID) grinding is a process that employs a metal-bonded superabrasive wheel together with an in-process dressing by means of an electrolytic action. The basic principle of the electrolysis is illustrated in Fig. 9.2. The electrolytic process continuously exposes new sharp abrasive grains by dissolving the metallic bond around the superabrasive grains, in order to maintain a high material removal rate and to obtain a constant surface roughness.

A key issue in ELID is, according to Chen and Li,[2] to balance the rate of removal of the metal bond by electrolysis and the rate of wear of the superabrasive particles. While the superabrasive wear rate is directly related to grinding force, grinding conditions, and workpiece mechanical properties, the removal rate of the metal bond depends on ELID parameters such as voltage, current, and the gap between the electrodes.

The rate of dissolution of the bond metal is highest at the metal-diamond interface particles. In other words: the tendency of electrolytic dissolution is to expose the diamond particles.[2] Consequently, the metal dissolution rate increases with concentration of the diamond particles.[2]

Figure 9.3 shows the basic mechanism of the ELID process, while Fig. 9.4 presents the progression of the dressing during the grinding process, which materializes in three stages: truing, dressing, and grinding/dressing.

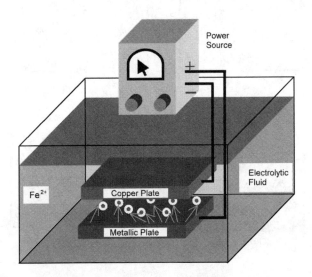

Figure 9.2. The principle of electrolysis.

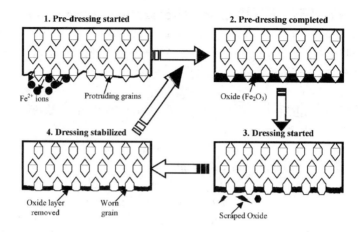

Figure 9.3. Mechanism of ELID.[27]

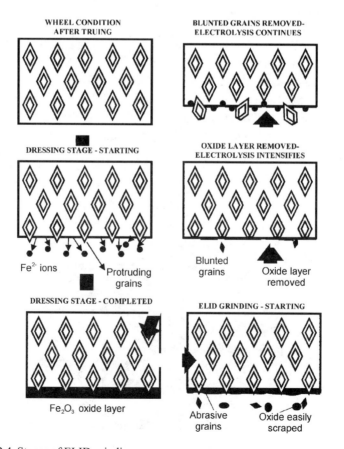

Figure 9.4. Stages of ELID grinding.

For a fixed gap and applied voltage, the current density does not change much with concentration of diamond particles.[2] Hence, in order to maintain a constant rate of metal removal, the applied electric field should be lower for a higher diamond concentration and, vice versa, the applied electric field should be higher for a lower concentration of particles. This electric field concentration effect is greatly reduced when the diamond particle is half-exposed.[3] The field sharply decreases from its highest value near the diamond-metal boundary, to a small value at a distance of the order of the diamond particle size.[3]

In a conventional grinding operation, the tool face is smooth and has no protrusion of diamond particles after truing.[3] Mechanical dressing opens up the tool face by abrasion with a dressing stone, which exposes the grits on the leading side while they remain supported on the trailing side. Laser and electro-discharge dressing opens up the tool face by thermal damage, producing craters, microcracks, and grooves. This degrades the diamonds because diamond undergoes a graphitization alteration at approximately 700°C. During electro-chemical dressing operations grits are exposed by dissolving the surrounding metal bonds.[3]

The stages of ELID grinding are depicted in Fig. 9.4 and presented below:

(i) *Truing* is carried out to reduce the initial eccentricity below the average grain size of the wheel and to improve wheel straightness, especially when a new wheel is first used or re-installed. Precision truing of a micrograin wheel is carried out to achieve a runout of less than 2–4 microns. In Fig. 9.5, details of the truing method are presented. A special electrical discharge (ED) truing wheel, made out of high temperature bronze-tungsten carbide alloy and insulated from its central shaft, is mounted on a three-jaw chuck, and connected to a negative pole. Both wheels rotate at rather low speeds and the ED-truing wheel reciprocates with the machine tool's saddle. Little or no coolant is supplied to prevent electrolysis and to obtain high truing precision. After truing, a pre-dressing operation is required prior to grinding.

(ii) *Pre-dressing* of the wheel by electrolytic means aims to increase protrusion of the abrasive grains. The procedure is performed at low speed and takes about 10–30 minutes.

(iii) Grinding is performed simultaneously with continuous in-process dressing by electrolytic means.

Due to the changes in the wheel surface condition, the electrolysis parameters change in the last two stages as shown in Fig. 9.6 and explained in Sec. 9.4.

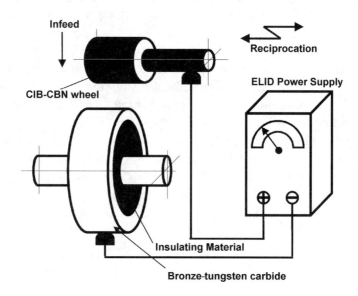

Figure 9.5. ELID truing mechanism.[26]

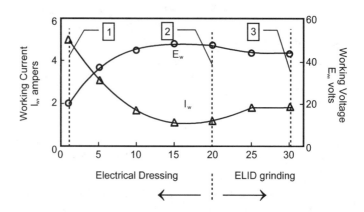

Figure 9.6. Current characteristics during ELID grinding.[6]

9.4 ELECTRICAL ASPECTS OF ELID GRINDING

The current, I_w, and voltage, E_w, vary during a complete ELID procedure. When the pre-dressing stage starts, the active surface of the wheel has a high electrical conductivity. Consequently, the current is high, while the voltage between the wheel and the electrode is low, as indicated by vertical Line 1 in Fig. 9.6. After several minutes, the cast iron bond material is removed by electrolysis, being transformed into Fe^{2+} ions. According to the chemical transformations shown in Fig. 9.2, the ionized Fe forms hydroxides of $Fe(OH)_2$ or $Fe(OH)_3$. The hydroxides further change into oxides Fe_2O_3 through electrolysis. The electro-conductivity of the wheel surface is reduced by the oxide that acts as an insulating layer (about 20 µm thick). The current decreases while the voltage increases as shown by vertical Line 2 in Fig. 9.6.

The grinding process can now commence with the protruding abrasive grains. As the grains are worn, the insulating oxide layer is also worn. This increases the electro-conductivity of the wheel so the electrolysis will intensify generating a fresh insulating layer, as depicted by vertical Line 3 in Fig. 9.6. The protrusion of the grains remain approximately constant. The layer of oxide has a larger flexibility and a lower retention characteristic than the bulk bond material.[4]

Figure 9.7 depicts the characteristics of the oxide film thickness required for different types of grinding operations: roughing or finishing. For rough grinding, a thin insulating layer is required so the abrasive grains can significantly protrude out of the bond and help increase the material removal rate, while for a mirror-finish ELID grinding, a relatively thick insulating layer is preferred because it will limit the real depth of cut of the abrasive grains. The thickness of the oxide layer can be controlled by modifying the characteristics of the electrical current output by the ELID power source. This creates the possibility of running both rough and finish grinding operations using the same setup and adjusting only the current characteristics of ELID and relative speed between the wheel and the work.

An important aspect during ELID is a small increase of the wheel diameter (or thickness) which occurs during ELID grinding due to the formation of the etched and oxide layers.[4] In Fig. 9.8, typical increases in wheel diameter due to insulating layer formation are shown for different types of electrolytes.

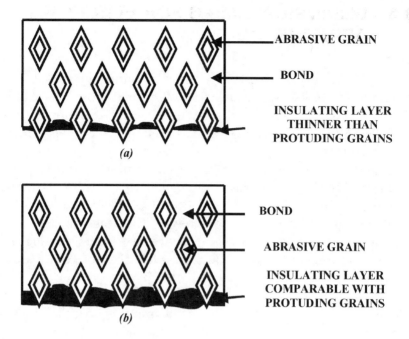

Figure 9.7. Ideal wheel conditions for *(a)* efficient grinding, and *(b)* mirror-surface finish.[20]

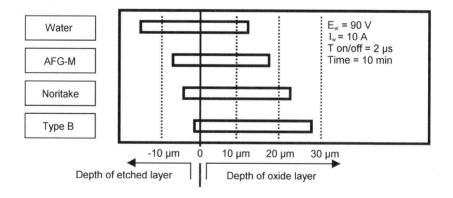

Figure 9.8. The depth of etched and oxide layers with different electrolytes.[4]

9.5 GRINDING WHEELS FOR ELID APPLICATIONS

The grinding wheels utilized for ELID-assisted abrasive processes have two main characteristics. They contain fine and very fine superabrasive grains, and they are made with electrically conductive bond. Usually, the combination of ultrafine superabrasive grains and metal bond makes the wheels prone to rapid dulling and very difficult to dress, unless special techniques, as is ELID, are employed to maintain their cutting ability throughout the process. Therefore, the grinding wheels for ELID applications include cast iron bonded and cast iron fiber-bonded wheels, with diamond or cubic boron nitride (CBN) superabrasives.

Cast Iron Bonded Diamond. These wheels are manufactured by mixing diamond abrasive, cast iron powders or fibers, and a small amount of carbonyl iron powder. The compound is shaped to the desired form under a pressure of 6–8 ton/cm^2, and then sintered in an atmosphere of ammonia. These wheels are unsuitable for continuous grinding for long periods of time, particularly for metals. This is because of the following:

(i)   A tough metal-bonded wheel is difficult to dress, so efficient and stable grinding cannot be achieved.

(ii)   High material removal rate wears the abrasive and requires frequent redressing.

(iii) The wheel becomes embedded with swarf during grinding of steels and other metals.

Cast Iron Fiber-bonded Diamond. These wheels provide high grinding ratio and high material removal rates.

Cubic Boron Nitride (CBN). Tough metal-bonded CBN wheels can be dressed during the grinding process using the ELID technique. This process can be used to control abrasive protrusion before and during the grinding of ceramics.

9.6 ELID GRINDING OF CERAMICS

In recent years, a number of publications confirm the merits of ELID grinding for common brittle materials, but also for BK-7 glass, silicon, and fused silica using fine-mesh superabrasive wheels.[5] Many of these publications report that the ELID system provides the ability to obtain spectacularly fine finishes on brittle material surfaces, down to the nanometer scale of 4 to 6 nanometers, after grinding operations. For some applications, this completely eliminates the need for loose abrasive lapping and/or polishing operations. ELID grinding has also been applied to the fabrication of large optical components, 150 to 250 millimeters in diameter. The data also suggest that ELID grinding can be successfully applied to very thin deposited substrates.

The US structural ceramics market was estimated at over $3,500M as compared with $20M in 1974, $350M in 1990 and $865M in 1995. Applications of ceramics are found in tool manufacture, automotive, aerospace, electrical and electronics industries, communications, fiber optics, and medicine (http://www.acers.org/news/factsheets.asp).

The properties of ceramic materials, as for all materials, depend on the types of atoms, the types of bonding between the atoms, and the way the atoms are packed together, also known as the atomic scale structure. Most ceramics are compounds of two or more elements. The atoms in ceramic materials are held together by a chemical bond. The two most common chemical bonds for ceramic materials are covalent and ionic, which are much stronger than metallic chemical bonds. That is why metals are ductile and ceramics are brittle.

The atomic structure primarily affects the chemical, physical, thermal, electrical, magnetic, and optical properties. The microstructure also affects the properties but has its major effect on mechanical properties and on the rate of chemical reaction. For ceramics, the microstructure can be entirely glassy (glasses), entirely crystalline, or a combination of crystalline and glassy. In the last case, the glassy phase usually surrounds small crystals, bonding them together.

The most important characteristics of ceramic materials are high hardness, resistance to high compressive force, resistance to high temperature, brittleness, chemical inertness, electrical insulation properties, superior electrical properties, high magnetic permeability, special optic and conductive properties, etc.

The interest in advanced structural ceramics has increased significantly in recent years due to their unique physical characteristics and due to significant improvements in their mechanical properties and reliability. Despite these advantages, the use of structural ceramics in various applications has not increased as rapidly as one might have expected, partly due to high machining cost. The cost of grinding ceramics may account for up to 75% of the component cost compared to 5–15% for many metallic components.[6]

The primary cost drivers in the grinding of ceramics are:

- Low efficiency machining operations due to the low removal rate.

- Highly expensive superabrasive wheel wear rate.

- Long wheel dressing times.[6]

A conventional grinding process applied to ceramic materials often results in surface fracture damage, nullifying the benefits of advanced ceramic processing methods.[7] These defects are sensitive to grinding parameters and can significantly reduce the strength and reliability of the finished components. It is, therefore, very important to reduce the depth of grain penetration to very small values so that the grain force is below the critical level for structural damage. The critical value of grain penetration depth for a hard ceramic is typically less than 0.2 microns. This small value of grain penetration depth is made possible using the ELID grinding technique with very fine grain wheels.

Although ELID grinding is good for workpiece accuracy, it is not necessarily beneficial to workpiece strength as the discussion of the effects of removal rates demonstrates.[8] Stock removal rate increases with the increasing number of passes, higher stock removal rates are obtained with a stiffer machine tool in the first few passes. For grinding wheels of a similar bond type, a larger stock removal rate is obtained for the wheels of larger grit sizes. Cast iron bonded wheels used during the ELID grinding allows a larger stock removal rate, yet a lower grinding force, than a vitrified bond grinding wheel used in a conventional grinding process. Machine stiffness has little effect on residual strength of ground silicon under multipass grinding conditions; this can be attributed to the effect of the actual wheel depth of cut on workpiece strength.[9] As the number of passes increases, the actual depth of cut approaches the set depth of cut, which means that regardless of machine tool stiffness, grinding force does not necessarily alter workpiece strength in a stable grinding process. Also, more compressive residual stress can be induced with a dull grinding wheel, with a grinding wheel of a larger grit size, or with a wheel that has a stiff and strong bond material. However, a larger grinding wheel grit size causes a greater depth of damage in the surface of the ground workpiece. As the number of passes increase, the normal grinding force also increases. This increase of force is steep initially and slows as the number of passes increases, a phenomenon more evident for a high stiffness machine tool. Due to machine tool deflection, the normal grinding force is initially smaller with lower machine stiffness.[9] Eventually, the normal force approaches a limit value, regardless of the machine stiffness characteristics.[9] To avoid damage to the workpiece, it is necessary to limit the grain penetration depth, which is more directly dependent on the removal rate than on the grinding force.

A little studied, yet controversial, aspect of the ceramic grinding process is the pulverization phenomenon that takes place in the surface layer of a ceramic workpiece during grinding.[10] Surface pulverization makes ceramic grains in the surface much smaller than those in the bulk, and gives the ground surface a smoother appearance.

9.7 MATERIAL REMOVAL MECHANISMS IN GRINDING OF CERAMICS AND GLASSES

In general, there are two approaches to investigating abrasive-workpiece interactions in grinding of ceramics:[7]

- *The Indentation-fracture mechanics approach*, which models abrasive-workpiece interactions by the idealized flaw system and deformation produced by an indenter.

- *The machining approach* involving measurement of forces coupled with scanning electron microscope (SEM) and atomic force microscope (AFM) observation of surface topography and of grinding debris.

The stock removal during grinding of ceramics is a combination of microbrittle fracture and quasiplastic cutting. The quasiplastic cutting mechanism, typically referred to as *ductile mode* grinding, depicted in Fig. 9.9a, results in grooves on the surface that are relatively smooth in appearance. By carefully choosing the values for the grinding parameters and by controlling the process, ceramics can be ground predominantly in this so-called ductile mode. On the other hand, *brittle mode* grinding, shown in Fig. 9.9b, results in surface fracture and surface fragmentation. Ductile mode grinding is preferred since negligible grinding flaws are introduced and structural strength is maintained.

As shown in Fig. 9.9 there is a plastically deformed zone directly under the grit. In brittle mode grinding, two principal crack systems are generated; median (radial) cracks and lateral cracks. Brittle mode removal of material is due to the formation and propagation of the lateral cracks.

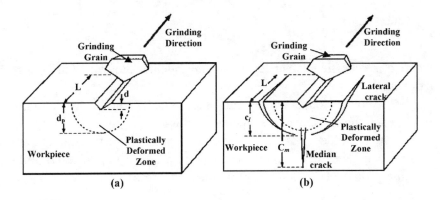

Figure 9.9. An abrasive grain depicts removing material from a brittle workpiece: *(a)* in ductile mode grinding, and *(b)* in brittle mode grinding.[7]

The specific depth at which a brittle-ductile transition occurs is a function of the intrinsic material properties of plasticity and fracture. The critical depth, according to Bandyopadhyay,[7] is:

Eq. (9.1) $$\frac{Plastic\ Flow\ Energy}{Fracture\ Energy} \approx \frac{E_p}{E_f} \approx d$$

where d is the critical depth of cut.

Although it is not always easy to observe microcracks produced by grinding, the depth of a median crack can be estimated using a formula from Inasaki:[11]

Eq. (9.2) $$l_{mc} = \left\{0.034(\cotan\psi)^{2/3}\left[(E/H)^{1/2}/Kc\right]\right\}^{2/3} F^{2/3}$$

where ψ is the indenter angle, F is the indentation load, E is the modulus of elasticity, and Kc is the fracture toughness of the material. Therefore, the depth of the median crack depends on the material properties, force, and grinding grit shape.

Indentation load, F, is estimated by dividing the grinding force by the number of active cutting edges in the contact area between the grinding wheel and the workpiece. This relationship applies above a threshold force F^*. The critical force, F^*, that will initiate a crack can be estimated by:

Eq. (9.3) $F^* = \dfrac{\alpha \cdot K_c^4}{H^3}$

where α is a coefficient that depends on the indenter geometry and H is the hardness.

In conclusion, crack size can be estimated, theoretically, from Eq. (9.2), when the force exceeds a certain critical value, determined by Eq. (9.3). Typical values of critical force to propagate sub-surface damage are presented in Table 9.1.[7] In order to cut in the presence of plastic deformation, the grain load should be, for SiC and Si_3N_4, less than 0.2 N and 0.7 N, respectively, for a typical grain shape.

Scanning electron microscope (SEM) and atomic force microscope (AFM) techniques can be utilized to evaluate surface and sub-surface fracture damage. Typical micrographs are shown in Figs. 9.10 and 9.11. From SEM and AFM micrographs, one can assess the difference between brittle mode and ductile mode material removal.

Table 9.1. Critical Force Required To Propagate Sub-surface Damage[7]

MATERIALS	H (GPa)	E (GPa)	K_c (kN/m$^{3/2}$)	F* (N)
SiC	24.5	392	3,400	0.2
Si_3N_4	14	294	3,100	0.73

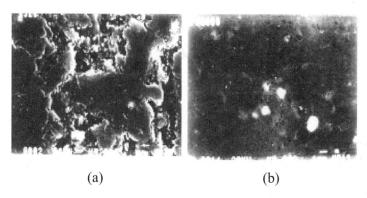

(a)   (b)

Figure 9.10. Scanning electron microscope (SEM) micrographs: *(a)* # 325, and *(b)* # 8000.[7]

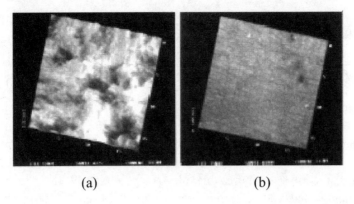

(a) (b)

Figure 9.11. Atomic force microscope (AFM) micrographs: *(a)* # 325, and *(b)* # 8000.[7]

9.8 COMPARISON BETWEEN ELID AND OTHER GRINDING TECHNIQUES

Significant reduction in grinding force has been reported with the application of ELID to workpieces ground both in the longitudinal and transverse directions.

ELID Grinding. Protruding grains abrade the workpiece. As a result, the grains and the oxide layer wear down. The wear of the oxide layer increases the electrical conductivity of the wheel. As a result, the electrical current in the circuit increases, leading to an intensification of the electrolysis. Consequently, protrusion of the abrasive grains increases and, during a short period, the thickness of the oxide layer recovers. The described electrical behavior is nonlinear due to the formation of this insulating oxide layer. The oxide layer has a beneficial lubricating role in the grinding process. The process of wear and recovery of the oxide layer follows in a rather stable manner during the entire ELID grinding operation.

Other In-process Dressing Technologies. Nakagawa and Suzuki[12] investigated various techniques of in-process dressing. The effects of in-process dressing using a dressing stick were studied. The wheel is dressed at the beginning of each stroke. Higher material removal rates were reported. An application of this procedure to side-grinding is difficult. However, use of a dressing stick accelerates wear of the superabrasive grains.

The technique of electro-chemical dressing was introduced by McGeough in 1974.[13] An electro-conductive metal-bond wheel forms the anode and a fixed graphite stick forms the cathode. The dressing process takes place by electrolysis. Welch, et al.,[14] employed sodium chloride electrolyte. However, sodium chloride is corrosive and harmful to the machine tool.

Another dressing technique is based on the principle of electrical discharge machining (EDM). The conductive grinding wheel is energized with a pulsed current. The flow of ions creates hydrogen bubbles in the coolant, creating an increasing electric potential. When the potential becomes critical, a spark is generated that melts and erodes the material that clogs the wheel. This procedure does not continuously provide protruding abrasive grains, and is considered unsuitable for ultra-fine grinding of materials, especially with micrograin size grinding wheels.

Other nonconventional machining processes based on electro-chemical metal removal are electro-chemical machining, electro-chemical grinding, electro-chemical polishing, etc.

9.9 APPLICATIONS OF ELID GRINDING

ELID grinding has been investigated for various materials including ceramics, hard steels, ceramic glass, ceramic coatings, etc., having a variety of shapes (plane, cylindrical external and internal, spherical and aspherical lens, etc.), and of greatly varying dimensions. Lately, new applications of the ELID principle were tested for biomedical materials.[15][16] Recent investigations reported include the following:

- ELID side-grinding.[4][8][9]
- ELID double-side grinding.[17]
- ELID lap-grinding.[18][19]
- ELID grinding of ceramics on a vertical rotary surface grinder.[6]
- ELID grinding of ceramics on a vertical grinding center.[20]
- ELID grinding of bearing steels.[21]
- ELID grinding of ceramic coatings.[22]

- ELID ultraprecision grinding of aspheric mirror.[23]
- ELID grinding of microspherical lens.[24]
- ELID grinding of large optical glass substrates.[5]
- ELID precision internal grinding.[25][26]
- ELID grinding of hard steels.[27]
- ELID mirror-like grinding of carbon fiber reinforced plastics.[28]
- ELID grinding of chemical vapor deposited silicon carbide.[4]

9.9.1 ELID Face-grinding

An ELID face-grinding[4][8][9] set-up is illustrated in Fig. 9.12. ELID grinding was found to offer advantages compared with conventional grinding operations applied to the same workpieces under similar conditions. The oxide layer obtained during ELID grinding modifies the nature of the contact between the wheel and workpiece; it is equivalent to changing the wheel bond. The oxide layer makes the wheel surface more flexible compared with the hard and highly-retentive cast iron bond in the conventional process.

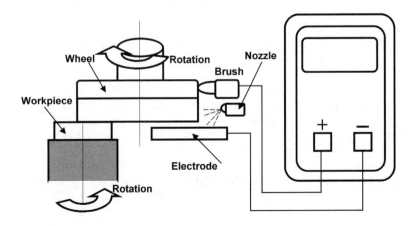

Figure 9.12. Principle of ELID face-grinding.[4][9][22]

The conclusions of the study were as follows:

- ELID grinding can be employed to produce mirror finishes.

- Surface roughness was slightly better.

- Fewer pits and sticking-out projections were produced on workpiece surfaces since ductile mode removal was found to be dominant.

- ELID grinding was particularly recommended for precision grinding of hard-brittle materials on conventional machine tools having low rigidity.

9.9.2 ELID Duplex (Double-sided) Grinding

An ELID duplex grinding[17] set-up is illustrated in Fig. 9.13. The conclusions for this ELID grinding operation compared with conventional grinding are presented below:

- ELID double-sided grinding (duplex grinding) can be employed to produce mirror-like surface finishes.

- Surface finish of workpieces ground with ELID was better than the roughness of the workpieces ground conventionally.

- The material removal mechanism for ELID duplex grinding was brittle mode for coarser wheels and ductile mode for finer wheels.

- The material removal rates for ELID duplex grinding was slightly higher than conventional grinding.

- ELID grinding is highly recommended in precision grinding of hard-brittle materials on conventional machine tools.

In Fig. 9.14, the evolution in time of the stock removal rate for ELID grinding is presented. In Fig. 9.15, a comparison between the topography of the ground surface with and without ELID is depicted.

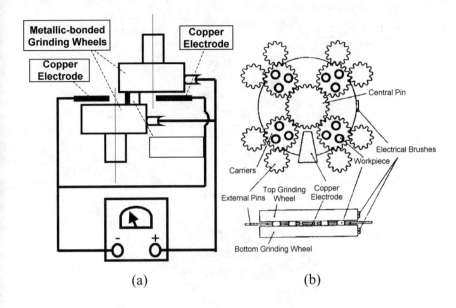

(a)   (b)

Figure 9.13. Set-up for ELID duplex (double-sided) grinding: *(a)* eccentric wheels[17] and *(b)* concentric wheels.

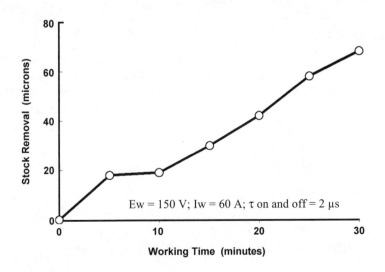

Figure 9.14. Stock removal vs time for ELID grinding.[17]

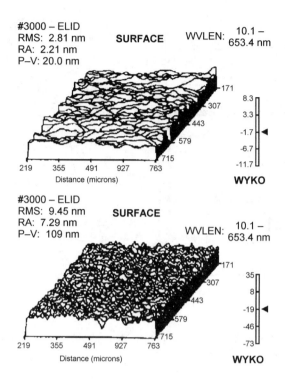

Figure 9.15. Surface topography of ground workpiece: *(a)* with ELID and *(b)* without ELID.[31]

9.9.3 ELID Lap-grinding

ELID lap-grinding[18][19] employs constant pressure and uses metal bonded abrasive wheels with ELID. The process was found to produce better results from the surface roughness and flatness viewpoints than conventional lap-grinding. ELID lap-grinding was found to be a perfect candidate for the following:

- Mirror finishes that cannot be obtained with constant feed grinding.

- Grinding with metal-bond wheels having grain-mesh size finer than JIS #10000.

- Employing a simple set-up mounted on an existing lapping machine.

The principle is illustrated in Fig. 9.16.

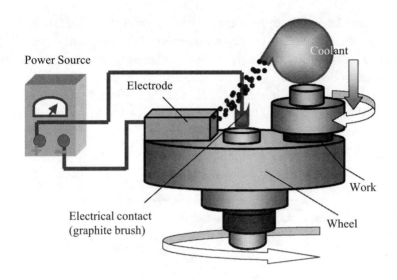

Figure 9.16. Principle of ELID lap-grinding.[19]

Workpieces (silicon or tungsten carbide in this specific investigation) are pressed against the lapping wheel. The lapping wheel is connected to the positive pole through a fine brush that smoothly contacts the wheel surface, while the electrode is connected to the negative pole. The gap between the two electrodes is 0.3 millimeters. Conclusions from the investigations were as follows:

- ELID lap-grinding of silicon with a JIS #4000 wheel was found to yield a mirror finish with a roughness of 3.8 nanometers, compared with the 7.4 nanometers roughness obtained after conventional lap-grinding.

- ELID lap-grinding of tungsten with a JIS #4000 wheel allowed a more stable removal rate than conventional lap-grinding over an 80 minute period.

- ELID lap-grinding of silicon was achieved in brittle mode for JIS #1200 and lower, and in ductile mode for JIS #4000 and finer mesh size.

- When silicon was ground together with tungsten carbide, surface roughness was lower compared with the surface finish obtained when silicon is ground alone; the brittle-to-ductile transition was observed for a JIS #4000 mesh size wheel.

In a later study by Itoh, et al.,[19] silicon and glass (BK7) were ELID lapped with metal-resin bond diamond wheels having grit sizes of JIS #8000, JIS #120000 and JIS #3000000. Material removal rate results are presented in Fig. 9.17.

ELID lap grinding using the JIS #3,000,000 metal-resin bond wheel produced very good quality ground surfaces of PV 2.8 nanometers for silicon and PV 2.5 nanometers for glass.

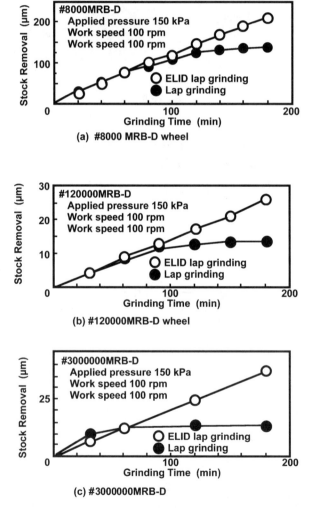

Figure 9.17. Grinding efficiency vs time for wheels of *(a)* #8000, *(b)* #120000, and *(c)* #3000000 mesh size.[19]

9.9.4 ELID Grinding of Ceramics on a Vertical Rotary Surface Grinder

Experiments were conducted on a vertical rotary surface grinder[6] having a 5.5 kW motor spindle. The workpieces were made from sintered reaction bonded silicon nitride (SRBSN) and cast-and-sintered silicon nitride (Si_3N_4). The grain size ranged between 0.3–0.4 μm (SRBSN) and 0.6–0.8 μm (Si_3N_4). Cast iron fiber-bonded diamond wheels of 200 millimeters in diameter were used. A Noritake AFG-M grinding fluid at a rate of 20–30 liters/minute was utilized as the electrolytic fluid.

The grain sizes for different JIS mesh sizes of diamond grinding wheels are presented in Table 9.2.

Table 9.2. Grain Sizes of Diamond Grinding Wheels[6][21]

MESH SIZE #	GRAIN SIZE (μm)	AVERAGE GRAIN SIZE (μm)
170	88–84	110
325	40–90	63.0
600	20–30	25.5
1200	8–16	11.6
2000	5–10	6.88
4000	2–6	4.06
6000	1.5–4	3.15
8000	0.5–3	1.76

A direct-current pulse generator was utilized as a power supply. The square-wave voltage amplitude was 60 volts with a peak current of 10 amps. The pulse width was adjusted to 5 microseconds on- and off-time. The conclusions of the investigation were as follows:

- Mirror finishes were obtained after grinding with a #4000 mesh size wheel, as shown in Figs. 9.18 and 9.19.

- Cutting speeds had no significant affect on the surface roughness of the workpiece, as shown in Fig. 9.20.

- Feed rate had no significant effect on the surface roughness, especially for mesh size wheels finer than JIS #325, as shown in Fig. 9.21.

- Lower surface roughness characterized SRBSN compared with Si_3N_4, especially with rougher wheels.

- Similar surface roughness characterized SRBSN and Si_3N_4 when finer wheels were used.

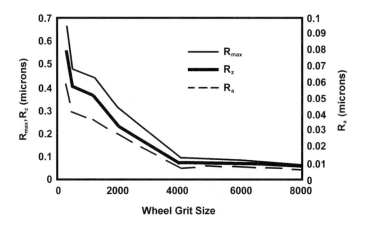

Figure 9.18. Effect of grit size on surface-roughness values.[6]

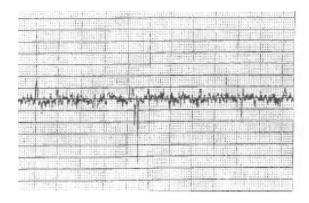

Figure 9.19. Typical surface-roughness traces for #4000 mesh size wheel.[6]

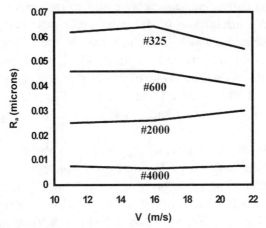

Figure 9.20. Effect of cutting speed on surface roughness.[6]

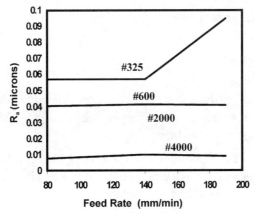

Figure 9.21. Effect of feed rate on surface roughness.[6]

9.9.5 ELID Grinding of Ceramics on a Vertical Grinding Center

The experiments were carried out on a vertical machining center.[20] The silicon nitride workpieces were clamped firmly in a vice and fixed onto the base of a strain-gauge dynamometer. The dynamometer was clamped onto the machining center table and a reciprocating grinding operation was performed.

A direct-current pulse generator was utilized as a power supply. The square-wave voltage was 60–90 volts with a peak current of 16–24 amps. The pulse width was adjusted to 4 ms on- and off-time. Different values between 0.01 and 0.05 mm for the depth of cut, and between 2 and 5 mm for the width of cut were explored. Material removal rates of 250 mm³/min and up to 8,000 mm³/min were obtained. Results obtained after conventional grinding were compared with the results output by ELID grinding.

A modified ELID dressing procedure was also investigated. The modified ELID dressing was performed in two stages: (a) at 90 volts for 30 minutes; the insulating oxide layer was mechanically removed by an aluminum oxide stick of #400 grit size at 300 rpm; (b) another dressing stage at 90 volts for 30 mins. The conclusions of the investigation were as follows (Fig. 9.22):

- ELID grinding favors high rates of material removal.

- ELID grinding is recommended for low-rigidity machine tools and low-rigidity workpieces.

- Grinding force increased continuously during conventional grinding.

- Grinding force was less in ELID grinding than in conventional grinding; this effect became more visible after 18 minutes of grinding.

- Voltage increased with duration of ELID grinding, while the grinding force was reduced; this effect became apparent after 18 minutes of grinding.

- The full potential of ELID grinding was achieved after 24 minutes of grinding when the grinding force stabilized at a low value.

- ELID and conventional grinding produced almost the same surface roughness after a rough grinding operation.

- The grinding force was constant and low after the application onto the wheel of the modified ELID dressing procedure.

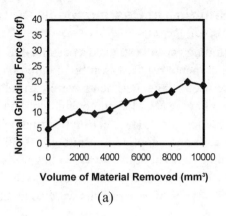

(a)

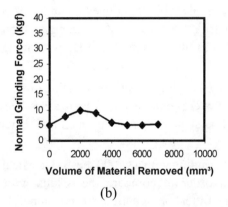

(b)

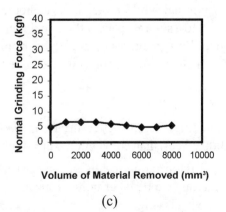

(c)

Figure 9.22. Relationship between the volume of material removed and the grinding force for *(a)* conventional grinding, *(b)* ELID grinding after ELID dressing, and *(c)* ELID grinding after modified ELID dressing.[20]

9.9.6 ELID Grinding of Bearing Steels

ELID grinding was investigated as a super-finishing technique for steel bearing components.[21] Cast-iron bond cubic boron nitride wheels (CIB-CBN) were used: the experiments were carried out on a common cylindrical grinder having a 3.7 kW motor spindle. The negative electrode was made from stainless steel. The electrolytic grinding fluid was Noritake AFG-M diluted 1:50 and supplied at a rate of 20–30 l/min. A dc pulse generator was employed. The square-wave voltage was 60–150 volts with a peak current of 100 amps. Pulse width was adjusted to 4 microseconds on- and off-time. Three types of experiments were conducted:

- Traverse and plunge-mode ELID grinding experiments were completed in order to evaluate the effects of grinding wheel mesh size and grinding method on surface roughness and waviness (as shown in Figs. 9.23 and 9.24).

- Traverse grinding experiments, using different mesh size wheels, were conducted in order to assess the influence of wheel mesh size on surface finish and material removal rate.

- ELID grinding with a #4000 mesh size wheel was carried out in order to compare the results with the results obtained after honing and electro-polishing. During electro-polishing, the workpiece is connected to the positive (or anodic) terminal, while the negative (cathodic) terminal is connected to a suitable conductor, and direct current (dc) is applied. Both positive and negative terminals are submerged in the solution, forming a complete electrical circuit.

The conclusions were as follows:

- ELID grinding offered a better surface finish (R_a=0.02 μm for #4000 wheel) than conventional grinding.

- Plunge ELID grinding results in coarser surface roughness than traverse ELID grinding, especially for rougher abrasive wheels (Fig. 9.25).

- Waviness of the ground surface is reduced for higher wheel mesh size (Fig. 9.26).

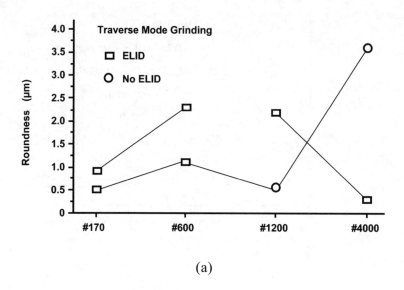

(a)

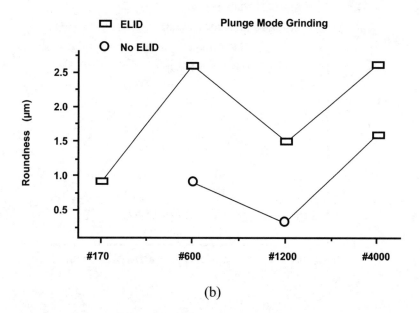

(b)

Figure 9.23. Comparison of roughness: *(a)* traverse, and *(b)* plunge.[21]

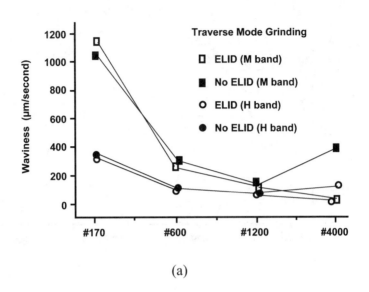

(a)

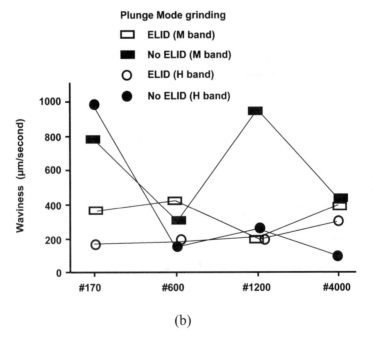

(b)

Figure 9.24. Comparison of waviness: *(a)* traverse, and *(b)* plunge.[21]

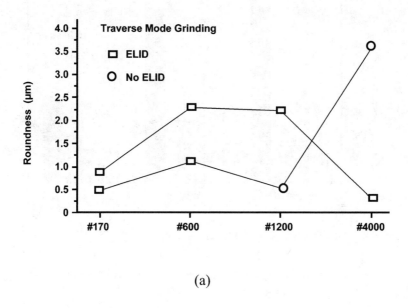

(a)

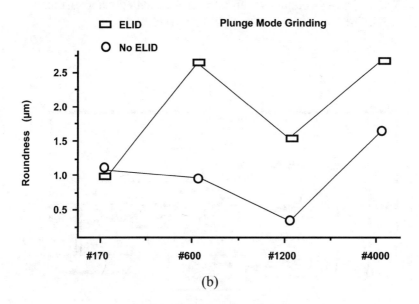

(b)

Figure 9.25. Comparison of roundness: *(a)* traverse, and *(b)* plunge grinding.[21]

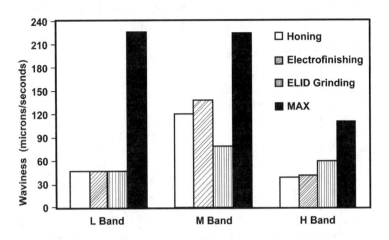

Figure 9.26. Surface waviness with different processes.[21]

- The ELID process was more stable for traverse grinding than for plunge grinding (Fig. 9.27).

- Roundness of the ground surface increased with wheel mesh number utilized.

- Out-of-roundness can be improved with greater stiffness of the machine tool.

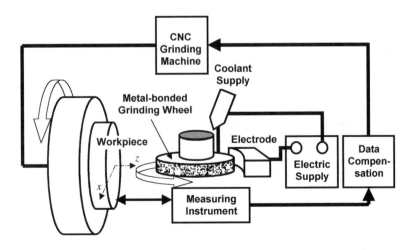

Figure 9.27. Aspherical form control system.[23]

- ELID traverse grinding tends to offer a more promising potential than plunge mode grinding.

- Three to four spark-out passes improved waviness and roundness of the ground surface.

- Effects of grinding parameters on surface roughness for ELID grinding and conventional grinding are comparable.

- Increased depth of cut and increased traverse rate produced a poorer surface roughness.

- ELID grinding gave better results than both honing and electric polishing.

- ELID ground surfaces have greater high band waviness than honed surfaces, due to the smaller workpiece-tool contact area.

- ELID grinding induced a lower compressive surface stress of about 150–400 MPa, than the 600–800 MPa compressive stress resulting from honing.

- The 10 μm depth of the compressive layer produced by ELID operation was half of the 15–20 μm depth produced by honing operation.

- The cycle time for ELID grinding was twice as long as the cycle time for either honing or the electric polishing operation. However, lower roughness and higher removal rates were achieved by ELID grinding with a coarser wheel and a faster traverse speed. It was concluded that ELID grinding was more cost effective for small batch production situations.

9.9.7 ELID Grinding of Ceramic Coatings

Ceramic coatings[22] include a large group of subspecies, such as chemical vapor deposited silicon carbide, plasma spray deposited aluminum oxide, and plasma spray deposited chromium oxide. In order to achieve the required quality, efficient machining of these ceramic coatings is still under development. Zhang, et al.,[22] made a comparative study of diamond grinding of ceramic coatings on a vertical grinder. Two types of dressing procedures were applied to a cast-iron bond diamond wheel having #4000 mesh size, alumina rotary dressing, and ELID dressing and grinding. Conclusions of the study were as follows:

- There is a critical current value for each electrolytic dressing system; when the current is smaller, the thickness of the insulating oxide layer increases with increase of current; otherwise, it decreases.

- Thickness and depth of the oxide layer depend largely on the coolant type.

- A small increase in the wheel diameter and/or thickness occurred after electrolytic dressing. Conversely, during rotary and other mechanical methods of dressing, there was a reduction of the wheel diameter.

- Workpiece roughness decreased more rapidly with rotary dressing than in ELID grinding in the first three minutes. After three minutes, roughness decreased constantly in ELID grinding but showed a wavelike model for rotary dressing.

- Wear of abrasive grains produced an unstable grinding performance for the rotary dressing technique, while grinding performance remained constant during ELID grinding.

- Surface roughness depends on material properties in both methods.

- All ceramic coatings, except sintered SiC, had a lower roughness after ELID than after rotary dressing.

- Plasma spray deposited chromium oxide is difficult to grind to an extremely fine roughness.

- For all dressing methods, the micrographs showed that the material removal mechanism involved both brittle fracture and ductile mechanisms. For ELID, the ductile mode was predominant, except for sintered SiC and for the rotary dressing, when the brittle fracture mode was predominant.

- Ductile mode grinding can be implemented even on a common grinder by controlling the wheel topography.

- For ELID grinding the interaction between the abrasive grain and the workpiece surface is modeled by a spring-damper system (because of the existence of the oxide layer), while for rotary dressing the contact is rigid and stiff. The oxide layer absorbs vibrations and reduces the actual exposed cutting edge of the abrasive grain.

9.9.8 ELID Ultra-Precision Grinding of Aspheric Mirrors

The quality of soft x-ray silicon carbide mirrors[23] influences the performance of modern optical systems (see Fig. 9.27). To accomplish a high precision of these aspheric mirrors, an ELID grinding system was employed.

A #1000 cast iron grinding wheel was mechanically trued and pre-dressed using electrical methods. The workpiece surfaces were concave spherical with a curvature of two meters. After grinding, the form was measured and the data was compared with the planned data by the mean least squares method. A form error was calculated and compensation data were generated. Accordingly, a new form was ground. This procedure was applied five times in order to exponentially decrease the errors from 2.6 μm to 0.38 μm.

9.9.9 ELID Grinding of Microspherical Lens

ELID grinding was investigated for production of microspherical lenses[24] in a ductile mode. The implementation of ductile mode cutting requires expensive items such as ultra-precision vibration-free rigid machine-tools, high resolution feed-motion control, submicron grit wheels and a clean work environment. The conclusions reached were:

- ELID grinding is stable, efficient, and economical.

- Some problems in achieving ductile or semiductile mode grinding of micro-optical components occurred with low grinding speed, instability of ultrafine abrasive and small-size wheels, difficulty in achieving precise and efficient truing and dressing of the wheels, and difficulty in obtaining precise and effective fixturing.

- Coarse grit size wheels (#325) do not show any difference in the final roughness when ELID is applied.

- Finer #4000 wheels, however, output lower surface roughness when ELID was employed.

- ELID high-precision grinding of microspherical lens with cup wheels (ELID-CG) achieved high spherical accuracy and low roughness around R_a 20 nanometers.

- ELID-CG can be successfully utilized to fabricate a microspherical lens with a more stable process, higher efficiency and better surface quality than conventional grinding.

9.9.10 ELID Grinding of Large Optical Glass Substrates

In this research, ELID grinding was employed to grind optical components[5] of 150–250 mm diameter. ELID grinding using fine mesh superabrasive wheels produced spectacularly low roughness of 4–6 nm R_a on brittle surfaces including BK-7 glass, silicon, and fused silica. For some applications ELID grinding eliminated polishing or lapping operations.

9.9.11 ELID Precision Internal Grinding

Little has been reported on mirror-finish internal grinding due to the limitation in abrasive grit size applicable to nonmetallic bond grinding wheels. A novel method to carry out ELID grinding of internal cylindrical surfaces on an ordinary grinding tool, named interval ELID, which is presented in Fig. 9.28, was developed.[25][26] The wheel was dressed at intervals (before each stroke) and the abrasive grains remained protruded. After a pre-dressing operation, the insulating oxide layer was 30 mm thick, and increased the external diameter of the grinding wheel.

The characteristics of the electrical current utilized for interval ELID grinding are shown in Fig. 9.29.

In internal grinding, abrasive wear occurs rapidly due to the smaller diameter of the wheel. The problems caused by this were overcome by an interval ELID grinding technique. Two new techniques for internal grinding, namely ELID II and ELID III, are also described in the literature, Fig. 9.30. For ELID II, a fixed cylindrical dressing electrode dresses the cast iron fiber-bonded cubic boron nitride (CIB-CBN) grinding wheel before each stroke. For ELID III, the metal resin bonded grinding wheel is connected to the positive terminal of the power supply, while the workpiece itself is connected to the negative pole. It was concluded that:

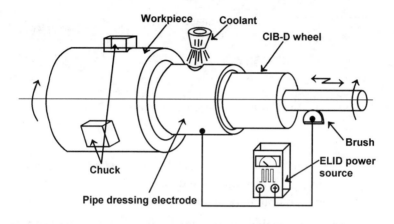

Figure 9.28. Schematic of interval ELID grinding.[25]

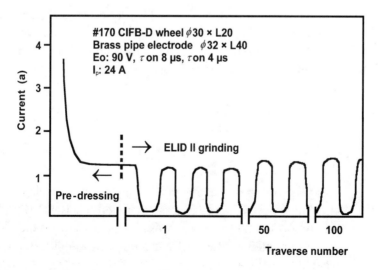

Figure 9.29. Current fluctuation in interval ELID grinding.[25]

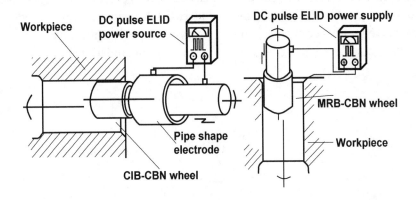

Figure 9.30. ELID II and ELID III internal grinding processes.[26]

- Due to the limitation on wheel diameter, wheel speed can be adjusted only within a small range. The effect of wheel speed on output parameters was not significant.

- A higher wheel speed, within a limited range, resulted in a finer surface roughness.

- The obtained values of grinding parameters after interval ELID grinding are similar to those obtained after conventional internal grinding.

- The surface quality of the ground workpieces was better for increased mesh size value of the wheel. Fine abrasive wheels can, therefore, be used to grind smooth surfaces without risking the loss of the stability of the surface roughness.

- Internal mirror-finish was possible for pieces made of bearing steel and alumina.

- Pipe-shaped dressing electrodes are superior to other shape electrodes.

- It is possible to achieve a mirror-finish with a coarse grinding wheel. The roughness obtained after ELID III grinding with #2000 grit size was almost the same as the roughness obtained after ELID grinding with a #400 CBN wheel.

- A combination of ELID II and ELID III can be employed for finishing operations, especially when small diameters are applicable.

- Rough and finish grinding can be performed on the same machine tool using ELID II and ELID III procedures.

9.9.12 ELID Grinding of Hard Steels

Hardened bearing steels such as M50 were ground to produce an optical quality surface, finer than 10 nm R_a, using a 76 μm CBN grain size and 500 μm depth of cut, as shown in Figs. 9.31 and 9.32. The final surface roughness was reduced by the burnishing action of the worn CBN grits. ELID grinding was employed to reduce surface roughness by maintaining the protrusion and sharpness of the CBN grits and to avoid the pull-out of carbides in the secondary finishing zone phenomenon.

Another grinding technique employed to minimize microcracking, surface burn, and phase transformation is low-stress grinding (LSG). However, LSG places special demands on machine tool stiffness, low and controllable vibration levels, low wheel speed, and frequent wheel dressing. Low-stress grinding is characterized by low removal rates, low grinding ratios, and significantly increased production costs. Also, it was found that some localized surface damage and surface roughness in the range of 100–200 nanometers R_a were obtained.

Onchi, et al.,[29] reported a roughness of 30 nm R_a achieved after grinding of SAE 52100 with a porous CBN wheel, yet with relatively low removal rates and very fine CBN grits.

Puthanangady, et al.,[30] after superfinishing hardened steel pieces with #500 grit size fused-alumina stones reported a surface finish parameter R_a of up to 60 nm.

Stephenson, et al.,[27] employed CBN grinding wheels having 30 μm grain size for roughing, 2 μm grain size for intermediate finishing and 0.7 μm grain size for the final mirror-finish.

A 100 mm diameter D151 electroplated diamond wheel at 3,000 rev/min was employed with a traverse rate of 5 mm/rev and an in-feed of 1–4 μm per pass. The electrical power supply parameters were: 60 volts, 10 amps peak current, 6 μs on-time, 2 μs off-time with a square pulse wave. Important findings are described below:

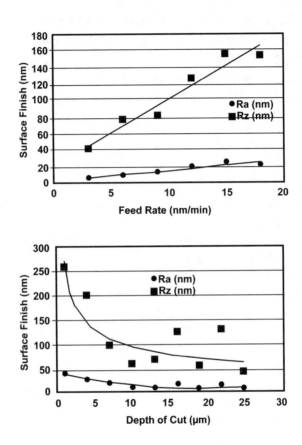

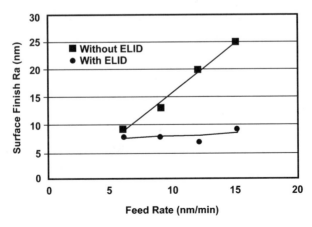

Figure 9.31. Surface finish vs feed rate and depth of cut.[27]

Figure 9.32. Surface finish vs feed rate for ELID and non-ELID grinding.[27]

- A repeatable surface roughness of less than 10 nm R_a was obtained with 75 µm CBN grit and 500 µm depth of cut.
- The lowest surface roughness of 2.3 nm R_a was obtained with a 200 µm depth of cut.
- Chip thickness was estimated at 1–10 nm.
- Carbide pull-out of the CBN grits could be avoided by employing ELID.
- Optical quality surfaces were considered to have been obtained by a combination of processes in the primary and secondary finishing zones of the cup-wheel, with the final surface finish enhanced by the burnishing action of worn CBN grits.

9.9.13 ELID Mirror Grinding of Carbon Fiber Reinforced Plastics

Carbon fiber reinforced plastics (CFRP) are used in the aerospace industry, and for machine tool spindles, power-transmission shafts, and robotic arms.[28] It was found that:

- Surface roughness of the CFRP improved substantially for grinding with diamond wheels of increasing mesh number up to #4000.
- For diamond wheels having mesh number greater than #4000, roughness did not noticeably improve.
- The upper limit of surface roughness for a #6000 wheel was R_{max} 0.65 µm.
- Surface roughness obtained for CFRP was finer than for brittle materials ground under similar ELID conditions. This was explained by the elastic deformation of this material.
- Mirror-finish is strongly dependent on the grinding direction. Grinding at 90° with respect to fiber direction favors the obtaining of best results.
- The spark-out effect on roughness was significant for rapid-feed grinding but small for creep-feed grinding.

- Mirror-finish is accompanied by a homogenization mechanism resulting from grinding heat and chip smearing.

9.9.14 ELID Grinding of Chemical Vapor Deposited Silicon Carbide

Chemical vapor deposited silicon carbide (CVD-SiC) is the second most ideal material for deflection mirrors used in short wavelength laser systems, being surpassed only by the crystalline diamond. Conventional polishing techniques were found to be incapable of finishing CVD-SiC mirrors. The study concluded that:

- ELID grinding can achieve extremely smooth surfaces.
- Surface roughness reduced with grain size.
- Fewer pits and whiskers sticking out were produced in ELID grinding during brittle fracture when the abrasive grains crush and plough the surface of the workpiece, compared with other dressing techniques. The number of pits and whiskers reduced with decreasing grain size.
- The ratio of the ductile versus brittle fracture mechanisms was higher for ELID grinding than for conventional dressing. The ratio increased with decreasing grain size.
- Ductile mode removal was realized with ELID even on a conventional less-stiff and less-precise grinder, by optimum control of depth and composition of the insulating oxide layer.

9.10 CONCLUSIONS

During the last decade, a number of publications have demonstrated the merits of ELID for abrasive grinding of brittle materials such as common advanced ceramic materials, BK-7 glass and fused silica, ceramic coatings, and hard steels, as well. Materials have been ground in various shapes and sizes: plane, cylindrical external and internal, and spherical and aspherical lens. For some applications ELID grinding eliminated polishing and/or lapping operations.

ELID grinding provides the ability to produce extremely fine finishes on brittle material surfaces, with surface roughness on the nanometer scale (4 to 6 nm). Yet, coarse grit size wheels (JIS #325 and coarser) show only slight, if any, difference in the final roughness when ELID is applied compared with conventional grinding. However, for finer wheels (JIS #4000 and finer) ELID gives finer surface roughness compared with conventional grinding.

ELID grinding is more stable than conventional grinding allowing high removal rates over a longer period. Material removal rates between 250 mm^3/min and 8000 mm^3/min were reported, with specific stock removal rates between 25 and 800 $mm^3 \cdot mm^{-1}$/min.

Cast-iron bond wheels utilization resulted in both a larger stock removal rate and a lower grinding force than vitrified bond grinding wheels.

ELID grinding can be completed in brittle-fracture mode for coarse wheels and in a ductile mode for finer wheels (JIS #4000 to #20,000, FEPA #1200 and finer). By carefully selecting the grinding parameters and controlling the process, ceramic materials can be ground predominantly in a ductile mode resulting in relatively smooth grooves on the surface.

Ductile mode grinding can be implemented even on a conventional grinder by controlling the wheel topography. ELID grinding can be implemented even on low-rigidity machine-tools and for low-rigidity workpieces.

Cutting speed and feed-rate were proven to have little effect on workpiece surface final roughness.

Grinding forces are lower during ELID grinding than during conventional grinding. Grinding force was found to be relatively constant and reduced in value after the first ELID stage was completed.

Fine ELID grinding induces compressive surface stress of about 150–400 MPa. The depth of the compressive layer produced during ELID grinding may be half that of the one produced during honing. For example, 10 μm depth was produced after ELID fine grinding compared with 15–20 μm produced after honing.

Hardened bearing steels can be ultra-precisely ground to produce an optical quality surface characterized by 10 nm R_a or less. The final surface roughness is enhanced by the burnishing action of the worn grits.

REFERENCES

1. Ohmori, H., and Nakagawa, T., Mirror Surface Grinding of Silicon Wafers With Electrolytic In-process Dressing, *Annals of the CIRP*, 39(1):329–332 (1990)

2. Chen, H., and Li, J. C. M., Anodic Metal Matrix Removal Rate In Electrolytic In-process Dressing I: Two-dimensional Modeling, *J. Appl. Phys.*, 87(6):3151–3158 (2000)

3. Chen, H., and Li, J. C. M., Anodic Metal Matrix Removal Rate in Electrolytic In-process Dressing II: Protrusion Effect and Three-dimensional Modeling, *J. Appl. Phys.*, 87(6):3159–3164 (2000)

4. Zhang, C., Ohmori, H., Kato, T., and Morita, N., Evaluation of Surface Characteristics of Ground CVD-SiC Using Cast Iron Bond Diamond Wheels, *J. Int. Soc. Prec. Eng. and Nanotech.*, 25:56–62 (2001)

5. Grobsky, K., and Johnson, D., ELID Grinding of Large Optical (Glass Substrates), Report of Zygo Corp. (1998)

6. Ohmori, H., Takahashi, I., and Bandyopadhyay, B. P., Ultra-Precision Grinding of Structural Ceramics by Electrolytic In-process Dressing (ELID) Grinding, *J. Mater. Process Tech.*, 57:272–277 (1996)

7. Bandyopadhyay, B. P., and Ohmori, H., The Effect of ELID Grinding on the Flexural Strength of Silicon Nitride, *Int. J. Mach. Tool Mfg.*, 39:839–853 (1999)

8. Zhang, B., Yang, F., Wang, J., Zhu, Z., and Monahan, R., Stock Removal Rate and Workpiece Strength in Multi-Pass Grinding of Ceramics, *J. Mat. Proc. Tech.*, 4:178–184 (2000)

9. Zhang, C., Kato, T., Li, W., and Ohmori, H., A Comparative Study: Surface Characteristics of CVD-SiC Ground with Cast Iron Bond Diamond Wheel, *Int. J. Mach. Tool Mfg.*, 40:527–537 (2000)

10. Zhang, B., and Howes, T. D., Material-removal Mechanisms in Grinding Ceramics, *Annals of the CIRP*, 43(1):305–308 (1994)

11. Inasaki, I., Speed-stroke Grinding of Advance Ceramics, *Annals of the CIRP*, 37(1):299 (1988)

12. Nakagawa, T., and Suzuki, K., Highly Efficient Grinding of Ceramics and Hard Metals on Grinding Center, *Annals of the CIRP*, 35(1):205–210 (1986)

13. McGeough, J. A., *Principles of Electro-chemical Machining*, Chapman and Hall Publ. House, London (1974)

14. Welch, E., and Yi, Y., Electro-chemical Dressing of Bronze Bonded Diamond Grinding Wheels, *Proc. Int. Conf. Mach. Adv. Mat.*, Gaithersburg, MD, NIST, 874:333–340 (1993)

15. Komotori, J., Mizutani, M., Katahira, K., and Ohmori, H., ELID Surface Treatment for Metallic Biomaterials by ELID Grinding, *Int. Workshop on Extreme Optics and Sensors,* Ikebkuro, Jpn (Jan. 14–17, 2003)

16. Mizutani, M., Komotori, J., Nagata, J., Katahira, K., and Ohmori, H., Surface Finishing for Biomaterials—Application of the ELID Grinding Method, *3rd Int. Conf. Adv. Mat. Development and Performance*, Daegu, Korea (Oct. 16–19, 2002)

17. Itoh, N., Ohmori, H., Kasai, T., Karaki-Doy, T., and Yamamoto, Y., Development of Double-sided Lapping Machine, Hicarion, and Its Grinding Performances, pp. 130–133, ELID Team Project.

18. Itoh, N., and Ohmori, H., Grinding Characteristics of Hard and Brittle Materials by Fine Grain Lapping Wheels with ELID, *J. Mat. Proc. Tech.*, 62:315–320 (1996)

19. Itoh, N., Ohmori, H., Moryiasu, S., Kasai, T., Karaky-Doy, T., and Bandyopadhyay, B. P., Finishing Characteristics of Brittle Materials by ELID-Lap Grinding Using Metal-resin Bonded Wheels, *Int. J. Mach. Tool Mfg.*, 38(7):747–762 (1998)

20. Bandyopadhyay, B.P., Ohmori, H., and Takahashi, I., Efficient and Stable Grinding of Ceramics by Electrolytic In-process Dressing (ELID), *J. Mat. Proc. Tech.*, 66:18–24 (1997)

21. Qian, J., Li, W., and Ohmori, H., Cylindrical Grinding of Bearing Steel with Electrolytic In-process Dressing, *Precision Eng.*, 24:153–159 (2000)

22. Zhang, C., Ohmori, H., Marinescu, I, and Kato, T., Grinding of Ceramic Coatings with Cast Iron Bond Diamond Wheels—A Comparative Study: ELID and Rotary Dresser, *Int. J. Adv. Mfg. Technol.*, 18:545–552 (2001)

23. Moryiasu, S., Ohmori, H., Kato, J., Koga H., and Ohmae, M., Ultra-precision Form Control of Aspheric Mirror with ELID Grinding, *Mfg. Sys.*, 27(4) (1998)

24. Ohmori, H., Lin, W., Moryiasu, S., and Yamagata, Y., Advances on Micro-mechanical Fabrication Techniques: Microspherical Lens Fabrication by Cup Grinding Wheels Applying ELID Grinding, RIKEN Review #34 (Apr. 2001)

25. Qian, J., Li, W., and Ohmori, H., Precision Internal Grinding With Metal-bonded Diamond Grinding Wheel, *J. Mat. Proc. Tech.*, 105:80–86 (2000)

26. Qian, J., Ohmori, H., and Li, W., Internal Mirror Grinding With a Metal/Metal-Resin Bonded Abrasive Wheel, *Int. J. Mach. Tool Mfg.*, 41:193–208 (2001)

27. Stephenson, D. J., Veselovac, D., Manley, S., and Corbett, J., Ultra-precision Grinding of Hard Steels, *J. Int. Soc. Prec. Eng. Nanotech*, 25:336–345 (2001)

28. Park., K. Y., Lee, D. G., and Nakagawa, T., Mirror Surface Grinding Characteristics and Mechanism of Carbon Fiber Reinforced Plastics, *J. Mat. Proc. Technol.*, 52:386–398 (1995)

29. Onchi, Y., Matsumori, N., Ikawa, N., and Shimada, S., Porous Fine CBN Stones for High Removal Rate Superfinishing, *Annals of the CIRP*, 44(1):291–294 (1995)

30. Puthanangady, T. K., and Malkin, S., Experimental Investigation of the Superfinishing Process, *Wear*, 185:173–182 (1995)

31. Ohmori, H., Influence of Electrical Conditions of Ultra-precision Mirror Surface Grinding With Electrolytic In-process Dressing (ELID), *Proc. 1st Int. Abrasive Tech. Conf., Kaitech, Seoul, Korea*, pp. 104–110 (Nov. 1993)

10

Grinding Wheel and
Abrasive Topography

Topography is concerned with defining and mapping the shape of a surface, in this case, the grinding wheel or other abrasive tools. Both macrotopography and microtopography are important, although in different ways. The basic wheel shape is part of the macrotopography and is important for the overall accuracy of the workpiece profile generated. Microtopography is also significant for maintaining the required workpiece profile, but additionally for the workpiece roughness, energy requirements, down-time for redressing, wheel life, removal rates, and so on.

10.1 BASIC WHEEL SHAPE

The basic wheel geometry is defined by the profile of its periphery and faces. The bulk shape is important for the achievement of the basic workpiece geometry, as illustrated in Fig. 10.1. Some basic causes of inaccuracy are shown in the following sections.

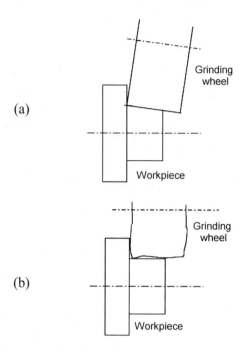

Figure 10.1. Examples of shape and misalignment effects: *(a)* Wheel misalignment causes conicity and dishing of a cylindrical workpiece, and *(b)* wheel shape errors replicated on the workpiece.

10.1.1 Misalignment

One form of shape error results from misalignment between the grinding wheel and the workpiece. Even where the wheel is a perfect cylinder, due to misalignment shown in Fig. 10.1(a), the wheel surfaces presented to the workpiece are incorrect. A cone error is introduced into the cylindrical workpiece surface. The workpiece face becomes the shape of a dish. Some of the basic causes of misalignment are as follows:

- Machine structure misalignment.
- Wheel mounting misalignment.
- Wheel spindle deflection due to the grinding load.
- Dressing-slide misalignment.

10.1.2 Profile Errors

In Fig. 10.1(b), the wheel shape has profile errors. It is not a perfect cylinder. The corner between the cylindrical surface and the face is not square and the surfaces are not straight. These shape errors will be directly replicated in the workpiece profile. Some basic causes of wheel shape errors are:

- Lack of straightness in a dressing slide.
- Irregular wear generated during grinding.
- Corner breakdown of the wheel due to concentrated grinding stresses.
- Wear of a single-point dressing tool during a dressing operation.

A frequently employed method of measuring wear depth and irregularities in the profile of a wheel is the so-called *razor-blade technique*. A part of the wheel not used for grinding is used as a reference surface for measurement. A thin sheet of steel is plunge-ground twice by the wheel, once before and once after experiencing wear in grinding. This procedure reproduces the wheel profile along the edge of the strip. The difference in the profile before and after grinding can then be compared to determine the depth of wear. Measurement is achieved by mounting the strip on a stylus or projection profile-measuring machine to produce an accurate magnification of the shape. The principle is illustrated in Fig. 10.2.

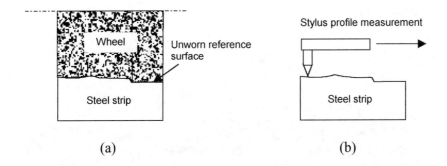

Figure 10.2. Principle of wheel wear measurement: *(a)* The profile and wear depth is ground onto a thin strip, and *(b)* the strip is measured using a suitable profile measuring system.

10.1.3 Vibrations

Waviness of the grinding profile tends to build up with time of grinding. This may start due to machine vibrations. Waviness tends to build up at particular frequencies due to a regenerative coupling between depth of cut variations, force variations, and consequent machine deflections. Waviness eventually has to be eliminated by truing the wheel. Vibrations during truing and dressing also account for many problems in grinding. The most basic cause of vibration is wheel unbalance. Wheel unbalance causes the wheel axis to run-out as illustrated in Fig. 10.3. The effect of wheel unbalance can be largely corrected by truing the grinding wheel at the same speed as employed for grinding. At Point A, the run-out towards the dressing tool is removed. When Point A arrives at A', the run-out is away from the diamond and less material is removed. The result is an eccentricity dressed into the wheel shape precisely correcting the balance run-out.

Sometimes, for the convenience of machine design, the wheel is dressed on the side opposite to the grinding point at 180°, or sometimes at 90°, or some other angle relative to the grinding point. Dressing at other points allows approximate removal of the run-out due to basic unbalance, but does not remove the run-out due to other vibration frequencies. This is illustrated for a vibration harmonic at twice the wheel speed (Fig. 10.4). The run-out due to the vibration causes an error to be dressed into the wheel at Point A. When the wheel has rotated 180°, the error will arrive at the grinding point. Point A will be at the new position, Point A'. For a double frequency vibration, Point A' will again be moving to the right as it was at Point A. The dressed error will, therefore, add to the vibration movement and the error machined on the workpiece will be doubled instead of being eliminated.

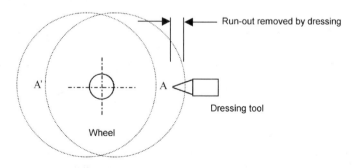

Figure 10.3. Run-out due to wheel unbalance can be balanced by dressing at grinding point.

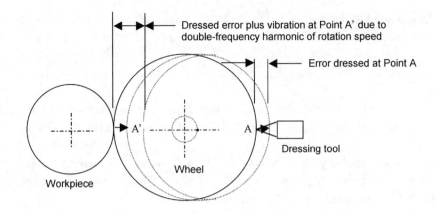

Figure 10.4. The effect of dressing opposite the grinding point. The error dressed into the wheel at Point A when it arrives at Point A' adds to vibration at twice the rotational speed.

Ideally, dressing should always take place at the grinding point and at the same speed of wheel rotation to minimize run-out errors. In practice, this is not always achieved, so the operator must be especially careful to avoid vibrations. The basic profile accuracy of the workpiece is dependent on the dressed geometry of the wheel, the extent of wheel wear, and the accuracy of the machine alignments and motions.

10.2 THE IMPORTANCE OF MICROTOPOGRAPHY

The microtopography of an abrasive tool determines the cutting efficiency of the process since the topography defines the shape, hence the sharpness of the cutting edges. The topography defines the spacing of the grains and reflects the porosity of the surface. After studying this topic, the reader should be aware of the effects of topography on the performance of an abrasive. The topography of an abrasive tool depends on the following:

- The abrasive tool structure.
- Truing and dressing of the tool.
- Process wear experienced by the abrasive.

An abrasive tool is composed of abrasive grains, bond material, and pores as illustrated in Fig. 10.5. The structure conforms to the constraint that

Eq. (10.1) $V_g + V_b + V_p = 100\%$

where V_g is the volume percentage of the abrasive grains, V_b is the volume percentage of the bond material, and V_p is the volume percentage of air, otherwise known as the porosity.

If a smaller proportion of abrasive grains is mixed into the structure, the result is increased porosity. Less bond material also increases porosity. Both methods reduce the strength of the structure.

Figure 10.6 shows examples of porosity viewed using a scanning electron microscope (SEM). A SEM gives the clearest view of a wheel surface. On the left is a high-porosity CBN wheel. On the right is a medium-porosity wheel. The two wheels have different grinding characteristics as a consequence of the different character of the surfaces.

The tips of the grains perform the abrasion of the workpiece. The bond material holds the grains together. The bond material also holds the grains apart and is responsible for grain spacing. Together, the bond material and grains provide the strength of the structure. Weak bond bridges allow grains to break out of the surface, the remaining cutting edges become more heavily loaded so that fracture wear increases. However, if fracture wear is too low, grains become very dull due to wear, grinding forces increase, and problems of wheel loading can result.

Pores provide clearance for the flow of grinding chips. Pores that are too small, quickly become clogged or loaded. By measuring the wheel topography, it is possible to obtain quantitative measures of performance of an abrasive. This is particularly valuable in deciding on requirements for an improved abrasive structure or for improvements required in the dressing process.

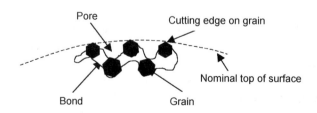

Figure 10.5. Abrasive structure comprised of grains, bonds, and pores.

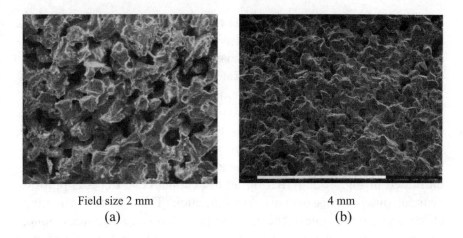

Field size 2 mm   4 mm
(a)   (b)

Figure 10.6. Scanning electron microscope (SEM) photographs of wheel surfaces: *(a)* high-porosity CBN wheel: B91-VR150, and *(b)* medium-porosity CBN wheel: B91-150V1.

A wheel specification includes parameters such as abrasive grain type, bond type, structure effective hardness, and mean grain size or mesh size. The wheel specification is not the whole description, however. The topography of the wheel surface provides additional information. The topography reflects wheel wear and effects of truing and dressing. Structure, dressing, and wear all affect the openness of the surface, the number of cutting edges available, and the sharpness or dullness of a tool.

Chapter 3 introduced the nature of cutting-edge distributions and the effect of truing and dressing on the grinding process. The purpose of this chapter is to consider in more detail, the definition of the topography of an abrasive tool and its measurement.

10.3 TOPOGRAPHICAL DEFINITIONS

Definitions are required for the measurement of topography. Figure 10.6 shows two different wheels, but the differences between them need to be determined.

The problems of describing the topography of a grinding wheel are found in Ref. 1. There are a large number of parameters that currently exist for characterizing surfaces.[2]–[4] Stout, et al.,[5] presented sixteen topographic parameters including amplitude parameters, spatial parameters, hybrid parameters, and function parameters for characterizing bearing and fluid retention properties. Given the large number of parameters that currently exist, the discussion in this chapter is centered on three parameters of fundamental importance for the grinding process.

Three parameters that can be measured are γ as the dullness of the cutting edges, C_a as the number of active cutting edges per unit area, and V_{pw} as the effective porosity of the wheel at the cutting surface.

Malkin has shown that increasing the wear-flat area is important for the increase of specific energy.[6] The importance of cutting edge shape is illustrated in Fig. 10.7. A flat tip on the cutting edge impedes upward material flow and discourages chip formation. This causes low cutting efficiency. A sharp edge on the tip of the grain acts like the prow of a ship. The cutting edge pierces the surface encouraging upward flow of the material. Upward flow encourages chip formation and high cutting efficiency.

10.3.1 Cutting Edge Dullness (γ)

Cutting edge dullness, γ, is defined as the ratio of the effective cutting edge radius, r_0, to the grain radius, r_g, as illustrated in Fig. 10.8. An alternative definition of cutting edge dullness may be obtained from the ratio of wear flat area per unit area. This latter definition is less sensitive to changes in the wear-flat area.

Eq. (10.2) $$\gamma = \frac{r_0}{r_g}$$

10.3.2 Cutting Edge Density (C_a)

Cutting edge density, C_a, is defined as the number of active cutting edges per unit area of the wheel surface. Changing the wheel structure or changing the depth of cut changes the number of active cutting edges per unit area.

It should be remembered that cutting efficiency does not only depend on the shape of the grain. The same grain may give rise to low-cutting efficiency or to high-cutting efficiency. The difference between the two cases illustrated in Fig. 10.9 is the grain depth of penetration. This was

Figure 10.7. Cutting efficiency depends on cutting edge shape. A sharp edge encourages chip removal. A dull grain impedes chip removal.

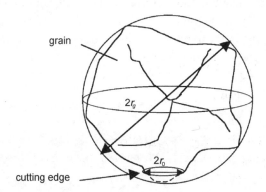

Figure 10.8. Cutting edge dullness is defined from the effective radius of the wear-flat areas on the grains and the mean grain size.

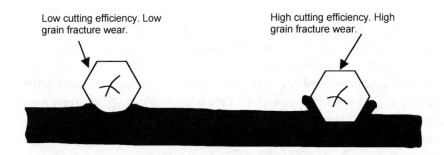

Figure 10.9. The cutting efficiency and grain fracture wear depend on grain penetration, hence on grain spacing. A large spacing increases grain penetration.

discussed in previous chapters as a size effect. With small grain penetration, the material flows mainly sideways. With large grain penetration, the material is encouraged to flow upwards and sideways in an oblique cutting action. As shown in Ch. 3, "Kinematic Models of Abrasive Contacts," the depth of grain penetration is affected by the grain spacing. For the same depth of cut, smaller grain spacing means a smaller depth of grain penetration.

König and Lortz[7] pointed out that not all the cutting edges statically measured in the wheel surface come into contact with the workpiece in grinding. Shaw and Komanduri[8] also pointed out that even edges at the same level do not necessarily cut, since it is possible that a preceding edge may have already removed the material in the path of a succeeding edge. Verkerk, et al.,[9] proposed that two parameters be used, C_{stat}, the total number of static cutting edges per unit area and C_{dyn}, the number of cutting edges per unit area when dynamics are included in the assessment. From a basic kinematic analysis, the dynamic cutting edge density for a given wheel topography should depend uniquely on the quantity $(v_w/v_s)/(a_e/d_e)^{1/2}$. The density of dynamic cutting edges can then be related to the static value on the basis of this relationship[10]

Eq. (10.3)
$$C_{dyn} = C_{stat} \left(\frac{v_w}{v_s} \right)^c (\theta')^c$$

where θ' is the wheel rotation angle into the arc of contact and the exponent, c, is a constant for a particular wheel and dressing conditions ranging between 0.4 and 0.8.

Cutting edge density is strongly affected by wheel-workpiece deflection. This is illustrated in Fig. 10.10. Real contact length is usually much larger than the theoretical geometric value.[11] This is because of local deflection of the wheel similar to deflection of a car tire against the road. Deflection brings more cutting edges from below the outer surface of the wheel into contact with the workpiece and greatly increases the active cutting edge density. Examination of the wheel surface shows many more cutting edges are involved in grinding than would be expected from static measurements.

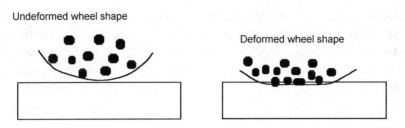

Figure 10.10. Local deflection of the wheel during grinding greatly increases the number of active cutting edges per unit area.

Cutting edge density may be determined by counting the number of active cutting edges per unit area in the surface of the wheel. Active grains experience wear that may be observed. This allows active grains to be counted, and also allows the depth of the active layer within the wheel to be assessed.

Cutting edge density is related to cutting edge spacing. Increasing cutting edge density reduces spacing and implies smaller grain penetration. Grain forces lessen with smaller grain penetration, but grinding is less energy efficient. This point is illustrated in Fig. 10.9 and is important for the effects of changes in topography on performance.

With reference to Fig. 3.14, cutting edge density is related to grain spacing according to

Eq. (10.4)
$$C = \frac{1}{L \cdot B}$$

where L is the mean spacing between active grains in the grinding direction and B is the mean spacing between active grains in the lateral direction.

10.3.3 Effective Porosity Ratio (V_{pw})

Effective porosity ratio concerns porosity at the wheel surface. Effective porosity is defined as the effective pore volume of the pores to the volume of the wheel on the wheel surface as shown in Fig. 10.11. The effective porosity is different from the overall porosity within the wheel structure defined by Eq. (10.1). The effective porosity ratio describes the openness of the wheel at the surface. Effective porosity within the active layer of the surface is defined as

Eq. (10.5) $$V_{pw} = \frac{V_p \big|_{z \le z_i}}{V_w \big|_{z \le z_i}}$$

where z_i is the depth of the active layer.

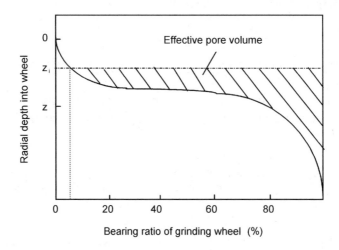

Figure 10.11. Effective pore volume decreases with increasing depth into the wheel surface.

The effective porosity decreases with depth into the surface as illustrated in Fig. 10.11. This type of diagram is known as a bearing area curve. Above the surface, there is 100% air. On entering the surface, the proportion of material to air at a particular level increases. This proportion is known as bearing ratio. In this figure, we are only concerned with pores at the surface. These are the pores that convey grinding fluid, into which chips may escape from the workpiece. Below the surface, the effective porosity becomes zero, hence, the bearing ratio is assumed to be 100%. Of course, cutting through the grains would reveal pores inside the structure. However, those pores are not considered to be part of the effective porosity of the surface. Clearly, effective porosity of the surface may be greater or smaller than the structural porosity of the wheel.

Figure 10.11 illustrates a problem of how to define the zero level of the surface. Conceptually, it is the first point of contact with the surface. In practice, measuring the first point of contact presents a difficulty in that an infinitesimal pointmust be detected; this introduces uncertainty. The first point of contact may be some small contamination of the surface, such as a particle of dust or hair or an unrepresentative grain or whisker of wheel loading from the workpiece. It is advisable not to place too much reliance on a measurement from a supposed zero level.

10.4 MEASUREMENT TECHNIQUES

Techniques for monitoring a grinding wheel during the process must, necessarily, be non-destructive. Some of the main techniques employed are:

- Stylus techniques.
- Optical techniques.
- Scanning electron microscopy.

Some relative merits of these techniques are discussed below.

10.4.1 Stylus Techniques

Stylus methods have been used to measure surface roughness since 1927.[11] A stylus with a very small radius on the tip is dragged horizontally across the surface to be measured. The vertical movements of the stylus are measured using a displacement transducer. The movements represent a two-dimensional record of the shape of the surface.

The principle of stylus measurement is illustrated in Fig. 10.12. The vertical magnification is high and the horizontal magnification is much lower. The vertical movements of the stylus are usually measured relative to the position of a skid sliding on the surface of the wheel. This produces a trace as shown. The stylus is not allowed to fall deep into the pores since this would produce extremely large movements when magnified 1000× or more. It would also risk damage to the delicate stylus mechanism. A typical measurement depth is less than 100 microns.

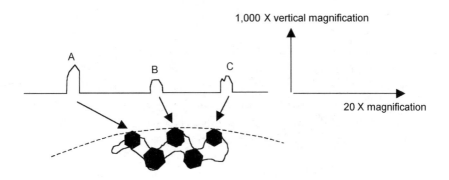

Figure 10.12. A 2-D stylus trace of a grinding wheel surface.

There are several problems with stylus measurements. Accuracy of the stylus measurement depends on the shape and size of the stylus tip. The larger the tip radius, the greater the inaccuracy of the profile trace. The abrasive is very hard and continued use of a stylus against an abrasive surface causes wear of the stylus. The stylus is relatively expensive and is very easily damaged. There is also the problem measuring a large wheel, i.e., how to locate the wheel under the instrument. This raises the need for making and measuring a replica of a part of the surface using appropriate replication materials. The replica introduces additional errors although reasonable accuracy has been achieved in practice with certain replication materials.

The trace across a grain often suggests multiple cutting edges as in Peak C in Fig. 10.12. It is necessary,therefore, to interpret the trace. Are there two grains or only one? In this case, it is only one grain and it would be assumed to represent only one cutting edge. It is tempting to take the separation between points on a 2-D trace as the basis for grain spacing and grain depth below the surface. However, this is misleading. In practice, cutting edges do not lie neatly in a straight line or coincide with the peaks of a trace.

The only way to map a surface with accuracy using a stylus is to take repeated traces. After each trace, the sample is displaced sideways by a small precisely defined distance. In this way, a small area can be scanned along a series of parallel lines and a picture built up in a similar way to the projection of a picture on a television screen. Feeding the data from the traces into appropriate software, a clear 3-D picture of the grain peaks is produced as in Fig. 10.13.

μm

- 65
- 60
- 55
- 50
- 45
- 40
- 35
- 30
- 25
- 20
- 15
- 10
- 5
- 0

Figure 10.13. 3-D topography of a wheel surface obtained by stylus measurement.

Three-dimensional data can be processed in various ways. Two-dimensional plan sections can be produced at various levels under the surface. Two-dimensional sections allow cutting edge density and effective porosity values to be extracted as a function of cutting depth. Examples of 2-D sections produced from 3-D data are shown in Fig. 10.14.

10.4.2 Microscopy

An optical microscope is not primarily a depth measurement instrument. An optical microscope image does not directly allow measurement of contours at a specified depth. Accurate measurement of the bearing area requires a method for precise depth resolution of the image provided by a stylus instrument or optical instruments with depth resolution capabilities. A stereoscopic microscope using a built-in vertical illuminator or a link to a computer partially overcomes this problem. An example of a fine-grain wheel viewed through an optical microscope is shown in Fig. 10.15. The wheel has been used for grinding, and wear flats can be seen on grain tips.

A stereo microscope is relatively inexpensive and invaluable for examination of a surface. Fine details of the surface can be clearly seen using a large magnification. When viewed normal to the wheel surface, flattened areas on the grain tips reflect light and appear shiny against a dark background. The number of distinct wear-flat areas per unit area on the wheel surface can be counted and can sometimes be interpreted as the cutting edge density.

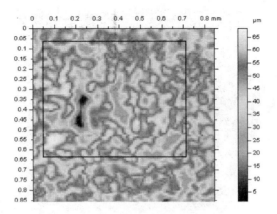

(a) 2D image measured by Tallysurf

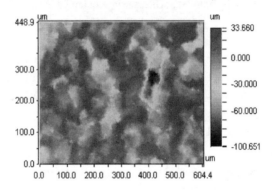

(b) 2D image measured by Wyko RST plus

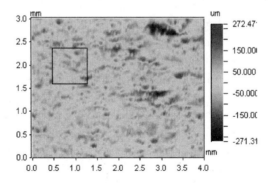

(a) 2D image measured by Uniscan OSP 100

Figure 10.14. Two-dimensional sections of an area of the active surface of a grinding wheel obtained by scanning using: *(a)* a stylus instrument, *(b)* interferometry, and *(c)* laser triangulation.

Figure 10.15. The surface of a fine-grain wheel viewed through an optical microscope showing light-colored striated wear flats.

At low magnifications, a relatively large field of view can be achieved, including a number of abrasive grains. It is possible to see grains and bond, although it is not always possible to distinguish between the two. It is possible to see workpiece material loaded on and between the grains and to see and measure wear flats on the peaks of the grains. The problems of interpreting the view of a wheel from an optical microscope are illustrated by the example in Fig. 10.15.

A detailed and clear picture may be obtained from a scanning electron microscope as shown in Fig. 10.6. However, a large SEM may not be available to many potential users. An SEM provides stereographic photographs and can also provide contour plots with high magnification and large depth of field, allowing cutting edge density to be observed at various depths of penetration. A limitation of most instruments is the need to employ removable segments or sections of a wheel small enough to fit into the observation area or the SEM chamber. Yet another problem is that replication materials tend to melt when being coated in preparation for measurement by SEM.

Two optical techniques make it possible to achieve quantitative characterization of surface topography including depth information. These are the Wyko optical interferometer and Uniscan laser triangulation instrument. These are highly specialized instruments and may not be available to everyone, due to cost. The equipment is provided with sophisticated software essential for topographical analysis.

Figure 10.16 shows a 3-D image of a surface obtained using the interferometer. The data obtained can be analyzed to provide the topographical parameters of interest.

Examples of 2-D sections obtained using the three methods are shown in Fig. 10.14. Measurements are shown from the same area of a wheel using a Talysurf (3-D) stylus instrument, a Wyko RST Plus interferometer, and a Uniscan OSP 100 laser triangulation instrument. In a color diagram, the grains and wear flats can be distinguished using both the Talysurf (3-D) and the Wyko RST Plus, but cannot be distinguished using the Uniscan OSP 100.

The interferometer with specialized software can be used for finer detailed examination of very small areas as in Figs. 10.14 and 10.16. The vertical range is limited, and the surface needs to have a reflectivity greater than 15%.

Laser triangulation employs a laser diode to create a structured light pattern and a CCD camera to capture the image. Typically, the camera angle is set up at a triangulation angle of 18° to the light source direction. With the aid of image analysis software, the detected contours are transformed into 3-D data. The laser triangulation instrument has an advantage over the interferometry instrument in that it is possible to achieve a relatively large measurement area similar to the stylus instrument. Laser triangulation also allows a large range of depths to be measured. Laser triangulation is, therefore, useful for obtaining statistical information from wheels.

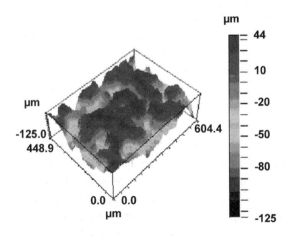

Figure 10.16. Image of 3-D topography produced by interferometry.

10.4.3 Comparison of Measurement Ranges

In the choice of an instrument for a specific application, knowledge of resolution and range is fundamental. Limiting factors are the vertical range and resolution of the system, the horizontal range and resolution, the horizontal datum, the probe size, and geometry. Another consideration is the effect of the color and texture of the wheels and replica surfaces for measurement. For example, CBN wheels usually have hard abrasives and dark colored bonds. Replica materials may be of different colors too.

From Table 10.1, the Wyko RST Plus offers more accuracy, but a smaller measurement area for 3-D analysis. The Talysurf (3-D) provides slightly lower accuracy, but a larger area for measurement. The Uniscan OSP 100 provides the lowest accuracy, but the largest area for measurement.

Table 10.1. Measurement Range and Accuracy

	STYLUS	INTER-FEROMETRY	LASER TRIANGU-LATION
Instrument	Talysurf (3-D)	Wyko RST Plus	Uniscan OSP 100
Measurement area (Typical value)	120 x 50 mm	8 mm x 6 mm (depends on variable objectives)	100 x 100 mm
Measurement depth	1.5 mm	0.1 nm – 500 μm	35 mm
Measurement accuracy	0.03 μm	0.01 μm	0.5 μm

Measurement errors vary according to the measurement technique. A measurement error includes not only an instrument error, but also an image processing error. In order to analyze the peripheral surface of a grinding wheel, the peripheral-curved shape needs to be fitted and regressed to a flat surface by the software. Inadequate data at the measurement border and inconsistency of the local form of the wheel surface introduces fitting errors. In addition, the raw data from the measurements usually include some bad data using interferometry. Bad data means the surface cannot be measured at that point due to a limitation of darkness, slope, or some other reason. Measurement software provides processed data for such points, usually by interpolation. This process introduces interpolation errors. Both fitting and interpolation errors depend on the uniformity of the wheel surface.

10.4.4 Replication Techniques

Replication is not a measurement technique in itself. Replication may be employed to overcome difficulties of making direct measurements on large wheels or of monitoring a wheel *in situ* when it is undesirable to remove and replace the wheel without having to redress the wheel surface. This situation applies when it is desirable to monitor wheel changes within a redress cycle.

A replica is prepared using a setting compound or soft metal. The replication material is poured or pressed onto the area of interest on the wheel. When cured, this provides a solid surface that ideally should be identical with the area of interest. The replicas can be measured and observed by stylus, optical, and microscope techniques. Investigations by Eckert[12] on a water-based plastic system and investigations by George[13] and James and Thum[14] on epoxy resin systems have shown that both types have the ability to replicate surface structure and have relatively low shrinkage.

Brough, et al.,[15] used soft polished lead that was pressed for thirty seconds into the surface of the wheel to the depth of the active layer or slightly greater. The control of the depth of penetration into the lead is achieved by careful control of the applied pressure and time of application. The force applied on the individual grains is less than the force experienced by the grains when cutting steel. This is essential to avoid damaging the topography. One of the advantages of using lead as a replica material was that it allowed a stylus instrument to be used directly on the replica. If the material used to take the initial replica is too soft to measure with a stylus instrument, a hard replica must be made. This can be achieved by making a hard replica of the soft replica. Inevitably, repeated replication leads to loss of detail in the surface.

Cai[1] used a two-part synthetic rubber compound known as Microset 101. This material allowed a resolution of 0.1 microns and exhibited high contrast characteristics for use with the laser triangulation and interferometry instruments. Microset 101 could not be used directly for stylus measurement due to its softness.

10.4.5 Image Processing

After the basic depth information is obtained, the data may be fed into an image processing software such as Matlab Image Processing for the analysis of wheel topography. Three-dimensional analysis allows measurement results to be presented in 2-D, working slice by slice from the top surface of the wheel. The cutting edge density can be evaluated against depth by counting the number of unconnected bearing areas by software. Bearing area curves can also be generated and the wear flat area calculated. Comparing cutting edge density and the wear flat area observed through the microscope with values determined using Matlab Image Processing, the difference was approximately 10%. Once the active cutting depth has been decided, the effective porosity ratio can be determined using Wyko software.

10.5 TOPOGRAPHY CHANGES IN GRINDING

The following section compares typical changes in topography for two different fine-grain vitrified CBN wheel specifications used for precision grinding. A high-porosity wheel, B64-150VX, and a medium-porosity wheel, B91-150V1, are compared. For CBN wheels, a B64 wheel has a smaller grain size than a B91. The measured changes in topography are compared in Fig. 10.17.

A surprising result is that the cutting edge density of the B64 wheel is much smaller than the cutting edge density of the B91 wheel. This means the grain spacing of the finer-grain wheel is actually larger than the spacing of the larger-grain wheel. The explanation was found by closer examination of the wheel; in the manufacture of very fine grain wheels, there is a greater tendency for several grains to agglomerate, effectively forming one much larger grain and reducing the number of cutting edges available.

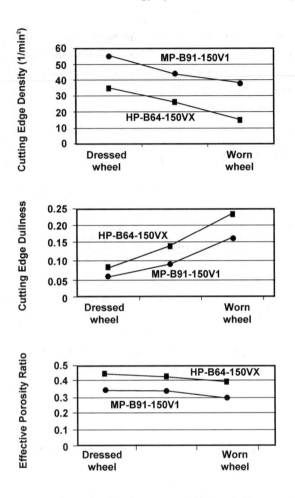

Figure 10.17. Changes in topography due to wear in the grinding process: a medium porosity B91 CBN wheel and a high-porosity B64 CBN wheel.

During grinding, the cutting density decreased with tim. For the high-porosity wheel, the cutting density by the end of grinding was less than half the value at the start of grinding. Wear reduced the number of separate grain tips measured and grain fracture removed cutting edges from the cutting action. The rate of wear was less pronounced for the medium-porosity wheel because the closely spaced grains were less prone to fracture. The conclusion was that, for an easy-to-grind workpiece material, medium porosity gives better wheel life than a high-porosity wheel.

Another surprise came from the evaluation of active cutting depth and cutting edge density. Active cutting depth was much greater than would be expected from the theoretical chip thickness based on the diameter of the wheel and the depth of cut. The depth of active cutting edges within the wheel surface was found to be 8 to 44 times greater than theoretical maximum chip thickness. The deflection of the wheel brought grains from deeper within the surface down into contact with the workpiece. Other contributions to this result were the statistical variations in grain spacing and depth discussed in Ch. 3, "Kinematic Models of Abrasive Contacts." However, the deflections contributed a large part of the discrepancy. The average active cutting depth for the MP-B91 wheel was 11 μm and for the HP-B64 wheel, 20 μm. These values were used to evaluate active cutting edge density, cutting edge dullness, and effective porosity ratio of the two wheels.

The results show that the dullness ratio increases as grinding proceeds. The dullness parameter is the ratio of wear flat size to mean grit size. For the finer grain wheel, the mean grit size is smaller so that for the same wear flat area, dullness ratio is increased. For the HP-B64 wheel, the wear flat size increases to almost 25% of the mean grit size.

As wear increases, the effective porosity reduces slightly. This is despite the reduced cutting edge density. The reduced effective porosity corresponds to increased bearing area as wear flat area increases even though the number of cutting edges is greatly reduced.

10.6 GRINDING INCONEL 718

Figures 10.18 and 10.19 show the effects of topographical changes when grinding a difficult-to-grind aerospace alloy, Inconel 718. Inconel 718 has the tendency to cause wheel-loading. The medium-porosity CBN wheel of Fig. 10.6 only managed to remove 60 mm^3/mm of material in grinding tests. After this, the loading was so severe and the quality so poor that grinding had to stop. The high-porosity CBN wheel of Fig. 10.6, however, was successful in grinding up to 500 mm^3/mm before the acceptable roughness level of 0.18 microns Ra was exceeded. The G-ratio was 2,000 with the high-porosity wheel, a high figure for this material. Increasing the wheel speed by 20% further doubled the G-ratio to 4,000, reduced the roughness to 0.13 microns Ra, and allowed 1,400 mm^3/mm to be removed. This shows the advantages of increased porosity and optimization of wheel speed for difficult-to-grind materials.

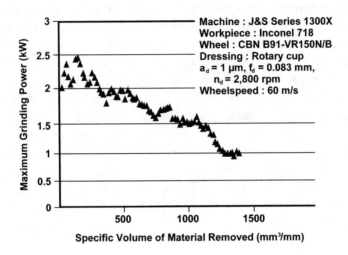

Figure 10.18. Effect of topographical changes on grinding energy.

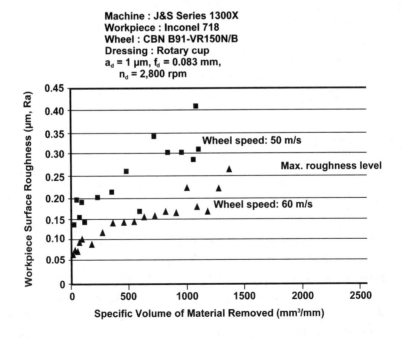

Figure 10.19. Effect of topographical changes on workpiece roughness in grinding.

Figure 10.18 shows grinding power reduced with grinding time as the cutting edge density reduced. This is despite the dullness of the grains. These results would be expected from the size effect, but not from the increasing measured dullness. This suggests grain dullness was relatively unimportant in this case compared with cutting edge spacing for grinding efficiency.

Three factors contribute to reduced wheel loading for high-porosity wheels. First, high-porosity helps transport coolant into the grinding zone. Improved fluid delivery reduces grinding temperature, improves lubrication, and provides more space to clear the swarf produced in grinding. Second, high-porosity causes grains to agglomerate, resulting in fewer active cutting edges and larger cutting edge spacing. The larger spacing leads to a larger mean uncut chip thickness and a larger grain depth of penetration. Based on the commonly observed size effect, the specific energy and grinding temperature decrease with a larger mean uncut chip thickness. Finally, high-porosity is associated with lower wheel hardness and increasing bond fracture. Increasing bond fractures results in a sharper grinding wheel.

Figure 10.19 shows the effect of wheel topography on workpiece roughness. Reduced cutting density increases roughness. Eventually roughness exceeds the specified tolerance and the wheel must be redressed. Higher wheel speed reduces workpiece roughness and, in this case, greatly increases the redress life. Higher wheel speed can also contribute to reduced wheel loading and longer wheel life. Higher wheel speed reduces grain depth of penetration leading to lower grinding force on each cutting edge. Adhesive loading decreases with lower grinding force.

REFERENCES

1. Cai, R., Assessment of Vitrified CBN Grinding Wheels for Precision Grinding, Ph.D Thesis, Liverpool John Moores University, Liverpool, England, UK (2002)

2. Peters, J., Vanherck, P., and Sastrodinoto, M., Assessment of Surface Typology Analysis Techniques, *Annals of the CIRP*, 28(2):539–553 (1979)

3. Nowicki, B., Multiparameter Representation of Surface Roughness, *Wear*, 102:161–176 (1985)

4. Whitehouse, D. J., Handbook of Surface Metrology, Institute of Physics Publ., Bristol (1994)

5. Stout, K. J., Sullivan, P. J., Dong, W. P., Mainsah, E., Luo, N., Mathia, T., and Zahyouani, H., The Development of Methods for the Characterization of Roughness in Three Dimensions, Report of the Commission of the European Community (1993)

6. Malkin, S., and Cook, N. H., The Wear of Grinding Wheels, Part 1– Attritious Wear, *Trans. ASME*, 93:1120–1128 (1971)

7. König, W., and Lortz, W., Properties of Cutting Edges Related to Chip Formation in Grinding, *Annals of the CIRP*, 24(1):231–236 (1975)

8. Shaw, M., and Komanduri, R., The Role of Stylus Curvature in Grinding Wheel Surface Characterization, *Annals of the CIRP*, 26(1):139–142 (1977)

9. Verkerk, J., et al., Final Report Concerning CIRP Cooperative Work on Characterization of Grinding Wheel Topography, *Annals of the CIRP*, 26(1):385–395 (1977)

10. Malkin, S., *Grinding Technol.*, Ellis Horwood, Publ. (1989)

11. King, T. G., Whitehouse, D. J., and Stout, K. J., Some Topographic Features of Wear Processes—Theory and Experiment, *Annals of the CIRP*, 25(1):351–357 (1977)

12. Eckert, J. D., Replica Techniques for the Study of Fracture Surfaces and Topography Study in General, *Prakt. Metallogr.*, 33:369–372 (1996)

13. George, A. F., A Comparative Study of Surface Replicas, *Wear*, 57:51–61 (1979)

14. James, P. J., and Thum, W. Y., The Replication of Metal Surfaces by Filled Epoxy Resins, *Precision Eng.*, 4:201–204 (1982)

15. Brough, D., Bell, W. F., and Rowe, W. B., Achieving and Monitoring High Rate Centerless Grinding, *Proc. 21st Machine Tool Design and Research Conf.*, Swansea U., Pergamon Press (1980)

11

Abrasives and Abrasive Tools

11.1 INTRODUCTION

Materials used as abrasives include both natural minerals and synthetic products. Abrasive materials can be considered as cutting tools with geometrically unspecified cutting edges that are characterized by high hardness, sharp edges, and good cutting ability. The sharpness of abrasive grains may be described in terms of edge radius and apex angle. As grain size increases, the percentage of sharp apex angles decreases, indicating a deterioration of grain cutting ability. In addition, cutting ability depends on specific features such as grain structure and cleavage, which are connected with the ability of cutting grains to regenerate new sharp cutting edges and points.

The choice of abrasive for a particular application may be based on durability tests involving impact strength, fatigue compression strength, dynamic friability, and resistance to spalling which occurs under the influence of single or cyclic thermal stress.

The abrasives industry is largely based on five abrasive materials; three are considered to be conventional abrasives, namely silicon carbide (SiC), aluminum oxide (alumina, Al_2O_3), and garnet. The other two, namely diamond and cubic boron nitride (CBN), are termed superabrasives.

There are dramatic differences in properties and cost between conventional abrasives and superabrasives. For example, values of hardness on the Knoop scale are:

Diamond: $7000 \, kg \, mm^{-2}$

CBN: $4500 \, kg \, mm^{-2}$

SiC: $2700 \, kg \, mm^{-2}$

Alumina: $2500 \, kg \, mm^{-2}$

Garnet: $1400 \, kg \, mm^{-2}$

A primary requirement of a good abrasive is that it should be very hard; but hardness is not the only requirement of an abrasive. The requirements of a good abrasive are discussed below. The decision to employ a particular abrasive will be based on various criteria relating to workpiece material, specified geometry, and removal conditions.

Cutting fluids should be used wherever possible in grinding to achieve high material removal rates coupled with low wear of the grinding wheel.

11.2 CONVENTIONAL ABRASIVE GRAIN MATERIALS

Basic properties of natural abrasives are listed in Table 11.1. Silicon carbide and aluminum oxide cost the least and are used the most. These two abrasives are capable of grinding most materials. Diamond is harder than CBN, and is very efficient for grinding carbides, but less efficient for grinding steels because of the chemical affinity of diamond for metals, which melts the abrasive and causes excessive wear. Similarly, alumina is more effective on most steels but less efficient on nonferrous metals and nonmetallic substances than on SiC.

Manufacturers use a variety of trade names for products based on silicon carbide and aluminum oxide. Table 11.2 lists some common trade names classified according to the type of abrasive. Some characteristics of typical alumina and SiC abrasives are presented in Table 11.3.

11.2.1 Aluminum Oxide (Al₂O₃)-based Abrasives

Figure 11.1 shows common aluminum oxide-based grains. Also called corundum, alumina ore was mined as early as 2000 BC in the Greek island of Naxos. Its structure is based on α-Al_2O_3, and various admixtures. Traces of chromium give alumina a red hue, iron makes it black, and titanium

Table 11.1. Natural Abrasives and Their Principal Characteristics

ABRASIVE	BASIC STRUCTURAL COMPONENT	CRYSTAL SYSTEM	HARDNESS (Mohs)	COLOR
Diamond	Carbon	Regular	10	Colorless, brown, gray, yellow
Aluminum Oxide	α-Al_2O_3 60–90%	Triagonal	9	Transparent, opaque, spotted
Garnet	Fe, Mg, or Mn $Al_2(SiO_4)_3$	Regular	7–7.5	Pink, red, brown, black
Quartz	SiO_2	Triagonal	7	Colorless, milky, white gray, transparent

Table 11.2. Trade Names and Common Names for the Two Types of Abrasives

TRADE NAMES AND COMMON NAMES	ABRASIVE TYPE
Abrasite, Adalox (Norton Company), Adamite, Alabrase, Alorix, Al_2O_3 (RPM Inc.), Aloxite, Ansandum, Ansolox, A-O-66, Baco, Barbalox, Boxite, Durexalo, Durexite, Electric-Abrasives, Garalum, Herculundum, Jewelox, Luminox, Lumnite, Metalite (Norton Company), Production (3M Co.), Three-M-ite(3M Co.), Usalox, Wasusite, Yellow-Stripe	Aluminum Oxide
Amunite, Ansilicon, Armourite, Blue Stripe, Carbicon, Carbonite, Carborix, Carborundum, Crystox, Cyrstolon, Durexail, Durite, Durundum, Jewelite, Sil-Carb, Tri-Mite, Vulcanite	Silicon Carbide

We found it difficult to determine the trademark owner for all names under which abrasives are sold on the market. Some of these are easy to pinpoint; others are already common names; still others are very difficult to identify as product brands of different companies.

Table 11.3. Summary of Mechanical Properties of Typical Alumina and SiC Abrasives

ABRASIVE	HARDNESS KNOOP	RELATIVE TOUGHNESS	SHAPE/ MORPHOLOGY	APPLICATIONS
Green SiC	2840	1.60	Sharp, angular, glassy	Carbides/ceramics precision
Black SiC	2680	1.75	Sharp, angular, glassy	Cast iron/ceramics/ ductile non-ferrous metals
Ruby Al_2O_3	2260	1.55	Blocky, sharp edged	HSS and high alloy steel
White Al_2O_3	2120	1.75	Fractured facets, sharp	Precision ferrous
Brown Al_2O_3	2040	2.80	Blocky, faceted	General purpose
Al_2O_3/10%ZrO	1960	9.15	Blocky, rounded	Heavy duty grinding
Al_2O_3/40%ZrO	1460	12.65	Blocky, rounded	Heavy duty/snagging
Sintered Al_2O_3	1370	15.40	Blocky, rounded, smooth	Foundry billet and ingots

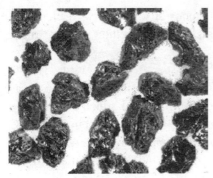

Green Silicon Carbide

White Alumina

Brown Alumina

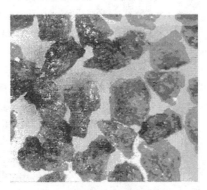

Pink Alumina

25% Zirconia-Alumina

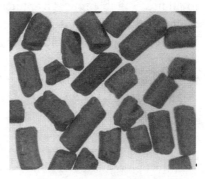

Sintered Extruded Alumina

Figure 11.1. Alumina grain types.[73]

makes it blue. Its triagonal system reduces susceptibility to cleavage. Precious grades of Al_2O_3 are used as gemstones, and include sapphire, ruby, topaz, amethyst, and emerald.

Charles Jacobs,[106] a principal developer, fused bauxite at 2,200°C (4,000°F) before the turn of the 20th century. The resulting dense mass was crushed into abrasive particles. Today, alumina is obtained by smelting aluminum alloys containing Al_2O_3 in electric furnaces at around 1,260°C (2,300°F), a temperature at which impurities separate from the solution and aluminum oxide crystallizes out. Depending upon the particular process and chemical composition there are a variety of forms of aluminum oxide. The poor thermal conductivity of alumina (33.5 W m^{-1} °K^{-1}) is a significant factor that affects grinding performance. Alumina is available in a large range of grades because it allows substitution of other oxides in solid solution, and defect content can be readily controlled.

For grinding, lapping and polishing bearing balls, roller races, and optical glasses, the main abrasive employed is alumina. Its abrasive characteristics are established during the furnacing and crushing operations, so very little of what is accomplished later significantly affects the features of the grains.

Aluminum oxide is tougher than SiC. There are four types of gradations for toughness. The toughest grain is not always the longest wearing. A grain that is simply too tough for an application will become dull and will rub the work, increasing the friction, creating heat and vibrations. On the other hand, a grain that is too friable will wear away rapidly, shortening the life of the abrasive tool. Friability is a term used to describe the tendency for grain fractures to occur under load. There is a range of grain toughness suitable for each application. The white friable aluminum oxide is almost always bonded by vitrification. It is the main abrasive used in tool-rooms because of its versatility for a wide range of materials. In general, the larger the crystals, the more friable the grain. The slower the cooling process, the larger are the crystals. To obtain very fine crystals, the charge is cooled as quickly as possible, and the abrasive grain is fused in small pigs of up to two tons. Coarse crystalline abrasive grains are obtained from 5–6 ton pigs allowed to cool in the furnace shell.

Alumina abrasives are derived either by electrofusion, or by chemical precipitation and/or sintering, as presented below.

Electrofusion. The raw material, bauxite, containing 85–90% alumina, $2-5\%$ TiO_2, up to 10% iron oxide, silica, and basic oxides, is fused in an electric-arc furnace at 2,600°C (4,700°F). The bed of crushed and calcined bauxite, mixed with coke and iron to remove impurities, is poured into the bottom of the furnace where a carbon starter rod is laid down. A couple of large vertical carbon rods are then brought down to touch and a heavy current applied. The starter rod is rapidly consumed, by which time the heat melts the bauxite, which then becomes an electrolyte. Bauxite is added over several hours to build up the volume of melt. Current is controlled by adjusting the height of the electrodes, which are eventually consumed in the process.

After cooling, the alumina is broken up and passed through a series of hammer, beater, crush, roller, and/or ball mills to reduce it to the required grain size and shape, producing either blocky or thin splintered grains. After milling, the product is sieved to the appropriate sizes down to about 40 μm (#400). The result is brown alumina containing typically 3% TiO_2. Increased TiO_2 content increases toughness while reducing hardness. Brown alumina has a Knoop hardness of 2090 and a medium friability.

Electrofused alumina is also made using low-soda Bayer process alumina that is more than 99% pure. The resulting alumina grain is one of the hardest, but also the most friable, of the alumina family providing a cool-cutting action. This abrasive in a vitrified bond is therefore suitable for precision grinding.

White aluminum oxide is one of the most popular grades for micron-size abrasive. To produce micron sizes, alumina is ball-milled or vibro-milled after crushing and then traditionally separated into different sizes using an elutriation process. This consists of passing abrasive slurry and water through a series of vertical columns. The width of the columns is adjusted to produce a progressively slower vertical flow velocity from column to column. Heavier abrasive settles out in the faster flowing columns while lighter particles are carried over to the next. The process is effective down to about 5 μm and is also used for micron sizing of SiC. Air classification has also been employed.

White 99% pure aluminum oxide, called monocorundum, is obtained by sulphidation of bauxite, which outputs different sizes of isometric corundum grains without the need for crushing. The crystals are hard, sharp, and have better cleavage than other forms of aluminum oxides, which qualifies it for grinding hardened steels and other tough and ductile materials. Finer-grain aluminum oxide with a good self-sharpening effect is used for finishing hardened and high-speed steels, and for internal grinding.

Not surprisingly, since electrofusion technology has been available for one hundred years, many variations of the process exist both in terms of starting compositions and processing routes. For example:

a) **Red-Brown or Gray Regular Alumina.** Contains 91–93% Al_2O_3 and has poor cleavage. This abrasive is used in resinoid and vitrified bonds and coated abrasives for rough grinding when the risk of rapid wheel-wear is low.

b) **Chrome Addition.** Semi-fine aloxite, pink with 0.5% chromium oxide, and red with 1–5% chromium oxide, lies between common aloxite, having less than 95% Al_2O_3 and more than 2% TiO_2, and fine aloxite, which has more than 95% Al_2O_3 and less than 2% TiO_2. The pink grain is slightly harder than white alumina, while the addition of a small amount of TiO_2 increases its toughness. The resultant product is a medium-sized grain available in elongated, or blocky but sharp, shapes. Ruby alumina has a higher chrome oxide content of 3% and is more friable than pink alumina. The grains are blocky, sharp edged, and cool cutting, making them popular for tool room and dry grinding of steels, e.g., ice skate sharpening. Vanadium oxide has also been used as an additive giving a distinctive green hue.

c) **Zirconia Addition.** Alumina-zirconia is obtained during the production process by adding 10–40% ZrO_2 to the alumina. There are at least three different alumina-zirconia compositions used in grinding wheels: 75% Al_2O_3 and 25% ZrO_2, 60% Al_2O_3 and 40% ZrO_2, and, finally, 65% Al_2O_3, 30% ZrO_2 and 5% TiO_2. The manufacture usually includes rapid solidification to produce a fine grain and tough structure. The resulting abrasives are fine grain, tough, highly ductile, and give excellent life in medium to heavy stock removal applications and grinding with high pressures such as billet grinding in foundries.

d) **Titania Addition.** Titania-aloxite, containing 95% Al_2O_3 and approximately 3% Ti_2O_3, has better cutting ability and improved ductility than high-grade bauxite common alumina. It is recommended when large and variable mechanical loads are involved.

e) **Single Crystal White Alumina.** The grain growth is carefully controlled in a sulphide matrix and is separated by acid leaching without crushing. The grain shape is nodular which aids bond retention, avoiding the need for crushing and reducing mechanical defects from processing.

f) Post Fusion Processing Methods. This type of particle reduction method can greatly affect grain shape. Impact crushers such as hammer mills create a blocky shape while roll crushers cause splintering. It is possible, using electrostatic forces to separate sharp shapes from blocky grains, to provide grades of the same composition but with very different cutting actions.

The performance of the abrasive can also be altered by heat treatment, particularly for brown alumina. The grit is heated to 1,100°C–1,300°C (2,015°F–2,375°F), depending on grit size, in order to anneal cracks and flaws created by the crushing process. This can enhance toughness by 25% to 40%.

Finally, several coating processes exist to improve bonding of the grains in the grinding wheel. Red iron oxide is applied at high temperatures to increase surface area for better bonding in resin cut-off wheels. Silane is applied for some resin bond wheel applications to repel coolant infiltration between the bond and abrasive grit and thus protect the resin bond.

Chemical Precipitation and/or Sintering. A limitation of electrofusion is that the resulting abrasive crystal structure is very large; an abrasive grain may consist of only 1 to 3 crystals. Consequently, when grain fracture occurs, the resulting particle loss may be a large proportion of the whole grain. This results in inefficient grit use. One way to avoid this is to dramatically reduce the crystal size.

The earliest grades of microcrystalline grits were produced as early as 1963[93] by compacting a fine grain bauxite slurry, granulating to the desired grit size, and sintering at 1,500°C (2,735°F). The grain shape and aspect ratio could be controlled by extruding the slurry.

One of the most significant developments since the invention of the Higgins furnace was the release in 1986, by the Norton Company, of seeded gel (SG) abrasive.[105][107] This abrasive was a natural outcome of the wave of technology sweeping the ceramics industry at that time to develop high strength engineering ceramics using chemical precipitation methods. This class of abrasives is often termed "ceramic." Seeded gel is produced by a chemical process. In a pre-cursor of boehmite, MgO is first precipitated to create 50 µm-sized alumina-magnesia spinel seed crystals. The resulting gel is dried, granulated to size, and sintered at 1,200°C (2,200°F). The resulting grains are composed of a single-phase α-alumina structure with a crystalline size of about 0.2 µm. Defects from crushing are avoided; the resulting abrasive is unusually tough but self-sharpening because fracture now occurs at the micron level.

As with all new technologies, it took significant time and application knowledge to understand how to apply SG. The abrasive was so tough that it had to be blended with regular fused abrasives at levels as low as 5% to avoid excessive grinding forces. Typical blends are now 5 SG (50%), 3 SG (30%), and 1 SG (10%). These blended abrasive grades can increase wheel life by up to a factor of 10 over regular fused abrasives although manufacturing costs are higher.

The grain shape of SG can be controlled to surprising extremes by the granulation processes adopted. The shape can vary from very blocky to very elongated.

In 1981, prior to the introduction of SG, the 3M Co. introduced a sol-gel abrasive material called Cubitron for use in coated-abrasive fiber discs.[9] This was a submicron chemically precipitated and sintered material but, unlike SG, had a multiphase composite structure that did not use seed grains to control crystalline size. The value of the material for grinding wheel applications was not recognized until after the introduction of SG. In the manufacture of Cubitron, alumina is co-precipitated with various modifiers such as magnesia, yttria, lanthana, and neodymia to control microstructural strength and surface morphology upon subsequent sintering. For example, one of the most popular materials, Cubitron 321, has a microstructure containing submicron platelet inclusions which act as reinforcements somewhat similar to a whisker-reinforced ceramic.[9]

Direct comparison of the performance of SG and Cubitron is difficult because the grain is merely one component of the grinding wheel. Seeded gel (SG) is harder (21 GPa) than Cubitron (19 GPa). Experimental evidence suggests that wheels made from SG have longer life, but Cubitron is freer cutting. Cubitron is the preferred grain in some applications from a cost/performance viewpoint. Advanced grain types are prone to challenge from a well-engineered, i.e., shape selected, fused grain that is the product of a lower cost, mature technology. However, it is important to realize that wheel cost is often insignificant compared to other grinding process costs in the total cost per part.

Seeded gel grain shape can be controlled by extrusion. Norton has taken this concept to an extreme and in 1999 introduced TG2 (extruded SG) grain in a product called ALTOS. The TG2 grains have the appearance of rods with very long aspect ratios. The resulting packing characteristics of these shapes in a grinding wheel creates a high-strength, light-weight structure with porosity levels as high as 70% or even greater. The grains touch each other at only a few points where bond also concentrates in the same way as a spot weld. The product offers potential for higher stock removal rates and higher wheelspeeds due to the strength and density of the resulting wheel body.[54]

With time, it is expected that SG, TG, Cubitron, and other emerging chemical precipitation/sintering processes will increasingly dominate the conventional abrasive market.

11.2.2 Garnet

Garnet includes a class of natural minerals of a wide chemical composition. Garnet crystallizes in the regular form of rhombic dodecahedron. Transparent or attractively colored garnets are used for jewelry, and when formed as large crystals, used for abrasives. As an abrasive, the value of garnet is determined by the friability rather than the hardness. The friability is determined by the presence of cracks along the cleavage planes.

Garnet is used in the abrasives industry for abrasive paper used for fine grinding of wood. Garnet is also used on glasses, leather, and varnished and painted surfaces. As a loose abrasive, garnet is used in finishing plate glass to a mirror finish and edging of lenses. Most bonding agents used for garnet are organic because of partial decomposition and melting at the temperatures employed in the firing of vitrified wheels.

11.2.3 Quartz

Quartz is a triagonal crystal composed of SiO_2 and impurities of CO_2, H_2O, NaCl, $CaCO_3$. Quartz is comprised of a large variety of rocks, permatites, and many hydrothermal formations. It is resistant to chemical and mechanical weathering. The main uses are in naturally bonded sandstones as coated abrasives for sanding wood, and as loose abrasive for polishing glass and for sandblasting.

11.2.4 Silicon Carbide (SiC)

Silicon carbide (SiC), initially called carborundum, was discovered in 1891 by Acheson.[104] Acheson electrically heated a mixture of sand and coke and found shiny black grains tough enough to scratch glass. Silicon carbide is manufactured in an Acheson resistance heating furnace at a temperature of around 4,350°F. The chemical reaction is:

Eq. (11.1) $SiO_2 + 3C \rightarrow SiC + 2CO$

A large carbon resistor rod is placed on a bed of raw materials to which a heavy current is applied. The raw material includes sawdust to add porosity to help release CO, and salt to remove iron impurities. The whole process takes about 36 hours and typically yields 10–50 tons of product. From the time it is formed, the SiC remains a solid; no melting occurs. The SiC sublimates above 2,750°C (5,000°F). After cooling, the SiC is sorted by color, from green 99% pure SiC, to black 97% pure.

The pieces are further broken, crushed, and separated into different size ranges by screening. The procedure also involves a modification of grain shape. Sharp, elongated grains are used for coated abrasives, blocky grains are used for foundry application, and the remainder for general work. Silicon carbide was initially produced in small quantities and sold for $1,940 per kg ($880/lb) as a substitute for diamond powder for lapping precious stones. In its time, it might well have been described as the first synthetic "superabrasive" compared with natural emery and corundum minerals otherwise available at the time. However once a commercially viable process of manufacture was established, the price fell sharply, and by 1938, SiC sold for $0.22 per kg ($0.10/lb). Today material costs have increased to about $1.75 per kg. ($0.80/lb).

Silicon carbide has a low thermal expansion coefficient, which decreases with increasing temperature. It has high hardness, and sharp crystal edges and therefore qualifies as a good abrasive. Green SiC with a high purity of 97–99%, but costly and hard to produce, is used for grinding sintered carbides. Impure black SiC, having less than 95% SiC, is used for grinding hard and friable materials with low tensile strength, such as cast iron, bronze, aluminum, and glass. Silicon carbide has a Knoop hardness of 2,500 and is very friable. The impurities within the black grade increases toughness to some extent but the resulting grain is still significantly more friable than alumina. Above 760°C (1,400°F), SiC shows a chemical reactivity towards metals with an affinity for carbon such as iron and nickel.

This limits its use to grinding hard, nonferrous metals. Silicon carbide also reacts with boron oxide and sodium silicate, common constituents of vitrified wheel bonds.

11.2.5 Polishing Abrasives

Polishing abrasives may be required to be hard or soft according to the desired material removal rates. Hard abrasives are employed for rough abrasive paste compositions; examples are pumice, beryllium oxide, chrome oxide, iron oxide, garnet, corundum, emery, quartz, silica carbide, and glass. Soft abrasives are employed for fine polishing paste compositions; examples include kaolin, chalk, barite, talc, tripoli, and Vienna lime.

- *Pumice:* A vitreous spongy compound formed after drying of volcanic lava. Pumice is used for polishing materials such as silver, wood, horn, and bone.

- *Beryllium oxide (BeO):* A white amorphous powder hardened above 1,730°C (3,150°F). Beryllium oxide is used for polishing sintered carbides.

- *Chromium oxide (Cr_2O_3):* A green fine-crystalline very hard material used for polishing hard materials such as chrome, platinum, and steel; the latter acquiring a better corrosion resistance in the process.

- *Iron oxide (Fe_2O_3):* Also known as jeweler's rouge and varies in color from bright red to violet, the darker the color, the higher the hardness. Iron oxide is used for polishing metals, glasses, and stones, and for very fine polishing of gold, silver, brass, and steel.

- *Kaolin:* Used in a fluid, doughy, or solid agent to polish metals, glasses, metallographic specimens, etc.

- *Chalk:* Obtained from crushed, rinsed and washed limestone, is white and very friable. Chalk is perfectly inert to CO_2 and moisture. It is used for polishing silver, brass, and steel.

- *Barite (Barium sulfate):* A white, gray, yellow, or red mineral, soft and friable, used as a polishing agent.

- *Talc:* A natural large agglomerate of hydrous silicate of magnesium, ranging in color from white, shades of green, to red and brown with a glossy sheen. This is the softest polishing agent and is used for polishing soft materials such as alabaster and marble.

- *Tripoli:* A white sedimentary mineral obtained after co-agulation of silica gels in laminae compacted into soft rock mass. Tripoli is easy to crush and is water-absorb-ing. It is used to polish wood, stone, and metals.

- *Vienna lime:* A white powdery composite of fine-crystal-line calcium and magnesium oxides, obtained after burn-ing dolomite. Vienna lime readily binds CO_2 and atmo-spheric moisture, losing its abrasive properties in the process.

- *Garnet*: A homogeneous mineral, which contains no free chemicals. All oxides and dioxides are combined chemi-cally to give $Fe_2O_3Al_2(SiO_4)_3$. The iron and aluminum ions are partially replaceable by calcium, magnesium and manganese. It is red-to-pink in color, with irregular, sharp angular particles. Garnet is inert, nonhygroscopic, slightly magnetic, friable to tough, with a melting point of 2400°F.

- *Corundum*: An aluminum oxide mineral with structural formula Al_2O_3. It is characteristic of A-rich and silica poor environments, and appears in association with other aluminous and/or silica-poor minerals. The structure con-sists essentially of a dense arrangement of oxygen ions in hexagonal closest-packing with Al^{3+} ions in two-thirds of the available octahedral sites, which makes the structure neutral. The sheets are held together by strong covalent bonds and this results in a very hard and dense structure. Crystals of corundum are commonly well formed, often rather large, showing hexagonal and rectangular sections, which are commonly bounded by hexagonal prisms or steep hexagonal di-pyramids resulting in barrel-shaped forms. Corundum is normally colorless, but colored crys-tals are common due to various element impurities. Crys-tals can be red, blue, pale, pink, green, violet, and yellow. Much of the gem quality corundum is found as individual isolated crystals in metamorphosed crystalline limestone. Fine grained corundum is often separated in the heavy fraction of sedimentary rocks and used as an abrasive. The name corundum is sometimes restricted to the nontransparent or coarser kinds.

- *Emery*: A fine-grained intimate mix of corundum and magnetite used for grinding and polishing. Emery is rather dark-colored: black or dark gray. The particle is blocky with sharp edges and has a Mohs hardness of 8. The high amounts of impurities and the weak internal structure as compared to naturally occurring minerals have restricted its wide utilization in new applications.

- *Quartz:* Belongs to the silicate class of minerals (SiO_2), being the most common mineral found on the face of the Earth. Its color is as variable as the spectrum, but clear quartz (rock crystal) is by far the most common color followed by white or cloudy (milky quartz). Purple (amethyst), pink (rose quartz), gray or brown (smoky quartz) are also common. Quartz is transparent to translucent, with a luster that is vitreous to glossy. Quartz crystals are piezoelectric: they produce an electric voltage when pressurized along certain directions of the crystal. Quartz is a 7 on Mohs scale and its fracture is conchoidal. With its trigonal crystal system and no cleavage, is used in ground form as an abrasive in stonecutting, sandblasting, and glass grinding

- *Silica carbide*: The most common source of silica in the "shop environment" is from sand paper that uses silica carbide as the abrasive material. Synthetic silica carbide abrasive materials (clear green sika) is used for sizing, roughing, smoothing, and polishing; using special polishing and smoothing machines as well as variable concave and convex profile NC machines.

- *Glass*: Processed recycled glass has been sold as an abrasive blasting medium under several brand names in different parts of the United States. Limited public information is available regarding the performance of finely sized crushed glass in abrasive applications because manufacturers have developed their own proprietary data. Lower health risk of recycled glass and the fact that it can be processed to physical characteristics close to silica sand represent a potential competitive market advantage. Finely-sized crushed glass has an angular grain shape. It performs well in preparing steel to a "white metal" condition, in dry blasting or combined with water for use in slurry blasting.

11.2.6 Abrasive Wheel Type-marking

An example of the type-marking employed for conventional wheels is presented in Fig.11.2. The first number is optional and represents the manufacturer's symbol indicating the exact quantity of abrasive utilized. The letter A indicates aluminum oxide or the letter C a silicon carbide abrasive material. The number to the right of the abrasive type letter indicates the abrasive grain size by a mesh number; a larger mesh number indicates a finer grain size. The mesh number specifies the number of wires per inch for the control sieve. Since each nominal grain size includes a range of abrasive grain sizes, the grain dimension corresponding to a particular grain size number may be characterized by an average value. The grain dimension, d_g, is quoted as equal to the aperture opening of the control sieve, or, alternatively, according to the approximate relationship:

Eq. (11.2) $$d_g = \frac{0.6}{M} \text{ (inches)}$$

where M is the mesh number.

Sometimes, the manufacturer adds a single digit after the mesh number to indicate the presence of a mixture of grain sizes. Grinding wheel specification is discussed further in Sec. 11.7.

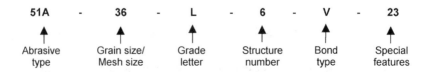

51A	-	36	-	L	-	6	-	V	-	23
Abrasive type		Grain size/ Mesh size		Grade letter		Structure number		Bond type		Special features

Figure 11.2. Wheel type-marking.

11.3 SUPERABRASIVES

The term superabrasive is used for either of the two very hard abrasives, diamond and cubic boron nitride. Diamond is the hardest known abrasive material. Cubic boron nitride is the most recent of the 4 major abrasive types, and the second hardest after diamond. Cubic boron nitride is also known as Borazon®, from General Electric, who discovered it, Amborite® and Amber Boron Nitride® after de Beers, or in Russian literature as Elbor, Cubonite, or β-BN.

The abrasive composite layer on a superabrasive wheel, such as diamond, is limited to a thin rim or layer on a plastic or metal hub in order to reduce the amount of costly diamond or CBN. It was soon realized that this structure also lends itself to the design of very high-speed wheels that can be used for high removal rates.

Not only have synthetic diamonds become a substitute for natural diamond, but also variations of the synthesis process have led to a large spectrum of grain types, having different friability characteristics, and having a large range of applications.

11.3.1 Natural Diamond

Natural diamond grows predominantly in an octahedral form that provides several sharp points optimal for single point diamond tools. It also occurs in a long stone form, created by the partial dissolution of the octahedral form, used in dressing tools such as the Fliesen® blade developed by Ernst Winter & Son. Long stone shapes are also produced by crushing and ball milling diamond fragments. Crushing and milling introduces flaws that significantly reduce strength and life. (See Fig. 11.3.)

Twinned diamond stones called maacles also occur regularly in nature. These are typically triangular in shape. The twinned zone down the center of the triangle is the most wear-resistant surface known. Maacles are used for sharpening chisels as well as for reinforcements in the most demanding form roll applications.

Natural diamond abrasive grains derive from crystals considered unsuitable for jewelry, having flaws, inclusions, and defects. Before use, the diamond is crushed and filtered through a series of mesh nets. The fragments obtained have random shapes, sharp cutting edges, and high strength or low friability. They are bonded into metal or electroplated bonds. The characteristic yellow color is due to nitrogen atoms dispersed in the lattice. The blocky shape of synthetic monocrystalline diamond contrasts with the highly irregular shape of natural diamond grit generated by crushing.

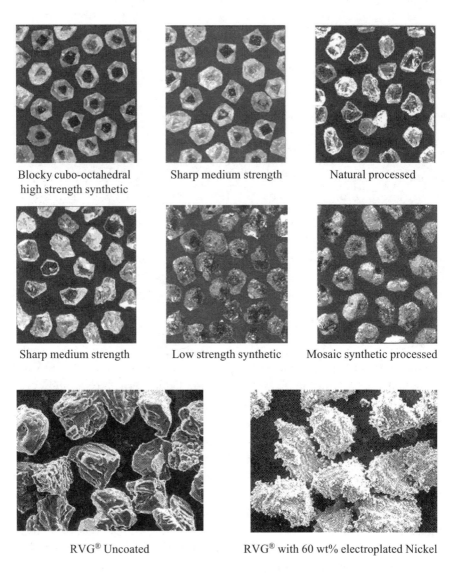

Blocky cubo-octahedral high strength synthetic   Sharp medium strength   Natural processed

Sharp medium strength   Low strength synthetic   Mosaic synthetic processed

RVG® Uncoated   RVG® with 60 wt% electroplated Nickel

Figure 11.3. Typical diamond grit shapes, morphologies,[29] and coatings.[38]–[40]

Natural diamonds were mined in India from 800 BC. Chemically, diamond is an allotropic form of carbon in a cubic system, with impurities of SiO_2, MgO, FeO, Fe_2O_3, Al_2O_3, TiO_2, graphite, etc. In air, diamond starts oxidizing at 800°C–900°C (1,500°F–1,700°F) and graphitizes at 1,000°C–1,100°C (1,900°C–2,000°F). Diamond is resistant to acids and bases, but dissolves in fused soda nitre. Diamond has a high thermal conductivity ($2,092$ W m^{-1} °K^{-1}) and a low electrical conductivity.

The shape of the crystals can be octahedral, dodecahedral or hexahedral. Diamond has very good cleavage parallel to the direction of the octahedron faces.

Diamonds may be used as rough diamonds, cut and polished diamonds, and diamonds for drill tools. Rough diamonds are used for dressing tools, cut and polished diamonds are used for honing tools.

Diamond holds a unique place in the abrasives industry. Being the hardest known material it is not only the natural choice for grinding the hardest and most difficult materials but it is also the only material that can effectively true and dress abrasive wheels. Diamond is the only wheel abrasive that is still obtained from natural sources. Synthetic diamond dominates wheel manufacture but natural diamond is preferred for dressing tools and form rolls. Diamond materials are also used as wear surfaces for end stops and work-rest blades on centerless grinders. In these types of applications, diamond can give 20–50 times the life of carbide.

Diamond is created by the application of extremely high temperatures and pressure to graphite. Such conditions occur naturally at depths of 250 km (120 miles) in the upper mantle of the earth's surface or in heavy meteorite impacts. Diamond is mined from Kimberlite pipes that are the remnant of small volcanic fissures typically 2 to 45 meters (5 to 150 feet) in diameter where magma has welled up in the past. Major producing countries include South Africa, West Africa (Angola, Tanzania, Zaire, Sierra Leone), South America (Brazil, Venezuela), India, Russia (Ural Mountains) Western Australia, and recently, Canada. Each area and even each individual pipe will produce diamonds with distinct characteristics. Production costs are high, on average 6 million kg (13 million lbs) of ore must be processed to produce ½ kg (1 lb) of diamonds. Much of this cost is supported by the demand for the jewelry trade. Since World War II, the output of industrial grade diamond has been far outstripped by demand. This spurred the development of synthetic diamond programs initiated in the late 1940s and 1950s.[65]

The stable form of carbon at room temperature and pressure is graphite that consists of carbon atoms in a layered structure. Within the layer, atoms are positioned in an hexagonal arrangement with strong sp^3 covalent bonding. However, bonding between the layers of graphite is weak. Diamond is metastable at room temperature and pressure and has a cubic arrangement of atoms with pure sp^3 covalent bonding. There is also an intermediate material called wurtzite or hexagonal diamond where the hexagonal layer structure of graphite has been distorted above and below the layer planes but not quite to the full cubic structure. The material is nevertheless almost as hard as the cubic form.

The principal crystallographic planes of diamond are the cubic (100), dodecahedron (011) and octahedron (111). The relative rates of growth on these planes are governed by the temperature and pressure conditions, together with the chemical environment both during growth and, in the case of natural diamond, during possible dissolution during its travel to the earth's surface. This, in turn, governs the stone shape and morphology.

The direct conversion of graphite to diamond requires temperatures of 2,200°C (4,000°F) and pressures higher than 10.35 GPa (1.5×10^6 psi). Creating these conditions was the first hurdle to producing man-made diamonds. General Electric achieved it through the invention of a high pressure/temperature gasket called the "belt" and announced the first synthesis of diamond in 1955. Somewhat to their chagrin, it was then announced that a Swedish company, ASEA had secretly made diamonds 2 years previously using a more complicated 6-anvil press. The Swedish company, ASEA, had not announced the fact because they were seeking to make gems and did not consider the small brown stones they produced the culmination of their program. De Beers announced their ability to synthesize diamonds shortly after GE in 1958.

The key to manufacture was the discovery that a metal solvent such as nickel or cobalt could reduce the temperatures and pressures required to more manageable levels. Graphite has a higher solubility in nickel than diamond, therefore, graphite first dissolves in the nickel and then diamond precipitates out. At higher temperatures, the precipitation rate is faster and the number of nucleation sites is greater. The earliest diamonds were grown fast at high temperatures and had weak, angular shapes with a mosaic structure. This material was released by General Electric under the trade name RVG® for *Resin Vitrified Grinding* wheels. Most of the early patents on diamond synthesis have now expired and competition from emerging economies have driven down the price to as low as $880 per kg ($400/lb.) although quality and consistency from some sources is questionable.

By controlling growth conditions, especially time and nucleation density, it is possible to grow much higher-quality stones with well-defined crystal forms: cubic at low temperature, cubo-octahedral at intermediate temperatures, and octahedral at the highest temperatures.

The characteristic shape of good quality natural stones is octahedral, but the toughest stone shape is cubo-octahedral. Unlike in nature, this can be grown consistently by manipulation of the synthesis process. This has led to a range of synthetic diamond grades, typified by the MBG® series from GE and the PremaDia® series from de Beers, which are the abrasives of choice for saws used in the stone and construction industry, and wheels for grinding glass.

Quality and price of the abrasive is governed by the consistency of shape and, also, by the level of entrapped solvent in the stones. Since most of the blockiest abrasive is used in metal bonds processed at high temperatures, the differential thermal expansion of metal inclusions in the diamond can lead to reduced strength or even fracture. Other applications require weaker phenolic or polyamide resin bonds processed at much lower temperatures and use more angular, less thermally stable diamonds. Grit manufacturers therefore characterize their full range of diamond grades by room temperature toughness (TI), thermal toughness after heating at, for example, 1,000°C (1,800°F) (TTI), and shape (blocky, sharp, or mosaic). In the mid-range, sharp grades include crushed natural as well as synthetic materials.

Diamond coatings are common. One range includes thick layers or claddings of electroplated nickel, electronless Ni-P and copper or silver at up to 60% wt. The coatings act as heat sinks, while also increasing bond strength and preventing abrasive fragments from escaping. Electroplated nickel for example produces a spiky surface that provides an excellent anchor for phenolic bonds when grinding wet. Copper and silver bonds are used more for dry grinding, especially with polyamide bonds, where the higher thermal conductivity outweighs the lower strength of the coating.

Coating can also be applied at the micron level either as a wetting agent or as a passive layer to reduce diamond reactivity with the particular bond. Titanium is coated on diamonds used in nickel, cobalt, or iron-based bonds to limit graphitization of the diamond while wetting the diamond surface. Chromium is coated on diamonds used in bronze-based bonds to enhance chemical bonding and reactivity of the diamond and bond constituents.

For electroplated bonds, the diamonds are acid etched to remove any surface nodules of metal solvent that would distort the plating electrical potential on the wheel surface leading to uneven nickel plating or even nodule formation. Etching also creates a slightly rougher surface to aid mechanical bonding.

Since 1960, several other methods of growing diamond have been developed. In 1970 DuPont launched a polycrystalline material produced by the sudden heat and pressure of an explosive shock (see Fig. 11.4). The material was wurtzitic in nature and produced mainly micron-size particles more suitable for lapping and polishing than for grinding.

11.3.2 Synthetic Diamond

Synthetic diamond may be monocrystalline or polycrystalline. *Monocrystalline grits* are utilized for particularly demanding applications. The easily recognizable cubic or cubo-octahedral shape reflects the characteristic crystallographic structure of diamond. The almost perfect crystals are obtained through a slow growth, low-density nucleation process, with limited metallic inclusion and little or low interaction between various grits that are growing within the melt.

Polycrystalline grits that are highly friable are produced by greatly accelerating the nucleation rate within the press so the diamond nuclei precipitate from the melt in a large number. Due to the crowding in the melt, the normal growth pattern is inhibited, so grits having undefined geometric shapes are obtained that look like an agglomerate of smaller crystals. These are very prone to fragmentation and partial break-up with dark spots within the grits that correspond to metallic inclusions. Weak crystals used in resinoid grinding wheels are free-cutting.

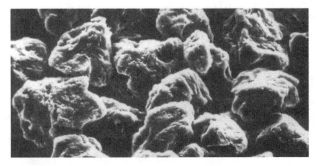

Figure 11.4. DuPont Mypolex® polycrystalline diamond produced by explosive shock.

Metal-clad polycrystalline grits provide the possibility of high stock removal rates. The surface tension of diamond grits worsens the contact characteristics between resin and diamond, making difficult the extraction of worn diamond grits from the bond. Conversely, there is a good interaction between resin bond and metal-cladding, increasing the effective surface of the grit.

Early synthetic diamonds were obtained by converting graphite at 1500°C–3000°C (2800°F–5500°F) and 5.2 to 10.35 MPa (0.75 to 1.5 × 10⁶ psi), in the presence of catalysts such as iron, chromium, nickel, titanium, radium, and cobalt. These are more friable and have coarser faces as compared to natural diamond. Due to this, synthetic diamonds are retained well in resinoid bonds and perform better than natural diamonds by 30% up to 100%.

Chemical vapor deposition (CVD) diamond forms as a fine crystalline columnar structure (see Fig. 11.5). There is a certain amount of preferred crystallographic orientation exhibited; more so than, for example, polycrystalline diamond (PCD) but far less than in single crystal diamond. Wear characteristics are therefore much less sensitive to orientation in a tool. The CVD diamond is not used as an abrasive but is very promising fabricated in the form of needle-shaped rods for use in dressing tools and rolls. Fabrication with CVD is slightly more difficult as it contains no metal solvents to help during electrical discharge wire-cutting machining. Diamond wetting also appears to be more difficult and must be compensated for by the use of an appropriate coating.

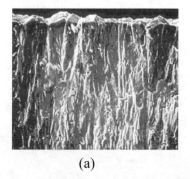
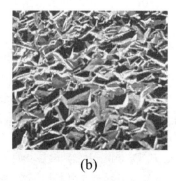

(a)   (b)

Figure 11.5. CVD diamond microstructure:[41] *(a)* cross section, and *(b)* as deposited surface.

In the last decade, increasing effort has been made to grow large synthetic diamond crystals at high temperatures and pressures. The big limitation has always been that press time and, hence, cost goes up exponentially with diamond size. The largest saw grade diamonds are typically #30/40. The production of larger stones in high volume, suitable for tool and form roll-dressing applications, is not yet cost-competitive with natural diamond. However, a recent exception to this, first introduced by Sumitomo, is needle diamond rods produced by the slicing up of large synthetic diamonds. The rods are typically less than 1 mm in cross section by 2 to 5 mm long (similar in dimensions to the CVD diamond rods discussed above), but orientated along the principle crystallographic planes to optimize wear and fracture characteristics.

Despite the dramatic growth in synthetic diamond, demand for natural diamond has not declined. If anything, the real cost of natural diamond has actually increased especially for higher quality stones. The demand for diamonds for jewelry is such that premium stones used in the 1950s for single-point tools are more likely now to be used in engagement rings. Very small gem-quality stones, once considered too small for jewelry and used in profiling dressing discs, are now being cut and lapped by countries such as India. With this type of economic pressure it is not surprising that the diamonds used by industry are those rejected by the gem trade because of either color, shape, size, crystal defects such as twins or naats, or inclusion levels, or are the processed fragments, for example, from cleaving gems. Although significant quantities of processed material are still used in grinding wheel applications, it is the larger stones used in single-point and form-roll dressing tools that are of most significance. Here, the quality of the end product depends on the reliability of the diamond source and of the ability of the toolmaker to sort diamonds according to requirements. The highest quality stones are as-mined virgin material. Lower-quality stones may have been processed by either crushing and/or ball-milling, or even reclaimed from old form rolls or drill bits where they have previously been subjected to high temperatures or severe conditions. (See Fig. 11.6.)

11.3.3 Chemical-Physical Properties of Diamond

Diamond, despite its great hardness and wear resistance, is not indestructible. The durability is a consequence of a very particular coincidence of a high thermal conductivity (2 to 6 times larger than copper), and a low thermal expansion coefficient. Diamond conducts heat by serving as a medium for phonons, which are small packets or quanta of vibrational energy. The phonon transfer is extremely efficient due to the low mass of carbon atoms and high bonding energy between them. Diamond, as a metastable form of graphite at normal pressures, tends to turn into graphite at relatively low temperatures: 700°C (1,300°F) in air and 1,500°C (2,700°F) when surrounded by an inert atmosphere. When subjected to heat for long periods, diamond degrades rapidly.

Since diamond is a form of carbon, interactions with ferrous materials are to be expected. The chemical affinity of carbon for low carbon steels can facilitate a chemical attrition that leads to formation of specific compounds such as Fe_3C, which tends to cause premature pull-out of diamond grits.

Premium dressing
stones

Lower quality/processed
dressing stones

Premium long stones

Lower quality/processed
long stones

Maacles

Ballas

Figure 11.6. Natural industrial diamonds.[44]

11.3.4 Cubic Boron Nitride (CBN)

Cubic boron nitride (CBN) is an allotropic crystalline form of boron nitride almost matching the hardness of diamond. Cubic boron nitride was first synthesized in 1957. The hexagonal crystals are obtained at 2,800°F–3,700°F and pressures of 3.45 to 6.2 MPa (0.5 to 0.9 × 10⁶ psi), in the presence of catalyzers such as alkali metals, antimony, lead or tin, lithium, magnesium, and nitrides. The popularity of CBN is due to its hardness, thermal resistance which is higher than diamond's, allowing work at 1,900°C (3,500°F), and the good chemical resistance of CBN to ferrous alloys. Cubic boron nitride is used mainly for grinding high-quality tool steels.

Because of its chemical nature, CBN has no affinity for low carbon steels, being widely employed for grinding high-speed steels. Cubic boron nitride has an excellent thermal stability; oxidation starts above 1,000°C (1,800°F) and becomes complete around 1,500°C (2,700°F). In addition, the boron oxide layer created protects against further oxidation. Therefore, high speeds of 30.5–61 meters per second (6,000 to 12,000 feet per minute) can be employed. However, the affinity of CBN to water vapor, which dissolves the protective oxide layer and creates the conditions for hydrolysis to take place with boric acid and ammonia as an output, is sometimes thought to be a drawback to wet grinding. In practice, wet grinding with CBN is much better than dry grinding. The most efficient coolant with CBN abrasive is straight sulphurated mineral oil, or, second best, emulsion of sulphochlorinated oils in water with a minimum concentration of 10%.

11.3.5 Boron Nitride (B₄N)

Boron nitride (B₄N) is a crystalline material synthesized from boric anhydride and pure low-ash carbon material in electric furnaces at 1,800°C–2,500°C (3,300°F–4,500°F). Its hardness is about 3,800 HV and it has a good cutting ability in the form of loose grains. However, a low oxidation temperature, of 430°C (800°F), prevents the use of boron nitride for grinding wheels. It is used exclusively in the form of pastes for sintered carbide lapping, or as grit for sandblasting.

11.4 STRUCTURE OF SUPERABRASIVES

Strength and hardness of superabrasives are a direct consequence of a particular crystallographic structure common to both diamond and CBN resulting from string tetrahedral covalent bonds between atoms. This is as surprising as the fact that diamonds are derived from their hexagonal equivalents, referred to as graphite and white graphite.

11.4.1 Structure of Diamond

Diamond and graphite are allotropic forms of carbon, which differ only in bond structure between the same carbon atoms. Graphite has a lamellar structure with strong covalent bonds between two atoms of carbon within a 2-D hexagonal layer, which explains its use as a lubricant agent, with a 1.42 Å interatomic distance. Between two hexagonal layers the bonding is much weaker, with an interatomic distance of 3.35 Å, so the sliding between layers is relatively facilitated. While in graphite each carbon atom has three principal partners within its own layer, diamond has a tetrahedral bonding with each carbon at the center of a tetrahedron formed by its four nearest neighbors. This four-fold bonding and the high degree of symmetry are responsible for the remarkable strength and hardness of the diamond lattice. The interatomic distance in diamond of 1.54 Å is comparable with the distance measured within a hexagonal building block in graphite, but much smaller than the distance between two layers. Hence, the large difference between the densities of diamond 3,500 kg/m³ (0.127 lbs per cubic inch) and graphite 2,200 kg/m³ (0.08 lbs per cubic inch). The tetrahedron structure with one carbon atom at the center and four carbon atoms at the four apexes explains why diamonds frequently occur as crystals with easily identifiable octahedral, cubic or cubo-octahedral shapes.

11.4.2 Structure of Cubic Boron Nitride (CBN)

The description given for the structure of diamond applies also to CBN if the central atom is considered to be replaced by boron and its neighbors by nitrogen atoms. The newly formed covalent compound has two allotropic forms: a hexagonal form (known as white graphite), and a cubic form known as CBN. Since the covalent bond between nitrogen and boron atoms is heteropolar, the hardness and density of CBN are lower than in the perfectly symmetrical homopolar diamond, where only carbon atoms bond together. Another notable difference concerns the availability of various allotropic forms; whereas diamond and graphite are natural products, CBN is a purely synthetic material.

The development of CBN was swiftly oriented towards increasing differentiation and specialization. The procedure for producing CBN is similar to the one for producing synthetic diamond. The raw material utilized is hexagonal cubic boron nitride or white graphite, derived from the pyrolysis of boron chloride-ammonia (BCl_3NH_3).

Boron nitride is made at room temperatures and at pressures using the reaction:

Eq. (11.3) $BCl_3 + NH_3 \rightarrow BN + 3 \cdot HCl$

The resulting product is a white slippery substance with an hexagonal layered atomic structure called HBN (or α-BN) similar to graphite but with alternating nitrogen and boron atoms. Nitrogen and boron lie on either side of carbon in the periodic table, and it was realized that high temperatures and pressures could convert HBN to a cubic structure similar to diamond. The first commercial product was released in 1969.

Both cubic (CBN) and wurtzitic (WBN or γ-BN) forms are created at comparable pressures and temperatures to those for carbon. The chemistry of BN is quite different to carbon; for example, bonding is not pure sp^3 but 25% ionic and BN does not show the same affinity for transition metals. The solvent/catalyst turned out to be any one of a large number of metal nitrides, borides, or oxide compounds of which the earliest commercially used one (probably with some additional doping) was Li_3N. This allowed economic yields at 6 MPa (870,000 psi), 1,600°C (2,900°F) and cycle times faster than 15 minutes.

As with diamond crystal growth, CBN shape is governed by the relative growth rates on the octahedral (111) and cubic planes. However, the (111) planes dominate and furthermore, because of the presence of both B and N in the lattice, some (111) planes are positive terminated by B atoms and some are negative terminated by N atoms. In general, B (111) plane growth dominates and the resulting crystal morphology is a truncated tetrahedron. Twinned plates and octahedrons are also common. Only by further doping and/or careful control of the P-T conditions can the morphology be driven towards the octahedral or cubo-octahedral morphologies.

As with diamond, CBN grit grades are characterized by toughness, TI and TTI, and by shape. Of these, TI and TTI dominate, controlled by doping and impurity levels. Surface roughness is also more of a factor than for diamond. For example, GE 1 abrasive is a relatively weak irregular crystal. The coated version GE II used in resin bonds has a simple nickel-plated cladding. However, GE 400 which is a tougher grain but with a similar shape has much smoother, cleaner faces. The coated version GE 420 is therefore first coated with a thin layer of titanium to create a chemically bonded roughened surface to which the nickel cladding can be better anchored.

Only a relatively few grades of CBN are tough and blocky with crystal morphologies shifted away from tetrahedral growth. The standard example is GE 500 used primarily in electroplated wheels. De Beers also has a material, ABN 600, whose morphology has been driven towards the cubo-octahedral.

The manufacture of CBN is dominated by GE in the USA, by de Beers from locations in Europe and South Africa, and by Showa Denko from Japan. Russia and Romania have also been producing CBN for over 30 years, and recently China has become a significant player. Historically, consistency has always been a problem with materials from these latter sources but, with intermediate companies such as ABC Abrasives, and St. Gobain Abrasives Div., they are becoming a real low-cost alternative to traditional suppliers. It is, therefore, expected that CBN prices will be driven down over the next decade offering major new opportunities for CBN technology. Currently CBN costs are of the order of $1,500/lb– $5,000/lb or at least 3–4 times that of the cheapest synthetic diamond.

As with carbon, wurtzitic boron nitride (WBN) can also be produced by explosive shock methods. Reports of commercial quantities of the material began appearing in about 1970, Nippon Oil and Fats,[74] but its use has again been focussed more on cutting tool inserts with partial conversion of the WBN to CBN, and this does not appear to have impacted the abrasive market.

11.4.3 PCD Diamond

The 1970s saw the introduction of polycrystalline diamond (PCD) blanks that consisted of a fine grain sintered diamond structure bonded to a tungsten carbide substrate. The material was produced by the action of high temperatures and pressures on a diamond powder mixed with a metal solvent. Since it contained a high level of metal binder it could be readily fabricated in various shapes using EDM technology. Although not used in grinding wheels it is popular for reinforcements in form dress rolls and for wear surfaces on grinding machines. Its primary use though is in cutting tools.

11.4.4 CVD Diamond

In 1976, reports began to come out of Russia of diamond crystals produced at low pressures through chemical vapor deposition (CVD). This was initially treated with some skepticism in the west even though Russia had a long history of solid research on diamond. However, within five years Japan was also reporting rapid growth of diamond at low pressures and the product finally became available in commercial quantities about 1992. The process involves reacting a carbonaceous gas in the presence of hydrogen atoms in near vacuum, to form diamond on an appropriate substrate. Energy is provided in the form of hot filaments or plasmas at temperatures higher than 815°C (1,500°F) to dissociate the carbon and hydrogen into atoms. The hydrogen interacts with the carbon and prevents any possibility of graphite forming while promoting diamond growth on the substrate. The resulting layer can form to thicknesses deeper than 1 mm.

11.5 GRIT SIZES, GRIT SHAPES, AND PROPERTIES

11.5.1 Grit Sizes

Grit sizes are traditionally measured by the fraction of bulk grains that can pass through a series of vibrating screens or meshes having a specified number of openings per square inch. The specifications do not call for 100% of the grains to be of nominal size. A 60–100 grit size allows the mix to include a small percentage of coarser grains and a somewhat higher percentage of finer grains.

Grain sizes relate to the screening mesh used for sizing. Mesh number is the number of wires per linear inch of the screen, through which the grains pass while being retained at the next, finer size screen. Grain dimensions are the dimensions of the smallest cubicoid that can be described on the given grain. The largest grain produced may be whatever passes through a quarter-inch screen. The smallest may be so fine it floats on water. The size of the grains reduces as mesh number increases. It is therefore important to know which measure is employed to classify grit size. The Federation of European Producers of Abrasives (FEPA) uses grain dimension for grit size and ANSI uses mesh number for grit size, as follows.

The particle size distribution of abrasive grains is defined by numerous international standards. All are based on sizing by sieving in the sizes typical of most grinding applications. In the case of the ANSI standard (B74.16 1995), mesh size is defined by a pair of numbers that correspond to sieves with particular mesh sizes. The lower number gives the number of meshes per linear inch through which the grain can only just fall, while staying on the surface of the sieve with the next highest number of meshes which is the higher number. The Federation of European Producers of Abrasives (FEPA) (ISO R 565–1990, also DIN 848-1988) gives the grit size in microns of the larger mesh hole size through which the grit will just pass. The two sizing standards are in fact very similar, FEPA has a somewhat tighter limit for oversize and undersize (5–12%) but no medium nominal particle size. ANSI has somewhat more open limits for oversize and undersize (8–15%) but a targeted mid-point grit dimension. FEPA is more attuned to the superabrasive industry, especially in Europe, and may be further size controlled by the wheel maker; ANSI is more attuned to conventional wheels and in many cases may be further broadened by mixing 2 or 3 adjacent sizes.

There is also a system called US grit size with a single number that does not quite correlate with either the upper or lower ANSI grit size number. This has created considerable confusion especially when using a single number in a specification. A FEPA grit size of 64 could be given as 280, 230, or 270 depending on the wheel manufacturers particular coding system. This can readily lead to error of a grit size when selecting wheel specifications unless the code system is well-defined. In Table 11.4, correspondence between different systems is presented.

11.5.2 Modern Grain Developments

Research is accelerating both into alumina-based grain technology and into new ultra-hard materials. In the area of ceramic-processed alumina materials, St. Gobain released SG in 1986 (US Patent 4,623,364) followed

Table 11.4. Correspondence Between Systems of Grit Size Classifications

FEPA Designation	ISO R 5665–1990 Aperture range (microns)	ANSI Grit Size	US GRIT Number
Standard			
1181	1180/1000	16/18	
1001	1000/850	18/20	
851	850/710	20/25	
711	710/600	25/30	
601	600/500	30/35	
501	500/425	35/40	
426	425/355	40/45	
356	355/300	45/50	
301	300/250	50/60	50
251	250/212	60/70	60
213	212/180	70/80	80
181	180/150	80/100	100
151	150/125	100/120	120
126	125/106	120/140	150
107	106/90	140/170	180
91	90/75	170/200	220
76	75/63	200/230	240
64	63/53	230/270	280
54	53/45	270/325	320
46	45/38	325/400	400

(Cont'd.)

Table 11.4. *(Cont'd.)*

FEPA	ISO R 5665–1990	ANSI	US GRIT
Designation	Aperture range (microns)	Grit Size	Number
Wide Range			
1182	1180/850	16/20	
852	850/600	20/30	
602	600/425	30/40	
502	500/355	35/45	
427	425/300	40/50	
252	250/180	60/80	

by extruded SG in 1991 (US Patent 5,009,676). More recently, in 1993, Treibacher released an alumina material with hard filler additives (US Patent 5,194,073).

Electrofusion technology has also advanced. Pechiney produced an Al-O-N grain (Abral™) produced by fusion of alumina and A/ON followed by slow solidification. It offers much higher thermal corrosion resistance relative to regular alumina while also having constant self-sharpening characteristics akin to ceramic-processed grain materials but softer acting.[80]

New materials have also been announced with hardness approaching CBN and diamond. Iowa State University announced in 2000 an Al-Mg-B material with a hardness comparable to CBN (US Patent 6,099,605). Dow Chemical patented in 2000 an Al-C-N material with a hardness close to diamond (US Patent 6,042,627). In 1992, the University of California patented some α-C_3N_4 and β-C_3N_4 materials that may actually be harder than diamond (US Patent 5,110,679).

Whether any of these materials will eventually prove to have good abrasive properties and can be produced in commercial quantities has yet to be seen. Nevertheless, there will undoubtedly be considerable advances in abrasive materials in the coming years.

11.5.3 Friability

The concept of friability is very important for the operation of diamond grinding wheels. Any diamond, regardless of shape and size has the same hardness. But the friability defines the tendency to break up into smaller fragments under pressure. Friability is a function of shape, integrity, and purity of the crystal; properties that are themselves a consequence of the crystal growth characteristics during synthesis. Due to its adjustable brittleness and friability, synthetic diamond enlarged the application range of diamond as an abrasive. When metallic lithium is added, the grits obtained are black and opaque, due most probably to an excess of boron in the crystals. When lithium nitride is utilized as a solvent/catalyst, the crystals obtained are yellow and translucent. Both crystals obtained are well geometrically shaped with sharp cutting edges.

The toughness or friability of the abrasive grains is not the only important characteristic of grits. Grain size, hardness, shape, specific gravity, and structure are important too.

11.5.4 Hardness

Hardness is usually defined as the resistance to penetration by another material. There are a number of hardness scales mostly based on the principle of the depth made in the tested material by a standard indenter under a standard pressure. However, because of the wide range of hardness in materials, no standard covers the entire spectrum of materials hardness. For abrasives, the scales used most are Rockwell-Cone and Knoop.

11.5.5 Grain Shape

Abrasive grain shape affects the number of cutting edges, and helps define the spatial distribution of cutting edges and points, relative to a workpiece and the magnitude of the apex angles. For grinding, there is usually no possibility to orient the grains during the wheel-making process, so the grain for wheels is blocky. For applications when high pressures are required, blocky grains withstand pressure better than elongated grains. For all coated abrasive products, the process includes a stage of orienting the elongated grain. Microscopic comparison of abrasive grains with standard samples is the most common test to determine the grain shape, but is difficult to accomplish within a shop-floor production process.

11.5.6 Specific Gravity

The specific gravity of SiC is 3.2 and of aloxide nearly 4.0. In comparison, water is 1.0, cast iron is 7.2, and gold is 19.3.

11.5.7 Porosity

Vitrified wheels have a bond structure porous enough to allow large quantities of coolant to wash the tool-workpiece interface. Porosity plays an important role in the performance of an abrasive. The effects of porosity on grinding wheel performance are discussed in greater detail in Ch. 10, "Grinding Wheel and Abrasive Topography."

11.5.8 Properties of Diamond

Mechanical Properties. The hardness of diamond is a difficult property to define for two reasons. First, hardness is a measure of plastic deformation but diamond does not plastically deform at room temperature. Second, hardness is measured using a diamond indenter. However, since the hardness of diamond is quite sensitive to orientation, a Knoop indenter is used. A Knoop indenter is a distorted pyramid with a long diagonal seven times the short diagonal. The indenter is orientated in the hardest direction so that reasonably repeatable results can be obtained. The following hardness values have been recorded:

(001) plane [110] direction 10,400 kg/mm^2

(001) plane [100] direction 5,700 kg/mm^2

(111) plane [111] direction 9,000 kg/mm^2

A more important measure than hardness though is mechanical wear resistance. This is also a difficult property to pin down because it is so dependent on load, material hardness, speed, etc. Wilks and Wilks[97] showed that when abrading diamond with diamond abrasive the wear resistance follows the hardness but the differences between orientations are far more extreme. For example, on the cube plane, the wear resistance between the (100) and (110) directions varies by a factor of 7.5. In other planes, the differences can be as great as a factor of 40 sometimes with only relatively small changes in angle. Not surprisingly, diamond gem lappers often talk of diamonds having *grain*. Factors regarding wear resistance of diamond on other materials in grinding must include all attritious wear processes including thermal and chemical.

Being so hard, diamond is also very brittle. It can be readily cleaved on its 4 (111) planes. Its measured strength varies widely due, in part, to the nature of the tests but also because it is heavily dependent on the level of defects, inclusions, and impurities present. Not surprisingly, small diamonds (with smaller defects) give higher values for strength than larger diamonds. The compressive strength of top quality synthetic diamond (#100) grit has been measured at 9.75 MPa (1.4×10^6 psi).

Chemical Properties. The diamond lattice is surprisingly pure; the only other elements known to be incorporated are nitrogen and boron. Nitrogen is present in synthetic diamonds up to 500 parts per million in single substitutional sites and gives the stones their characteristic yellow/green color. Over extended time, at high temperature and pressure, the nitrogen migrates and forms aggregates, and the diamond becomes the colorless stone found in nature. Synthetic diamond contains up to 10% included metal solvents while natural diamond usually contains inclusions of the minerals in which it was grown (olivine, garnet, spinels).

Diamond is metastable at room temperatures and pressures; it will convert to graphite given a suitable catalyst or sufficient energy. In a vacuum or inert gas, diamond remains unchanged up to 1,500°C; in the presence of oxygen it will begin to degrade at 650°C (1,200°F). This factor plays a significant role in how wheels and tools are processed in manufacturing.

Diamond is also readily susceptible to chemical degradation from carbide formers, such as tungsten, tantalum, titanium, and zirconium, and true solvents of carbon including iron, cobalt, manganese, nickel, chromium, and the Group VIII platinum and palladium metals. This chemical affinity can be both a benefit and a detriment. It is a benefit in the manufacture of wheels and tools where the reactivity can lead to increased wetting and therefore higher bond strengths in metal bonds. For diamond tool manufacture, the reactant is often part of a more complex eutectic alloy (for example, copper-silver, copper-silver-indium, or copper-tin), in order to minimize processing temperature, disperse and control the active metal reactivity, and/or allow simplified processing in air. Alternatively, tools are vacuum brazed. For metal-bonded wheels, higher temperatures and more wear-resistant alloy bonds are used but fired in inert atmospheres.

The reactivity of diamond with transition metals such as nickel and iron is a major limitation to the use of diamond as an abrasive for these materials. Some researchers[89][90] showed that during single-point turning of mild steel with diamond, chemical wear was excessive and exceeded abrasive mechanical wear by a factor of 10^4. During single-point turning of pearlitic cast iron, though, the chemical wear rate exceeded the abrasive mechanical wear rate only by a factor of 10^2. Furthermore, the wear on pearlitic cast iron was actually 20 times less than that measured using CBN tools. Much less effect was seen on ferritic cast iron which, unlike the former material, contained little free carbon; in this case, diamond wear increased by a factor of 10, turning workpieces of comparable hardness.

It is generally considered, as the above results imply, that chemical-thermal degradation of the diamond prevents diamond from being used as an abrasive for steels and nickel-based alloys but that, under certain circumstances, free graphite in some cast irons can reduce the reaction between diamond and iron to an acceptable level. For example, in honing of automotive cast iron cylinder bores, performed at similar cutting rates and speeds of 1.2 m/min (4 feet per minute) to those used in the turning experiments mentioned above, diamond is still the abrasive of choice, out-performing CBN by a factor of 10. However, in the case of cylindrical grinding of cast iron camshafts, at the higher speeds of 1.35 m/min typically (16,000 feet per minute) and temperatures the case is reversed.

Thermal Properties. Diamond has the highest thermal conductivity of any material with a value of 600 to 2,000 Wm^{-1}°K^{-1} at room temperature falling to 70 Wm^{-1}°K^{-1} at 700°C (1,300°F). These values are 40 times greater than the thermal conductivity of alumina. However, an abrasive can have an extremely high thermal conductivity but if the heat capacity of the material is low it will simply get hot faster. Thermal models such as those based on Jaeger,[50] have therefore identified a composite thermal property as key to temperatures in transient situations in grinding. Please note for grain temperatures in grinding, the steady-state thermal conductivity is much more important than the transient thermal property. (See the Hahn model in Ch. 6, "Thermal Design of Processes.") The thermal property is $\beta = (k\rho c)^{1/2}$ where k is the thermal conductivity, ρ is the density, and c is the specific heat capacity. Please note that thermal diffusivity, α, is not the same as the transient thermal property β (see Ch. 6); $(k\rho c)^{1/2}$ for diamond is 600 mm^2 s^{-1} compared to 0.30 to 150 mm^2 s^{-1} for most ceramics, including alumina and SiC, and steels. Copper has a value of 370 mm^2 s^{-1} due in part to a much higher heat capacity than that of diamond. This may explain its benefit as a cladding material. The linear thermal expansion of diamond is (1×10^{-6})/K at 200°C increasing to (4.8×10^{-6})/°K at 870°C (1,600°F). The thermal expansion coefficients are of significance to bonded wheel manufacturers who must try to match thermal expansion characteristics of bond and grit throughout the firing cycle.

11.5.9 Properties of CBN

Mechanical Properties. The hardness of CBN at room temperature is about half that of diamond but double that of conventional abrasives. The differences in abrasion resistance however are much more extreme. Difference by a factor of 2 translates into a difference of 100–1,000 times or even greater in abrasion resistance depending on the abrading material. As with diamond, the key is the total wear resistance to all attritious wear processes.

Like diamond, CBN is brittle, but differs in having 6 (110) rather than 4 (111) cleavage planes. This gives a more controlled breakdown of the grit especially for the truncated tetrahedral shape of typical CBN grains. The grain toughness is generally much less than that of blocky cubo-octahedral diamonds. This, combined with its lower hardness, provides one very useful advantage. The CBN wheels can be dressed successfully by diamond (rotary) tools.

Chemical Properties. Cubic boron nitride (CBN) is thermally stable in nitrogen or vacuum to at least 1,500°C (2,700°F). In air or oxygen, CBN forms a passive layer of B_2O_3 on the surface that prevents further oxidation up to 1,300°C (2,400°F). However, this layer is soluble in water, or more accurately high temperature steam at 900°C (1,650°F), and will allow further oxidation of the CBN grains following the reactions in the following equation:[20][100]

Eq. (11.4)
$$2BN + 3H_2O \rightarrow B_2O_3 + 2NH_3$$
$$BN + 3H_2O \rightarrow H_3BO_3 + N_2 > 900°C \ (1,650°F)$$
$$B_2O_3 + 3H_2O \rightarrow 2H_2BO_3 > 950°C \ (1,750°F)$$
$$4BN + 3O_2 \rightarrow 2B_2O_3 + 2N_2 > 1,000°C \ (1,800°F)$$

This reactivity has been associated with reduced life grinding in water based coolants compared with straight oil coolants. However the importance of this reaction is not so clear cut, as discussed in the chapter on coolants (Ch. 8, "Fluid Delivery").

Cubic boron nitride (CBN) is also reactive towards alkali oxides, not surprisingly in the light of their use as solvents and catalysts in CBN synthesis. The B_2O_3 layer is particularly prone to attack or dissolution by basic oxides such as Na_2O GE Superabrasives[39] by reaction:

$$B_2O_3 + Na_2O \rightarrow Na_2B_2O_4$$

Such oxides are common constituents of vitrified bonds and the reactivity can become extreme at temperatures above 900°C (1,650°F) affecting processing temperatures for wheels.[99] Cubic boron nitride does not show any significant reactivity or wetting by transition metals such as iron, nickel, cobalt, or molybdenum until temperatures in excess of 1,300°C (2,370°F). The low reactivity is reflected in low rate of wear grinding these materials in comparison with diamond.

Cubic boron nitride (CBN) shows marked wetting by aluminum at only 1,000°C (1,900°F), while with titanium as demonstrated in wetting studies of low temperature silver – titanium eutectics, CBN reacts readily at 1,000°C (1,830°F) to form TiB_2 and TiN.[16] This provides an explanation of why in grinding aerospace titanium alloys such as Ti-6Al-4V, CBN wheels wear typically 5 times faster than diamond wheels.[57] By comparison the wear rate using SiC abrasive is 40 times greater than CBN.

Pure CBN is colorless although commercial grades are either black or amber depending on the level and type of dopants. The black color is believed to be due to an excess of boron doping.

Thermal Properties. The thermal conductivity of CBN is almost as high as that of diamond. At room temperature its value is 200–1300 $W\ m^{-1}\ ^{\circ}K^{-1}$, depending on purity. Cubic boron nitride has a transient thermal diffusivity of $\beta = 480\ mm^2\ s^{-1}$. The thermal expansion coefficient of CBN is about 20% higher than the diamond's characteristic value.

11.6 BONDS

The primary function of a bond is to hold the abrasive grits during the grinding process. The maximum grit retention capability is not synonymous with high performance since the bond should fulfill several functions:

- Provide adequate grit retention, without premature pull-out from the surface of the abrasive.
- Allow controlled bond erosion, leading to gradual exposure of new cutting points.
- Provide sufficient strength to ensure optimal transfer of the grinding forces from the spindle to workpiece.
- Provide adequate heat dissipation.

Grit friability and grit retention are the two most crucial factors at the grinding interface. Bond systems can be divided into two types: those holding a single layer of abrasive grain to a solid core, and those providing a consumable layer many grains thick with the abrasive held within the bond. The latter may be mounted on a resilient core or produced as a monolithic structure from the bore to the outer diameter. There are also segmented wheels used for special purposes.

There are three major bond types used for all abrasives plus a fourth single-layer group used for superabrasives. There are also coated abrasives which are dealt with in Sec. 11.9. The three major types are:

- Resin bonds.
- Vitrified bonds.
- Metal bonds.

Common to all types of bonds is the fact that during the bonding process the bonding agent becomes fluid and wets to some degree all other components present in the rim. Mechanical properties for the three major bond types are given in Table 11.5.

Table 11.5. Mechanical Properties for Most Utilized Bonds

MECHANICAL PROPERTY	RESIN BOND	VITRIFIED BOND	METALLIC BOND
Brinell Hardness, (HB)	228	380	278
Rupture Strength, (psi)	1,046	1,243	2,073
Elasticity Modulus, (psi)	173,500	599,500	792,000

11.6.1 Bonds for Conventional Abrasives

Most bonds for conventional abrasives are either vitreous or resin. Vitreous bonds and resin bonds are also employed for superabrasives but in these situations the bond requirements differ. There are also other bonds used for specialized machining applications as detailed below.

Vitreous Bonds. Vitrified alumina wheels represent nearly half of all conventional wheels and the great majority of precision high production grinding. Vitrified superabrasive technology, especially CBN, is the fastest growing sector of the precision grinding market. The nature of an abrasive depends both on the processing of the grit and the vitrification process.

Vitreous bonds are essentially glasses made from the high temperature sintering of powdered glass frits, clays, and chemical fluxes such as feldspar and borax. Their attractions are their high temperature stability, brittleness, rigidity, and their ability to support high levels of porosity in the wheel structure. The mixture of frits, clays, and fluxes are blended with the abrasive and a binder, such as dextrin and water, and pressed in a mold usually at room temperatures. The binder imparts sufficient green strength for the molded body to be mechanically handled to a kiln where it is fired under a well-controlled temperature/time cycle in the range of 90°C–1300°C (1100°F–2400°F) depending on the abrasive and glass formulation. The frit provides the glass for vitrification, the clays are incorporated to provide green strength up to the sintering temperature, while the fluxes control/modify the surface tension at the abrasive grain-bond interface. The clays and flux additions therefore control the amount of shrinkage and, except for the very hardest of wheel grades, are kept to a minimum. The pressing stage in wheel manufacture is performed either at a fixed pressure or a fixed volume, providing a controlled volume of porosity after firing.

In addition to the grit type and size discussed above, it can be seen that two other factors are key to the wheel specification: grade or hardness designated by a letter, and structure designated by a number. To understand how these factors relate to the physical properties of the wheel consider first how loose abrasive grains pack. If grains with a standard size distribution are poured into a container and tamped down they will occupy about 50% by volume. It will also be noticed that each grain is in contact with its neighbors resulting in an extremely strong and rigid configuration. Now consider the effect of adding the vitrified bond to this configuration. The bond is initially a fine powder and fills the interstices between the grains. Upon sintering, the bond becomes like a viscous liquid that wets and coats the grains. There is usually actual diffusion of oxides across the grain boundary resulting in chemical, as well as physical, bonding. If, for example, 10% by volume of vitrified bond has been added then a porosity of 40% remains. The size and shape of individual pores are governed by the size and shape of the grains. The percentage of abrasive that can be packed into a given volume can be increased to greater than 60% by broadening the size distribution. The volume of abrasive can also be reduced to as low as 30% while maintaining mutual grain contact by changing the shape of the abrasive. For example, long, needle-shaped (high-aspect ratio) abrasive grains have a much lower packing density than standard grain.[32]

However, consider the situation where the grit volume of a standard grit distribution is now reduced from 50%. The most obvious effect is that immediately some of the grains stop being in contact. The integrity and strength of the whole can now only be maintained in the presence of the bond that fills the gaps created between the grains and provides the strength and support. These points are called bond posts and become critical to the overall strength and performance of the wheel. As the grit volume is further reduced the bond posts become longer and the structure becomes weaker. Not surprising that the abrasive volume percentage is a critical factor and is designated by the wheel manufacturers as *Structure number*. Each supplier will use slightly different notations and these are not generally reported for commercial reasons.

With the abrasive volume defined, the remaining volume is shared between the bond and the porosity. The bond bridges can be made stronger by increasing the amount of bond to make them thicker. The more bond there is, the lower the porosity and the harder the wheel will act. The definition of grade again varies from supplier to supplier. For some, it is simply a direct correlation to porosity, for others it is a more complicated combination of porosity % P and structure number. Malkin[66] gives one supplier's system where the grade letter is designated so that:

Eq. (11.5) $\text{Grade} \propto 43.75 - 0.75P + 0.5S$

This designation makes grinding performance change more predictably from one grade letter to the next in relation to burn, dressing forces, power, etc.

In addition to the size of the bond-bridge, fracture mode is also critical. The bond must be strong enough to hold the grains under normal grinding conditions, but under higher stress it must allow the grain to fracture in a controlled way. However, the bond material should not be too strong as compared to abrasive grit strength, because otherwise it promotes the glazing of the abrasive grains, which, in turn, can lead to their burn.

One method to regulate fracture mode is by adding fine quartz or other particles to the bond to control crack propagation. Another is to recrystallize the glass creating nucleation centers that act in a similar fashion.

The primary attraction of producing wheels with high structure numbers is to achieve high levels of porosity while still maintaining structural integrity. High porosity provides for good coolant access and chip clearance in the grinding process. However, it is more difficult to maintain initial strength before firing and maintain integrity of the pores during manufacture of the wheel without using additives to act as structural supports or *pore formers*. Pore formers typically consist of hollow particles such as bubble alumina, glass beads, or mullite that remains an integral part of the wheel structure but break open at the grinding surface. Fugitive materials such as naphthalene, sawdust, or crushed walnut shells that burn out in the firing process may also be used as pore formers. Hollow particles maintain a strong and coherent wheel structure, while fugitive fillers create a structure with high permeability that allows coolant to be carried deep into the wheel.

Wheel manufacturers such as Universal have taken this concept further and produced wheels with multiple pore former size distributions to create both macroporosity for high permeability, and microporosity for controlled fracture of the bond. This type of wheel, with trade names such as Poros 2™, has proved very effective for creep-feed grinding.

Extruded seeded gel (Fig. 11.7) needle-shaped grains has provided an opportunity for creating extremely porous and permeable structures. The natural packing density of grains with an aspect ratio of 8:1 is about 30%. Norton has recently developed a product called Altos™ with a totally inter-linked porosity as high as 65–70%. Even though the structure contains only a few percents of bond, it is very strong because the bond migrates and sinters at the contact points between grains acting analogously to *spot welds*. The high structural permeability allows prodigious amounts of coolant to be carried into the grinding zone. The shape can be varied from very blocky to very elongated as illustrated in photographs below. This type of wheel probably gives the highest stock removal rates of any vitrified wheel, higher even than those possible with vitrified CBN, together with excellent G-ratios for a conventional abrasive. Thus, it becomes profitable to identify the major applications for grinding of difficult and burn-sensitive materials like the nickel based alloys, which are commonly utilized for components in aerospace and power generation industries, for instance.

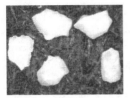  

Figure 11.7. Seeded gel abrasive grain shapes, (St. Gobain Abrasives).

In broad terms, wheel grades E to I are considered soft and are generally used with high structure numbers (11–20 for induced porosity) for creep feed and burn-sensitive applications. Grades J to M are considered medium grade and are usually used with lower structure numbers for steels and regular cylindrical and internal grinding. Very hard wheels are produced for applications such as ball-bearing grinding. These wheels are X or Z grade and contain as little as 2% porosity. Specifications of this hardness are produced by either hot pressing or by over-sintering such that the bond fills all the pores. As such, their structural numbers can vary from 8 up to 24.

Pores may also be filled with lubricants such as sulfur, wax, or resin by impregnating regular wheel structures after firing. Sulfur, a good high temperature EP lubricant, is common in the bearing industry for internal wheels although becoming less popular due to environmental concerns.

Resin Bonds. Resin bonds cover a broad range of organic bonds fabricated by hot pressing at relatively low temperatures, and characterized by the soft nature of cutting action, low temperature resistance, and structural compliance. The softest bonds may not even be pressed but merely mixed in liquid form with abrasive and allowed to cure. Concepts of grade and structure are very different from vitrified bonds. Retention is dependent on the localized strength and resilience of the bond surrounding the grain and very sensitive to localized temperatures created in the grinding zone and the chemical environment. For example, the bond is susceptible to attack by alkali components in coolants.

Resin bonds can be divided into three classes based on strength/ temperature resistance; these are plastic, phenolic resin, and polyamide resin.

Plastic Wheels. These are the softest of wheels made using epoxy or urethane-type bonds. With conventional abrasives, they are popular for double disc and cylindrical grinding. At one time, prior to the introduction of vitrified CBN, they were the primary wheels for grinding hardened steel camshafts where they gave both a very soft grinding action and a compliance that helped inhibit the generation of chatter. They are still popular in the knife industry and job shops for grinding burn-sensitive steels. Manufacturing costs and cycle times are low so pricing is attractive and delivery times can be short. For superabrasive wheels, plastic bonds appear to be limited to ultra-fine grinding applications using micron diamond grain for the glass and ceramics industries. Its compliance offers an advantage of finer finish capabilities but wheel life is very limited.

Phenolic Resin Wheels. Phenolic bonds represent the largest market segment for conventional grinding wheels after vitrified bonds, and dominate the rough grinding sector of the industry for snagging and cut-off applications. The bonds consist of thermosetting resins and plasticizers that are cured at around 150°C to 200°C (300°F to 400°F). The bond type was originally known as *Bakelite* and for this reason still retains the letter "B" in most wheel specifications. Grade or hardness is controlled to some extent by the plasticizer and use of fillers.

Unlike vitrified wheels, most resin wheels are used under controlled pressure and not with fixed infeed systems. Resin wheels are very often used at high speed. The abrasive size is therefore usually used to control recommended grade. Finer grit wheels remove material faster for a given pressure but wear faster, and are used to avoid excessive porosity required in a coarse wheel to obtain an effective cutting action. Porosity reduces burst speed and allows grits to be easily torn out. With higher available pressures coarser grit sizes can be used. Glass fibers are also added to reinforce cut-off wheels for higher burst strength.

Resin bonds are also used for precision applications where resilience provides benefits of withstanding interrupted cuts and better corner retention. One such area is flute grinding of steel drills where the wheel must maintain a sharp corner and resist significant side forces. The most recent product, Aulos™, introduced by St. Gobain in 2001 is capable of grinding at high material removal rates while still producing several drills between dresses. This is one example of where the advances in conventional engineered abrasives are competing very successfully and actually out-performing emerging CBN technologies.

For superabrasive wheels, phenolic resin bonds represent the earliest, and still most popular, bond type for diamond wheels and especially for tool-room applications. The bonds were originally developed for diamond with the introduction of carbide tooling in the 1940s. Their resilience made them optimal for maintaining tight radii while withstanding the impact of interrupted cutting typical of drill, hob, and broach grinding. To prevent localized temperature rise, the abrasive is typically metal coated to act as a heat sink to dissipate the heat. In addition, high volumes of copper or other metal fillers may be used to increase thermal conductivity and heat dissipation.

Phenolic resin bonds were quickly adopted upon the introduction of CBN in 1969, and it remains the dominant bond for the steel tool industry.[25] Because the basic technology is mature, the number of wheel makers is too numerous to list. However, the marking system for wheels is covered by standards such as ANSI B74-13 for the USA or JIS B 4131 for Japan.

Many wheel makers are located close to specific markets to provide quick turn-around. Alternatively, many are obtained from low-cost manufacturing countries. The key to gaining a commercial advantage in this type of competitive environment is application knowledge either by the end-user developing a strong data base and constant training, or using the knowledge of the larger wheel makers with strong engineering support.

Polyamide Resin Wheels. Polyamide resin was developed by DuPont in the 1960s originally as a high temperature lacquer for electrical insulation. By the mid 1970s it had been developed as a cross-linked resin for grinding wheels giving far higher strength, thermal resistance, and lower elongation than conventional phenolic bonds. The product was licensed to Universal Diamond Products and sold under the trade name of Univel™, where it came to dominate the high-production carbide grinding business especially for flute grinding. Polyamide has 5–10 times the toughness of phenolic bonds and can withstand temperatures of 300°C (575°F) for 20 times longer. Its resilience also allows it to maintain a corner radius at higher removal rates or for longer times than phenolic resin.

In recent years, alternate sources for polyamide resin have become available. They are significantly less expensive than the DuPont based process but to date have not quite matched the performance. However, the price/performance ratio is attractive, making the bonds cost-competitive relative to phenolic resin bonds in a broader range of applications. In some applications, the Univel™ product has proved so tough in comparison to regular phenolic bonds that induced porosity techniques from vitrified bond technology such as hollow glass beads have been used to improve the cutting action.

Other Bond Systems. There are several traditional bond systems used with conventional abrasives. These include the following:

- *Rubber.* Introduced in the 1860s, rubber bonds are still used extensively for regulating wheels for centerless grinding and some reinforced grades for wet cut-off grinding. Manufacturing rubber bonds is an increasing problem for environmental reasons. Alternatives are being sought where possible.

- *Shellac.* Shellac (or "elastic") bonded wheels were first made in 1880, and due to a combination of elasticity and resilience, probably still represent the best wheel for producing fine, chatter-free finishes for grinding steel rolls for the cold strip-steel mills and paper industries. Shellac comes from fluid exuded by insects as they swarm Cassum or lac trees in India. As such it is highly variable both in availability and properties depending on the weather conditions and species. On occasion, a single wheel maker can use 10% of the entire world's production. Not surprising, wheel makers have sought alternate solutions for grinding applications.

- *Silicate.* Silicate bonds were first produced around 1870 by mixing wet soda of silicate with abrasive, tamping in a mold, drying, and baking. It is still popular in certain parts of the world by reason of simplicity and low cost of manufacture. The wheels are generally used for large face wheels.

11.6.2 Bonds for Single-layer Superabrasives

Single-layer wheels are generally limited to superabrasives because of the economics of wheel life. Single-layer wheels can be subdivided into two groups: electroplated wheels fabricated at essentially room temperatures, and brazed wheels fabricated at temperatures as high as 1,000°C (1,800°F).

Electroplated. Electroplated wheels consist of a single layer of superabrasive grains bonded to a precision-machined steel blank using nickel deposited by an electroplating or occasionally electronless plating process. The plating depth is controlled to leave about 50% of the abrasive exposed (Fig. 11.8).

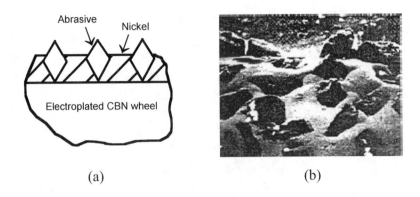

(a)   (b)

Figure 11.8. *(a)* Schematic of a plated CBN wheel section,[4] and *(b)* the surface of an electroplated CBN wheel.[78]

Brazed Single-layer Bonds. Electroplating is a low temperature process (below 100°C) which holds the abrasive mechanically. Consequently, the plating depth required to anchor the abrasive needs to be higher than 50% of the abrasive height. The alternative process is to chemically bond the CBN to the steel hub by brazing, using a relatively high temperature metal alloy system based on, for example, Ni/Cr,[21][22][63][78] with trade names such as MSL™ (metal single layer) from St. Gobain Abrasives (Fig. 11.9).

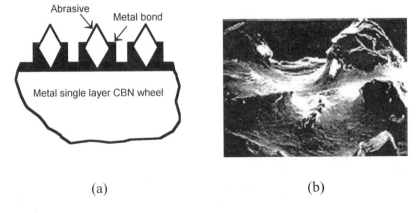

(a)   (b)

Figure 11.9. A brazed wheel;[4] *(a)* schematic and *(b)* the surface of a brazed wheel.

Use of a chemical bonding method allows a much greater exposure of the abrasive and hence an increased usable layer depth. It also gives greater chip clearance and reduced grinding forces. However, brazing occurs at temperatures up to a 1,000°C (1,800°F) that can degrade the grit toughness and distort the steel blanks. Braze also wicks up around the grit placing it under tensile stress upon cooling and thus further weakening it. Consequently, the use of brazed wheels tends to be for high stock removal roughing operations of materials such as fiber glass, brake rotors, and exhaust manifolds, or applications with form tolerances larger than 2 microinches.

11.6.3 Bonds for Multilayer Diamond

The first bonded diamond wheel made was apparently developed in the early 1930s with a Bakelite resin. Resin bond predominates in the grinding of tungsten carbide, where a good balance between wheel life and removal rate is required.

Resin Bonds. Resinoid diamond wheels contain a polycrystalline, low strength, metal clad diamond grit, allowing efficient self-sharpening via partial fragmentation. The range of conditions in which resin bond wheels can be utilized efficiently is much larger than in the case of metal bond wheels. An abundant supply of coolant, a prerequisite to grinding performance, makes the utilization of resin bonded grinding wheels very efficient. A resin bonded wheel is weakened by exposure to excessive heat.

Sintered Metal Bonds. Metal bonds hold strong blocky monocrystalline grit more firmly than resin bonds. During sintering the metallic binder softens and wets the diamond grit. The excellent grit retention and the high abrasive wear resistance produce wheels suited to form grinding and brittle materials such as glass and ceramics. Self-sharpening is possible only under high grinding forces. Compared with resin bonded wheels, these high pressures are possible only with stable, stiff machine tools. A disadvantage of metal bond wheels is the high sintering temperature necessary to melt the metallic phase, which leads to high manufacturing costs. Metal bonds can be used dry only in the case of small contact areas.

Vitreous Bonds. Vitreous bonds for diamond have different requirements from conventional wheels. Materials ground with diamond tend to be hard, nonmetallic, brittle materials. Therefore, there are fewer problems with wheel loading with grinding debris, and wheel porosity can be relatively low. On the other hand, hard workpiece debris is likely to cause much greater bond erosion than other workmaterials. Therefore, either the bond erosion resistance must be higher or, more practically, a lot more bond must be used.

Diamond does not show significant chemical bonding with the components of a vitreous bond. The bond must therefore rely on mechanical bonding. Diamond reacts with air at temperatures above 650°C (1,200°F). Therefore, diamond wheels must either be fired at low temperatures, or in an inert or reducing atmosphere. The lowest temperature bonds however were traditionally very prone to dissolution in water that limited shelf life and made air firing unattractive. A simple method was developed to manufacture wheels at higher temperatures by hot pressing using graphite molds. The graphite generated a reducing carbon-rich atmosphere locally. Since the mold strength was low the bond had to be heated above regular sintering temperatures to limit pressing pressures. Consequently, the bonds were fully densified with less than 2% open porosity. Pockets could be generated in the wheel by adding soft lubricating materials such as graphite or HBN that wore rapidly on exposure to the wheel surface. Fugitive fillers were also added to burn-out during firing. Nevertheless, the wheels had a major limitation—their bond content was so high they could not be automatically dressed using diamond tools. As such, they fell into the same category as metal and resin bonds that had to be trued and then subsequently conditioned. This type of structure was used extensively in applications such as double disc and PCD grinding.

The grinding forces with diamond can be very high and efforts are made to limit them by reducing the number of cutting points by significantly lowering the volume of diamond from 50%. This introduces the term *concentration*, a measure of the volume of superabrasive per unit volume of wheel. A 200 concentration is equivalent to 8.8 ct/cm^3 by weight or 50% by volume. Most diamond wheels are typically 12–100 concentration.

Recent years have seen a revolution in the development of porous cold-pressed vitrified diamond bonds driven by the increased use of PCD and carbide for cutting tools, and the growth of engineering ceramics. Vitrified diamond bonds, much of which are used in conjunction with micron sizes of diamond grit, are used for edge grinding of PCD and PCBN cutting tools, thread grinding of carbide taps and drills, and fine grinding and centerless grinding of ceramics as for seals and some diesel engine applications.

11.6.4 Bonds for Multilayer CBN

When CBN was introduced into the market in 1969 its cost naturally lent itself to being processed by wheel makers experienced in manufacturing diamond abrasives. Bonding systems included dense hot-pressed vitrified systems as described above for diamond. Unfortunately these lacked properties, such as chip clearance and dressability, required for high production grinding of steels where CBN would prove to be most suited. A range of bond types is available for CBN as well as for diamond and conventional abrasives.

Vitreous Bonds. Early vitreous bonds used by conventional wheel makers were so reactive that they dissolved the CBN into the bond converting it into boric oxide. Grit suppliers tried to counter this by producing CBN grains with thin titanium coatings. Not surprisingly, it took 10 years and numerous false starts before porous vitreous bonds with the capability of being dressed automatically became available. While some manufacturers still pursued hot-pressed bonds with high fugitive or other filler content,[62] the majority developed reduced reactivity cold-pressed bonds using methods common to processing of conventional vitrified wheels. Just as with conventional abrasives, it was possible to modify the bond formulations to obtain just sufficient reactivity and diffusion to create strong wetting and bonding.

The demands of vitreous bonds for CBN differ from the demands for either conventional or diamond bonds. Typical wheel supplier specifications, in compliance with standard coding practice, are shown in Table 11.6.

The wheel specification format is dictated by the standard practices of the diamond wheel industry. As such, the hardness is expressed as a grade letter but wheel structure is often not given or is described in the vaguest of terms. As with vitrified diamond, concentration plays a key role in controlling the number of cutting points on the wheel face. Concentrations for CBN wheels tend to be higher than in diamond wheels, at up to 200 concentration (50% by volume), especially for internal and many cylindrical grinding applications. This limits the structure number to a small range.

Early CBN wheels were very dense structures with porosity levels of the order of 20%. With the high cost of CBN, performance was focused on obtaining maximum wheel life. With the development of cylindrical grinding applications for burn-sensitive hardened steels in the 1980s, porosity levels rose to 30%. Recently, with the rapid expansion of CBN into aerospace and creep-feed applications, porosity levels have risen to the order of 40%. A comparison with porosity values shows clearly that the grade of CBN wheel for a given application, even with the development of higher porosity structures, is 5–10 points harder than for conventional wheels. This is not surprising, as CBN is so expensive it must be held for a much greater period of time, even if higher levels of wear flats are created. This is possible because of the high conductivity of CBN relative to both aloxide and most workpiece materials. A number of wheel manufacturers do try to match up grades of CBN wheels to be close to those of conventional wheels for the purpose of helping end users more familiar with conventional wheels specify a given wheel for an application. However it must be understood that dressing forces will be much higher, both because of the grain hardness and because of the additional bond, while hydrodynamic forces from the coolant will also be higher because of the lower porosity. This places additional challenges on the systems stiffness and creates a need for new strategies for achieving part tolerances.

Table 11.6. Vitrified CBN Commercial Bond Codes

EXAMPLE	EXAMPLE
B 200 N 150 V BA – 3.0 **(1) (2) (3) (4) (5) (6) (7)**	**49 B126 V36 W2J6V G 1M** **(1) (2) (3) (4) (5) (6)**
(1) Grit type	(1) Grit type
(2) Grit size (US Mesh)	(2) Grit size (µm)
(3) Hardness (grade)	(3) Concentration (V36 = 150 concentration)
(4) Concentration	(4) Vitrified bond system
(5) Bond type	(5) Grade
(6) Bond feature	(6) Internal coding
(7) Layer depth (mm)	

The glass bonds and manufacturing techniques used for vitrified CBN wheels are proprietary and there is rapid development still in progress. For example, General Electric[40] in 1988 recommended bonds not be fired over 700°C (1,300°F). Subsequently, Yang[100] found the optimum firing temperature to be 950°C (1,750°F) for a nominally identical bond composition. An important factor is to match thermal expansion characteristics of the glass with the abrasives,[8] or to optimize the relative stress developed between bond and grain in the sintering process. Yang, et al.,[100] reported this could be readily optimized by adjustments to minor alkali additives primarily sodium oxide. As with bonds for conventional abrasives, bond strength can be improved by the introduction of micro-inclusions for crack deflection either in the raw materials or by recrystallization of the glass.[94] With the far greater demand for life placed on CBN vitrified bonds and the narrower working range of grades available, quality control of composition and particle size of the incoming raw materials, the firing cycles used to sinter the bonds are critical. It has often been process resilience, as demonstrated by batch to batch consistency in the finished wheels, which has separated a good wheel specification from a poor one.

Sintered Metal Bonds. Metal represents the toughest and most wear resilient of bond materials, and is almost exclusively used with superabrasives. Much of this is for stone and construction, glass grinding, and honing.

Metal bonds for production grinding tend to be based on bronze in the alloy range of 85:15 to 60:40 with various fillers and other small alloy components. They are the most resilient and wear resistant of any of the bonds discussed but also create the highest grinding forces, and the most problems in dressing. Their use has been limited to thin wheels for dicing and cut off, profile grinding, fine grinding at low speeds, and high-speed contour grinding such as the Quickpoint ™ process.[53] This latter process is dependent on maintaining a well-defined point on the wheel and therefore the maximum wheel life. However, in many cases involving CBN, metal has been replaced by vitrified bond, even at the sacrifice of wheel life, in order to improve the ease of dressing.

Metal bronze bonds become more brittle as the tin content is increased. In the 1980s, brittle metal bond systems began to emerge with sufficient porosity that profiles could be formed in the wheel automatically by crush dressing using steel or carbide form rolls. Yet, the dressing process did not leave the wheel in a free-cutting state and therefore the surface had to be subsequently conditioned using dressing sticks or brushes. Also, the biggest problem was the extreme forces involved in dressing. Where use has been reported, such as OSG in Japan,[103] high stiffness grinders have had to be built specifically for this type of wheel. As such its use has been limited awaiting advances in standard production machine tool stiffness. Vitrified technology has been substituted in most cases.

The concept of a porous, brittle metal bond has been taken further by increasing the porosity level to the point of having inter-connected porosity in a sintered metal skeleton and vacuum impregnating the pores with resin. This is sold under trade names such as Resimet™ from Van Moppes (a St. Gobain Group Co.). This type of bond has been extremely successful for dry grinding applications on tool steels and carbide. It is freer cutting and gives longer life than resin, requires no conditioning, while the metal bond component offers an excellent heat sink.

11.7 DESIGN AND SPECIFICATION OF GRINDING WHEELS

11.7.1 Conventional Grinding Wheels

An abrasive wheel is an agglomeration of abrasive grains, bonding material, and air, carefully chosen for a particular application, thoroughly mixed, fired at strictly controlled temperatures, and finally trimmed to finished shape and size. It has many layers of abrasive in random orientation. Many grains do not have the best orientation for cutting, but there are many grains oriented in a random manner. Due to the high speed of the wheel and the large number of hard cutting edges brought into contact with the workpiece, the wheel forms an effective cutting tool.

Most abrasive wheels are relatively incompressible from the edge of the hole to the outer diameter, which means the wheels are well suited to producing close-tolerance work. However, vitrified and resinoid abrasives are elastic compared to steels and more susceptible to local fractures. This softness of these types of abrasive tools offers a limited degree of protection against minor sharp contacts with the workpiece when the wheel is rotating. However, it must always be remembered that a grinding wheel is a very dangerous tool if mishandled.

Traditional grinding wheels are solid with an arbor hole for mounting. Segmented wheels are formed by bonding abrasive segments onto the surface of a wheel. Segments may be bonded to the face of a wheel or onto the periphery.

There are five main factors in the specification of an abrasive tool as shown in Fig. 11.2 and Table 11.7; these are:

- Abrasive type.
- Grit size.
- Grade.
- Structure.
- Bond.

Table 11.7. Typical Wheel Specifications for Conventional Abrasives

ABRASIVE	MESH NUMBER/ GRIT SIZE		GRADE	STRUCT -URE	BOND
A-Aloxide	**Coarse**	**Medium**	Soft: A-I	Closed to open: 1-15	V - vitrified
C-SiC	10, 12, 14, 16, 20, 24	30, 36, 46, 54, 60	Medium: J-P		B - resinoid
					BF resinoid reinforced
	Fine	**Very fine**	Hard: Q-Z		E-shellac
	70, 80, 90, 100, 120, 150, 180	220, 240, 280, 320, 400, 500, 600			R-rubber RF-rubber reinforced

The following features are marked on grinding wheels:

- Geometric shape.
- Principle dimensions of the tool.
- Physical and mechanical properties.

Grade. The hardness of an abrasive tool is represented by a grade letter related to the strength of the bond. It is based on the percent of bond and abrasive grain in the wheel mix. Hardness is defined in terms of resistance of the bond against abrasive grain extraction due to grinding forces. Ideally, if the wheel has the correct grade, the abrasive grains will be pulled out as soon as they reach a certain degree of dullness. To this extent, hardness determines the self-sharpening characteristic, although friability of the grit is also very important for a self-sharpening effect. Wheel hardness is not equivalent to grain hardness. Harder grade wheels contain more bond than softer grade wheels.

A hard wheel retains dulled grains in the bond for a longer time compared to a soft wheel. Hard wheel grades are used for soft material and vice-versa. Harder wheels are used for smaller contact areas and interrupted cuts. Sometimes, the grinding conditions can make the wheel appear to have a softer grade. For example, increased workspeed, decreased wheelspeed, and reduced wheel diameter all make the wheel grade appear softer. The presence of vibrations during grinding requires a higher-grade wheel than usual.

Structure. The structure number relates inversely to concentration. Concentration is the percentage volume of the abrasive to the overall tool volume. The wheel may be open/wide, medium, or close/dense. More open structures have a lower concentration and higher structure number. Open structures are used when there is a high risk of wheel loading or workpiece burning. Vitrified bond wheels, having medium and open structure, have smaller mass than dense structures. Wide grain spacing helps the coolant to be brought into the grinding area, and the chips to be contained and removed from the area. If the wheel is too dense, the chips are retained in the active face of the wheel, a condition called loading. Wide grain spacing especially for vitrified wheels, is obtained by adding ground walnut shells during the wheel-making process, which burn out during the high temperature operation, leaving small pores.

Bond Type. The bond type specifies the type of material holding the abrasive grains in the tool in a specific shape. Bond types include vitreous, resin, and metal bonds as described in Sec.11.6. Bond strength determines the maximum safe speed of a grinding wheel.

Vitrified abrasives for precision work use primarily medium and fine abrasive grains. Vitrified tools are relatively inelastic and have good chemical stability and high temperature resistance. Vitreous bonds are made out of clay, feldspar, quartz, and other minerals. These bonds can be fusible, utilized with aluminum oxide, or sintered, when the abrasive is silicon carbide. Most vitrified wheels carry the color of the abrasive: black or green for SiC, and white, gray or tan for aloxide. The presence of chromium oxide in small quantities produces a pink wheel. The presence of vanadium oxide create a green aloxide color. The additions have little effect on the abrasive characteristics of the grains, and are used mainly for identification and marketing purposes.

Resinoid wheels are mostly black because of the color of the bond. Resin bonds, made from synthetic resins that bakelize above 450°F, are elastic and have good mechanical strength. Resinoid wheels can be simply designed for relatively high peripheral speeds. Soft grade wheels are less likely to scratch the work than vitrified wheels. The hardness can be lowered by adding alkalis to the grinding fluid. Resin is used with coarser grains for heavy stock removal.

Rubber bonds, natural or synthetic, are used for cut-off wheels since they are more durable in cut-off applications. Rubber bond cut-off wheels can be operated at higher speeds than resinoid wheels. Conversely, they have a poor resistance to high temperatures.

Shellac bonds, made out of natural resin with shellac, are used for finishing wheels, having a high elasticity and good polishing action.

The magnesite bond is composed of caustic magnesite and magnesium chloride. They wear quickly being used for rough grinding.

The silicate bond is used for large wheels for grinding of fine-edge cutting tools. They tend to wear quickly but have a cool cutting action important in this type of application.

Metal bonded wheels are used for bonding superhard abrasives such as diamond and CBN. The bonds are electroplated or sintered from powders of copper, tin, aluminum, nickel, etc.

The coarser the grain, the higher the stock removal, and the finer the grain size, the lower the roughness. There are exceptions. For an exceptionally hard material, a coarser grain will not remove significantly more stock than a fine grain. Sometimes, a mix of two size ranges of grains will be employed in a wheel.

11.7.2 Multilayer Superabrasive Grinding Wheels

Diamond wheels are usually metal or resin bonded, and rarely vitrified. For CBN wheels, metal, resin, and vitreous bonds are used. About half of all diamond grinding wheels are resinoid. For superabrasive wheels, the concentration must be matched to the intended application; as the concentration increases so does the efficiency of machining and wheel durability. Concentrations of 50 to 75 tend to be used for short-run production with fine-grain wheels, while concentrations of 100 to 150 tend to be used for metal-bond diamond wheels. Vitrified wheels tend to use concentrations of around 100, and resinoid wheels below 100. However, there are wide variations from these guidelines for special applications.

Because diamond and CBN abrasives are expensive, superabrasives are molded in a thin layer, up to 1/4 inch thick, around the core. A typical marking system and typical specification ranges are given in Table 11.8 for diamond wheels.

Table 11.8. Typical Diamond Wheel Specifications

ABRASIVE	GRAIN SIZE	WHEEL GRADE	CONCEN-TRATION		BOND	DEPTH OF DIAMOND LAYER
MD	**180**	**R**	**3**		**BL**	**1/8**
D Un - processed	40, 60, 80, 100 Coarse	H, J, L soft	4	100	B, M Resin metal	1/32
BD-blocky	120, 150, 180, 220 Medium	M, R Medium	3	75	D, G, K, L, N, P, W, Q Indicate bond modifi-cations	1/16
ED-processed blocky	270, 320, 400, 500 Fine	P, Q, R, S, T Hard	2	50		1/8
MD-resinoid diamond			1	25		1/4
ND-nickel coated diamond						
CD-copper coated						
ID-metal bonding diamond						
XD-steel grinding						
Abrasive type: D for diamond and B for cubic boron nitride						

Core Material. Some manufacturers add another element to the wheel marking to indicate the type of core material utilized.

11.7.3 Electroplated Single-layer Superabrasive Grinding Wheels

The accuracy and repeatability of grinding with electroplated wheels is dependent on many factors. The blank must be machined to a high accuracy and balanced; ideally blank profile tolerances are maintained to within 2 µm and wheel run-out maintained within 5 µm.[68] The abrasive is generally resized to provide a tighter size distribution than that used in other bond systems. This is to avoid any high spots and better control of aspect ratio. The abrasive is applied to the blank by various proprietary methods to produce an even and controlled-density distribution. For tight tolerance applications or reduced surface roughness, the wheel may also be post-conditioned, also termed *dressing*, *truing*, or *shaving*, where an amount equivalent to approximately 5–7% of the grit size is removed to produce a well defined grit protrusion height above the plating. With good control of plate thickness this helps control and/or define a usable layer depth.

The hardness, or more accurately the wear resistance, of the nickel plating is controlled by changes to the bath chemistry. Nitride coatings, similar to those used on cutting tools, have also been reported to improve the wear resistance of the nickel but data have been mixed indicating the performance parameters are not fully understood.[19][51]

The standard precision values for plated wheel form capability are presented in Table 11.9. For conditioned plated wheels, the standard values are presented in Table 11.10.

The size of the grit must be allowed for when machining the required form in the blank. The allowance differs from the nominal grit size and depends on the aspect ratio of the particular grit type.

The tables give values for typical form holding capabilities and finish as a function of grit size for standard precision plated and post-plated conditioned wheels. The finish values will vary to some extent dependent on workpiece hardness. The values indicated are those for grinding aerospace alloys in the hardness range of 30–50 HRC using CBN abrasive.

A major attraction of plated wheels is the fact that they do not require dressing, and therefore eliminate the need for an expensive form diamond roll and dressing system. However, they do present certain challenges, for example, the changes in power, finish, and wheel-wear with

Table 11.9. Precision Standards for Plated-wheel Form Capability

GRIT SIZE	GRIT SIZE	MIN. RADIUS	PRE-CISION ALLOW-ANCE BLANK	PRE-CISION TOLER-ANCES (+/-)	PRE-CISION FINISH	PRE-CISION BREAK IN DEPTH
(FEPA)	(US MESH)	(mm)	(microns)	(microns)	R_a	(microns)
B852	20/30	2	920	100		
B602	30/40	1.5	650	90		
B501	35/40	1.3	650	80		
B427	40/50	1	500	60	160	63
B301	50/60	0.8	400	40	125	
B252	60/80	0.7	290	30	85	40
B213	70/80	0.6	260	30	75	
B181	80/100	0.5	220	30	63	35
B151	100/120	0.4	190	30	38	
B126	120/140	0.3	165	25	35	
B107	140/170	0.26	140	25	32	
B91	170/200	0.23	130	25	32	20
B76	200/230	0.2	100	25	28	
B64	230/270	0.18	90	20	25	
B54	270/325	0.15	75	20	22	
B46	325/400	0.12	65	20	20	10

Table 11.10. Precision Standards for Conditioned Plated-wheel Form Capability

GRIT SIZE	GRIT SIZE	MIN. RADIUS	PRECISION ALLOWANCE BLANK	PRECISION TOLERANCES (+/-)	PRECISION FINISH	PRECISION BREAK IN DEPTH
(FEPA)	(US MESH)	(mm)	(microns)	(microns)	R_a	(microns)
B852	20/30	2	850	10		
B602	30/40	1.5	600	10		
B501	35/40	1.3	600	10		
B427	40/50	1	450	8	80	
B301	50/60	0.8	360	6	70	
B252	60/80	0.7	250–280		60	
B213	70/80	0.6	240–250		50	
B181	80/100	0.5	200		40	<5
B151	100/120	0.4	165	5	32	
B126	120/140	0.3	140–150	5	28	
B107	140/170	0.26	110		25	
B91	170/200	0.23	95		22	
B76	200/230	0.2	80			
B64	230/270	0.18	70	4	NA	
B54	270/325	0.15	60	4		
B46	325/400	0.12	50	3		

time for a typical precision plated CBN wheel when grinding aerospace alloys. Initially, the roughness is high as only the tips of the grits are cutting. The power then rises rapidly together with an associated rapid rate of wheel wear and a drop in roughness. The process then generally stabilizes, with wear flat formation being balanced by fracture, unless the grinding conditions are too aggressive. This leads to a much more protracted period of time when the rates of change of all three variables are reduced by up to a factor of 10. Failure eventually occurs either from power levels becoming so high that burn occurs, or by stripping of the plating or grits from the core. This latter effect is particularly concerning because in most cases it cannot yet be detected in advance, or predicted easily except by empirical data from production life values from several wheels.

The number of variables for a given wheel specification that can make a significant impact on performance are quite limited. The plating thickness is held within a narrow band. The homogeneity of the plating is controlled by the plating rate and anode design. Care should be taken to avoid nodule formation especially around tight radii. Nickel-phosphorus is sometimes used in these circumstances to give the most even and hardest surfaces. The biggest variable is the grit itself and how it wears under the prevailing grinding conditions. If the grit is too weak then fracture and rapid wheel wear occur. If the grit is too tough, wear flats build up and burn ensues. Another major factor is coolant. When grinding aerospace alloys with CBN in oil, the high lubricity of the coolant ensures a slow but steady build-up of wear flats. In a water-based coolant the lubricity is much lower causing more rapid wear of the grain tips. However, water has much higher thermal conductivity than oil and induces thermal shock in the abrasive leading to weakening and fracture. When used wheels are examined, it is found that those used to grind in oil may last several times longer than those used to grind in water and still have a high proportion of their layer depth remaining. Similar wheels grinding in water wore down completely to the nickel substrate. One conclusion is to use a tougher grit when grinding in water but a slightly weaker grit in oil.

In most production applications, oil coolant is required to obtain the necessary life to make a plated process competitive over other methods.

Plated CBN has proved remarkably cost effective in aircraft engine manufacture with low batch volumes of parts requiring profile tolerances in the 0.0004 inch to 0.002 inch as well as for high speed rough grinding of camshafts and crankshafts.

Used plated wheels are generally returned to the manufacturer for stripping and re-plating. The saving is typically about 40% and the steel core can be reused with care 5–6 times.

11.7.4 Wheel Shape and Tolerances

Wheel dimensions are usually expressed and marked on the wheel with three numbers:

Eq. (11.6)   Outer dia. (D) · Thickness/width (T) · Hole dia. (H)

For superabrasives, the layer thickness (X) is added afterwards. Conventional wheels are typically sold as standard stock sizes although they can be cut to size and bushed in the bore to order. They can also be pre-profiled for certain applications such as worm gear grinding. Many superabrasive wheels, especially resin bond, also come in standard stock sizes but many are custom built, often with complex pre-molded profiled layers. The cost of this is readily offset against the savings in abrasive and initial dress time.

Wheel dimensional tolerancing is dependent on the application and supplier. Some typical manufacturer guidelines are given below. Conventional cored wheels should not have a tight fit due to the serious risk of cracking due to differential thermal expansion with the steel wheel mounts. On the other hand steel-cored superabrasives for high speed require the best possible running truth to avoid problems of vibration.

Conventional roughing wheels	H13 bore fit
Conventional precision wheels	H11 bore fit
Standard superabrasive wheels	H7 bore fit
High Speed CBN wheels	H6–H4 bore fits

Side and outer diameter run-out of wheels vary from one manufacturer to another and depend on the application. Typical side run-out tolerances for superabrasive wheels are presented in Table 11.11. Outer diameter tolerances may be considerably larger, as long as concentric running truth is maintained. It is important to note that wheel run-out is also affected by wheel balance.

Table 11.11. Typical Side Run-out Tolerances for Superabrasive Wheels

WHEEL DIAMETER	STANDARD SUPER-ABRASIVE WHEELS	HIGH SPEED CBN
(mm)	(µm)	(µm)
<250	20	5
>250–400	30	10
>400–600	50	15
>600–750	70	20

11.7.5 Wheel Balancing

Balance is closely associated with run-out. Balance tolerances will depend on application. For instance, the Japanese code, JIS B4131, gives the balance tolerances in terms of center of gravity displacement. However, most balancing is performed either statically or, more frequently, by dynamic balancing at a fixed speed. For high speed, the balance requirements are significantly more stringent than those shown above. For example, a high-speed CBN wheel for camshaft grinding of 350 mm diameter (13.8 inches) weighing 11.35 kg (25 lbs) operating at 5,000 rot/min (rpm) must be balanced to less than 0.45 grams (0.015 oz.) in order to prevent visual chatter—even when used on a grinder with a high stiffness hydrostatic spindle. This is almost an order of magnitude tighter than current standards.

Chatter is visible down to displacements of 254 nanometers (10 micro-inches). Achieving displacements below this, even with the most sophisticated hydrostatic wheel spindle bearings, is an increasing challenge as wheelspeeds increase. For regular hydrodynamic or ball-bearing–based spindles, some form of dynamic balancer mounted within the machine is essential. Automatic balancers are mounted on the spindle nose or within the spindle assembly and function by adjusting the position of eccentric weights in a similar fashion to static balancing. A separate sensor detects levels of vibration. This type of balancer may need to have the wheel guarding altered to accommodate the assembly although automatic balancing is standard on many new cylindrical grinders.

The response times of automatic balancers have managed to keep pace with higher speed requirements and many can operate at 10,000 rot/min (rpm) or higher. They have expanded to cover not only large cylindrical wheel applications but smaller wheels down to 150 microns (6 inches) for use on multipurpose machining centers. Others have the vibration sensor built into the balance head and can also be used as a crash protection and acoustic dress sensor.

Automatic balancers have been developed for balancing in two planes for compensation of long wheels such as for centerless grinding and for balancing complete wheel/spindle/motor assemblies.

Coolant is also a key factor for maintaining balance. A grinding wheel can absorb a considerable quantity of coolant. For example, a 220# WA 1A1 wheel can hold up to 16 wt%. When the wheel is spinning the coolant is not released spontaneously but may take several minutes, even hours, depending on the viscosity. The primary problem arises when coolant is allowed to drip on a stationary wheel or a stationary wheel is allowed to drain vertically. This will throw the wheel into an out-of-balance condition when next used. The effect is unlikely to cause actual wheel failure but will generate prolonged problems of vibration even when constantly dressing the wheel. It is also an issue with vitrified CBN wheels at wheelspeeds greater than 60 m/s even though the porous layer may only be a few millimeters thick.

11.7.6 Design of High-speed Wheels

Wheelspeeds have risen significantly in the last 10 years particularly for vitrified CBN. In 1980, 60 m/s (1200 ft/min) was considered high speed, by 1990 80 m/s (1600 ft/min) was increasingly commonplace in production. More recently, speeds of 125–150 m/s (25,000–30,000 ft/min) were introduced. Nowadays, several machines for using vitrified-bond grinding wheels have been reported entering production for grinding cast iron at 200 m/s (40,000 ft/min). Speeds of up to 500 m/s (100,000 ft/min) have been reported experimentally with plated CBN.[55] Such wheelspeeds place extreme demands on both the wheel manufacturer and machine tool builder. The wheel must be capable of running safely in an "overspeed" test of 150% maximum speed. The machine must have adequate guarding to contain the flying debris and to cope with the energy released in the event of a wheel failure. The spindle must be capable of the high running speeds without overheating or seizure. The whole system must operate with low vibration levels at the operating speeds.

Traditionally, vitrified wheels of conventional design generally default to a maximum wheelspeed of 22.5–35 m/s (4,500–7,000 ft/min) depending on bond strength and wheel shape. Exceptions exist such as for thread and flute grinding wheels, 40–60 m/s (8,000–12,000 ft/min) and internal wheels up to 40 m/s (8,000 ft/min).

To achieve higher speeds requires an understanding of how wheels fail. Vitrified bonds are brittle, elastic materials that fail catastrophically when the localized stresses exceed the material strength. These stresses can occur from clamping of the wheel, grinding forces, acceleration/deceleration forces, from changes in wheelspeed in out-of-balance or thermal stresses. However, with skilled use and proper handling of the wheel, the greatest stress is the maximum hoop stress imposed at the bore of the wheel due to radial expansion at high rotational speeds.[10][11]

The stresses and displacements created in a monolithic grinding wheel can be readily calculated from the classical equations for linear elasticity. These can be solved using finite difference approximations to give radial displacements at any radius of the wheel. Wheel failure, in line with this analysis, occurs from cracks generated at or near the bore where the stress is highest. The failure is catastrophic with 4 or 5 large pieces being thrown off tangentially. To increase the burst speed either the overall strength of the bond must be increased (e.g., finer grit size, lower porosity, and better processing methods to eliminate large flaws) or the strength must be increased where the stress is highest. For conventional wheels, this has often been achieved with a two-component vitrified structure where the inner portion is higher strength.

For vitrified CBN wheels, higher speeds are achieved by substituting the inner section of the wheel with a high strength material such as aluminum, and especially steel. Wheel manufacture consists of epoxy bonding or cementing a ring of vitrified CBN segments to the periphery of the steel. The segmented design serves several purposes. First, it produces a much more consistent product than a continuous or monolithic structure because of the limited movements required in pressing segments of such small volume. This is especially true when, as in the examples shown above, a conventional backing (white) layer is added behind the CBN to allow the use of the full layer of the abrasive. This gives both a better consistency in grinding and a higher Weibull number for strength consistency. Second, it allows a wheel to be repaired in the event of being damaged which provides a considerable cost saving for an expensive CBN wheel. Third, the segments provide stress relief, acting as *expansion joints* limiting hoop stress due to segment expansion as wheelspeed increases and also as the steel core expands due to centrifugal forces.

Trying to model segmented wheels using the traditional laws of elasticity has proved difficult because of complex effects within and around the adhesive layer. Finite element analysis (FEA)-based models are now more common with much of the groundwork having been done by Barlow and Rowe, et al.[10]–[12] Both hoop stresses and radial stresses can lead to wheel failure. Hoop stress is dependent on the expansion of the core and the segment length. Radial stress is dependent on the expansion of the core but also, more importantly, on the mass and therefore thickness of the segment. It was deduced that there is an optimum number of segments. Higher segment numbers give rise to additional stresses at the joint edges because as the wheel expands in a radial direction it must contract in the axial direction.

For thin segments the major stress is circumferential, but for thicker segments the dominant stress shifts to radial. For this reason, abrasive segments no thicker than 10 millimeters are, usually, utilized for high-speed grinding wheels. This immediately places limitations on profile forms allowed. The key factor is the mass of the segment, and its impact on radial stress is also important when considering the effect of wheel radius on burst speed. As the wheel radius is reduced, the centrifugal force (mv^2/r) must increase which will directly increase the radial stress. Ideally, the calculated burst speed should be at least twice the maximum recommended operating speed. For the 0.2 inch CBN layer, the burst speed levels off at 60,000 ft/min (300 m/s). The most striking factor about this graph is that the burst speed drops rapidly as the wheel diameter is reduced. For a wheel diameter of 150 mm (6 inches) the maximum recommended wheelspeed would be only 100 m/s (20,000 ft/min). Not surprisingly, high-speed wheels operating in the range of 100 m/s to 200 m/s (20,000 to 40,000 ft/min) tend to be larger than 300 mm (12 inches) in diameter with flat or shallow forms.

Segmental wheel research first began with conventional wheels in the 1970s as part of an effort to evaluate the effect of high speed.[1][5][98] However, the labor-intensive manufacturing costs were not competitive for the economic gains in productivity. The recent development of ultra-high porosity specifications (and therefore low-density) using extruded abrasives has allowed the development of wheels with thick layers of conventional abrasive capable of operation at up to 175 m/s (35,000 ft/min).

The last and most important benefit of segmental designs is safety. The results of a failure in a large conventional wheel can be extremely destructive of the machine tool. With this level of energy released the situation is potentially dangerous even when well guarded. By comparison the energy released by a segment failure, and the level of damage accompanying it is very small. Any failure though is still unacceptable to the wheel maker and to the wheel user.

Wheels are speed tested by over-spinning the wheel by a factor prescribed by the appropriate safety code for the country of use. In the USA, ANSI B7.1 specifies that all wheels must be spin tested with an over-speed factor of 1.5 times the operating speed. The theory behind this reverts back to conventional wheel research where the 1.5 factor was proposed to detect pre-existing flaws that might otherwise cause fatigue failure in the presence of moisture or water based coolants during the expected life of the wheel.[43] In Germany, for the highest speed wheels, the DSA 104 code requires the wheel design be tested to withstand 3 times the operating stress (giving an over-speed factor of 1.71). However, all production wheels must be tested at only 1.1 times the operating speed. This code was due to the result of research in Germany that indicated that high over-speed factors could induce flaws that could themselves lead to failure. This discrepancy in the safety laws has a major impact on transfer of technology in the context of the global economy.[82][83]

Spin testing in itself however is insufficient. Higher stresses actually occur in the epoxy bond rather than in the abrasive. Fortunately, the bond strength of epoxy is about ten times greater than the abrasive. However, epoxy is prone to attack by moisture and coolant and will weaken over time. Efforts have been made to seal the bond from the coolant,[59] but these are generally ineffective and the wheel maker must have life data for his particular bonding agent in coolant. Since wheels may have a life of several years on the machine with spares held in stock for a comparable time this data takes considerable time to accrue.

Another problem is how to hold the wheel body on the machine spindle. Centripetal forces cause the wheel to expand radially both on the outer diameter and the bore. It must, therefore, also contract axially. The problem is to prevent movement of the wheel on the hub either by minimizing bore expansion/contraction and/or by maintaining sufficient clamping pressure on the wheel to resist torsional slippage.

Blotters are required for conventional wheels so as to negate the effects of micro-asperities in the grit structure of the vitrified body that would otherwise give rise to local stress concentrations. They are made of either paper or plastic (polyester) with a thickness of typically 0.4 milimeters (0.015 inches). The flanges are fixed together with a series of bolts tightened to less than 27 meter-Newtons (20 ft-lbf) torque. Optimum torque values can lead to lower rotational stresses and higher burst speeds. Overall clamping pressures need to be kept up to 7 GPa (1,000 psi) and usually are considerably below this value. However, over-torquing will cause distortion of the flange leading to a high stress peak that can readily exceed 14 GPa (2,000 psi) and lead to wheel failure.[71] De Vicq[30] recommended using tapered flange contact faces to compensate. Unfortunately this is impractical except for dedicated machines, and the accuracy of torquing methods is not always sufficient to ensure correct flange deflection. In the absence of any significant axial contraction, it is relatively straight-forward to calculate clamping forces required to prevent rotational slippage.[70]

Clamping Force to Compensate for the Mass of the Wheel. The clamping force must be sufficient to cope with the accelerations and decelerations of the wheel including the various forces, e.g., the gravity force. For example,

Eq. (11.7) $\qquad F_m = mg/\alpha$

where m = mass of wheel, α = coefficient of friction for blotter, paper blotter = 0.25, and plastic blotter = 0.15.[30]

Clamping Force for Wheel Out of Balance.

Eq. (11.8) $\qquad F_m = \delta v^2/gr_0^2\alpha$

where δ = out of balance (force·distance), v = wheelspeed, and r_0 = wheel radius.

Clamping Force for Motor Power Surge. Assuming electric motors can develop a surge torque of 2.5 times their rated torque before stalling:

Eq. (11.9) $\qquad F_s = 2.5W\, r_f/V\alpha\, r_o$

where r_f = flange average radius and W = spindle motor power.

Clamping Force for Reaction of Wheel to Workpiece. Again, assuming a motor surge capability of 2.5 times rated torque:

Eq. (11.10) $\qquad F_n = 2.5W/V\alpha\mu$

where μ = grinding force ratio, typically 0.2–0.8.

In addition to these forces, there will be effects of accidental vibration and shocks, possible compression of the blotters, and the increased clamping force required as the wheel wears when holding constant surface footage. Practical experience leads to another factor of 2 on forces. The total clamping force required becomes:

Eq. (11.11) $F_{total} = 2(F_m + F_m + F_s + F_n)$

Knowing the number of bolts, tables are available giving torque/ load values for the required clamping force. Calculations then need to be made to determine the flange deflection.

With high-speed steel-cored wheels the need for blotters is eliminated. Clamping is therefore steel-on-steel and not prone to the same brittle failure from stress raisers. Nevertheless, there is the uncertainty of wheel contraction and its effect on clamping. One solution is to eliminate the flanges entirely. Landis Gardner Co., a division of UNOVA, Inc., located in Waynesboro, PA, developed a one-piece wheel hub where the entire wheel body and tapered mount is a single piece of steel with the vitrified CBN segments bonded on to the periphery.[79]

The straight one-piece hub was found to minimize bore expansion and is the design currently used in production. The *turbine* or parabolic profile minimizes outer diameter expansion but at the expense of additional bore diameter expansion. It is also considerably more expensive to machine. This one-piece concept proved extremely successful in crankshaft pin grinding and camshaft lobe grinding for speeds in the range of 60–120 m/s (12,000–24,000 ft/min). It has eliminated the need for automatic balancers, and allows fast change-over times for lean manufacturing with minimal or no redress requirements.

Electroplated CBN wheels have been developed for considerably higher wheelspeeds than vitrified CBN. The plated layer can withstand greater expansion of the hub. Research was reported as early as 1991, by Koenig & Ferlemann,[55] at 500 m/s (100,000 ft/min) using wheels manufactured by Winter, while Tyrolit recently offered a product design rated, in its literature, for 440 m/s (88,000 ft/min).

The wheel design described by Koenig & Ferlemann[55] has several novel features including the use of light-weight aluminum alloy for the hub material, a lack of a bore hole to further reduce radial stress, and an optimized wheel body profile based on turbine blade research to give the minimum wheel mass for uniform strength. Although this technology has been available for 10 years, there are very few machines in production running at over 200 m/s (40,000 ft/min). The plated CBN layer can withstand the expansion at high speed. The radial expansion at 500 m/s (100,000 ft/min) is about 150 microns. But the expansion is rarely perfectly uniform. Even a 1% difference due to any anisotropy in the hub material will lead to build-up of regenerative chatter and performance issues well before the expected life of the wheel.

Aircraft grade, aluminum alloys are used as hub materials for some high-speed vitrified CBN wheels. The obvious attraction is the lower density relative to steel. Various grades are available with tensile strengths of 0.8–1 MPa (115–145 psi). However, they have higher thermal expansion and appear to give more size and stability problems.

In order to compensate for bore expansion, two other methods have been developed. The first is by Erwin Junker Maschinenfabrik who developed a patented bayonet-style (cam and follower) three-point type mount. The design assures 50 micro-inch run-out repeatability at speeds up to 140 m/s.[5] The second method is the incorporation of the increasingly popular HSK tool holder shank. The HSK system was developed in the late 1980s at Aachen, Germany as a hollow tapered shaft capable of handling high speed machining. It became a shank standard in 1993 with the issuance in Germany of DIN 69893 for the hollow 1:10 taper shank together with DIN 69063 for the spindle receiver. This system has seen a rapid growth, especially in Europe, as a replacement for the various 7:24 steep taper shanks known as the CAT or V-flange taper in the US (ASME-B5.10 – 1994), the SK taper in Germany (DIN 69871), and the BT taper in Japan (JISC-B-6339/BT.JISC). These tapers previously dominated the CNC milling industry. Muller-Held[72] reported that the HSK system could provide maximum position deviations as low as 0.3 microns compared with 2 microns for an SK taper. Lewis[61] reported it was 3 times more accurate in the X and Y planes and 400 times better in the Z axis. Bending stiffness was 7 times better, while the short length of the taper allowed faster tool change times. The important detail though for this discussion is the fact that the design has a hollow taper that expands under centrifugal load at a greater rate than the holder. Consequently, unlike other mounting systems discussed, the taper tightens more as the rotational speed increases. Currently, the primary hesitation in broader adoption of this technique is the cost, availability, and delivery, due to the limited number of capable high-precision manufacturing sources.

11.7.7 Chatter Suppression

Chatter is an ever-present problem for grinding accuracy. Many claims have been made that the design of the wheel, especially the use of hub materials with high damping characteristics used in conjunction with superabrasive wheels, can suppress its occurrence.[17][92]

In the absence of damping, energy is exchanged without loss during the course of motion at particular natural frequencies and the vibration amplitude will increase over time depending on the rate of input of energy. Damping absorbs energy either through internal friction of the particular material or more often in joints and seams. Prediction of the damping of a machine is extremely difficult to calculate although it may be determined empirically by use of a hammer test and measuring the decay. The source of the energy that creates chatter can be either external, leading to forced vibration, or inherent in the instability of the grinding process, leading to self-excited vibrations.

Forced vibrations can be eliminated in three ways. First is to eliminate the energy at its source. Wheels, motors, belts, and workpieces should all be balanced as should the three-phase power supply. Ultra-precision bearings and ball-screws should be used and properly maintained in work and wheel spindles and slides. Second, the grinder should be insulated from sources of vibration such as hydraulic and coolant pumps, and vibrations carried through the foundations. Third, where resonance cannot be eliminated, the machine dynamics must be modified. Where a particularly prominent frequency exists (for example, in a motor or canti-levered member), tuned mass dampers consisting of a weight with a damped spring can be fitted at the point where the vibration needs to be reduced. This may often consist of a weight attached to the member with a rubber sheet sandwiched between them. The sizes of the mass, spring, and damper are selected so that the mass oscillates out of phase with the driving frequency and hence dissipates energy. Judging by the size of mass often required for this approach it seems difficult to accept a wheel hub could have a sufficiently high damping factor to absorb this level of energy.

Self-excited chatter occurs only during grinding and the amplitude increases with time. In wheel-regenerative chatter, a small perturbation, due to instability in the system, starts a regular variation in grinding forces that in turn creates an uneven level of wear around the wheel. The process is thus regenerative. There are several methods available for suppressing this form of chatter. First, the system stiffness and damping can be increased. Second, the grinding conditions can be continuously varied by changing the workspeed, wheelspeed, work support compliance, or by periodically disengaging the wheel from the workpiece. For example, Gallemaers, et al.,[37] reported that by periodically varying the workspeed to prevent lobe build up on the wheel they could increase the G-ratio by up to 40% and productivity up to 300% by extending the time between dresses. Third, the stiffness of the contact area can be reduced to shift the state of the system towards a more stable grinding configuration.[88] Finally, various filter effects can be used to reduce the wavelength to less than the contact width. The workspeed can be slowed to the point where the chatter lines merge. Alternatively, the frequency of the chatter can be increased to the point that the grinding process itself acts as a filter to absorb the vibration energy. For this reason, the natural frequency of wheels are targeted at values higher than 500 Hz or, ideally, higher than 1,000 Hz.

Most scientific studies on "damped" wheel designs are based on suppressing self-excited chatter. Furthermore they all use hub materials that not only have good damping characteristics but are also considerably more compliant and light-weight than "standard" hub materials such as steel. Sexton, et al.,[85] reported excellent results in reducing chatter grinding steel with resin CBN wheels by the use of a *Retimet* nickel foam hub material with a radial stiffness of 0.5 N/mm·mm. This was compared with values of 4–10 N/mm·mm for standard phenolic aluminum and Bakelite hubs. McFarland, et al.,[69] used polypropylene with a radial stiffness of 1.56 N/mm·mm and a natural frequency of 1169 Hz. This was compared to a radial stiffness for an aluminum hub of 24 N/mm·mm. By comparing the performance of a resin bonded diamond wheel on a flexible phenolic aluminum composite hub with a similar bond on an aluminum hub for grinding SiN it was demonstrated an improvement in life of over 70%. FEM analysis of the contact zone revealed over twice the radial deflection with the flexible hub. It appears that compliance in the hub can be transferred through a resin superabrasive layer and significantly increase contact width.

In conclusion, light-weight flexible hubs can provide benefit in grinding by limiting self-excited chatter with superabrasive wheels. Damping may also be an issue, but hub compliance and frequency responses are more likely to be the controlling factors. The concept is unlikely to be effective where significant forced vibration is present although it is sometimes difficult to differentiate between the two.

Compliance is much higher in conventional wheels. The effect of contact width for suppressing chatter is particularly pronounced when using plastic bonds for camshaft grinding or shellac bonds for roll grinding.

Some benefit is even seen using rubber inserts in the bores of vitrified Al_2O_3 wheels for roll grinding. Even with a resinoid diamond wheel, Busch[18] was able to show a 300% improvement in life merely by placing a rubber sleeve between the wheel and flange to increase compliance. Further research is likely to be focused on this aspect of wheel design as superabrasive technology targets applications such as roll and centerless grinding. It should be noted that attempts have been published regarding introducing micro-elasticity into vitrified diamond and CBN bonds for commercial products.[42] A more comprehensive review of grinding chatter, excluding centerless grinding is given by Inasaki, et al.[49]

11.8 ABRASIVE PASTES

During loose abrasive processes, such as lapping and polishing, the abrasive compound is a paste or suspension of abrasive in a medium or binder. Depending on the kind of abrasive, the intrinsic grain dimensions of abrasive particles used in lapping are as follows:

- Silicon carbide and aloxide: 5–1 µm.
- Boron carbide: 60–5 µm.
- Diamond: 5–0.5 µm.
- Chromium oxide: 2–1 µm.

For polishing pastes, the intrinsic grain dimensions are around 1 µm. Abrasives include chromium oxide, iron oxide, Vienna lime, alumina, chalk, and talc.

11.8.1 Binders for Abrasive Pastes

The binder is the medium in which the abrasive grains are suspended to form the lapping or polishing suspension or paste. During machining, the abrasive grains are retained on a plate surface by the binder. To prevent excessive consumption, the binder should fuse on light contact with the polishing plate and should adhere well to the surface. The greater the difference between the fusion temperature and vaporization temperature of the binder, the better will be the quality of the paste. Binders for pastes include stearin, oleic acid, paraffin, animal fat, wax, and petroleum jelly.

- *Stearin*. A white, solid, crystalline substance that melts at 140°C (160°F), which intensifies the polishing process and transmits hardness and cohesion to the paste.

- *Oleic Acid*: An unsaturated fatty acid that melts at 15°C (57°F), that accelerates the polishing process by dissolving metal oxides and thins the paste.

- *Paraffin*: A waxy crystalline complex mixture of white high fatty hydrocarbons that melts at 111°F. Paraffin is an excellent binder since it is not susceptible to charring and resinification, and which endows cohesion, elasticity, hardness, and adhesion to polishing pastes.

- *Fats:* Organic fusible glycerides of saturated solid fats and unsaturated organic oils, often used instead of stearin.

- *Waxes*: Including carnauba plant wax, beeswax and montan wax, are solid, unctuous or liquid fatty acid esters with higher fatty alcohols. Waxes provide the qualities of hardness and cohesion to a paste.

- *Petroleum jelly*: Obtained from asphaltless paraffin-base crude oils reduces the hardness of a polishing paste.

Surface-active substances and emulsifiers are sometimes added to polishing pastes to intensify the machining and to transmit a higher durability to the abrasive compound. Thixotropic substances including aluminum soaps, aluminum alcoholates, complex bentonite, and fine talc powder below 1 μm, are added to fluid pastes to increase viscosity.

11.9 COATED ABRASIVES AND ABRASIVE BELTS

11.9.1 Coated Abrasive Tools

Coated abrasive tools consist of coated abrasive grains bonded onto a backing material in the form of paper, cloth, or fiber strips. Abrasive paper and cloth in the form of belts and endless belts are used for machining metals. Silicon carbide coats are used for soft metals, while aluminum oxide coats are used for steel.

Coated abrasives are described in terms of shape and size, kind of abrasive, grit size, type of base (which can be paper, cloth, and paper reinforced with cloth and fiber), and type of glue (which can be either hide glue or synthetic resin glue).

Abrasive belts can be classified in terms of structure, single-coat and multiple-coat belts, mode of abrasive grain mounting, using gravitation or electrostatics, and bonding technology. The early development of abrasive belts was delayed by the lack of waterproof adhesives to allow the use of coolant with abrasive belts, and lack of stability and rigidity in the machines.

A coated abrasive belt is a length of coated abrasive material spliced together on the bias to make as smooth a joint as possible. It is mounted on two or more rolls: a contact roll to push the belt against the work, and one or more rolls to tension and rotate the belt. Abrasive belts can remove significant stock and the belts do not need dressing. The belt is used until it is necessary to change it due to deterioration in machining performance. The comparison between dressing time in conventional grinding and belt-changing time tends to favor belt grinding for suitable operations. Belt grinding is limited to straight form grinding, although limited contouring is possible with formed contact rolls.

For coated abrasives, virtually all the abrasive grain employed is either silicon carbide or aluminum oxide augmented by a little natural garnet or emery for woodworking and small amounts of diamond or CBN.

The aloxide grains are elongated, pointed, and sharp compared with grains employed in grinding wheels. Silicon carbide is regarded as harder and sharper, even though not as tough, as aloxide. Silicon carbide is used to grind low-tensile metals, glass, plastics, leather, and other soft materials. It penetrates and cuts fast under light pressure, which is beneficial for operations where appearance is more important than close tolerances.

11.9.2 Backings Materials for Coated Abrasives

There are four types of backings: paper, cloth, vulcanized fiber, and combinations of these laminated together:

- *Paper backings:* The least costly and are used when strength and pliability of the backing is not important.
- *Cloth backings:* Made from a special woven type of cotton cloth.

- *Vulcanized fiber:* A very heavy, hard, strong, and stiff backing utilized, mainly, for resin bonded discs.

11.9.3 Adhesives for Abrasive Belts

The function of the adhesive is to attach the abrasive grain to the backing. If the product is intended for dry use, the adhesive can be glue. Using coolants, the adhesive may be one of the liquid phenolic resins. The resins can be modified to provide longer or shorter drying times, greater strength, more flexibility, or other properties required.

11.9.4 Comparison Between Grinding Wheels and Abrasive Belts

Abrasive belts are often used for finishing wide surfaces, particularly where appearance is more important than close tolerances. In this type of application, abrasive belts tend to be more economic and give better performance than grinding wheels. For very high precision with strictly controlled tolerances, the rigidity of a grinding wheel gives better performance than an abrasive belt.

The cutting ability of a grinding wheel can be restored by a dressing operation, while the cutting ability of an abrasive belt declines continuously with use, and cannot be regenerated, even though a partially worn belt may produce lower surface roughness than a fresh belt.

A new abrasive belt has an advantage compared to a grinding wheel due to the grain orientation on a belt. Grinding wheels act harder in grinding than abrasive belts, with a typical hardness of 52 HRC. Abrasive belts are used extensively to grind hardened steels and to grind gray and malleable cast irons and nonferrous alloys. But, generally, the limit factor governing utilization is set by the nature of the abrasive grain rather than by the form of the tool.

11.9.5 Abrasive Belts in Furniture Production

Abrasive belts are widely used for *sanding* materials used in furniture production. The selection and use of coated abrasive belts has a significant impact on the ability to produce a good quality finish that maintains color, consistency, and clarity from unit to unit. The belts are mounted on a sanding machine and fed through at a constant speed in contact with the workpiece.

For optimum performance, it is important to select the best and most cost-effective abrasive product. The abrasive product needs to be optimal in terms of grain size, grain type, adhesive, and backing material for the particular application.

11.9.6 Abrasive Grains for Abrasive Belts

There are several abrasive grain types available and the type of grain selected has a direct relationship to product life and workpiece roughness. As a rough guide, aluminum oxide is good, zirconia-alumina is better and seeded gel (SG) aluminum oxide is considered best.

The difference in abrasive performance with SG alumina can be traced to the way the seeded gel grain is manufactured. The abrasive is made using a process that creates ceramic aluminum oxide crystals containing billions of abrasive particles per grain. This "microstructure" allows each grain to remain sharp during sanding by continually exposing fresh, sharp cutting points. Comparing belt life and finish consistency using SG versus standard aluminum oxide, productivity increases of 300% to 600% are reported.

The grit size of the abrasive grain must be sufficiently fine to produce roughness within the specified range and coarse enough for machining efficiency. It is important not to specify too coarse a grit size. Coarse grit products create patterns that must be completely removed as the surface roughness approaches the required value for the sanded surface. A series of grit sizes are employed in sequence to bring roughness down to the acceptable level. It is not recommended to skip more than one grit size in a grit sequence when finishing upgrading. Skipping two or more grits in a sequence can become costly, because an excessive number of finer grit belts must be used to remove the pattern left by the previous coarse grit.

11.9.7 Backings for Abrasive Belts

Belt selection includes the choice between cloth-backed and paper-backed products. On sanding machines, cloth-backed belts are normally used on the primary contact roll heads because of the potential for breakage when thick and tapered parts are fed into the sander. Cloth-backed coated abrasive belts withstand shock better caused by such pieces.

Paper-backed belts are generally recommended for use on platen finishing heads. Compared to cloth belts of the same grit size, paper belts are less costly and typically produce a higher quality finish. And, with today's designs, paper belts can produce a belt life nearly as good. In many applications, SG paper-backed belts have out-produced cloth belts by a factor of 3 to 1. When testing paper versus cloth on belt applications, it is advisable to decrease the belt tension gauge pressure by approximately 20 to 25%. The reason being that cloth stretches more than paper.

Both cloth- and paper-backed coated abrasives are now available with "antistatic" abrasive technology, which neutralizes the static charges of dust particles, allowing these particles to be exhausted more efficiently. In this way, antistatic technology minimizes dust on the workpiece and sanding equipment, and prevents premature loading of the abrasive belts. Because the workpiece stays cleaner, belt life and cutting rate are improved and a better overall finish results.

When approaching a sanding problem, it is important to understand "cause and effect" relationships in the process. A machine adjustment may temporarily solve the problem at hand; however, it may create a different problem entirely. This is particularly true when making frequent adjustments on multihead wide belt machines.

For example, through-feed speed has a direct relationship to stock removal rate. Slower through-feed speeds allow greater stock removal, and vice versa. However, it is important to remember that slower through-feed speeds can generate higher temperatures, as described in Ch. 6, "Thermal Design of Processes." Finish also has a direct relationship to through-feed speed. Faster through-feed-speeds produce a finer finish with lower temperatures. The opposite is true using slower speeds.

These examples demonstrate how an understanding of the sanding and finishing process can help you to identify, or better still, avoid sanding problems. To optimize sanding results, it is advisable to work with the coated abrasives manufacturer to ensure optimal product selection and usage.

REFERENCES

1. Abdel-Alim, A, Hannam, R. G., and Hinduja, S., A Feasibility Analysis of a Novel Form of High Speed Grinding Wheel, *21st Int. Machine Tool Design & Research Conf.*, Swansea, pp. 305–311 (Sep. 1980)

2. *Abrasive Magazine*, pp. 20–28 (Mar./Apr. 1999)

3. *AES Magazine*, front cover, 31(1) (1990)

4. Anon, CBN at High Speeds, *Am. Mach*, p. 50 (Mar. 1991)

5. Anon, High-speed Plunge Grinding, *Mgf. Eng.*, pp. 67–69 (Jun. 1979)

6. Anon, What is Single Point OD Grinding?, *Mod. Mach. Shop,* pp. 62–69 (Dec. 1997)

7. Bailey, M. W., and Juchem, H. O., *IDR*, Pt. 3, pp. 83–89 (1993)

8. Balson, P. C., *Vitreous Bonded Cubic Boron Nitride Abrasive Article,* US Patent 3,986,847

9. Bange, D. W., and Orf, N., *Tool Prod.,* pp. 82–84 (Mar. 1998)

10. Barlow, N., and Rowe, W. B., Discussion of Stresses in Plain and Reinforced Cylindrical Grinding Wheels, *Int. J. Mach. Tool Design & Research,* Pergamon Press, 23(2/3):153–160 (Oct. 1983)

11. Barlow, N., Jackson, M. J., Mills, B., and Rowe, W. B., Optimum Clamping of CBN and Conventional Vitreous-Bonded Cylindrical Grinding Wheels, *Int. J. Tools Mfg.,* 35/1:119–132 (1995)

12. Barlow, N., Jackson, M. J., and Hitchiner, M. P., Mechanical Design of High-speed Vitrified CBN Grinding Wheels, *Mgf. Eng: 2000 and Beyond IMEC, Conf. Proc.* (1996)

13. Barnard, J. M., Creep-Feed Grinding Using Crushform and Dressable Superabrasive Wheels, *Superabrasives '85 SME Conf. Chicago, IL,* MR85-292 (1985)

14. Barnard, J. M., Crushable CBN and Diamond Wheels, *IDR,* (1):31–34 (1989)

15. Bauery, R., and Caughel, C. M., *Use of Sol-Gel Technology for High-Performance Abrasives.*

16. Benko, E., *Ceram. Int.,* 21:303–307 (1995)

17. Broetz, A., Innovative Grinding Tools Increase the Productivity in Mass Production: Grinding of Crankshafts and Camshafts with Al_2O_3 and CBN Grinding Wheels, Prec. Grind & Finish in the Global Economy, Oak Brook, IL, *1st Conf. Proc., Gorham* (Oct. 2001)

18. Busch, D. M., Machine Vibrations and Their Effect on the Diamond Wheel, *IDR,* 30/360:447–453 (1970)

19. Bush, J., Advanced Plated CBN Grinding Technology, *IDA Diamond & CBN Ultrahard Materials Conf.,* Windsor, Canada (Sep. 1993)

20. Carius, A. C., Modern Grinding Technology, *SME,* Novi MI (Oct. 10, 1989)

21. Chattopadhyay, A. K., and Hintermann, H. E., Performance of Brazed Single-layer CBN Wheel, *Annals of the CIRP,* 43(1):313–317 (1994)

22. Chattopadhyay, A. K., Chollet, L., and Hintermann, H. E., Performance of Chemically Bonded Single-layer CBN Grinding Wheel, *Annals of the CIRP,* 39(1):309–312 (1990)

23. Coes, L., Jr., *Applied Mineralogy 1 – Abrasives,* Springer Publ. (1971)

24. Collie, M. F., *The Saga of the Abrasives Industry,* GWI/AGA Publ. (1951)

25. Craig, P., The Age of Resin Isn't History, *Cut Tool Eng.,* pp. 94–97 (Jun. 1991)

26. Marinescu, D., (ed.), Manufacturing Engineering–2000 and Beyond, *SME Publ.,* pp. 568–570

27. Daniel, P., Crushform Wheels Can Be Formed in Your Plant, *IDR,* (Jun. 1983)

28. De Beers Ind. Diamond Div., Monocrystal Diamond Product Range, commercial brochure (1993)

29. De Beers Ind. Diamond Div., Premadia Diamond Abrasives, commercial brochure (1999)

30. De Vicq, A. N., An Investigation of Some Important Factors Affecting the Clamping of Grinding Wheels Under Loose Flanges, *Mach. Tool Int. Res. Assoc.,* Macclesfield, UK (Aug. 1979)

31. DeVries, R. C., *Cubic Boron Nitride: Handbook of Properties,* GE Report #72CRD178 (Jun. 1972)

32. DiCorletto, J., Innovations in Abrasive Products for Precision Grinding, Prec. Grind & Finish in the Global Economy, *Conf. Proc., Gorham Advanced Materials,* Oak Brook, IL (Oct. 2001)

33. Engineer, F., Guo, C., and Malkin, S., Experimental Measurement of Fluid Flow Through the Grinding Zone, *J. Eng. Ind.,* 114:61–66 (Feb. 1992)

34. Field, J. E., *Diamond–Properties and Definitions,* Cavendish Lab. (1983)

35. Finishing in the Global Economy, *Conf. Proc., Gorham Advanced Materials,* Oak Brook, IL (Oct. 2001)

36. Frank, H., Mewes, D., and Schulz, S., Festigkeit von Schleifporen aus gebundenem Schleifmittel, *Ceramic Forum Int.,* Berichte, DKG, 75(1/2):44–49 (1998)

37. Gallemaers, J. P., Yegenoglu, K., and Vatovez, C., Optimizing Grinding Efficiency with Large Diameter CBN Wheels, *Int. Grind. Conf.,* Philadelphia, USA SME86-644 (Jun. 1986)

38. GE Superabrasives Report, Understanding the Vitreous Bonded Borazon CBN System (1988)

39. GE Superabrasives Report, Understanding the Vitreous Bonded Borazon CBN System, *Appl. Dev. Ops.,* Worthington (1988)

40. General Electric Borazon® CBN Product Selection Guide, commercial brochure (1998)

41. Gigel, P., *Finer Points,* 6(3):12–18 (1994)

42. Graf, W., *CBN- und Diamantschleifscheiben mit mikroelastischer Keramikbindung*

43. GWI, *Fatigue Proof Test Procedure for Vitrified Grinding Wheels,* Grinding Wheel Institute (1983)

44. Henri Polak Diamond Corp., commercial brochure (1979)

45. Heywood, J., *Grinding Wheels and Their Uses,* Penton Co. Publ. (1938)

46. Hitchiner, M. P., Systems Approach to Production Grinding with Vitrified CBN Superabrasives '91 Conf., *Proc SME,* 1.1–1.16 (1991)

47. Hitchiner, M. P., and Wilks, J., *Wear,* 93:63–80 (1984)

48. Hitchiner, M. P., Technological Advances in Creep Feed Grinding of Aerospace Alloys, 3rd Int. Grind. & Mach. Conf., *SME,* pp. 627–652, Cincinnati (Oct. 1999)

49. Inasaki I., Karpuschewski, B., and Lee, H. S., Grinding Chatter–Origin and Suppression, *Annals of the CIRP,* Keynote, 50(2) (2001)

50. Jaeger, J. C., *Proc. R. Soc.,* New South Wales, Vol. 76 (1942)

51. Julien, J., *Titanium Nitride and Titanium Carbide Coated Grinding Tools and Method Thereof,* US Patent 5,308,367 (May 1994)

52. Junker Group Int. Commercial Co., presentation.

53. Junker Maschinen, A New Era in the Field of O. D. Grinding, trade brochure (1992)

54. Klocke, F., and Muckli, J., High Speed Grinding with Microcrystalline Aluminum Oxide, *Abrasive Mag.,* pp. 24–27 (Jun./Jul. 2000)

55. Koenig W., and Ferlemann, F., CBN Grinding at Five Hundred m/s, *IDR,* 2/91:72–79

56. Koepfer, C., Grit, Glue–Technology Tool, *Mod. Mach. Shop,* p. 53 (Dec. 1994)

57. Kumar, K. V., 4th Int. Grinding Conf., *SME,* MR90-505, Dearborn, MI (1990)

58. Kumar, K. V., et al., Superabrasive Grinding of Superalloys, *SME Superabrasives '91,* MR91-216 Chicago, IL (Jun. 1991)

59. Kunihito, K., and Masuda, Y., *Segmented Grinding Wheel,* EP 433 692 A2 (Jun. 1991)

60. Landis Internal Report (Sep. 23, 1997)

61. Lewis, D. L., HSK Status Report, *Mgf. Eng.,* pp. 63–68 (Jan. 1996)

62. Li, R., *Improved Vitrified Abrasive Bodies,* WO Patent WO 95/19871 (Jul. 27, 1995)

63. Lowder, J. T., and Evans, R. W., *Process for Making Monolayer Superabrasive Tools,* US Patent 5,511,718, 11.4.94 (See also US 3,894,673 & US 4,018,576)

64. *Machine and Tool Blue Book,* pp. 69–72 (Apr. 1986)

65. Maillard, R., (ed.) *Diamonds – Myth, Magic and Reality,* Crown Publ. (1980)

66. Malkin, S., *Grinding Technology Book,* Ellis Horwood Publ., p. 36 (1989)

67. Malkin, S., *Grinding Technology,* Ellis Horwood Publ., Ltd (1989)

68. McClew, D., Technical and Economic Considerations of Grinding Aerospace Alloys with Electroplated CBN Superabrasive Wheels, *Precision Grinding '99, Gorham Advanced Materials,* Ambassador West Hotel, Chicago, IL. (Jun. 1999)

69. McFarland, D. M., Bailey, G. E., and Howes, T. D., The Design and Analysis of a Polypropylene Hub CBN Wheel to Suppress Grinding Chatter, *J. Mfg. Sci. E., Trans. ASME,* 121(1):28–31 (Feb. 1999)

70. Menard, J. C., Document WG6 – 4E *Calculations Based on Studies Conducted at the Technical High School of Hanover, Germany* (1963)

71. Meyer, R., Consultant for ANSI B7.1, *Safe Clamping of Cylindrical Grinding Wheels.*

72. Muller-Held, B., Development of a Repeatable Tool-holder Based on a Statically Deterministic Coupling MIT/RWTH Aachen Proj. Rep. (Feb. 1998)

73. Naxos Union Information, Grinding Wheels Production and Application, commercial brochure (1990)

74. Nippon Oil and Fats Co. Ltd., WURZIN (WBN) Tool, commercial brochure (1981)

75. Noichl, H., What is Required to Make a Grinding Wheel Specification Work? IGT Grinding Forum, U. of Bristol, n.d.

76. Norton, S. A., Project Altos, commercial brochure (1999)

77. *Crushable Wheels - Case Histories,* Part 2, pp.176–178 (Apr. 1989)

78. Peterman, L., ATI Techview PBS® versus Electroplating, trade paper.

79. Pflager, W. W., *Finite Element Analysis of Various Wheel Configurations at 160 m/sec.*

80. Roquefeuil, F., *ABRAL:A New Electrofused Alon Grain for Precision Grinding.*

81. Schacke, P., and Graf, W., Grinding and Honing Machine in One, *Swiss Quality Prod.*, pp. 16–19 (Jul. 2000)

82. Service, T., Safe At Any Speed, *Cut Tool Eng.* pp. 99–101 (Jun. 1991)

83. Service, T., Rethinking Grinding Wheel Standards, *Cut Tool Eng.*, pp. 26–29 (Dec. 1993)

84. Service, T., Superabrasive Safety, *Cut Tool Eng.*, pp. 22–27 (Jun. 1996)

85. Sexton, J. S., Howes, T. D., and Stone, B. J., The Use of Increased Wheel Flexibility to Improve Chatter Performance in Grinding, *Proc. Inst. Mech. Eng.*, (196):291–300, London, UK (1982)

86. Shaw, M. C., *Principles of Abrasive Processing*, Oxford U. Press (1996)

87. Slocum, A., *Precision Machine Design,-* SME Publ., pp. 336–341 (1992)

88. Snoeys, R., *Grinding Chatter*, pp. 415–424

89. Thornton, A. G., and Wilks, J., *Wear*, 53(1):165–187 (1979)

90. Thornton, A. G., and Wilks, J., *Wear*, 65(1):65–74 (1980)

91. Tymeson, M. M., *The Norton Story*, Norton Co. Publ. (1953)

92. Tyrolit Schleifmittel Swarovski, Grinding disk comprises an intermediate vibration damping ring which is made as a separate part of impregnated high-strength fibers, and is glued to the central carrier body and/or the grinding ring, Patent DE 20102684 (Apr. 19, 2001)

93. Ueltz, H. F. G., First Sintered Alumina Abrasive, US Patent 3079243

94. Valenti, A., Drofenik, M., and Petrovic, P., Glass-Ceramic Bonding in Alumina/CBN Abrasive Systems, *J. Mat. Sci.*, 27:4145–4150 (1992)

95. VSI-Z-*Special Werzeuge*, pp. 72–74 (Aug. 1992)

96. Wellborn, W., Modern Abrasive Recipes, *Cut Tool Eng.*, pp. 42–47 (Apr. 1994)

97. Wilks, E. M., and Wilks, J., *J. Phys. D., Appl. Phys.*, 5:1902–1919 (1972)

98. Yamamoto, A., A Design of Reinforced Grinding Wheels, *Bull. JSPE*, 16(4):127–128 (1972)

99. Yang, J., The Change in Porosity During the Fabrication of Vitreous Bonded CBN Tools, *J. Korean Ceramic Soc.*, 9:988–994 (1998)

100. Yang, J., et al., Effect of Glass Composition on the Strength of Vitreous Bonded CBN Grinding Wheels, *Ceramic Int.*, 19:87–92 (1993)

101. Yang, J., *J. Korean Ceramics Soc.*, 35(9):988–994 (1998)

102. Yang, J., and Kim Ho-Young Kim, D. Y., Effect of Glass Composition On the Strength of Vitreous Bonded CBN Grinding Wheels, *Ceramic Int.,* 19:87–92 (1993)

103. Yoshimi, R., and Ohshita, H., *Crush-formable CBN Wheels Ease Form Grinding of End Mills.*

United States Patents

104. US 492,767, & 1896 US 560,291, first SiC synthesis, Acheson, E. G. (1893)

105. US 4,623,364, SG abrasive, Cottringer, T., Van de Merwe R. H., and Bauer, R. (1986)

106. US 659,207, first electrofused alumina synthesis, Jacobs, C. F. (1900)

107. US 4,314,827, SG abrasive, Leitheiser, M. C., and Sowman, H. G. (1982)

108. US 2,947,617, first CBN synthesis, Wentdorf, R. H. (1960)

American National Standards Institute

109. American National Standards Institute, ANSI B74.3, Specifications for shapes and sizes of diamond or CBN abrasive products.

110. American National Standards Institute, ANSI B7.1, Safety requirements for the use, care and protection of abrasive wheels.

111. American National Standards Institute, ANSI B74.13, Bond specifications for diamond or CBN grinding wheels.

112. American National Standards Institute, ANSI B74.16, Particle size distribution of abrasive particles.

Japanese Industrial Standards

113. Japanese Industrial Standard, JIS B4130, Grain sizes of diamond and cubic boron nitride.

114. Japanese Industrial Standard, JIS B4131, Diamond or cubic boron nitride grinding wheels.

International Standards Organization

115. International Standards, ISO 6104, Abrasive products– Diamond or cubic boron nitride grinding wheels and saw–general: Survey, designation and multilingual designation.

116. International Standards, ISO 6106, Abrasive products–grain sizes of diamond or cubic boron nitride.

117. International Standards, ISO 6168, Abrasive products–diamond or cubic boron nitride grinding wheels–dimensions.

118. International Standards, ISO R565 1990, FEPA size designation for abrasive particles.

Grinding Wheel Manufacturers' Literature

Cincinnati Milacron	Diamond and CBN grinding wheels (1990)
	Grinding wheel application guide (1983)
DWH	Precision Tools in diamond and CBN (n.d.)
EHWA	Product Catalog (n.d.)
FAG (Efesis)	Superabrasives – CBN and diamond with vitrified bond – rotating dressing tools, Publication # SK56 001 EA (n.d.)
Kinik	Grinding wheels Catalog #100E (n.d.)
Krebs	Vitrified diamond and CBN grinding wheels (n.d.)
Meister Abrasives	Master Vitrified CBN (n.d.)
Noritake	Vit CBN wheel ultimate solution (1998)
	Vitrified bond CBN wheels (n.d.)
Norton	Project Optimos – grind in the fast lane (1999)
	Bonded abrasives specification manual (1999)
	Diamond vitrified VRF bond for roughing and finishing polycrystalline tools (2000)
	Norton CBN wheels precision production grinding (n.d.)

Osaka Diamond	Vitmate® series vitrified bond CBN wheels (n.d.)
Radiac Abrasives	An introduction to Radiac grinding wheels and their applications (1987)
Tesch	Diamond and CBN grinding wheels for mechanical industries (1995)
TVMK	General Catalog (2001)
Tyrolit	Grinding with Tyrolit (n.d.)
	Diamant-nitrure de bore cubique (CBN) (n.d.)
Universal	Poros Two (n.d.)
	Universal vitrified grinding wheels (1997)
	Universal stock catalog & price list (1994)
Universal Superabr.	Diamond & CBN wheels – Univel, vitrified, resin metal, plated (1998)
Van Moppes-IDP	'Crushform' diamond and CBN wheels (n.d.)
	Resimet® (n.d.)
Wendt	Diamond/CBN Grinding and Dressing Tools H/S/G-1 General Information (n.d.)
	Vit - CBN CBN grinding tools in vitrified bond (1991)
Winter	Grinding with diamond and CBN Book (1988)
	Winter HP™ High Precision Electroplated Wheels, Brochure B00A-38
Winterthur	Precision grinding wheels (1998)

12

Conditioning of Abrasive Wheels

12.1 INTRODUCTION

This chapter deals with the operations to prepare abrasive tools for effective machining. First, there is the preparation of a wheel before grinding and then the maintenance of the active surface of the wheel as grinding proceeds.[1] These operations that are auxiliary to the grinding process are divided into two distinct groups: the preparation and the conditioning of the grinding wheels.

12.2 GRINDING WHEEL PREPARATION

Preparation of abrasive and superabrasive grinding wheels includes the following operations:

- *Ringing.* Check the integrity and structural uniformity of a new abrasive wheel.

- *Mounting.* Consists of clamping the abrasive wheel between plane-parallel flanges and the subsequent mounting on a spindle. Clamping normally requires use of elastic layers between the wheel and the flanges. The procedure requires the uniform distribution of the clamping force to avoid stress concentrations.

- *Balancing.* Essential for safety and for effective grinding. An initial check is performed for *static balance* of the new abrasive wheel before it is mounted on the main spindle of the grinding machine. After mounting the wheel on the grinding machine, *dynamic balancing* is required particularly for high wheel speeds. Further dynamic balancing may be necessary during the use of the wheel, and is often performed after each conditioning of the wheel and at the same time and/or before starting a finishing grinding process.

Safety regulations must be observed during mounting of an abrasive wheel according to national health and safety regulations. The personnel who undertake this task should be appropriately trained and certified.

12.3 GRINDING WHEEL CONDITIONING

12.3.1 Introduction

Conditioning includes *cleaning-up* a wheel, to eliminate loading and restore active grain distribution on the wheel surface. Conditioning operations also include *truing* to achieve an accurate form and *dressing* to restore cutting efficiency of the grinding wheel. Since the same process is usually employed for all three purposes, the term dressing is a commonly used term for most conditioning operations on conventional vitrified wheels. However, for superabrasive wheels, it is necessary to distinguish dressing from other conditioning functions described below. Conditioning may sometimes include the application of *reduced removal rate grinding* or other techniques to open-up a superabrasive grinding wheel.

Ideally, the grinding process will take place with optimum technological parameters established as a result of commissioning tests, so that the specified tolerances can be maintained for a long time due to a slow uniform wear process of the active surface of the grinding wheel during machining. At the same time, a self-dressing process takes place due to the presence of friable abrasive grains in the active surface of the wheel. This *self-dressing process* occurs due to micro and macrofracture of the cutting edges of the grains. As the grains wear and start to become dull, fracture restores sharp cutting edges (Fig. 12.1a). A self-dressing process allows the working parameters to be maintained in the optimum range and workpiece geometric tolerances to be achieved without need for frequent reconditioning.

The normal grinding force (F_n) depends primarily on the ratio between the peripheral surface speed of the wheel (v_s) and the workpiece material removal rate (Q_w).

Eq. (12.1) $\qquad F_n = K \cdot v_s / Q_w$

where K is an abradability constant, which depends on the processed workpiece material and the sharpness of the grains.

In practice, the aim is usually to increase removal rate and, at the same time, to maintain roughness and other geometric deviations within the tolerances. Unfortunately, the aim of increasing removal rate tends to act to the detriment of the *wheel redress life*. High removal rate accelerates wear of the abrasive layer. A high wear rate is characterized by breaking and/or by pulling out of whole abrasive grains from the active surface. It is also characterized by thermal degradation of the bond and by transfer wear of workpiece material onto the active surface of the tool, a process known as *loading* (Fig. 12.1b). Rapid fracture wear of the wheel surface reduces forces, but greatly increases workpiece roughness, and is therefore likely to be unacceptable. Loading of the wheel usually increases grinding forces, but also worsens workpiece quality.

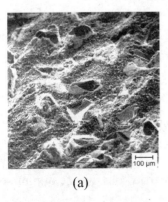

(a)   (b)

Figure 12.1. Scanning electron microscopy (SEM) photographs of active layers on resin-bonded CBN-wheels used for HSS-tools and cutter grinding: *(a)* linear evolution of grinding with self-dressing of the abrasive layer, and *(b)* dulled and loaded surface caused by extended grinding; a conditioning operation must be performed immediately.

Glazing is the opposite of intensive wheel wear and can occur as a consequence of employing low removal rates to obtain very low roughness. It also tends to occur when the wheel being employed is one or two grades too hard for the particular workpiece material and removal rate. Glazing occurs when there is a failure to achieve a self-dressing action and the grains wear flat during grinding. It is possible to see that the wheel surface is glazed because it becomes more light reflective. Glazing increases the grinding force and may reduce workpiece roughness. However, the high forces can easily lead to problems such as wheel loading and vibrations. The wheel should be redressed when it is found that the workpiece roughness or any other quality control measure exceeds the action control limit, or the forces and power have increased beyond an acceptable level. The progression of the forces, with time and their dependence on the dressing process in constant feedrate grinding, are discussed in Ch. 5, "Forces, Friction, and Energy."

In a force-controlled abrasive machining process, such as lapping or polishing, a constant force is usually applied to the tool rather than a set feedrate. The same tribological principles apply to the tool-wear process as discussed previously, but in a constant-force case it is found that the abrasive removal rate tends to reduce with machining duration and with tool wear. This is because the abrasive grains tend to become dull and there is less likelihood of a self-dressing action. The removal rate tends to reduce exponentially from its initial value.

Eq. (12.2) $Q_w = Q_{w,o} \cdot e^{-\beta K F_n \cdot v_s \cdot t}$

If the removal rate is allowed to diminish to half its initial value before action is taken to condition the tool,[2] the machining time between two successive conditioning processes is,

Eq. (12.3) $T_{1/2} = 0.69/\beta K F_n v_s$

A high removal rate corresponds to a high normal force and the tool-life is correspondingly reduced by tool wear. A high wear rate of the abrasive tool reduces grinding productivity.

To summarize, conditioning is carried out for one or more of the following reasons:

- Loss of wheel shape accuracy due to wear or wheel loading.

- High energy consumption and consequent temperature-rise during the grinding process.

- Loss of workpiece surface quality.

- Low accuracy of finished workpieces.

- Reduced productivity.

The purpose of conditioning a tool is to restore productivity to its initial value. A conditioning process aims to assure a clean abrasive surface of appropriate microtopography for subsequent abrasive machining operations. On average, only 10% of the active volume of the wheel is worn away by grinding. The remaining 90% is removed by wheel conditioning operations.[3]

Tool redress life, for repeated batch production, can be experimentally determined on the basis of systematic commissioning tests. Determination of the tool redress life is part of a process of optimizing the dressing and grinding conditions for a new product. Determination of tool redress life constitutes an aspect of ongoing process improvement in modern industrial practice.

12.3.2 Cleaning-up a Wheel

Cleaning-up in this context represents the removal of debris, the worn layer, and any loading at the wheel surface. The need for clean-up can be understood from the following features of the grinding process:

- Wear changes the distribution of grains in the active layer. The number of abrasive contacts in the active layer may be either increased or reduced. A heavy wear process reduces the number of contacts by pulling out grains and by macrofracture. Cleaning-up the wheel restores the number of active grains at the surface. To clean-up the wheel may require removal of a relatively large depth from the abrasive surface that may be equivalent to approximately one average grain diameter.

- When machining with superabrasives, great care is taken to select machining and conditioning process parameters as far as possible, to avoid the problem of loading. Loading due to transfer wear accumulates within the pores of the abrasive layer and at the grain tips. Loading tends to be greater with difficult-to-grind materials.[4] It is also more likely to be a problem in dry grinding or with the use of process fluids lacking a *detergent property*. Loading is a particular problem where there is a chemical affinity between the workpiece material and the abrasive composite. In these circumstances, the particles of workpiece material adhere to the abrasive grains of the grinding wheel. This process is known as *adhesive loading*. Some materials form long continuous chips in grinding. These have the tendency to block the pores of the abrasive surface. This process is known as *pore loading*. Pore loading and adhesive loading are both increased by high contact temperatures and ductile material flow conditions. To fully restore machining efficiency, it is necessary to completely remove the loading. This requires machining away a relatively large depth of the abrasive layer. With superabrasive tools, due to the high cost of the superabrasives, it may not be economic to completely clean up the loaded layer.

• Metallic transfer layers may interact with *friction polymer films* formed on the working surface of the wheel. The synthesis of polymer films may occur due to the presence of organic free radicals in the process fluids. These organic radicals form by tribopolymerization of an adherent film on the active surface of the tool in an intensive grinding process. (See Ch. 15, "Tribochemistry of Abrasive Machining.") Metallic and organic layers formed on the wheel surface change the nature of contact of the tool/workpiece friction couple elements. The results of this change are a reduced real pressure in the interface and the onset of metal-to-metal contact. This defective contact is both a result and a cause of adhesive loading, as illustrated in Fig. 12.1b.

12.3.3 Truing

As stated previously, cleaning-up a wheel is often carried out simultaneously with truing. However, truing describes the machining of the correct profile for the active layer of the wheel. Therefore, truing may need to be carried out in a slower and more carefully controlled operation than cleaning-up. Truing of the wheel is intended to make the achievement of the workpiece profile within the specified workpiece tolerances possible. Truing aims to restore the specified peripheral form, roundness, and concentricity of the abrasive wheel for the particular state of wheel-balance. Truing must always be performed if the wheel is re-balanced and must be performed after balancing. Truing may also be essential to achieve other technical requirements, such as parallelism for machining side faces and the perpendicularity of these planes on the rotation axis. The requirements of the macrotopography of a wheel were discussed in Ch. 10, "Grinding Wheel and Abrasive Topography."

12.3.4 Dressing

In a strict sense, dressing is performed to achieve a specific microtopography on the working surface of the wheel. Both the morphology and the distribution of abrasive grains in the active layer define the grinding wheel microtopography, as described in Ch. 10. The abrasive wheel topography determines grinding forces and temperature, and influencing productivity and quality of the process. The topography of the abrasive tool varies during the grinding process. A dressing process generates a specific microgeometry of the abrasive surface. This process aims to achieve an optimum microtopography, which is specific for the particular process conditions. Dressing takes place at the commencement of operation with a new wheel and/or at periodic intervals during the grinding process. The dressing process is responsible for the sharpness of the wheel. The sharpness is reflected in the specific energy in grinding, as discussed in Ch. 5, "Forces, Friction, and Energy." The dressing operation also affects the wear behavior of a grinding wheel and can therefore be considered to influence the effective hardness properties of the wheel. It is obvious that we have not only a modification of the microtopography by dressing, but also an essential transformation of the microstructure of the working layer. A conditioning process modifies the main features of the qualification of the grinding tool,[5] i. e., the E-modulus and the hardness of the abrasive composite layer.

There are some problems with terminology in the conditioning of grinding wheels, since for superabrasive wheels it is necessary to carry out a dressing process after truing. For vitrified grinding wheels with conventional abrasives, it is often assumed that "the act of truing a wheel, also dresses it."[6] The combined cleaning-up, truing, and sharpening of vitrified wheels is generally referred to as dressing.[7] Accordingly, for vitrified grinding wheels with conventional abrasives, the cleaning-up, truing, and dressing stages can often be amalgamated into one operation.

For other types of abrasive tools, conditioning has to be executed in separated stages. It is very important to realize and respect this requirement; otherwise acceptable results will not be obtained during the grinding process either from technological or from economic considerations.

12.4 DRESSING TOOLS

12.4.1 Introduction

Conditioning of grinding wheels by conventional methods require special dressing tools. These dressing tools have an active cutting surface containing hard or ultra-hard materials of a special type and size (carats).

A diversity of designs of dressing tools is available, including single- and multigrain tools. Dressing tools can be classified into one of the following subgroups:

- Metal cutters.
- Single/multiset diamond and diamond matrix traverse dressers.
- Superabrasive wheels and points.
- Block and roll dressers.
- Crushing roll dressers.

The type of dressing tool can have profound effects on the grinding performances.

12.4.2 Ultra-hard Materials for Dressing Tools

In most situations, natural or synthetic diamonds form the basic component for the active cutting surface of a dressing tool. The application of HP/HT diamonds synthesized at 1600°C and 7 GPa has been highly successful over the 25 years or more for dressing tools.[8] In addition to the classical SD-grits, there are also MCD and CVD logs, PCD dressing pieces, etc. The choice of diamond is an expensive constructional solution, but essential for increased wear resistance of the dressing tool. The working behavior of a dresser depends greatly on the nature and structure of its active part. The following types of ultra-hard materials are preferred for the manufacture of dressing tools:

- *Natural diamond (ND) gemstones.* ND single crystals are ultra-hard grains classified with a quality-index from "A–D." The grains result as a sub-product following preliminary selection of the cropped natural diamond. In fact, the attention of the selector is focused on producing crystals with a perfect form, purity, and transparency for use in jewelry. The crystals remaining after the preliminary selection are used for industrial purposes. These are then classified based on the criteria of quality, morphology, and size (carats).

  - The main criterion for *quality* is the number of grown points on the ND stones. (See Table 12.1.)

  - After selection based on quality, a further selection is made according to *morphological criteria,* that is corresponding to the initial form of the ND crystals. (See Table 12.2.)

  - Finally, ND gemstones are distributed into *size-groups* ranging from 0.25 to 4.00 carat.

  Different combinations may result from applying criteria for selecting industrial diamonds. The appropriate sort, type, and dimensions of different dresser types have to be established according to the basic characteristics of the ND grains for particular application domains.

Table 12.1. Quality of Selected Industrial Diamond Gemstones

QUALITY RANGES (INDICES)	NUMBER OF GROWN POINTS
Super (A)	5
Special (B)	3 and 4
Standard (C)	1 and 2
One-way (D)	1

Table 12.2. Morphology and Applications of the Industrial ND Gemstones and Grains

MAIN MORPHOLOGY	ND DRESSER TYPES	APPLICATIONS
Octahedrons	Single stone or profiled traverse dressers	High accuracy dressing and profiling
Points or needles	Multi-stone traverse dressers	Traverse dressing
Maacles	Multi-stone plain and roll dressers	High accuracy profiling
Broken (ND-grits)	Multi-grain dressers	Traverse dressing and profiling

- *Broken natural diamond (ND) grits.* Fragments or microcrystals, cropped or resulting from mechanical processing of larger ND grains. These ultra-hard grits derive from tough diamonds and are selected according to their granulation. The ND grits are used to fabricate sintered or electroplated dressing disks, cups and points, block-dressers, as well as profiling roll-dressers.

- *Synthetic diamond (SD) grits.* "Man-made" diamond powders, usually belonging to the SD tough class of friability. These grits are selected according to their grain size. Sometimes, the SD microcrystals are chemically treated and/or coated in order to increase adhesion between the grit and the binding material. The SD grits are used in the same field of application as the ND grits.

- *Synthetic monocrystalline diamond (MCD) logs.* Small prisms with a square base. The MCD logs are made from a single crystal of synthetic diamond, having a 8,000 to 9,000 Knoop hardness. These extra-hard crystalline small pieces are used for the manufacture of single- and multistone traverse dressers. Unlike natural diamonds, MCD logs have a constant section, structure, and hardness along the entire height of the single crystal. Table 12.3 gives basic dimensional sets of MCD logs available on the world market.[9]

- ***Diamond chemical vapor deposition (CVD) logs.*** CVD diamond logs are produced in a vacuum by lying down several polycrystalline SD-layers on the working surface of a prismatic support (log). The prismatic support is made from sintered carbide (SC) or from a very tough ceramic. The wear resistant coating consists of a thick polycrystalline SD multilayer applied without a binding matrix. The main CVD log sizes are presented in Table 12.3. One of the best applications of CVD logs is for the manufacture of dressing rolls having a complicated shape.[10]

- ***Polycrystalline diamond (PCD) pieces.*** These are dressing composites with different shapes and sizes, depending on the HP/HT- diamond synthesis technology. These ultra-hard composite PCD pieces consist of SD micrograins (size 2 to 25 μm), dispersed in a cobalt matrix. The hardness of PCD pieces ranges from 5,000 to 5,500 Knoop. A main application domain for PCD is for the active surface of dressing rolls. [11]

Table 12.3. Size of SD Dressing Logs (Diamond Monocrystalline MCD and Polycrystalline CVD Logs)

DIMENSIONS (mm)	CROSS-SECTIONAL AREA (mm^2)
0.4 × 0.4 × 4.0	0.16
0.6 × 0.6 × 4.0	0.36
0.8 × 0.8 × 5.0	0.64
1.2 × 1.2 × 5.0	1.44

12.4.3 Traverse Dressers

A profile may be generated on a wheel by traverse feeding a diamond tool across the surface using a pantograph[12] or a CNC system.[13][14] Traverse dressing tools are used for conditioning in an intermittent mode for vitrified grinding wheels with conventional abrasives. Dressers are available in various shapes and configurations for this purpose (Table 12.4).

The detailed specification of traverse dressers requires the following information:

- The type of the tool, number and carats of diamond single crystals, or the specification of the polycrystalline compacts inserted in the single- or multistone tool.
- The grain size and/or the carat value of the diamond crystals included in the active part of a multigrain dresser.
- The geometric characteristics of the ultra-hard active part of a profiled and polished single stone.
- A drawing for the feeding-tool fixture.

Great care must be taken in the conditioning of superabrasive wheels. Performing the conditioning operation using a dressing tool on superabrasive grains reduces the efficiency of grinding. The superabrasive grains are dulled and damaged by a truing process. For electroplated superabrasive wheels, it is generally considered inadvisable to true the wheel with a dressing tool. It is, therefore, very important that such wheels are supplied with an acceptably low concentricity in the mounting process.

A worn-out dresser will result in the production of very poor quality workpieces. Therefore, the active part of the dressing tool has to be monitored and controlled periodically, and replaced as necessary. Due to the high cost of a monocrystal-type dresser, reconditioning is recommended, where possible, by re-setting the flattened diamond stone and/or by re-pointing it.

12.4.4 Roll Dressers

The principal types of *diamond roll dressers* consist of diamond disk, roller, cup and point.[15][16] Various methods exist for production of diamond-roll dressers;[17] Table 12.5 lists these systematically.

Table 12.4. Types and Specifications of Different Diamond Dressers

DRESSER TYPE	DIAMOND GRIT TYPE	GRIT SET	PROPERTIES OF DIAMOND DRESSER
Single-stone	-ND stones, Quality A to C Size: 0.25 to 4.00 ct; -MCD or CVD logs; -PCD pieces	By hand	-Care by starting impact; -Cooling before contact; -Drag inclination 5 to 15°; -Natural active point; -Re-setting or re-pointing
Profiled diamond (polished)	-ND stones, Quality A or B Size: 0.50 to 2.50 ct; -MCD logs; -CVD logs.	By hand	-High precision and expensive dressing tools; -Drag inclination 5 to 15°; -Care by impact and cooling; -Exact edge geometry; Point-radius 0.25 to 0.75 mm Angle 40 to 60°; -Re-setting and/or re-pointing
Multi-stone	-ND stones (small sized), Quality C or D Size: 0.02 to 0.35 ct on to five layers; -MCD or CVD logs; -PCD pieces.	By hand or randomly	-Insensitive to starting impact; -For dressing the large and wide wheels; -For rough and heavy dressing; -Decrease dressing costs
Multi-grain: cylindrical, plate, block or nib	-Broken ND or SD-grits -Grain size D 650-D 1,100	Randomly	-Good dressability; -For dressing the centerless- and surface-grinding wheels; -For dressing the fine-grain conventional and CBN wheels

Table 12.5. Types and Specifications of Different Diamond Roller Dressers

PROCESS USED IN PRODUCTION OF THE ROLLER DRESSER	DIAMOND SETTING	BOND TYPE OF DIAMOND LAYER	PROPERTIES OF THE ROLLER-DRESSER
Reverse process*	Randomly distributed ND or SD grains	Electroplated*	High precision; Statistical distribution; Bigger concentration of diamond by different grain size
		Sintered-infiltrated**	
	Hand-set ND or SD grains	Electroplated*	Highest precision; Setting pattern; Variation of diamond concentration
		Sintered-infiltrated**	
	Hand-set PCD pieces, MCD or CVD logs.	Electroplated*	Highest precision; Setting pattern; Increased dressing productivity
		Sintered-infiltrated**	
Direct process*	Randomly distributed ND or SD-grains	Electroplated*	Precision depends on grain size and shape tolerances; More economical; The multi-layer makes possible the recondi-tioning of the dresser
		Sintered-pressed** (multi-layer)	

* Dressing tool can be reground and recoated; ** Dressing tool cannot be recoated.

A *crushing roll dresser* is made from materials that are softer than the abrasive grains of the wheel but harder than the composite abrasive-bond layer. Materials employed for crushing rolls include hardened tool steel, tungsten carbide, or boron carbide.[18] Crushing-rolls are mounted on the rotational axis of a profiling device.

12.5 TECHNOLOGIES FOR CONDITIONING VITRIFIED CONVENTIONAL WHEELS

12.5.1 Conditioning by Traversing the Diamond Tool Across the Wheel

The kinematics of the conventional conditioning of vitrified wheels is similar to turning on a lathe. During this operation, the volume of abrasive material removed from the wheel surface in each pass is proportional to the dressing depth, a_d (Fig. 12.2). The longitudinal lead of the feed motion is the dressing lead, f_d. The lead is related to the dressing traverse feedrate according to:

Eq. (12.4) $$f_d = \frac{\pi \, d_s v_{fd}}{v_s}$$

where v_{fd} is the dressing traverse feedrate, v_s is the peripheral velocity of the wheel, and d_s is the initial diameter of the abrasive wheel.

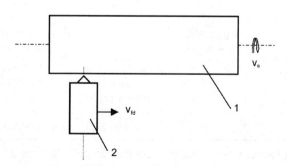

Figure 12.2. Dressing by traversing a tool across the wheel: *(1)* vitrified grinding wheel, and *(2)* traverse dresser.

The efficiency and quality of the conditioning process performed on a vitrified grinding wheel, depends mainly on the dressing lead and the dressing depth. Sometimes additional *spark-out* dressing passes may be executed with zero infeed at the end of the dressing operation, with the view of obtaining a smoother and more accurate wheel microtopography and implicitly of achieving better workpiece quality. While this procedure can have advantages in producing lower roughness in grinding, it carries the disadvantages that the abrasive grains are dulled and the wheel life is reduced.

Some effects of dressing kinematics were discussed in Chs. 3 and 5. A further important parameter is the *overlap ratio*, U_d, defined by Eq. (5.18). A small value of overlap ratio, less than 2 to 3, is used to achieve an open wheel surface with low grinding forces, for example, for rough grinding domain. A higher value of 7 to 10 is used to achieve the lowest roughness at the expense of higher values for grinding forces, for example, for finish grinding.

Employing a traverse feedrate, which is too low, may often cause grinding problems. This results in numerous interactions between the dressing tool and individual grits, causing damage to them.

Some typical values of the dressing parameters are presented in Table 12.6. These values provide a starting point for the optimization of the dressing process. However, due to the process complexity, optimization of the conditioning processes for a new grinding set-up will require systematic commissioning tests.

Table 12.6. Dressing Parameters for Conventional Vitrified Wheels, Using Traverse and Roller Dressers

DRESSER TYPE	DRESSING LEAD (f_d) (mm)	DRESSING DEPTH (a_d) (µm)
Single-stone	0.05 to 0.3	0.05 to 25
Chisel-shaped multi-stone	0.10 to 0.50	2.0 to 50
Cluster-shaped multi-stone	0.30 to 2.00	2.0 to 50
Diamond roller	—	0.1 to 2
Crushing roller	—	2.0 to 20

The conditioning parameters using a traverse tool need to be selected for a particular wheel and workpiece material couple operating under particular grinding conditions. This process may only be carried out in one direction, either from left to right or the reverse. The main characteristics and nature of the conditioning process remain the same, however, irrespective of whether a wet or dry working process is employed and of the specification of the process fluid.

12.5.2 Diamond Roller Dressing

The kinematic arrangement for traverse conditioning is similar to the external diameter grinding process; the roller dresser contacts a strip of the rotating abrasive wheel and is traversed along its surface. Both cylindrical parts of the wheel/roller friction couple are in rotational movement with different rotational speeds. The peripheral speed of the grinding wheel, v_s, and that of the dresser, v_d, are illustrated in Fig. 12.3. The dressing depth is achieved by means of an infeed mechanism.

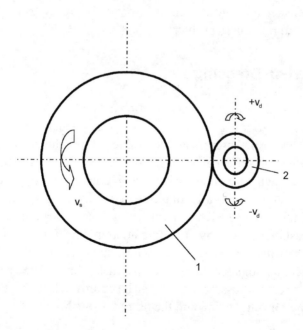

Figure 12.3. Roll dressing of a grinding wheel: *(1)* vitrified grinding wheel to be conditioned, and *(2)* roller dresser.

An important criterion for roll conditioning is based on the ratio of the dressing speed, v_d, to the abrasive wheel speed, v_s. [16] The dressing speed-ratio may be either positive or negative, depending on whether the speeds at the contact point act in the opposite direction or in the same direction. If the roller velocity acts in the same direction as the wheel speed, the roller velocity is required to be up to four times greater than if the roller velocity acts in the opposite direction. When the roller velocity acts in the same direction as the wheel velocity, the dressing process is similar to crushing. The abrasive grains in the wheel experience more fracture damage, and subsequent workpiece roughness will be high. For precision grinding, it is more useful for the velocities to act in the opposite direction, in order to achieve low surface roughness.[8] A general guide for the roller-dresser velocity is:

Eq. (12.5) $v_{d,max} = 0.8\ v_s$

At the end of a conditioning cycle, finishing passes may be performed without further dressing infeed, as in the case of traverse dressing. The same cautionary comments apply. It is also important that the dressing roller speed is selected to avoid a machine resonant frequency, as this will lead to vibration and roundness problems.

12.5.3 Crush Dressing

Crush dressing is a fast operation using a crushing roller to open a wheel, reduce grinding forces, and increase productivity. The crushing process is performed at intermediate values of peripheral roller speed and at relatively high contact pressures, typically 60 N/mm². The high working pressure can generate concentricity deviations (*run-out*) of conditioned abrasive wheel, adversely affecting system stability and machined workpiece quality. These adverse effects of crush dressing can be minimized by using helical crushing tools or by operating at optimum values of contact width between the wheel and the roller. Decreasing the clearance of the grinding spindle bearings can sometimes reduce instability of the crush dressing system. The crush dressing process can be improved by using two crushing rollers, one for roughening and the other for finishing.

Process fluids for crush dressing need to have good tribological properties. Ideally, the fluid will be neat oil with EP additives. In some cases, water-based synthetic solutions with highly active rust inhibitors have been used satisfactorily.

Crush dressing is not a process used in precision grinding. Precision grinding results are generally better, using a diamond roller dresser for profiling and dressing operations, than those obtained by crush dressing. The improved results obtained through diamond roller dressing are due to the smaller contact between the roller dresser and the wheel and through the increasing use of high wheel speeds for conditioning or dressing.

12.5.4 Continuous Dressing (CD) Using A Diamond Roller

Continuous dressing (CD) is a development beyond traditional periodic dressing. The CD process is performed at the same time as grinding. The CD process permits very high removal rate capability by effective control and constancy of the real contact pressure between the roller dresser and the wheel. The main advantage of this technique is the ongoing control of the grinding wheel topography during the grinding process. The continual maintenance of an optimum microtopography permits high quality and productivity to be achieved with a wide range of advanced materials. To develop this advanced technology it is necessary to employ rigid high-precision machine tools equipped with adequate CNC equipment.[19] This expense is necessary because the CD process is efficient only if a strict control of the grinding depth, dressing depth, and speed-ratio are achieved. It is necessary to continuously compensate the infeed position of the wheel during the grinding process to allow for the reducing wheel size as the roller dresser is advanced.

Continuous dressing has been notably employed for deep creep-feed grinding operations, where it is essential to maintain a sharp cutting surface on the wheel to avoid forces and thermal damage problems[15] (see also Sec. 5.4, Ch. 5, "Forces, Friction, and Energy").

12.6 TECHNOLOGIES FOR CONDITIONING SUPERABRASIVE WHEELS

12.6.1 Introduction

As mentioned previously, the various stages of conditioning of superabrasive wheels are performed separately. At each stage, it is necessary to employ appropriate tools, devices, and technologies. The truing and dressing procedures are specific for the nature of superabrasive grit and upon the type and other characteristics of the active layer bond. Ideally, superabrasive wheels are ultra-wear–resistant. However, if the wheels are abused either in the conditioning operation or in grinding, rapid wear and damage can occur. Given the high cost of superabrasive wheels, these wheels are employed in applications where it is worth spending a little extra time to ensure the conditioning is performed satisfactorily. With a little extra care and attention, the user will benefit from a long wheel life and high process productivity.

12.6.2 Truing Superabrasive Wheels

Truing to produce the macrogeometric profile of the active layer is performed with one of the numerous tools and devices specially designed for superabrasive wheels. Commonly employed devices include the following:

- A brake-controlled truing device.[20] This is a simple contrivance, which is directly driven by the trued grinding wheel. The active part may be a diamond roll or possibly a vitrified conventional abrasive roll having dimensions $\Phi 75 \times 25$ mm, hardness K or L and conventional abrasive CSi (green, mesh size 80 to 150). In the case of brake-controlled truing, replacing a silicon carbide roll with a diamond wheel greatly increases the tool cost, but allows much better accuracy to be achieved. The peripheral speed of the roll is much lower than the wheel speed due to a braking mechanism in the truing device. This truing technology, when used with a conventional abrasive truing roll, has a relative low accuracy due to the high wear rate of the roll.

- A diamond impregnated truing block or diamond truing nib. These dressers contain SD-grains of sizes D 126 to 301, at a concentration of 200. The grains are embedded in a metallic bond (bronze).[21] There is a risk of drastic damage to the tool and to the wheel if a diamond truing nib is brought into harsh engagement with the wheel.

- A new dressing technology, called *swing-step dressing,* is used for profiling CBN wheels. This process is designed for high-speed profile grinding and for high-efficiency deep grinding. Applying diamond dressers to condition superabrasive wheels requires an automated control system for wheel-contact detection. The best results for this are obtained with an acoustic emission device.[22]

- A very different method utilizes a peripheral milling dresser (PCD) for truing the resin-bonded CBN wheels. Peripheral milling dresser blanks are inserted in the rotating dresser with geometrically defined cutting edges and with average structural grain size of 25 µm.[23]

- A metal-bond rotary diamond impregnated truer, having a disk, cup wheel, or rod form. The best quality devices are driven separately from the grinding spindle by a hydraulic or electric motor.

- A SC crushing roll may be used for profiling the superabrasive grinding wheels.[18] The wheel specification needs to assure high friability properties for the composite material of the active layer.

The last three truing technologies are applied on modern grinders equipped with CNC or for automatic operation.[24] These methods permit high-profile accuracy to be achieved with superabrasive wheels.

As mentioned earlier, initial truing produces a dull and closed aspect of the active surfaces of a superabrasive wheel, irrespective of the type of truing technology employed. It is emphasized that truing is only a first stage of conditioning for superabrasive wheels. Further conditioning is required to achieve efficient cutting-surface properties on a superabrasive wheel.

12.6.3 Dressing Superabrasive Wheels

Dressing is the operation of restoring the active surface microtopography of a superabrasive tool after truing. Various dressing techniques and tools are described below:

- One technique is to bring an abrasive stick into contact with the wheel. This is often termed as *sticking*. Sticking is usually performed using CSi or Al_2O_3 dressing blocks.[25] As a rule, the stick contains a fine conventional abrasive powder vitrified with a ceramic bond (K or L hardness group). The selection of a dressing stick specification depends on the nature of the wheels on which the dressing operation is to be performed. For example, sticks that contain silicon carbide are used to dress diamond grinding wheels, while aluminum oxide is used to dress CBN wheels. The grit size of the conditioning stick is governed by the mesh size of the superabrasive grit of the wheel. Dressing can be done either manually or by machine using a clamping device. Performed manually, stick dressing introduces profile errors into the wheel. This unsophisticated method does not allow high accuracy. As previously mentioned, sticking also has the disadvantage that it wears the cutting edges of the grains and is not very effective in opening up the surface, particularly for fine-grain wheels.

- Good geometrical shape accuracy of superabrasive grinding wheels can be obtained by applying the method of roller dressing.[7]

- A combined conditioning method for CBN grinding wheels was developed in Japan.[26] This technique employs a special technology of conditioning with a prismatic diamond rotary dresser.

- Conditioning by grinding mild steel dressing blocks or rolls is another possibility. At the beginning, infeeds of 0.005 mm are used, increasing to 0.035 mm at the completion of the conditioning process. Free-machining mild steel blocks allows the surface of the wheel to be opened up. The main advantage of this operation is minimizing the abrasive layer to be removed. The conditioning process takes advantage of tribochemical oxidation reactions described in Ch. 15, "Tribochemistry of Abrasive Machining." The steel swarf resulting during the conditioning operation is heavily oxidized. The abrasive steel swarf (a mixture of $Fe_2O_3 + FeO$ powder) enhances the dressing process by wearing away the bond material surrounding the grains.[21]

- Dressing of CBN wheels with loose grits. Two main applications of this dressing technology are known:

  - Roll dressing with loose abrasives. An auxiliary steel roller and a secondary coolant system with a flow of slurry are used. The slurry is supplied directly through the slot between elements of the working couple. The elements of the pair have the same peripheral speed.

  - Jet sharpening of the trued wheel surface. The attack angle of the jet erodes the bond from the active layer surface.[27]

12.7 NONCONVENTIONAL TECHNOLOGIES FOR WHEEL CONDITIONING

Improvements in CD processes have been further obtained or are being sought by combining conventional technologies with the less conventional electrical, physical, and chemical technologies. Examples of technology used for dressing the active layer include:

- Electro-contact discharge dressing, which uses the principles of ECDM.[28]

- Electrolytic in-process dressing (ELID).[29] This process is presented in detail in Ch. 9, "ELID Grinding and Polishing."

- LASER dressing is still in the research stage.[30]

These unconventional conditioning processes may take place on the working surface of the abrasive tool, simultaneously with the grinding process, thus combining in-process truing and dressing.

Electrolytic in-process dressing conditioning, mentioned previously, has been particularly successful in increasing productivity and quality when using a metal-bonded diamond wheel.[31] The process can not only provide control of the dressing process but makes it possible to control a layer of lubricating oxide on the wheel surface. The process is extremely efficient for producing very low roughness and optical finishes. It also permits the continuous conditioning of the wheel simultaneously with performing grinding. Electrolytic in-process dressing may be performed either continuously or with intermittent action.

Performing electro-physical and chemical operations on a high-precision grinding machine requires appropriate precautions to be taken for the health and safety of the operators due to electrical, optical, or chemical hazards, and of the machine due to corrosion attack.

12.8 REMOVAL MECHANISMS IN CONVENTIONAL CONDITIONING

12.8.1 Introduction

Conditioning aims to transform the cutting surface from an initial condition to a final condition where:

- The requirement to start conditioning is signaled by a worn and loaded active working surface no longer capable of removing chips efficiently. Simultaneously, there is nonconformity to the quality assurance specifications for the grinding process. See Fig. 12.4 for an example of a worn and loaded wheel.

- Satisfactory conditioning has been completed when a good trued grinding wheel with an efficient surface has been achieved.

Difficulties tend to occur during wheel conditioning, if there is insufficient knowledge of the mechanics of the microcutting and fracture processes. These processes take place at the surface of the abrasive, but the effects propagate within the structure of the surface layer in an unpredictable manner.[32] Some general conclusions are summarized below.

Figure 12.5 shows SEM pictures of the microtopography of a conventional vitrified wheel compared with a SD-wheel. Both wheels have been conditioned correctly following use in grinding processes. Comparing these pictures, it may be appreciated that:

- The distribution of the abrasive grains is uniform and the surface concentration of the grains is that determined for the wheel specification, with respect to grain size and concentration.

- The grains situated in the active surface layer have the exterior edges well sharpened and uncovered, and the rest of them are well fastened into the bond.

- The open microtopography of the active surface permits easy removal of grinding swarf.

- A clean working surface may be achieved only by removing the swarf, the tool wear debris, and the degraded process fluid.

Figure 12.4. Scanning electron microscopy picture of the worn out working surface of a vitrified Al_2O_3-wheel used in the HSS-tools and cutter grinding process.

(a)   (b)

Figure 12.5. Scanning electron microscopy pictures of the dressed working surface: *(a)* vitrified grinding wheel with conventional abrasives (Al_2O_3), and *(b)* resin-bonded grinding wheel with coated superabrasives (SD).

12.8.2 Mechanism of Dressing Conventional Vitrified Wheels

Dressing of vitrified wheels is taken to include all three stages of wheel conditioning. The complex dressing process relies on the dynamic behavior of the friction couple. The dresser, which contains ultra-hard materials, and the wheel, which consists of a hard abrasive and the bond, forms the tool/wheel couple. The principal mechanisms, involved in the dressing process, include microcutting and fracture of the grains and of the bond bridges (Fig. 12.6).

If the conditioning processes are carried out well they ensure the optimum form and topography for the wheel. Conditioning processes can be better understood taking into consideration the functional aspects of the dresser/wheel friction couple:

- The contact pressures at the couple interface during a dressing process exceed the hertzian level due to macro- and microgeometry of the active part of the dresser, as well as due to the very high hardness of both materials which are in contact. In spite of the high contact pressures, the strains generated during conditioning are not significantly accumulated in the ceramic material structure and are insignificant in affecting the shape of the profile generated. However, the forces will lead to local microcracking mechanisms, accompanied by breaking and tearing in the surface structure of the abrasive composite.

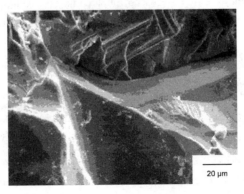

20 μm

Figure 12.6. Scanning electron microscopy picture of the microcutting and breaking processes of the abrasive grains and of the bond bridges by traverse conditioning of the vitrified Al_2O_3 wheel.

- In the operation of dressing by feeding a single-stone dresser across the wheel, the size of particles removed is usually small compared with the original abrasive grain size. To protect the integrity of the wheel with its breakable structure, the optimum value of dressing depth should be smaller than the grain size of the wheel.

- During dressing, abrasive grains fracture due to their rigidity. The breaking of the single or polycrystalline structure of the abrasive grain leads to further breaking of the intergranular bond bridges.[16] The ratio of grain fractures breaking to intergranular bond bridge fractures is directly proportional to the abrasive grain size and the hardness grade of the wheel.[33] The weight of fractured intergranular bridge material is increased for smaller abrasive grain size and with low hardness grade.

- Viewed under a microscope, dressing debris produced by traverse dressing of vitrified wheels is remarkably similar to debris produced by other conventional dressing tools. It can be inferred that the removal mechanisms are the same, irrespective of the type of dresser or of the technology employed.

12.8.3 Mechanism of Conditioning Superabrasive Wheels

The active structure of a superabrasive wheel differs from a conventional wheel, due to the nature and properties of the materials used for the superabrasives and the bonds. Superabrasive materials used for wheel fabrication are, in general, relatively inert in relation to the bond material. Methods used to achieve a better adherence between the superabrasive grains and the bond include:

- Metal coating for realization of a larger external contact.
- Physical-chemical treatment of the grains.
- Selection of a form and structure for the superabrasive grains to increase the contact surface area with the bond.
- Use of ingredients to increase bond adherence.

The high cost of superabrasive tools obliges the user to require both high productivity in the grinding process and long tool-life for the wheels. The conditioning process strongly affects the tool-life of the superabrasive wheels. Therefore, the following rules for operation are recommended:

- After truing, the volume of the bond near the working surface is usually too thick to allow high material removal rates in grinding. The pores are too small for the technological needs of the process. The wheel must be further conditioned to wear away the bond and to expose the abrasive grains more deeply.

- There is an optimum grain protrusion height, which assures a minimal grinding force, a best quality for the finished workpiece surface, and a reduced level of the wheel wear.[34] Achieving this optimum is the main goal of conditioning:

  - Too low of a protrusion height increases grinding forces and loading tendencies.

  - Too large of a protrusion height increases fracture wear of the bond.

- It is necessary to adopt very fine dressing cuts for SD or CBN wheels, compared with dressing cuts for conventional wheels. The aim is to limit the effects of conditioning to a very thin layer of the working surface of the wheel. Unless this procedure is followed, the structure of the superabrasive layer may be profoundly altered and newly generated clefts will develop extending deeper into the layer. This procedure leads to a noncontrolled evolution of the sharpening and wear processes, resulting in damage or even destruction of a superabrasive wheel.

- The wheel conditioning process may sometimes be carried out by temporarily modifying the grinding process parameters, rather than by a dressing operation. Gentle grinding conditions, or a free grinding material, may be employed initially to allow opening-up of the final grinding microtopography of a wheel.[35] This *post-dressing conditioning* is necessary when using fine-grain vitrified CBN-wheels for the first time or after a cleaning-up operation has been employed. A similar operation may be undertaken for resin-bonded wheels, where it is necessary to burn out the resin bond at the surface before the wheel can cut efficiently.

- A dressing effect can also be achieved by grinding initially at a lower removal rate. As the grinding forces fall during initial grinding, the removal rate can be increased. Alternatively, but less effectively, an abrasive dressing stick may be applied to dress the wheel surface. However, the use of a dressing stick rapidly wears the sharp cutting points of the superabrasive grains and reduces wheel redress life. After the wheel has been opened-up at reduced removal rate, it is important not to close up the wheel surface again by dressing with larger dressing depth.

- For fine-grain superabrasive wheels, a problem of *touch dressing* is commonly employed where only 1 to 5 μm of superabrasive layer is removed in each dressing operation.[35] This keeps the wheel surface open, but prevents a heavy cleaning-up operation from being performed since this would close-up the wheel surface again. Therefore, it is imperative that loading is avoided. Loading with difficult-to-grind materials can sometimes be avoided by using higher jet velocities for the additional flush process fluid and higher wheel speeds to reduce the volume of material removed per grain.[34] It is also important to use the most appropriate process fluid and coolant application to impede the heating of the dressing tool.[36] For conditioning, a minimum fluid specific flow-rate of 1 l/min · mm is required. In order to be able to undertake touch dressing, it is usually necessary to be able to detect contact between the dressing tool and the wheel without exceeding the required dressing depth. The touch contact between the dresser and the abrasive wheel is, therefore, detected using an acoustic emission sensor.

- In the special case of resin-bonded wheels, the conditioning process gives rise to a simultaneous self-sharpening action, where the coated diamond grains used have a friable character.

- For metal-bonded superabrasive wheels long duration and elevated working values characterize the dressing process. A nonconventional conditioning method may be undertaken and the electrolytic in-process dressing (ELID) operation may be employed. See Ch. 9, "ELID Grinding and Polishing."

- It is important that conditioning and grinding processes are performed without giving rise to unnecessary thermal and mechanical shocks. Mechanical shocks create a risk of fracturing the superabrasive grains from the active layer. Sudden increases in temperature or very high temperatures generate thermal shocks and tribochemical wear phenomena in the wheel and dresser working area.[37] These situations should be carefully avoided.

- Conditioning of superabrasive wheels depends on the type of wear mechanisms involved. Abrasive processes act both on the diamond or CBN grains and on the bond. Other wear processes on the active layer, such as tribochemical and surface fatigue wear, are less dominant than the physical wear process. However, due to the low bond hardness, in comparison with the ultra-hard constitutive, the superabrasive grains become exposed by abrasion. There is the risk that superabrasive grains may be pulled out and wasted.

12.9 MICROTOPOGRAPHY OF CONDITIONED WHEEL

12.9.1 Introduction

The purpose of conditioning is to generate an appropriate microtopography on the active surface for a particular grinding operation. The importance of this aspect is well known. However, a method is not yet available for continuous digital representation of the active surface topography of a wheel. Such a technique would be useful for active control and optimizing the grinding and conditioning processes. At present, abrasive manufacturing processes are directed in an empirical way, i. e., maintaining careful control of:

- Specification of the dresser for the particular wheel and workpiece requirements.

- Optimization of grinding and conditioning process conditions.

- Specification of the process-fluid, nozzle, velocity and flowrate.

The measurement of the active layer microtopography is still at the research stage in universities and a few companies. The examination of microtopography of the active layer in industry is uncommon, unless the manufacturing process goes out of control. Detailed information on measurement of wheel topography is given in Ch. 10, "Grinding Wheel and Abrasive Topography."

12.9.2 Experimental Methods for Representation of Wheel Microtopography

Direct Methods. Direct methods for wheel microtopography are shown below:

- Physical methods of investigation include:

  - Profilometry of the active layers.

  - Optical and/or electronic methods for investigating the abrasive surface.

- Special experimental techniques include:

  - Investigation of the active layer of the wheel after the grinding has stopped, and after removal from the machine; cleaning, and drying.

  - The taking of small-sized test pieces, which contain full information on the active layer microtopography; these test-pieces can be obtained by replicating, imprinting, or scratch methods.[38] The replicas are made without removing the wheel from the machine.

Generally speaking, an analysis begins with the characterization of the type of the abrasive grit, bond, and structure of the composite active layer. Only after that can the investigation of the active layer be properly affected.

Indirect Methods. Measuring and recording the power, forces, and temperatures[39] during the grinding process are indirect methods to investigating abrasive machining processes. For this purpose, various transducers to continuous measuring and recording forces, temperature, and energy during the machining process are set up on the grinding machine.[40] Exceeding a specified limit value is an indication of degradation of the microtopography and a signal that the active layer requires conditioning. It is considered that the conditioning process has been satisfactorily achieved when measured values return to the initial values. The need to determine when conditioning is required can be avoided for repeated batch production by *in-process dressing* or by *continuous diamond roll dressing.* This is made possible by providing the grinding machine with the necessary instrumentation and controls to maintain constant process conditions.

A statistical method of determining and comparing the grinding wheel microtopographies, before and after effectuation of the dressing process, is presented in the Bohlheim's report.[41]

12.10 WEAR OF THE DRESSING TOOLS

12.10.1 Introduction

Despite the very hard nature of ND grain, wear of the dressing tip starts from the first moment of use for conditioning. Important factors influencing tool wear are the grinding wheel specification, the nature of the dresser-active part, and duration of contact between the dressing tool and the wheel during dressing. One of the main tasks of the grinding machine operator is to evaluate wear of the dresser, and ensure that it does not exceed the limit value for the wear spot diameter on the ND single stone. Exceeding this limit results in a poor dressing action and various problems ensue in subsequent grinding. Problems may include high forces, vibrations, high roughness value, spiral marks on the workpiece, and catastrophic fracture of the dressing tool.

12.10.2 Influence of the Wheel Specification

Wear of the dressing tool is strongly influenced by the hardness and toughness of the abrasive or superabrasive grit in the wheel. Thus, the conditioning of wheels, containing tough abrasives such as CBN or CSi type, cause more intensive dresser wear than Al_2O_3 abrasives. A large abrasive grain size also has a negative influence on dresser tool life.

Increasing the size of an abrasive wheel increases dresser wear due to the increased duration of contact during conditioning. Recommendations have been developed to take into account the various influences on dresser wear. For example, increasing the diamond stone weight (size) can offset a high wear-rate of dresser. Equation (12.6) gives the required weight of inserted diamond grains (with a maximum error of +/- 20%) as a function of the diameter and width of the abrasive wheel.

Eq. (12.6) $$W_d = \frac{d_s + (2\, b_s)}{250}$$

where W_d is the weight of diamond grains (ct), d_s is the wheel diameter (mm), and b_s is the wheel width (mm).

12.10.3 Influence of the Ultra-hard Components of the Dresser

New synthetically produced ultra-hard materials have been developed for the same purpose as traditional ND gemstones.[8] The effectiveness of conditioning using natural and synthetic ultra-hard materials may be compared as follows.

- *Wear of ND gemstones.* Despite the compact and apparently strictly organized structure of natural diamond, there are significant differences from the theoretical crystallographic model. Apart from the size, shape, and color of the grain, there are important structural differences between various grains. These deviations from the typical crystallographic structure make the existence of flaws in industrial ND possible, including:

- Inclusion of impurities into the crystallographic net.
- External and internal cracks.
- Crystallographic faults, like naats and twins.
- Structural deformations, etc.

In addition to the structural deformations mentioned above, giving rise to foreseeable consequences, surprising differences in performance can occur due to anisotropy of physical and tribological properties of the ND stones. See Ch. 16, "Processed Materials." Ideally, the axis, or the plane of symmetry of the tool support should tally with the direction of maximum wear-resistance of the diamond. It is necessary to pay special attention to the orientation of the diamond stones by setting. As the exterior appearance of the ND grain seldom provides stable criteria for orientation, x-ray diffractometry may be used to inform the setting process.

Wear of an ND grain is the result of the total action of all the basic mechanisms of wear. Thus, although dressing is carried out under severe abrasion, other mechanisms of wear are also involved. The growth of the wear spot diameter (d_w) on the tip of an ND grain against dressing time (t_d) is similar to the standard wear curve for the diamond cutting tools.[42] We can observe that the standard wear curve[43] illustrates the progressive character of wear of a dressing tool used for conditioning (Fig. 12.7).

Under dressing conditions, the wear process of an ND single grain evolves successively through three characteristic stages:

- **Stage I.** *Initial wear.* This stage is characterized by rapid wear of the tool, until a small wear spot is formed. Under appropriate dressing conditions, the initial wear stage should be of relatively short duration.

- **Stage II.** *Steady wear.* The steady wear stage can be recognized by a linear rate of wear. This stage corresponds to the working duration of dresser application, known as the dresser *tool life* (T_d). The low and approximately steady wear rate being similar to the classic *laws of wear*.[44] The end of the dresser tool-life is marked by the size of the wear spot diameter corresponding approximately to the value given by Eq. (12.7).

- **Stage III.** *End of tool-life.* This final stage is characterized by a sudden increase of the wear rate. Wear has an exponential tendency as the tool approaches catastrophic failure.

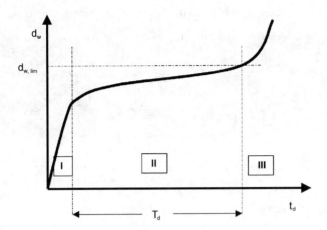

Figure 12.7. Increases of ND dresser wear (d_w) with duration (t_d) by conditioning an Al_2O_3 wheel: *(I)* initial wear, *(II)* wear in normal use, and *(III)* catastrophic wear at the end of the dresser tool-life.

An empirical guide for the maximum wear spot on the dresser tip is:

Eq. (12.7) $d_{w,lim} \leq 0.6$ mm

Throughout conditioning, the size and nature of the contact between the dresser and the wheel are continuously changing. Figure 12.8a shows a wear spot generated at the end of the initial wear stage. It can be seen that the surface of the spot has a well-marked relief. This relief has the appearance of wear caused by joint action of abrasion and fatigue. Figure 12.8b shows the wear spot at the mid-point of the steady wear stage. It can be seen that the active tip of the ND stone has developed a larger wear-flat due to abrasive wear. Figure 12.8c shows the wear spot at the end of the application stage. The surface of the spot is much rougher compared to the previous images. It demonstrates the intensification of the wear process due to imperfections in the ND structure and due to fatigue. Cracks appear on the surface of the wear spot and swiftly spread into the grain volume, due both to the increase in normal force and of vibrations. Exceeding the limit value of the wear spot diameter leads to complete failure of the ND grain.

 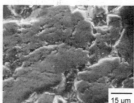 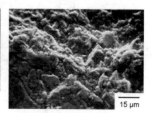

Figure 12.8. Scanning electron microscopy images of the wear spot on the tip of the ND single stone during traverse conditioning of a conventional Al_2O_3 wheel: *(a)* end of the initial wear stage ($d_w = 0.42$ mm), *(b)* middle of the steady wear stage ($d_w = 0.52$ mm), and *(c)* end of the normal dressing tool-life ($d_w = 0.61$ mm).

- **Wear of synthetic diamonds.** The following summarizes the main types of synthetic diamond and the characteristic behavior when conditioning grinding wheels:

  - *SD grits.* The wear processes of the active layer by these composite dressers are similar to the wear phenomena presented formerly for grinding wheels. See Chs. 5 and 15.

  - *MCD logs.* The use of MCD dressing logs for conditioning has advantages; the ultra-hard mono-crystalline material has a controlled and perfectly isotropic structure, with chemical purity and a steady dressing area.

The optimum area of the MCD-log depends on the dimensions of the grinding wheel to be conditioned. The optimum area also depends on the grain size of the abrasive and the characteristics of the bond. For example, a wheel of large grain size requires a larger contact area of the dresser. An optimum conditioning process yields a long and linear steady wear graph (Fig. 12.9). The graph of normal dressing tool life corresponds to Stage II of the standard wear curve. The initial Stage I wear may be neglected.

The wrong type of sintered synthetic diamond (SSD) log and/or incorrect dressing parameters shortens the dresser tool life and leads to dresser failure (see Stage III of the wear curve on Fig. 12.9). Irrespective of the dressing conditions, when the high of the monocrystal (h_w) reduces to less than 1 mm, dresser failure can take place at any time, by break-up or pull-out of the MCD log.

- *CVD diamond logs.* Wear behavior of CVD logs is similar in nature to MCD; both being synthetic products. However, the tool-life of a CVD dresser is shorter. Chemical vapor deposition (CVD) tool-life depends mainly on the adherence of the SD polycrystalline layer to the support.[45] Detachment of the ultra-hard layer is influenced by the relative thermal expansion of the support and of the diamond grains. The rate of expansion depends on the specific heat capacity and the coefficient of thermal expansion. The CVD tool wear is mainly caused by thermal shocks due to the temperature variations. Thermal shock loads on the dresser directly influence all four of the wear mechanisms. The best results are obtained using CVD logs having SC supports and thick diamond layers. An effective intermediate layer on the support improves the adherence of a CVD layer.[46]

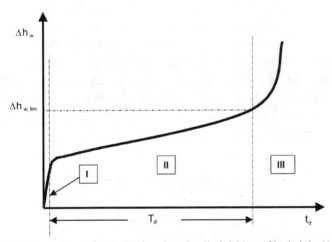

Figure 12.9. Linear wear of an MCD log, i.e., the diminishing of its height (Δh_w) against duration (t_d) by traverse-conditioning of an Al_2O_3-wheel: *(I)* initial wear, *(II)* steady wear, and *(III)* failure.

- *PCD pieces.* This has superior resistance to mechanical shocks than other products mentioned above. On the other hand, these materials are very sensitive to the thermal shocks experienced in the conditioning process. Faults first appear when the contact temperature exceeds 500°C.

Above this temperature, the cobalt binder leaches from the composite structure. This damages the diamond phase leading to the appearance of a *bloom defect* on the surface (Fig. 12.10).

These apparently shallow defects accentuate the friability of the active part and lead to crazing of the entire working surface. The development of crazing leads to failure of the whole PCD dresser.

Preliminary results from laboratory tests performed on these new ultra-hard materials are remarkable. The behavior during the conditioning process of synthetic dressing products is still the subject of research and development.[45][46]

12.11 CONCLUSIONS

It is necessary to replace an ND single stone dresser when the maximum allowable size of wear-flat is reached. Due to the high cost of ND dressers, it is recommended to recondition a blunt diamond through re-setting and/or re-pointing of the ND grain in the sintered metal matrix.

A traverse ND tool is not recommended for truing superabrasive wheels either from technical, or especially from economic considerations.

A specific and reproducible topography for the working surface of the wheel is best obtained using roller dressing technology and optimum dressing parameters established through testing.

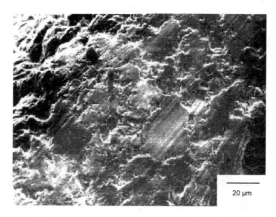

20 µm

Figure 12.10. Emergence of blooms by loss of cobalt binder on the working surface of a PCD dresser.

Single-layer superabrasive wheels cannot normally be trued or dressed by the user without serious damage. Conditioning is limited to ensuring that the surface is kept clean of grinding swarf. The open structure of a single-layer wheel usually helps to keep the wheel clean. The profile accuracy and topographical properties for single-layer wheels depends on the wheel manufacturer.

Various truing and dressing tools, devices, and technologies are used in the conditioning of superabrasive wheels. Common abrasives (SiC, Al_2O_3, etc.), and superabrasives (SD, PCD, CVD, MCD) are used for dressers, either metal-bonded or on a ceramic support. Free-machining steel may also be used to open up a wheel in a conditioning-by-grinding operation.

Abrasive processes used to condition wheels can be stimulated by nonconventional methods employing electrical processes or tribo-electro-chemical reactions (i. e., the nonconventional conditioning methods).

REFERENCES

1. Salije, E., Begriffe der Schleif- und Konditioniertechnik, Vulkan Vlg., Essen, Germany (In German) (1991)

2. Coes, L., Jr., *Abrasives,* Springer-Verlag, NY (1971)

3. Sen, P. K., Wheel Dressing and Natural Diamond, *IDR*, 51(1):32–38 (1991)

4. Cai, R., Rowe, W. B., and Morgan, M. N., Grinding Performance of High-Porosity Fine-Grain Vitrified CBN Wheels for Inconel 718; *Advances in Manufacturing Technology*, (K. Chen and D. Webb, eds.), Professional Eng. Publ., Ltd., London (2002)

5. Moser, M., Microstructures of Ceramics, *Akademiai Kiado*, Budapest (1980)

6. Doyle, L. E., Morris, J. L., Leach, J. L., and Schrader, G. F., *Manufacturing Processes and Materials for Engineering*, Prentice-Hall, Inc., Englewood Cliffs, NJ (1962)

7. Salmon, S. C., *Modern Grinding Process Technology*, McGraw-Hill, Inc., NY (1992)

8. Phaal, C., Neue Diamanttechnologien – Moeglichkeiten und Grenzen, *IDR,* 27(1):2–4 (In German) (1993)

9. Sen, P. K., Synthetic Diamond Dresser Logs:Serving the Future Needs of Industry, *IDR,* 62(3):194–202 (2002)

10. Lierse, T., and Kaiser, M., Dressing of Grinding Wheels for Gearwheels, *IDR,* 62 (4):273–281 (2002)

11. Cassel, M. F., PCD Boosts Productivity of Dressing Rolls, *IDR,* 52(1):4–7 (1992)

12. Hallewell, E. N., and Classey, D. S., Pantograph Diamond Dressers, *Proc. Ind. Diam. Conf.,* Chicago, IL (1969)

13. Kushigian, A., The Advantages of CNC Dressing, *Proc. Int. Grind. Conf.,* Dearborn, MI (1990)

14. Franke, S., CNC-gesteuertes Abrichten von profilierten Schleifscheiben, *Proc. 5th Int. Colloq.,* Braunschweig-Germany (In German) (1987)

15. Andrew, C., Howes, T. D., and Pearce, T. R. A., *Creep Feed Grinding,* Holt, Rinehart, & Winston, London (1985)

16. Meyer, H. R., and Klocke, F., New Developments in the Dressing of CBN and Conventional Grinding Wheels with Rotary Dressers, *Proc. Int. Grind. Conf.,* Dearborn, MI (1990)

17. Helletsberger, H., and Thumbichler, M., Diamant Abrichtrollen – eine wesentliche System-komponente beim Profilschleifen, Schleifen und Trennen, 119:2–8 (In German) (1993)

18. Dennis, P., and Kaiser, M., Profilschleifen mit abrichtbaren Diamant und CBN-Schleifscheiben, *IDR,* 33(1):21–26 (In German) (1999)

19. Ott, H. W., Basics in Grinding Technique, Ott Vlg., Pfaeffikon-ZH (Switzerland) (1993)

20. Hughes, F. H., Diamond Grinding of Metals, Diamond Res. Lab., Johannesburg, So. African Rep. (1990)

21. Fielding, E. R., Optimum Conditioned State for CBN Wheels, *Advances in Ultrahard Materials Application Technology,* (P. Daniel, ed.), Vol. 3, de Beers Ind. Diamond Div., Charters, England (1984)

22. Meyer, H. R., and Koch, N., Highly Productive and Cost-Effective Grinding Processes Require Mastery of Dressing Technology for Al_2O_3, Diamond and CBN Grinding Wheels, *Proc. 5th Int. Grind. Conf.,* Cincinnati, OH (1993)

23. Warnecke, G., Gruen, F. J., and Geis-Drescher, W., PCD in Rotary Dressers—An Appraisal, *IDR,* 48(3):106–110 (1988)

24. Liverton, J., Major Improvements to Precision Grinding Brought About by the Application of CNC, *Proc. 4th Int. Grind. Conf.,* Dearborn, MI (1990)

25. Ueno, N., et al., In-Process Dressing of Resin-Bonded Diamond Wheel, *Bull. Jpn. Soc. Prec. Eng.,* 24(2):112–117 (1990)

26. Yokogawa, M., Optimum Dressing Conditions for a Prismatic Diamond Rotary Dresser for Vitrified CBN-Wheels, *Proc. 5th Int. Grind. Conf.,* Cincinnati, OH (1993)

27. Salije, E., and Mackensen, H. G., Strahlschaerfen einer CBN-Schleifscheibe, Industrie Anzeiger, 107(25):11–18 (In German) (1985)

28. Friemuth, T., and Lierse, T., Electro-contact Discharge Dressing (ECDD) of Diamond Wheels, *IDR*, 58(2):57–61 (1998)

29. Kato, T., Ohmori, H., Zhang, C., Yamazaki, T., Akune, Y., and Hokkirigava, K., Improvement of Friction and Wear Properties of CVD-SiC Films with New Surface Finishing Method ELID-Grinding, *Key Eng. Mater.*, 196:91–102 (2001)

30. Hoffmeister, H. W., and Timmer, J. H., Laserkonditionieren von CBN-Schleifscheiben, *IDR*, 31(1)P:40–46 (In German) (1997)

31. Shore, P., ELID for Efficient Grinding of Super Smooth Surfaces, *IDR*, 53(6):318–322 (1993)

32. Malkin, S., and Anderson, R. B., Active Grains and Dressing Particles in Grinding, *Proc. Int. Grinding Conf.*, Pittsburg, PA (1972)

33. Malkin, S., *Grinding Technology:Theory and Application of Machining with Abrasives*, Ellis Horwood Series, John Wiley & Sons, NY (1989)

34. Syoji, K., et al, Studies on Truing and Dressing of Diamond Wheels (1st report) – The Measurement of Protrusion Height of Abrasive Grains by Using a Stereo Pair and the Influence of Protrusion Height on Grinding Performance, *Bull. Japan Soc. of Prec. Eng.*, 24(2):124–129 (1990)

35. Chen, X., Rowe, W. B., and Cai, R., Precision Grinding Using CBN Wheels, *Int. J. Machine Tools & Mgf.*, 42:585–593 (2002)

36. Kucher, K., Innengekuehlte Abrichtfliesen aus einen Guss, *IDR*, 36(4):304–305 (In German) (2002)

37. Uhlman, E., and Bruecher, M., Untersuchungen zum Oxidationsverschleiss von diamantbasierten Schneidstoffen, *IDR*, 35(2):127–134 (In German) (2001)

38. Nakayama, K., Takagi, J., Irie, E., and Okuno, K., Sharpness Evaluation of Grinding Wheel Face by the Grinding of Steel Ball, *Annals of the CIRP*, 28(1):227–231 (1980)

39. Nakayama, K., Takagi, J., Irie, E., Fukuda. T., and Ochiai, K., Sharpness Evaluation of Grinding Wheel Face by Coefficient of Friction, *Proc. Int. Conf. Prod. Eng.*, Tokyo (1980)

40. Nordmann, K., Sensoren zur prozessbegleitenden Ueberwachung beim Schleifen, Abrichten und Hartdrehen, *IDR*, 28(2):102–107 (In German) (1994)

41. Bohlheim, W., A Method for Determining Grinding Wheel Topography, *IDR*, 57 (2):58–62 (1997)

42. Moore, D. F., Principles and Applications of Tribology, Pergamon Press, Oxford (1975)

43. Czichos, H., and Habig, K. H., *Tribologie Handbuch:Reibung und Verschleiss*, Vieweg Vlg., Braunschweig (In German) (1992)

44. Crompton, D., Hirst, W., and Howse, M. G. W., The Wear of Diamond, *Proc. Royal Soc.,* A(333):435–454, London (1973)

44. Uhlmann, E., and Bruecher, M., Untersuchungen zum abrasiven Verschleiss von CVD-Diamant, *IDR,* 36(3):232–241 (In German) (2002)

45. Uhlmann, E., and Bruecher, M., CVD Diamond Tool Wear Caused by Thermal Shock Loads, *IDR,* 629(2):97–106 (2002)

46. Cook, M. W., Verschleisseigenschaften und Triboanwendungen von PKD, *IDR,* 329(1):28–34 (In German) (1998)

13

Loose Abrasive
Processes

13.1 INTRODUCTION

A range of loose abrasive processes are used for machining. Loose abrasive processes are mainly employed in high-precision finishing processes used to generate surfaces of desired characteristics. Abrasive finishing uses a larger number of multipoint or random cutting edges for effective material removal at smaller chip sizes than in the finishing methods that use cutting tools with a defined edges.[18] Abrasive finishing processes are accepted in a wide range of material applications and industries; typical examples are finishings of various components used in aerospace, automotive, mechanical seals, fluid handling, and many other precision engineering industries.

Advanced abrasive powders available in nanometer sizes are used in a variety of processes including loose abrasives (lapping, polishing), bonded abrasives (grinding wheels), and coated abrasives. Because the signs of the abrasives are of the order of submicrons, Taniguchi (1974) applied, for the first time, the term *nanotechnology* in relation to fine abrasive processes (or ultraprecision processing).[5] Consequently, he

presented a general classification of machining processes in conventional machining (CM), precision machining (PM), and ultraprecision
machining (UPM). In his opinion, by the year 2000 A.D., machining
accuracy in conventional machining would reach 1 mm, while PM and UPM
would reach 0.01 mm and 0.001 mm, respectively.[5]

The finishing abrasive processes using loose abrasives can be
approximately classified into the following categories: lapping, polishing, and
field-assisted processes. These types of operations are capable of producing
fine finishes on both ductile and brittle materials.

Lapping and polishing are precision finishing processes which
involve different mechanical arrangements. Lapping is a slow material-
removal operation. Though lapping tends to decrease the original surface
roughness, its main purpose is to remove material and modify the shape.
Lapping is used primarily to improve form accuracy rather than to reduce
surface roughness. Form accuracy includes, for example, flatness of flat
objects and sphericity of balls. In contrast, the term polishing implies better
finish with little attention to form accuracy (Table 13.1.)

Lapping and polishing are used for many workmaterials including
glass, ceramic, plastic, metals and their alloys, sintered materials, stellite,
ferrite, copper, cast iron, steel, etc. Typical examples of the diverse range
of processed components are pump parts (seal faces, body castings, rotating
valves, impellers), transmission equipment (spacers, gears, shims, clutch
plates), cutting tools (tool tips, slitter blades), hydraulic and pneumatics
(valves plates, seals, cylinder bodies, castings, slipper plates), aerospace
parts (lock plates, gyro components, seals), inspection equipment (test
blocks, micrometer anvils, optical flats, surface plates), and stamping and
forging equipment (spacers, type hammers, bosses, and a variety of other
tools with complex shapes).

The relative speeds in lapping and polishing are much lower than in
grinding. Consequently, the concentration of energy in the contact area is
much lower. The benefit is that average temperatures tend to be lower than
in grinding and may be negligible; the disadvantage is that specific energy is
higher, although, since the volumes of material to be removed are small, this
is not usually important.

Table 13.1. Surface Roughness Results for Loose Abrasive Processes[5]

PROCESS	WORKPIECE	R_a (nm)	R_q (nm)	R_{max} (nm)	REFS.
FIELD ASSISTED					
Magnetic fluid grinding					
w/SiC abrasives	Si_3N_4 ball alumina plates	60		100	Umehara (1994)
w/diamond abrasives	Si_3N_4 balls	25			Childs (1994b)
Magnetic float polishing					
w/Al_2O_3 abrasives	Si3N4 balls	35.8			Komanduri (1996b)
w/Cr_2O_3 abrasives		9.1			
w/CeO_2 abrasives		4.0		40	Jiang (1997b)
w/SiC abrasives	acrylic resin			40	Tani (1984)
Magnetic abrasive finishing					
w/diamond abrasives	Si_3N_4 rollers	40			Shinmura (1990)
Magnetic abrasive finishing	stainless steel rods	7.6			Fox (1994)
	brass tubes	40			Umehara (1995)
Dicing using electrophoretic deposition + diamond blade	Lithium Niobate			20	Ikeno (1991)
Polishing using electrophoretic deposition + elastic tool	BK7	40			Suzuki (1994)

(Cont'd.)

Table 13.1. *(Cont'd.)*

PROCESS	WORKPIECE	R_a (nm)	R_q (nm)	R_{max} (nm)	REFS.
LAPPING/POLISHING					
Lapping w/sintered cast iron plate with CBN abrasives	tungsten carbide			100	Hagiuda (1981)
Lapping w/tin lap with diamond abrasives	poly Mn-Zn ferrite	0.9			Touge (1996)
Lapping w/tin lap + various fine particles	single crystal sapphire			<1	Namba (1978)
	single crystal Mn-Zn			<1	
	poly Mn-Zn			5	
Lapping wheel w/low bonding strength	Single crystal Si			20	Tani (1986)
			2–3		Kawata (1993)
	BK7, lime glass, AlN	<10			
Synthetic fabric-faced lap w/diamond paste	Electroless Ni		<1		Parks (1994)
Teflon lap					
w/Al$_2$O$_3$ abrasives	BK7		<0.1		Leistner (1993)
w/SiO$_2$	Single crystal Si		<0.1		
Lap w/Mylar film + foamed SiC substrate	Black filter glass		1		Parks (1997)
Lapping w/fluorocarbon foam polisher w/SiO$_2$ powder	Fused silica			0.3	Kasai (1990)

(Cont'd.)

Table 13.1. *(Cont'd.)*

PROCESS	WORKPIECE	R_a (nm)	R_q (nm)	R_{max} (nm)	REFS.
LAPPING/POLISHING					
Polishing w/flexible tool w/elastic layer	fused silica		0.13		Ando (1995)
	CVD-SiC		0.15		
	CaF$_2$		0.12		
Mechano-chemical polishing	(100) Si-oxide surface		0.8–1.1		Sakata (1993)
	(100) Si/oxide interface		1.1–2.8		
Chemo-mechanical polishing	polySi	1.7–1.9			Yasseen (1997)
	SiC		0.47–0.6		Yamoaka (1994)
	CdS	1.3			Vitali (1996)
Mechanical polishing w/0.33 μm diamond abrasives	Ni-Zn ferrite		6.19		Xie (1996)
	pure Cu		8.22		
Polishing w/cloth polisher + fine particles	AlSi 304 stainless steel			2	Namba (1980b)
Float polishing	Fused silica, glass ceramic		0.1–0.2		Namba (1987)
	Natural quartz		0.2		Soares (1994)
Float polishing w/fluorocarbon tool	GaAs			2	Kasai (1988)

13.2 TWO-BODY AND THREE-BODY ABRASION (MECHANISMS)

Wear is the term that represents gradual material removal from a surface due to a mechanical movement and/or chemical process.[41] One of the most important mechanisms, due to its frequency, is abrasive wear, which means detachment of material from surfaces in relative motion, caused by protrusions and/or hard particles between the opposing surfaces or fixed in one of them.

The common classification of abrasive wear into the categories of two-body abrasion and three-body abrasion is widely used by most researchers in the field. The primary meaning of the two-body/three-body concept (Fig. 13.1) is to describe whether the abrasive particles are bound (two-body) or free to roll or slide (three-body). So, the term three-body abrasion refers to wear caused by free and loose abrasive particles existing as interfacial elements between a solid body and a counterbody, while two-body abrasive wear is caused by abrasive particles or asperities which are rigidly attached (embedded) in the second body.[9] Because of that, the abrasive particles in a two-body mechanism are able to cut deeply into the workpiece material, whereas in the case of three-body abrasion, the abrasive grains may spend only part of their time cutting into the material. Therefore, two-body abrasion is considered to produce wear rates three times bigger than the second mechanism using the same loading condition.[27]

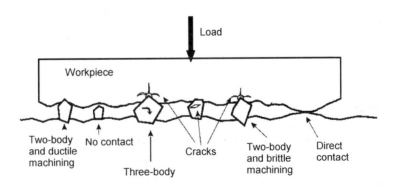

Figure 13.1. Schematic showing two-body and three-body abrasion.[46]

In 1998, Gates illustrated that this categorizing of two-body and three-body abrasion is ambiguous because there are situations when these two concepts could create misinterpretations. In his opinion, this classification can be used only to describe whether abrasive grains are rigidly held or free to roll. From a tribological point of view, one should take into account the severity of wear behavior: mild, severe, and extreme. Consideration should be given to the specific situation: gouging abrasion, high-stress (or grinding) abrasion, and low-stress (or scratching) abrasion. The difference between high-stress and low-stress abrasion is whether or not the abrasive grains are broken during abrasion. This is important since fracturing can create sharp cutting edges and give higher wear rates. Gates proposes this new classification (Table 13.2 and Table 13.3) as an improvement (without being considered precise).

Three important finishing abrasive operations are grinding, lapping, and polishing. Grinding is a two-body abrasive process because the grain is fixed in the grinding wheel. Lapping is primarily considered a three-body abrasive mechanism due to the fact that it uses free abrasive grains that can roll or slide between the workpiece surface and the lapping plate, although some grains become embedded in the lap leading to two-body abrasion. Other finishing processes that use loose abrasive grains could be a combination of two- and three-body abrasion. This is the case in polishing where the grains are temporarily fixed in a soft polishing pad (rolling as well as fixed abrasive grits participate in the material removal). Because of the complexity of the system, polishing involves two-body and three-body abrasion at the same time.

Table 13.2. Possible Situation-based Classification for Abrasive Wear[27]

ABRASIVE PARTICLES	CONTACT STRESS		
	LOW (PARTICLES DO NOT FRACTURE)	HIGH (PARTICLES FRACTURE)	EXTREME (GROSS DE-FORMATION)
Free	Low-stress free-abrasive	High-stress free-abrasive	
Constrained	Low-stress fixed-abrasive	High-stress fixed-abrasive	Extreme-stress fixed-abrasive

Table 13.3. Proposed Severity-based Classification for Abrasive Wear[27]

TYPICAL[a] SITUATIONS	ABRASIVE WEAR MODE		
	MILD	SEVERE	EXTREME
Particle size	Small	Moderate	Large
Constraint	Unconstrained	Partially constrained by counterface	Strongly constrained
Particle shape	Rounded	Sharp	Sharp
Contact stress	Low-insufficient to fracture particles	Moderate-sufficient to fracture particles	Very high-may cause macroscopic deformation or brittle fracture of material being worn
Dominant[b] mechanism	Microploughing	Microcutting	Microcutting and/or microfracture
Equivalent terms[c]	Low-stress abrasion	High-stress abrasion	Gouging abrasion
	Scratching abrasion[d]	Grinding abrasion[d]	High-stress two-body[e]
	Low-stress three-body[e]	High-stress three-body[e]	
		Low-stress two-body[e]	

[a] – Not all aspects of the "typical" situation necessarily apply simultaneously, [b] – Debris-removal mechanism is highly material-dependent, [c] – It has already been demonstrated at length that these alternative terminologies are ambiguous, therefore, no fully reliable correspondence with the proposed new terms can be expected. [d] – Term not favored even within the alternative classification scheme, [e] – Dominant sense of two-body/three-body distinction.

Even if mechanisms in two-body and three-body abrasion are the same, there are some obvious differences between the two methods. In two-body abrasion, the abrasive grains are constrained against the abraded surface and higher pressures can be exercised by them. Another difference is the effect of particle (or protuberance) size on wear rate.

In three-body abrasion, the distribution of grains in the contact area is subject to greater uncertainty. With an ample supply of abrasive, the average pressure on the grains is likely to be lower than in a two-body process. The pressures exerted by an abrasive particle also tend to depend on the grain size. The pressures are likely to be higher with large grain sizes. This affects the scratch depths on the workpiece surface. With low pressure and fine particle sizes, the scratches will be very small. For this reason, polishing, which is a predominantly two-body process, can produce very low surface roughness.

13.3 THE LAPPING PROCESS

Processes where the surface characteristics get maximum attention are called ultraprecision machining operations. They are often used to alter the surface attributes such as roughness, waviness, flatness, roundness, etc., without significant material removal from the workpiece; typical examples are lapping and polishing of optical lenses, computer chips or magnetic heads, honing of cylinder liners, microfinishing of bearing races, etc.

Lapping is a loose abrasive machining process that combines abrasive particles within an oil or aqueous medium depending on the material being finished. Fine abrasive is applied, continuously or at specific intervals, to a worksurface to form an abrasive film between the lapping plate and the parts being lapped. Each abrasive grain used for lapping has sharp irregular shapes and, when a relative motion is induced and pressure applied, the sharp edges of the grains are forced into the workpiece material. Each loose abrasive particle acts as a microscopic cutting tool that either makes an indentation or causes the material to cut away very small particles. Even though the abrasive grains are irregular in size and shape, they are used in large quantities and thus a cutting action takes place continuously over the entire contact surface (Fig. 13.2).

The depth of the marks and scratch grooves will determine the roughness of the surface, which is usually measured with a surface analyzer and described in terms of parameter such as R_a, R_3, and R_t in microns or micro-inches. Larger and harder grains produce higher surface roughness. Conversely, smaller abrasive grains tend to produce lower surface roughness. There is often a limit to the smoothness that can be obtained by lapping, even when very fine abrasive grains are used.

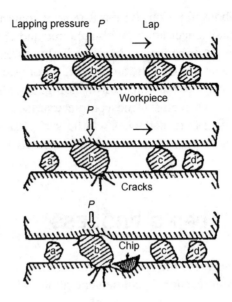

Figure 13.2. Model of the lapping process for hard, brittle materials.[4]

The main factors influencing the lapping characteristics are the type of lap, the type and size of the abrasive grains, the lapping fluid type, the lapping pressure, and the lapping speed.[4]

13.3.1 The Lap

The type of lapping plate is important since a workpiece can be badly scratched and contaminated with abrasives if the lap plate is too hard.[16] The composition of the lapping plate is of great importance because it can affect the results of the lapping process. A hard lapping plate resists being embedded with abrasive particles. Therefore, the grains roll more than slide so that most of the material removal is by stress-induced microfracture. Also, the grains are more likely to become embedded in the workpiece. A softer lapping plate allows abrasives to partially embed themselves in the lap, resulting in more sliding motion and material removal by ploughing. The result is a finer surface finish from soft plates, but less planarity. To obtain a better surface finish with respect to planarity, modern lapping often uses a hard plate and very fine grit abrasives.

To produce a plane surface it is important that the abrasive slurry is distributed uniformly between the workpiece and the lap. There is a tendency for abrasive particles to build up in some areas, increasing the local pressure and, hence, increasing the removal rate in those areas. Also, the leading edge of the lap will tend to engage the grains more quickly, causing a barreled lap shape. Therefore, consideration should be given to rotating the lap and using the three-lap technique described in workshop technology texts to maintain the planarity of the tools as well as the workpieces.

To produce a perfectly smooth surface, free from scratches, the lapping plate should be charged with a very fine abrasive. When the entire surface of the lap is charged, one should examine the lap for bright spots; if there are any visible bright spots, the charging will continue until the entire surface has a gray appearance. After a lap is completely charged, it should be used without applying more abrasives until it ceases to cut. If a lap is overcharged and an excessive amount of abrasive is used, there is a rolling action between the workpiece and the lapping plate that results in inaccuracy.

A large variety of lapping plates is accessible for almost any application: cast iron (for general engineering purposes), ceramic (for ceramic and other hard materials), glass (for electro-optic materials), aluminum/stainless steel, and many more.

Serrated or grooved lapping plates are best for flat surfaces with large areas, also for flat areas with holes in the surface. Laps with no serration or grooves are preferred for cylindrical lapping.

13.3.2 The Abrasive

There are a variety of abrasives that can be used for lapping: aluminum oxide (for general lapping with low surface roughness), silicon carbide (fast stock removal for hard or soft materials), boron carbide (for use with ceramic, carbide, and other hard materials), calcined alumina (for use with metals, optics, silicon wafers, and other semiconductor materials), and diamond slurries and pastes (available in a wide variety of micron sizes, concentrations, and emulsions).

Abrasives are either natural (diamond, corundum, emery, garnet, and quartz) or artificial crystalline forms (diamond, silicon carbide, aluminum oxide, cubic boron nitride, boron carbide, etc.). The second category can be divided into two groups:

- Fused abrasives which are the result of very high electric furnace temperatures that produce hard crystals).

- Unfused abrasives which are the result of lower temperatures and chemical additives. Unfused abrasives are not as hard as the fused abrasives.

The most widely used abrasives in industry include:

- *Diamond* – The hardest and sharpest known abrasive. It is both a natural and man-made synthetic abrasive which on the Mohs scale of hardness (Table 13.4), is numbered 10 (the hardest material). Diamond is best suited for tungsten carbide and other very hard materials. When a plate is embedded with the diamond abrasive, it cuts fast and produces fine finishes.

- *Cubic boron nitride (CBN)* – A synthetic abrasive almost as hard as diamond on the Mohs scale (9.9). It is recommended for lapping ferrous metals and especially for lapping 52100 bearing steel, cast iron, tool steel, stellite, super alloys and, in some cases, ceramic materials.

- *Silicon carbide (SiC)* – A fused, hard crystalline abrasive, 9.5 on Mohs scale. Fast cutting is achieved with good crystal breakdown to maintain abrasive sharpness when used to lap either high or low tensile strength materials. It is well-suited for rough lapping operations, forged or hardened gears, valves, tool room work and general maintenance where polish is not essential.

- *Aluminum oxide* (Al_2O_3) – A fused abrasive with a very hard crystal structure that is hard to fracture. It is utilized for lapping high tensile strength materials, rough lapping operations, hardened gears, ball bearing grooves, or lapping operations where pressure can be exerted to break down the crystals.

- *Corundum* – A natural abrasive found in the earth with a softer crystalline structure than silicon carbide or fused Al_2O_3. It is used for lapping a great variety of medium hard metals (Rockwell C 35-45).

Table 13.4. Mohs Scale of Hardness[46]

ABRASIVES	MOHS SCALE
Diamond	10.0
Cubic Boron Nitride (Borazon CBN)	9.9
Silicon Carbide	9.5
Aluminum Oxide	9.0
38 White Aluminum Oxide	9.0
Corundum	9.0
Chromium Oxide	8.5
Garnet	8 to 9
Quartz	7
Aluminas (hydrates)	5 to 7

Observations:

➢ Abrasives, though of equal or nearly equal hardness on Mohs scale, do not have equal cutting, lapping, or metal abrading power, nor do they produce the same lapped finish.

➢ Crystalline shapes, lines of cleavage, friability, chemical composition, etc., are responsible for lapping variables.

➢ Aluminas have a softer lapping action.

• *Unfused alumina* (hydrate-calcined) – Relatively soft and used for polishing. Calcined aluminas are produced by heat treatment and the degree of calcination determines the characteristics of the product. The terms soft, medium, and hard relate to abrasives resulting from mild, medium, and high degrees of calcination. Unfused alumina abrasives are recommended for the lapping and polishing of harder metals (Rockwell C 45-63). The shape, unlike the blocky crystals, is composed of flat or "plated" crystals with a thickness about one-sixth the diameter. Unfused aluminas allow more equal pressure to be distributed over a large surface area than the fused ones because of their plated shape. The disc shaped particles work with a shaving action rather than the rolling and gouging action of blocky abrasives and are less likely to produce deep scratches on the workpiece.

The size and size distribution of the abrasive are important factors in the surface obtained by lapping. The size of the abrasive is directly proportional to the material removal rate and surface roughness.

Larger grain sizes have a higher material removal rate than smaller abrasive grains but the smaller abrasives produce a lower surface roughness than larger abrasive sizes as shown in Figs. 13.3 and 13.4.

Another important factor is the concentration of abrasive (the amount of abrasive per volume of carrier) which represents the number of grains in contact with the surface of the workpiece material. When the concentration varies, the load distribution changes. This means that with an increase in the number of grains, the effective load per grain decreases because of the larger number of contact points. If the size of the abrasive grains decreases below sub-micron sizes, the number of grains increases accordingly.

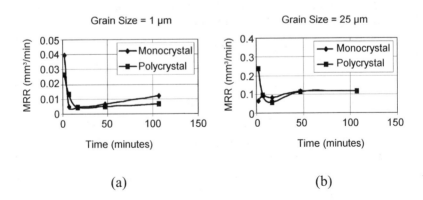

(a)   (b)

Figure 13.3. Material removal rate vs time (flow rate: 1.5 ml/min, rotation of the lapping plate: 56 rpm, slurry concentration: 1.4 g/500 ml): *(a)* diamond grain size 1 μm, and *(b)* diamond grain size 25 μm.

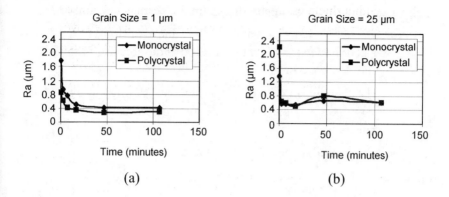

Figure 13.4. Roughness vs time (flow rate: 1.5 ml/min, rotation of the lapping plate: 56 rpm, slurry concentration: 1.4 g/500 ml): *(a)* diamond grain size 1μm; and *(b)* diamond grain size 25 μm.

13.3.3 The Lapping Fluid

Abrasive grains are transported to the lapping zone suspended in an oil or aqueous medium with the aim of achieving a continuous and even distribution across the lapping plate. This liquid carrier is available in various viscosities to cover almost any process situation. The carrier's function is to lubricate the two surfaces. Lubrication is necessary to reduce friction between the abrasive and the workpiece, to help disperse the abrasive product uniformly across the lap plate, and to remove the abraded debris from the work zone. Probably the most important characteristic of the carrier is its ability to suspend and uniformly disperse abrasive particles throughout the interface between the workpiece and the lap.[16] Environmental considerations of waste disposal are pushing lapping processes in the direction of water-based carriers. During operation, the abrasive grains in the slurry suffer a process of rounding and comminution, which reduces the grain size and the effectiveness of the abrasive as a machining tool. With wear replacement of the abrasive, fresh slurry is essential to maintain constant cutting conditions. If the fluid is always delivered to the same position on the lap, the increased concentration of slurry in that area can lead to uneven wear rates, a situation that is best to avoid.

In summary, the lapping process has the following characteristics, which vary in degree according to the particular system and equipment:

- The rate of material removal is low due to the low cutting speeds and the shallow penetration of the abrasive grains into the worksurface.

- The lapping process does not generate significant temperatures; it is considered a cool process that does not cause thermal damage.

- The general shape of the parts worked by lapping is mainly limited to flat, cylindrical, and spherical surfaces. However, irregularly-shaped parts can be processed.

- The accuracy of form produced by lapping can be very good with appropriate techniques, especially for flat surfaces:
  - Flatness to less than one light band: (He) 0.0000116"/0.3 microns.
  - Roughness of less than one micro-inch R_a: 0.025 microns R_a.
  - Size control to less than 0.0001"/2.5 microns.
  - Parallelism within 0.00005"/1.3 microns.

- Lapping may be successfully applied to brittle materials and fragile parts because a relatively uniform pressure is exerted on the workpiece.

- The lapping operation produces a gray "mat" finish (a nonreflective surface) due to the configuration of the randomized nondirectional pattern left by the rolling abrasive grains.

In lapping, the material removal mechanism includes rolling abrasive, sliding abrasive, and microcutting abrasive. It is a very complex mechanism that involves abrasive phenomena, brittle fracture, and plastic deformation. The predominant mode is dependent mainly on the load of the abrasive (whether the load exceeds the threshold for fracture). Chandrasekar and Bhushan[45] found that the depth and degree of surface deformed layers and the amount of material removal increase with higher normal force or contact pressure. The mechanical and thermal loads from frictional heating cause residual stresses to form on the surface. The residual stresses were found to be compressive, indicating that they were mechanically, not thermally, induced.

Lapping is a three-body abrasion mechanism because the abrasive particles may roll or slide rigidly between the lapping plate and the workpiece surface. Sometimes the particles only indent one or both of the surfaces. In a sliding mode, the wear process is basically that of two-body abrasion but usually for a short period of time. In a rolling mode, the wear processes involves ploughing damage and wear debris formation from the exposed lips of deformed material adjacent to the grooves. The abrasive particle shape influences the sliding mode and the rolling mode. If the particles are round and the same size, the probability of rolling is increased. If the particles are rectangular, with larger width than thickness, the probability of sliding is enhanced.

The abrasive wear rate is expected to increase with contact pressure and this means an increased number of contact points per unit area and a deeper penetration of the abrasive grains leading to deeper groove formation. Also, the contact pressure influences the fracture of the abrasive grains. At lower pressures, particle fracture may not be possible and the potential for wear is low while higher contact pressure causes some particle fracture (new sharp edges are produced on a greater number of particle) and the potential for wear may increase.

13.3.4 Lapping Types

One-Side Lapping (Single-Side Lapping). This is the best-known and most used method for producing flat surfaces. Basically, single-side lapping machines have a rotating annular-shaped wheel (lapping plate) and the workpieces are applied to the flat rotating wheel. Small lapping machines usually have three conditioning rings, while larger machines work with four conditioning rings. The workpieces to be lapped are placed within the conditioning ring and a load is applied (Fig. 13.5).

In this way, the parts are pressed against a film of abrasive slurry that is continuously dripped onto the rotating lapping plate. There are three important issues that should be controlled during the lapping: keeping the lap flat, applying a uniform and predictable pressure, and applying and maintaining a constant and consistent flow of the abrasive.

Double-Side Lapping. The most accurate method in terms of parallelism and uniformity of size is double-side lapping which means machining both sides of the workpiece simultaneously. With double-side processing there is less chance that foreign particles will be introduced into the process to settle between the workpiece and the load.

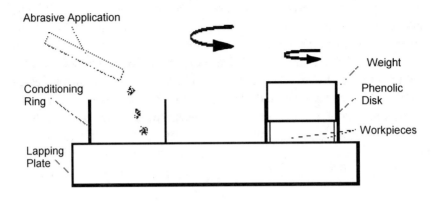

Figure 13.5. Single-side lapping principle.

Double-side lapping machines have two opposed lapping plates of equal diameter; the lower plate is mounted at a fixed level, while the upper plate can be power-raised and lowered usually by action through a hydraulic system (Fig. 13.6). Both lapping plates are driven and usually rotate in opposite directions. The rotation of the lower plate can be reversed when needed for particular reasons. With variable speed drive motors, different plate rotation speeds can be produced.

The lapping process may be carried out using cast-iron lapping plates and an abrasive compound that is kept in continuous circulation. As an alternative method, a bonded abrasive lapping wheel may be installed when the use of a liquid abrasive compound is not desirable, or when special effects on the workpiece are required.

A characteristic of the process is that the working pressure may be varied, making it possible to carry out the process with a gradual build-up of the force with which the lapping plates bear against the workpiece. Varying the working pressure can substantially reduce the lapping time and ease the development of the required roughness.

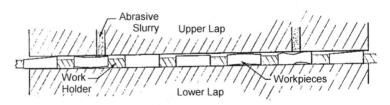

Figure 13.6. Position of the workpieces during double-side abrasive machining.[17]

Under ideal conditions, the results of the double-side lapping are optically flat surfaces (the workpiece has plane and parallel opposite surfaces) and micro-inch surface roughness.

Double-side lapping increases lapping efficiency, because the abrasive machining of two surfaces simultaneously takes approximately the same time as it takes to machine one surface alone.

A typical process cycle consists of the following sequence:

- Loading the workpieces (the upper wheel swings from the load position at one side to center over the lower wheel).

- The top wheel is lowered gently to rest on the top of the workpieces.

- The two wheels and the carrier drive units start slowly, automatically or controlled by the operator.

- The abrasive slurry is fed automatically through the top wheel during the cycle in order to provide the necessary abrasive action on both sides of the workpieces.

- The pressure of the upper wheel against the parts to be lapped is initially light (as the high points are machined away). Then the pressure is gradually increased until the optimum pressure, best suited for the job, is reached.

- The cycle continues until the desired size is obtained, at which time the rotation stops, the top wheel lifts and swings to the side for unloading the finished workpieces.

Double-side flat lapping offers some advantages:

- The efficiency of processing two sides of a workpiece in the same time as required to machine one side.

- A large number of parts can be lapped simultaneously.

- Best available method to obtain close tolerances for flatness, parallelism, and size.

- Removing stock from both sides of a workpiece at the same time helps to relieve internal stress of the part, thus making it easier to achieve flatness.

- Cut rate is uniform and repeatable. No dulling of the abrasive takes place because fresh sharp abrasive particles are gradually and continuously fed to the lapping area during the process cycle.

- Simple workholder design with no need to clamp or rigidly hold the part eliminates stresses in the workpiece, hence improving tolerances for flatness, parallelism, and size.

Cylindrical Lapping. Cylindrical parts or cylindrical portions of a particular component can be finished to a high degree of geometric accuracy, with low surface roughness, by means of lapping on a lapping machine equipped with a holder. The work holder has openings of the form and size corresponding to the dimensions of the workpiece.

Machines for cylindrical lapping utilize two annular lapping plates, each mounted on a vertical spindle. One or both of the lapping plates rotate; parts are placed in a workholder that guides them between the lap faces to produce an abrading action. The workholder is guided in the center by a rotating pin that can be adjusted to move eccentricly to the center of the lower lap. The cylindrical parts are placed in slots, the centerline of which are tangent to a circle in the center of the workholder. The rolling action of the parts causes the workholder to rotate. Controlled lapping occurs as the parts slide and slip during rolling, caused by the nonradial position of the workholder (Fig, 13.7).

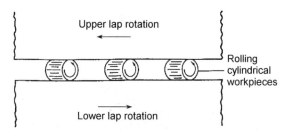

Figure 13.7. Schematic showing cylindrical lapping method.[17]

13.4 POLISHING PROCESS

Polishing is one of the oldest manufacturing technologies used as a surface-smoothing operation; it mainly consists of removing or smoothing out grinding or lapping lines, scratches, and other surface defects in order to decrease the surface roughness. Although there are many operations to make a surface smooth, polishing remains the best method to obtain the finest surface. High accuracy and ultraprecision technology are indispensable ingredients of modern polishing.

Traditionally, polishing was meant to involve best finish without regard for shape and form accuracy. It was also practiced for removing the damage (such as microcracks, voids, etc.) caused by previous manufacturing operations such as grinding.

The function of polishing is pictured in Fig. 13.8 of a plane-parallel polishing machine.

In conventional polishing the surface final attributes are decided by mechanical mechanisms, consequently it is important to utilize an abrasive harder than the material to be polished in order to obtain characteristic roughness levels of <1 mm. When the surface quality is required to be less than 0.1 mm, ultraprecision polishing methods are applied. In this case, the abrasive must be softer than the workpiece material; chemical and chemical-mechanical polishing methods can be used to accomplish material removal and surface roughness.[28]

The main removal mechanisms, scratching and microchipping, are characterized by abrasive grains that are temporarily embedded in a relatively soft polishing wheel (Fig. 13.9). The depth of the grain penetration into the workpiece surface causes elastic and plastic deformation phenomena. With increasing penetration depth, microchipping and microcracking take place.

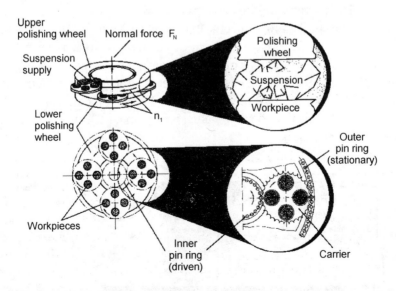

Figure 13.8. Principle of two-wheel polishing with guided workpieces.[28]

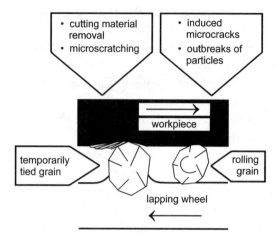

Figure 13.9. Polishing of a plane surface.[28]

It is not possible tp prevent scratches in traditional polishing since the abrasive always contains a range of grain sizes. This problem can be overcome by adapting the polishing method.

Figure 13.10 illustrates the factors that influence the polishing operation. It is very important that appropriate parameters selected for the process operation are associated with the parameters related to the material of the workpiece.

The primary difference between polishing and lapping is in the intent, rather than in the process. Because of this, high rates of material removal are not necessary. The aim of polishing is to achieve a desired surface finish. While lapping produces a gray nonreflective surface, polishing will generate a reflective or shiny surface. The degrees of reflectivity will vary in relation to the type of process used. Typical processes include hard and soft pad polishing, diamond abrasive/composite plate polishing, and optical pitch polishing.

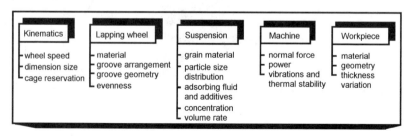

Figure 13.10. Parameters influencing the polishing process.[28]

The polishing process is carried out after lapping and is designed to smooth the surface while maintaining the precision of form obtained previously by lapping. Both soft polishers and fine abrasive grains can be utilized in order to accomplish the polishing process requirements. Relatively low pressures are used in polishing. The differences between lapping and polishing consist of the size of the abrasive grains and their holding method.

The polishing process is based on the principle of applying a fine abrasive powder to a polishing tool, with a relatively slow motion between the workpiece and the tool, to reduce the surface roughness of the part. The processing occurs due to the fine scratch effect carried out with extremely fine cuts into the workpiece by the grains kept elastically and plastically in the soft polisher. There are none of the tiny cracks that happen in lapping. Polishing creates very small irregularities in the worked surface and a small worked layer. This is why, by polishing, it is possible to obtain a higher degree of precision in dimensions and shape and greater surface smoothness than with lapping.

To minimize surface damage, gentle polishing conditions are required; low levels of force should be used, and the abrasive should not be considerably harder than the workmaterial.

A variety of polishing machines can be used, machines which have almost the same structure as a lapping device (polishing often uses much of the same set-up as lapping). The major difference in equipment is that where a lapping plate is usually made of bare metal for hardness and rigidity, a pad made of a soft material, such as a polymer, covers a polishing plate. For example, a cast iron lapping plate can be replaced with a disk of bronze or aluminum for polishing. Like lapping, polishing utilizes progressively smaller abrasive sizes to produce an extremely fine surface finish.

The conventional technique of polishing involves the use of a fine abrasive in a liquid carrier on a polishing pad. In this way, a rough surface having irregularities visible to the naked eye may be smoothed.[35] Unlike lapping, the carrier is chosen more for lubricity than for cooling and swarf removal. This is because of the low material removal rates and the desire for a superior surface finish. Polishing is subject to the same surface integrity problems as lapping, but because of the fine abrasives and surface finish, the dead layer is very thin.

There are several polishing methods used in industry; they can be classified into three important categories:

- Close contact (pitch polishing).

- Semi-contact polishing (conventional mechanical chemical polishing of Si wafers).

- Noncontact polishing (float polishing, elastic emission machining).

By changing the type of contact (from close contact to noncontact), one can alter the mechanism of polishing from mechanical to chemical.[5]

Polishing operations can be due to mechanical action only or due to chemical action alone (Fig. 13.11).

The combined actions accelerate material removal, which has to be small in order not to create profile errors, but reduce surface roughness and eliminate sub-surface damage.

Preston's equation is very useful in evaluating a polishing operation.

$$\frac{dT}{dt} = kp\frac{ds}{dt}$$

where k is the Preston coefficient and p is the polishing pressure. This equation implies that the velocity of the surface recession, dT/dt, is related to the relative velocity between the workpiece and the lap, ds/dt, through elasticity. This model assumes that a population of uniform-size spheres penetrates the ideal plane of the part in elastic Hertzian fashion and, in shear, ploughs a groove having a cross-section that is proportional to that of the penetration.[35]

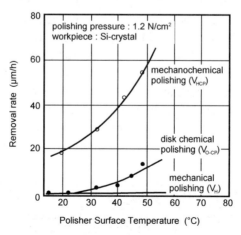

Figure 13.11. Material removal by mechanical, chemical, and chemical mechanical polishing.[35]

Several models have been presumed: a chemical model, a polishing-pad model, and a fluid-based model.

Figure 13.12 shows Cook's chemical model for wafer polishing, which takes into consideration the following reactions:

- The formation of hydrogen bonding between the solute (both water and slurry particles) and the solvent in the slurry.

- The creation of hydrogen bonding between solvated wafer and particle surfaces.

- Formation of molecular bonding between surfaces.

- The elimination of the bonded wafer surface as the slurry particles move away.

In Fig. 13.13, the model for a polishing pad is illustrated (Yu's model, 1993). The contact between the workpiece surface and the polishing pad is considered as statistically distributed hemispheres from which the contact surface can be derived. The mechanical removal can be calculated easily using Preston's equation.

The fluid-based erosion model (Runnels model, 1994) is pictured in Fig. 13.14. It is used for processes utilizing a hydrodynamic slurry layer. It takes into account Preston's equation, the simulation, the fluid dynamics, and the hydrodynamic action of the slurry.

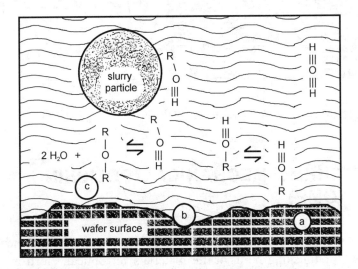

Figure 13.12. The chemical model.[43]

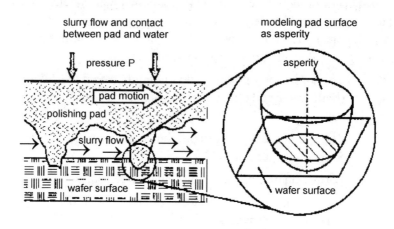

Figure 13.13. The polishing-pad model.[44]

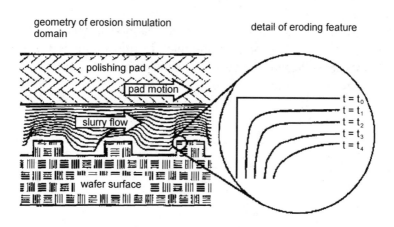

Figure 13.14. The fluid-based erosion model.[45]

13.5 CHEMO-MECHANICAL POLISHING (CMP)

Chemo-mechanical polishing (CMP) has become an indispensable process for finishing hard, brittle materials for optical, electronic, and structural applications.

In order to differentiate the particular factor that plays a major role in the material removal process, some researchers[5] have used several terms to describe chemo-mechanical polishing:

- *Chemo-mechanical polishing*: The main factor for material removal is a chemical mechanism followed by mechanical action for the removal of the reaction products.

- *Mechano-chemical polishing*: The mechanical action occurs first, and then the chemical action takes place.

- *Tribochemical polishing*: Due to friction, a reaction layer is formed without the need for particular abrasive particles.

All these terms are used to describe the same polishing process; they only indicate that material removal is caused by different actions.

The most used term is chemo-mechanical polishing (CMP) and it is a very important nanoscale finishing process for manufacture of computer disks and integrated circuits. This operation depends both on the chemical and the mechanical effectiveness of the abrasive and the environment with respect to workmaterial (Fig. 13.15). In CMP, the selected abrasive is softer or nearly of the same hardness as the workpiece material. For that reason, the damages due to mechanical action are minimized. The chemical reaction products, which have been formed between the workmaterial, the abrasive, and the environment, are eliminated from the work zone by subsequent mechanical action of the abrasive.[37] This results in the production of a very smooth and damage-free surface of the workpiece.

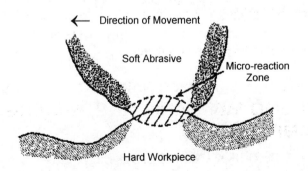

Figure 13.15. Schematic of the chemo-mechanical action between abrasive, work material, and the environment.[5]

The chemo-mechanical action takes place when a certain threshold pressure and temperature is applied in the contact zone and can initiate a chemical reaction between an appropriate abrasive, workmaterial, and environment. In CMP, material removal is accomplished by chemical reaction stimulated by frictional heat and the contact pressure at the contact area between the material of the workpiece and the abrasive. This process is successfully applied for finishing brittle materials because no scratching or groove formation is expected. Chemo-mechanical polishing (CMP) produces a weak reaction product compared to either the abrasive or the workmaterial. Hence, a higher removal rate can be obtained without causing damage due to brittle fracture of the brittle workmaterial.

For the CMP process, it is very important that the reaction products are taken away from the work zone; otherwise, they may influence the performance and the efficiency of the part in subsequent use.

Some researchers developed this method in order to planarize multilevel metal interconnect structures for micro-electronic chips.[38] Planarization by CMP was used to eliminate the peaks and valleys that are created when a dielectric film is deposited over the metallic features on a layer on the chip, before the following layer of gates and interconnects can be laid down.

Many researchers (Komanduri, Jiang, Bhagavatula) studied the chemo-mechanical polishing of Si_3N_4 bearing balls using different abrasives: B_4C (boron carbide), Al_2O_3 (aluminum oxide), Cr_2O_3 (chromium oxide), SiC (silicon carbide), etc. They investigated the abrasive effectiveness of the magnetic float polishing (MFP) method. Figure 13.16 shows the apparatus used for polishing Si_3N_4 balls.

The effectiveness of an abrasive for CMP depends on the chemical reaction with the workpiece material and the kinetic action involving the removal of the reaction product from the workmaterial. The chemical reaction will continue after the passivating layers are removed.[37] The reactions between the Si_3N_4 workmaterial and various abrasives are:

$$Si_3N_4 + Fe_2O_3 \rightarrow SiO_2 + FeO + FeSiO_3/ FeO \times SiO_2 + Fe_4N + N_2 \text{ (g)}$$

$$Si_3N_4 + 2Cr_2O_3 \rightarrow 3SiO_2 + 4CrN$$

$$Si_3N_4 + ZrO_2 \rightarrow SiO_2 + ZrSiO_4/ZrO_2 \times SiO_2 + ZrN + N_2 \text{ (g)}$$

$$Si_3N_4 + CeO_2 \rightarrow SiO_2 + CeO_{1.72} + CeO_{1.83} + Ce_2O_3 + N_2 \text{ (g)}$$

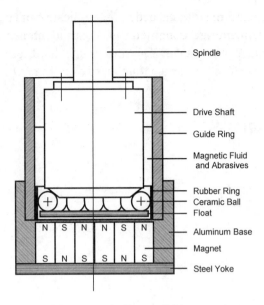

Figure 13.16. Schematic of the magnetic float polishing apparatus used for polishing Si_3N_4 balls.

Two types of reactions are considered: oxidation-reduction reaction and exchange reaction. Chemical reactions with Si_3N_4 workpiece materials are feasible with the abrasives used and water from the water-based magnetic fluid. Silicon dioxide (SiO_2) is the main reaction product leaving the surface of the workpiece material.

This method is based on the magneto-hydrodynamic behavior of a magnetic fluid that can float, and abrasives suspended in it by a magnetic field. Several permanent magnets, arranged alternatively, are put in an aluminum chamber filled with a magnetic fluid and abrasive. The forces applied by the abrasive on the ceramic balls are very low (1 N/ball) and highly controllable. When a magnetic field is activated, the magnetic particles in the magnetic fluid are attracted to the area where the magnetic field is higher. All nonmagnetic materials (abrasive grains and the acrylic float) are attracted in the area of the lower magnetic field.

The ceramic balls come into contact with the drive shaft and the abrasive grains, under the action of the magnetic buoyancy levitational force, polish them when the spindle rotates. The acrylic float produces a uniform polishing pressure. Using this technique, the material removal rate is 50 times higher than in conventional polishing because MFP uses higher polishing speeds (1,000–10,000 rpm instead of 50 rpm in lapping). The damage on the surface and subsurface is diminished by the application of very low polishing loads. The surface finish obtained with boron carbide abrasive (1–2 μm grain size) was 20 nm R_a and 200 nm R_t. Using another hard abrasive, SiC (1 μm grain size), the values for surface roughness were 15 nm R_a and 150 nm R_t.

REFERENCES

1. Schwartz, M., *Handbook of Structural Ceramics*, New York (1992)

2. Matsunaga, M., Fundamental Studies of Lapping, Report of the Inst. of Ind. Sci., U. of Tokyo, 16(2)

3. Walsh, R. A., *Machining and Metalworking Handbook*, McGraw-Hill, Inc. (1994)

4. Somiya, S., *Advanced Technical Ceramics*, Academic Press, Inc., San Diego, CA (1984)

5. Komanduri, R., Lucca, D. A., and Tani, Y., Technological Advances in Fine Abrasive Processes, *Annals of the CIRP*, 46/2:545–596 (1997)

6. Uhlmann, E., and Ardelt, T., Influence of Kinematics on the Face Grinding Process on Lapping Machines, *Annals of the CIRP*, 48/1:281–284 (1999)

7. Lambropoulos, J. C., Zhou, Y., Funkenbusch, P. D., Gillman, B., and Golini, D., Brittleness & Grindability of Brittle Workpieces, *Supertech-Superabrasives Technol. Conf. Proc.*, pp. 195–200, Livermore, CA (Nov. 7–8, 1996)

8. Verspui, M. A., and de Witt, G., Three-Body Abrasion: Influence of Applied Load on Bed Thickness and Particle Size Distribution in Abrasive Processes, *J. European Ceramic Soc.*, 17:473–477 (1997)

9. Trezona, R. I., Allsopp, D. N., and Hutchings, I. M., Transition Between Two-Body and Three-Body Abrasive Wear: Influence of the Test Conditions in the Microscale Abrasive Wear Test, *Wear*, pp. 205–214 (1999)

10. Buijs, M., and Korpel-Van Houten, K., *A Model for Lapping of Glass*, pp. 3014–3020, Chapman & Hall (1993)

11. Chen, C., Sakai, S., and Inasaki, I., Lapping of Advanced Ceramics, *Materials & Mfg. Processes*, 6/2:211–226, Marcel Dekker, Inc. (1991)

12. Narayan, P. B., Brar, A. S., and Sharma, J. P., Lapping and Polishing of Ceramics: Some Concerns and Solutions, *Solid State Technol.*, pp. 151–153 (Apr. 1988)

13. Buijs, M., and Korpel-Van Houten, K., Three-Body Abrasion of Brittle Materials as Studied by Lapping, *Wear*, 166:237–245 (1993)

14. Fang, L., Kong, X. L., Su, J. Y., and Zhou, Q. D., Movement Patterns of Abrasive Particles in Three-Body Abrasion, *Wear,* 162–164:782–789 (1993)

15. Chauhan, R., Ahn, Y., Chandrasekar, S., and Farris, T. N., Role of Indentation Fracture in Free Abrasive Machining of Ceramics, *Wear,* 162–164:246–257 (1993)

16. Millar, J., Lapping & Polishing Technology, *Abrasive Eng. Soc.,* 30(4):9–13 (1991)

17. Corsini, A. M., Abrasive Micromachining of Advanced Materials to Precision Tolerances, *Grinding & Machining of Advanced Materials,* Pittsburgh, PA (Oct. 11–13, 1995)

18. K. Subramanian, T. K. Puthanangady, Diamond Abrasive Finishing of Brittle Materials-an Overview, *Supertech-Superabrasives Technol.* (1996)

19. Gatzen, H. H., and Maetzig, J. C., Nanogrinding, *Precision Eng.,* 21:134–139 (1997)

20. Allor, R. L., and Jahanmir, S., Current Problems and Future Directions for Ceramic Machining, *The Amer. Ceramic Soc. Bull.,* 75(7):40–43 (1996)

21. Korman, R. S., Addressing Contamination Issues Raised by CMP Slurries, *Micro,* pp. 47–53 (1997)

22. Young, J. F., and Shane, R. S., *Materials and Processes; Part B: Processes,* Marcel Dekker, Inc., New York (1985)

23. Abrahamson, G. R., Duwell, E. J., and McDonald, W. J., Wear and Lubrication as Observed on a Lap Table with Loose and Bonded Abrasive Grit, *J. Tribology,* 113:249–254 (1991)

24. Chang, Y. P., Hashimura, M., and Dornfeld, D. A., An Investigation of the AE Signals in the Lapping Process, *Annals of the CIRP,* 45/1:331–334 (1996)

25. Ohmura, E., Koike, N., and Eda, H., Ultra-Precise Lapping of Magnetic Memory Disks with Automatic Conveyance, *Advancement of Intelligent Production,* pp. 515–520 (1994)

26. Budinski, K. G., *Surface Engineering for Wear Resistance,* Prentice-Hall, Inc., Englewood Cliffs, NJ (1988)

27. Gates, J. D., Two-Body and Three-Body Abrasion: A Critical Discussion, *Wear,* 214:139–146 (1998)

28. Klocke, F., Gerent, O., and Wagemann, A., Polishing of Advanced Ceramics, *Machining Sci. and Technol.,* 1(2), pg. 263–273 (1997)

29. Zhong, Z., and Venkatesh, V. C., Semi-Ductile Grinding and Polishing of Ophthalmic Aspherics and Spherics, *Annals of the CIRP,* 44/1:339–342 (1995)

30. Sun, J. J., Taylor, E. J., Gebhart, L. E., Zhou, C. D., Eagleton, J. M., and Renz, R. P., Investigation of Electro-chemical Parameters into an Electro-chemical Machining Process, *Technical Papers of the N. Amer. Mfg. Research Inst. of SME*, pp.63–68 (1998)

31. Rabinowicz, E., *Friction and Wear of Materials*, 2nd Ed., John Wiley & Sons, (1995)

32. Rabinowicz, E., Abrasive Wear Resistance as a Materials Test, *Lubrication Eng.*, 33:378–381 (1977)

33. Lucca, D. A., Brinksmeier, E., and Goch, G., Progress in Assessing Surface and Subsurface Integrity, *Annals of the CIRP*, 47/2:669–688 (1998)

34. Tönshoff, H. K., Karpuschewski, B., and Mandrysch, T., Grinding Process Achievements and their Consequences on Machine Tools Challenges and Opportunities, *Annals of the CIRP*, 47/2:651–666 (1998)

35. Ong, N. S., and Venkatesh, V. C., Semi-Ductile Grinding and Polishing of Pyrex Glass, *J. Materials Processing Technol.*, 83:261–266 (1998)

36. Liang, H., Kaufman, F., Sevilla, R., and Anjur, S., Wear Phenomena in Chemical Mechanical Polishing, *Wear*, 211:271–279 (1997)

37. Jiang, M., Wood, N. O., and Komanduri, R., On Chemo-Mechanical Polishing (CMP) of Silicon Nitride (Si_3N_4) Workmaterial with Various Abrasives, *Wear*, 220:59–71 (1998)

38. Larsen-Basse, J., and Liang, H., Probable Role of Abrasion in Chemo-Mechanical Polishing of Tungsten, *Wear*, 233–235:647–654 (1999)

39. Xie, Y., and Bhushan, B., Effects of Particle Size, Polishing Pad and Contact Pressure in Free Abrasive Polishing, *Wear*, 200:281–295 (1996)

40. Haisma, J., Spierings, G. A. C. M., Michielsen, T. M., and Adema, C. L., Surface Preparation and Phenomenological Aspects of Direct Bonding, *Philips J. Research*, 49:23–46 (1995)

41. Bozzi, A. C., and Biasoli de Mello, J. D., Wear Resistance and Mechanisms of WC-12%Co Thermal Sprayed Coatings in Three-Body Abrasion, *Wear*, 233–235:575–587 (1999)

42. Runnels, S., Feature-Scale Fluid-Based Erosion Modelling for Chemical-Mechanical Polishing, *J. Electro-chem. Soc.*, 141(7):1900–1904 (1994)

43. Cook, L. M., Chemical Processes in Glass Polishing, *J. Noncryst. Solids*, 120:152–171 (1990)

44. Yu, K. T., Yu, C., and Orlowski, M., Statistical Polishing Pad Model for CMP, *Proc. IEEE Int. Electron Devices Meeting*, Washington, DC (1993)

45. Chandrasekar, S. K. K., and Bhushan, B., Influence of Abrasive Properties on Residual Stresses in Lapped Ferrite and Alumina, *J. Amer. Ceramic Soc.*, 73(7):1907–11 (1993)

46. Grinding and Lapping Compounds, publ. United States Products, Co.

14

Process Fluids for Abrasive Machining

14.1 PROCESS FLUIDS AS LUBRICANTS

The purpose of this chapter is to review the main types of process fluids used in abrasive machining and their characteristics. In abrasive machining, liquid fluids are usually preferred to gaseous fluids. Besides cooling, a process fluid has the important function of lubricating. Lubrication means interposing a layer of low shear-strength in the interface between the elements of the friction pair. The main role of a lubricant is to prevent or, at least, to diminish solid contact between the elements to reduce adhesion wear.

14.1.1 Alternative Lubrication Techniques

Since liquid fluids are employed almost exclusively, other methods of lubrication tend to be neglected. Some examples of techniques that may be overlooked are:

- Spray, mist, or active chemical vapor lubrication.
- Surface layers with self-lubricating properties, produced by deposition or conversion on a half-finished product before grinding.

- Introduction of solid lubricants into the bond of the abrasive tool and/or into the composition of the workpiece material. This method requires the lubricant to be incorporated in the tool or workpiece at the time of initial fabrication. Self-lubricating materials may be introduced to enhance the effects of a process fluid or as an alternative to a process fluid.

One of the above techniques may be adopted as a move towards ecological protection against pollution by process fluids. Legislation at national and international levels increasingly aims to avoid potential health hazards and to protect the environment. The effect of complying with legislation leads to rising costs associated with the use and disposal of process fluids. The effect is to bring pressure to reduce or avoid the use of grinding fluids (Fig. 14.1).[1] However, before eliminating fluids from abrasive machining processes, it is important to consider how process fluids can be employed safely and efficiently. In spite of recent efforts to reduce the use of process fluids,[2] the possibility of completely eliminating process fluids is still remote for most applications.

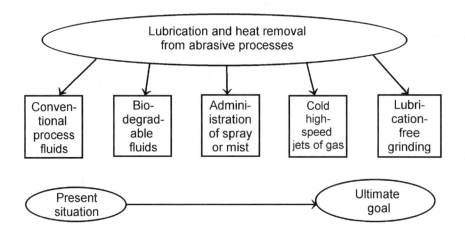

Figure 14.1. Avoidance of grinding fluids.[1]

14.1.2 Benefits of Lubricants in Abrasive Machining

Abrasive machining takes place at numerous cutting edges of indeterminate shape and at extremely small areas of real contact. These conditions involve a substantial proportion of rubbing and sticking friction as described in Chs. 4, 5, and 6. Because of the conditions involved, the use of a fluid strongly influences process efficiency and work quality. The presence of a process fluid in the couple interface greatly reduces friction forces and reduces abrasive tool wear.[3] The term "coolant" understates the functions of a process fluid.[4] In addition to process and bulk cooling, the fluid provides the additional functions of lubrication at the abrasive contact surfaces and flushing of swarf from the contact area. Process lubrication is important because it helps to achieve a high grinding rate and long wheel life.[5] A further function of the fluid is to protect the workpiece and the machine against corrosion.[6]

14.1.3 Three Main Groups of Fluids

Process fluids used for abrasive machining may be divided into three main groups:

- Gaseous (air, carbon dioxide, or inert gases) fluids that can be cooled beforehand.
- Water-immiscible (neat oil).
- Water-soluble and/or miscible (water-based).

Dry abrasive machining takes place exclusively in air. (This can be confusing because in tribology a "dry process" can mean functioning in high vacuum; however, this is almost never the case in abrasive machining.) In addition to acting as a coolant, the presence of air produces valuable lubricating effects. These effects include decreasing the friction coefficient, increasing the abrasive wear compared to adhesive wear, and reducing the tendency for seizure between the couple components.

Dry abrasive machining in the absence of liquid tends to be characterized by increased frictional drag.[7] Using only air as a process fluid, it is not usually possible to satisfy all the technical and economic requirements of a modern industrial process. For this reason, most machining processes are conducted with liquid lubrication. Neat oils or water-based fluids avoid many of the problems experienced in dry machining.[8] Improved lubrication results when using process liquids in the presence of air.[9]

14.1.4 Demands Arising from New Materials and Applications

Most fluids and additives are aimed at processing ferrous workpieces.[10] However, the 20th century saw the development of many new materials to be processed. These materials included new alloys, ceramics, polymers, and composites intended for restricted or even unique fields of application. Unfortunately, there is limited information on process fluids for many of the new materials. For example, there has been relatively little work on fluids for abrasive machining of titanium alloys, advanced ceramics, and/or composites, as well as for other nonferrous materials used in optics and electronics.[11] There is clearly a need for further information to support the machining of new materials.

The tribological conditions in machining are some of the most severe encountered anywhere.[12] The demands on the process fluid are made greater by the requirement for very high quality workpieces. Exacting quality norms imposed for finished workpieces require, in turn, the fulfilment of stringent technical requirements. In some cases, demanding quality norms may drastically limit permissible macro- and microgeometrical deviations of the workpiece surface. This imposes strict conditions for a uniform material composition and structure at the surface of the machined workpieces.[13] In such applications, all aspects of the process must be carefully considered including the selection and treatment of the process fluid.

14.2 LUBRICATION REGIMES

14.2.1 The Stribeck Curve

More than a century ago, Stribeck[14] represented the variation of the friction coefficient (μ) for a sliding bearing as illustrated in Fig. 14.2. The friction coefficient was represented as a function of the generalized Sommerfeld number (S_0) sometimes known as the Hersey number:

Eq. (14.1) $S_0 = \eta \cdot v/p_m$

where η is the fluid dynamic viscosity, v is the sliding speed, and p_m is the couple loading pressure.

A physical interpretation of the Stribeck curve was suggested based on distinct functional regimes emphasized on the curve $\mu = f(S_0)$. The regimes are *hydrodynamic lubrication* on the right-hand side, and *boundary/mixed lubrication* on the left-hand side of the diagram.

The minimum point of the Stribeck curve represents a transition between the purely hydrodynamic region and the boundary/mixed lubrication region.

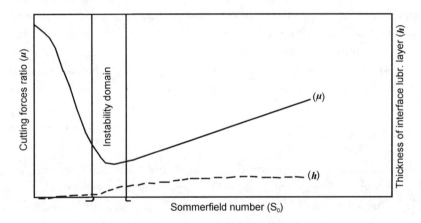

Figure 14.2. The Stribeck diagram.

14.2.2 Hydrodynamic Lubrication

At the beginning of the 20th century, hydrodynamic lubrication was well advanced, based on the work of Navier and Stokes. Reynolds' equation (1886) had been widely validated for continuous liquid films acting under restricted conditions. These restricted conditions do not apply in many practical situations, including rubbing and abrasive machining contact. The restrictions required are:

- The flow of the fluid has a Newtonian character.
- The fluid is incompressible.
- The viscosity of the fluid is constant.

- The contact surfaces of the solid bodies against which the flow process takes place are considered very smooth.
- The contact surfaces of the couple are separated by a fluid film operating under laminar conditions.
- The relative speed of the fluid in direct contact with a surface is zero.
- The fluid is inactive from the chemo-physical point of view against the solid bodies it contacts.

In a sliding bearing, the lubricant is represented in the form of a fluid film compelled to flow through the contact interface. The fluid film is thick enough to prevent physical contact between the friction couple elements. With hydrodynamic lubrication, the couple elements may function without wear. The only undesirable modifications of the surfaces occur at start and stop, when the relative sliding speed is very small or zero, and the fluid film has not had time to develop.

14.2.3 Boundary/Mixed Lubrication

Mixed lubrication is defined as a mixture of solid contact and liquid lubrication. This means that part of the area of contact is lubricated with liquid, but in other parts, the liquid film breaks down and lubrication depends on the lubricity of a very thin film of lubricant. Mixed lubrication is the term applied to the transitional region between full hydrodynamic or elastohydrodynamic lubrication and boundary lubrication. The liquid may still play a role in the areas where liquid lubrication breaks down due to the various slubrication mechanisms to be described in more detail throughout this chapter.

With a decrease of the S_0 beyond the minimum point of the curve, there is a rapid growth of the friction coefficient. This region represents the boundary lubrication regime. This domain is characterized by molecular thickness of the boundary lubrication film. The small film size is insufficient to prevent a wear process due to stochastic contacts between asperities in relative movement. The behavior of the lubricating layer under these conditions of partial solid contact is determined by physico-chemical properties of the process fluid. These are summarized by the following broad descriptions:

- *Fluid lubricity* is "the ability of a fluid to reduce wear and friction, other than by its purely viscous properties."[26] The lubricity of the thin lubricant films, generated at the couple interface, is evidenced by a marked reduction of the friction coefficient, μ. Kajdas and Colab[27] state "The lower the friction the higher the lubricity." Lubricity is increased by use of friction modifiers (FM) and antiwear (AW) additives (see Secs. 14.12.1 and 14.12.2).

- *Extreme pressure (EP)* properties of the process fluid are related to the load-carrying capacity of the thin lubricating layer under severe operating conditions.[28] This lubrication ability, achieved by use of EP-additives, tends to reduce scuffing and to prevent seizure (see Sec. 14.12.3).

Boundary layers appear as a result of tribophysical processes and/ or from tribochemical reactions between active substances from the process fluid, the environment, and the tool/workpiece pair.[29][30] The composition of the workmaterial has a major influence on the performance of process fluid.[31] A favorable aspect of the tribochemical role of the process fluid is the *softening effect*. This is the possibility of mechanical removal of the hard surface layer formed on the processed surface before or during the abrasive machining.[32] (See also the "Rehbinder Effect" in Ch. 15, "Tribochemistry of Abrasive Machining.") The formation of a boundary layer reduces friction and explains increased grindability in the presence of tribological additives.

14.2.4 Elastohydrodynamic Lubrication (EHL)

Elastohydrodynamic lubrication (EHL) is the typical regime for friction pairs with elastic contact such as ball bearings and gears. Elasto-hydrodynamic lubrication is a development of hydrodynamic lubrication to take into account the elastic deflection of the surfaces in contact. Film thicknesses are much smaller than in conventional hydrodynamic lubrication. Under these conditions, a continuous hydrodynamic film can only be achieved through elastic deflections; therefore, it is important to take the deflections into account.

The theory of the EHL regime was established in the latter half of the century, notably by Dowson and Higginson,[15] and further by Winer and Cheng.[16] On the Stribeck curve, the EHL regime is situated near the transition between full hydrodynamic lubrication and boundary lubrication.[17] Due to very high pressures at the contact interface, the EHL regime also involves elastic smoothing of the surface micro-asperities.[18] By this mechanism and by an increase in viscosity due to extreme pressure, EHL ensures continuity of the fluid film in a bearing contact.

14.2.5 Lubrication in Abrasive Machining Contacts

In abrasive machining, the situation is completely different from a conventional lubricated bearing. The abrasive grains of the tool have a crystalline structure and very high hardness. The grains coming in contact with the workpiece are subject to small elastic deformation. In comparison, the workmaterial is subjected to large plastic deformation. Due to the plastic deformation process and the viscous properties of the fluid at high temperature and pressure, only a lacunose lubricating film between the two moving surfaces is created.[24] By this, it means that lubrication plays an important part in small critical regions of the abrasive contact, but this contribution to lubrication is far from a uniform liquid film throughout the area of grain-workpiece contact. The modeling of elasto-plasto-hydrodynamic lubrication for real abrasive processes represents a future objective for research.[25]

14.3 VISCOSITY

In hydrodynamic and elastohydrodynamic lubrication, viscosity characteristics strongly influence fluid behavior in the contact area.

Viscosity is a fluid property representing the internal resistance to shear in a fluid layer. Viscosity is defined in terms of dynamic viscosity or kinematic viscosity.

14.3.1 Dynamic Viscosity, η

According to Newton's law of viscosity,[19] dynamic viscosity is the coefficient of proportionality between the shear stress, τ, of the fluid and the rate of shear du/dy:

Eq. (14.2) $\tau = \eta \cdot du/dy$

In the SI system, the basic unit of absolute dynamic viscosity, η, is the poise and is equivalent to units of Ns/m^2.

14.3.2 Kinematic Viscosity, ν

Kinematic viscosity is defined as dynamic viscosity divided by fluid density, ρ:

Eq. (14.3) $\qquad \nu = \eta/\rho$

The basic unit of kinematic viscosity, ν, is the Stoke equivalent to units of m^2/s.[20] In industry, other relative units are also used for fluid viscosity, for example, Engler degrees (°E), Saybolt seconds (SUS), and Redwood seconds.

14.3.3 Dependence of Viscosity on Temperature

In the hydrodynamic regime, viscosity of oils is strongly influenced by fluid temperature. For neat oils, the dependence of viscosity on temperature[21] is satisfactorily represented by Walther-Ubbelohde's empirical relationship:

Eq. (14.4) $\qquad \log \log (\nu + c) = a \cdot \log T + b$

where a, b, and c are constants, specific to a particular type of oil, and T is the absolute temperature.

In a double-logarithmic diagram of viscosity against temperature, the result is a fan of straight lines. The lines meet at a point for a particular oil. In order to simplify the presentation of viscosity values at different temperatures, manufacturers provide nomograms for different fluid groups. However, these viscosity-temperature relationships change due to oxidation-ageing of the lubricating oils.

14.3.4 Viscosity Index for Temperature

A viscosity index provided by a manufacturer is used to correct the viscosity of an oil depending on the viscosity-temperature relationship compared with those of two standard oils. In industry, Dean and Davies viscosity indexes (VI) are often used. The viscosity index compares the kinematic viscosity of an oil (v) with those of two reference oils (v_1 and v_2). For details about the method for calculating viscosity indexes, see the ASTM Standard on petroleum products (D567-53).

Eq. (14.5) $VI_{(v)} = 100 \cdot [(v_2 - v)/(v_2 - v_1)]$

Additives known an "VI-improvers" are used to reduce the influence of temperature on viscosity (see Sec. 14.8.2). Values of viscosity index for process fluids typically range between 0 and 120.[22]

14.3.5 Viscosity Index for Pressure

In EHL, viscosity is greatly increased by pressure and the effect of temperature is relatively unimportant. The dependence of viscosity on pressure has an exponential relationship expressed by the Barrus equation:[23]

Eq. (14.6) $\eta_p = \eta_o \cdot e^{\alpha p}$

where α is the pressure-viscosity coefficient and η_o and η_p are the dynamic viscosity values at normal and high pressure, respectively.

Weibull expressed the viscosity-pressure relationship in logarithmic form, where P stands for pressure:

Eq. (14.7) $\ln \eta_p = \alpha \cdot P + \ln \eta_o$

14.4 FRICTION COEFFICIENT IN MIXED/ BOUNDARY LUBRICATION

Friction between the tool and workpiece in the elasto-plasto-hydrodynamic lubrication regime that applies in abrasive contacts derives from three main causes.[33] The friction coefficient for wet abrasive machining can be expressed as a sum of several components.

Eq. (14.8) $\mu_a = \mu_{\text{fluid}} + \mu_{\text{solid}} + \mu_{\text{deformation}}$

However, adding friction coefficients is mathematically wrong, as argued in Sec. 5.10.2, Ch. 5, "Forces, Friction, and Energy." The friction forces add up to make the total friction force and the normal forces add up to make the total normal force. Mathematically, $(A + B + C)/(E + F + G)$ is not equal to $A/E + B/F + C/G$. Therefore, the individual friction coefficients should not be added directly. For this reason, Eq. (14.8) is included for qualitative explanation only:

- μ_{fluid} is determined by internal friction from the fluid film, characteristic of hydrodynamic lubrication. The hydrodynamic friction coefficient tends to be extremely low compared to the other two terms. The fluid friction depends on the value of the generalized Sommerfeld number. The hydrodynamic effect is maximized by high values of the combined surface speed $(u_1 + u_2)$ and of the fluid viscosity, and a low level of fluid contact pressure. The unintended creation of hydrodynamic lubrication in wet abrasive machining due to the very high peripheral speeds of the grinding wheel may cause major difficulties in processing. The hydrodynamic force tends to push the abrasive tool away from the workpiece and this force must, therefore, be reduced.[34] Limiting the hydrodynamic pressure generated in the contact area is very important for workpiece quality in grinding.[35] It is often necessary to limit the supply of fluid to the contact area in internal grinding where this problem tends to be most significant. Internal grinding is susceptible to the generation of hydrodynamic pressure due to the conformal converging wedge shape at the inlet to the grinding contact.

- μ_{solid} refers to the friction force generated by the mechanical contact between the asperities of the constituent elements of the friction couple. This is the main term for rubbing between two surfaces and is much higher than the hydrodynamic coefficient, but lower than the deformation coefficient.

- $\mu_{deformation}$ depends on external and internal solid friction, due to the elasto-plastic and shear deformation processes of the softer workpiece material, as it comes into contact with the hard asperities of the abrasive tool. The coefficient for the main deformation is higher than the other two. An energy-efficient cutting process, therefore, correlates with a high force ratio as explained in Ch. 5 "Forces, Friction, and Energy."

The fluid tends to lose its uniform and isotropic properties, due to temperature gradient and high pressures in the contact interface. It becomes difficult to determine the properties of the interjacent film within the mixed/boundary regime, as it depends on the degree of direct contact between couple elements.

Consequently by diminishing S_0, the friction coefficient gradually increases to the value in dry abrasive machining. The most disputed point of the Stribeck curve is its intersection with the ordinate axis. Depending on the properties of the couple elements and on the machining parameters, this point may be situated in any of these regimes, i.e., dry, boundary, or mixed.[36]

14.5 CLASSIFICATION OF PROCESS FLUIDS

Fluids can be classified according to the field of application, such as grinding, profile grinding, cut-off, honing, superfinishing, preservation-fluids, etc. We can subdivide these classes[37] by also specifying the processed materials, i.e., steel, superalloys, titanium, ceramics, glass, minerals, etc.

In abrasive machining, the use of gases (air, argon, helium, nitrogen, freon, or carbon dioxide) as the process fluid is very rare.[38] Most process fluids are liquids and are classified according to the nature of the base liquid, i.e., water-immiscible fluids and water-based fluids (Table 14.1).

14.6 NEAT OILS

Neat oils contain base oils and a mixture of mono- and/or polyfunctional additives. Neat oils can be subdivided according to the nature of the base oil. The nature and specific properties of base oils are discussed according to their origin (Table 14.1).

Table 14.1. Classification of Process Fluids, According to the Basic Liquid

CLASSES	SUB-CLASSSES	GROUPS
Neat oils (water-immiscible oils)	Natural fatty oils	Animal oils, Vegetable oils, Fish oils
	Mineral oils	Paraffinic oils, Olefinic oils, Aromatic oils
	Synthetic oils	Hydrocarbon structure oils, Complex structure oils
Water-base fluids	Water solutions	Mineral salt solutions, Synthetic solutions
	Water emulsions	Oil emulsions (soluble oils), Semisynthetic emulsions

14.6.1 Natural Fatty Oils

Natural fatty oils include animal, vegetable, and fish oils. These oils are made up of different esters (triglycerides) of fatty acids to approximately 95% by proportion. The remaining 5% consists of free fatty acids. Natural fatty oils have good lubricating qualities under relatively reduced loads. The stability or the service life of these fluids depends on the nonsaturation degree (the iodine index), which is proportional to their reactivity. Thus, fish or vegetable oils have a high degree of nonsaturation, being characterized by the highest iodine indexes and have lower stability than animal oils.

Oxidation of natural fatty oils occurs quite easily and results in the emergence of deposits and corrosive fatty acids. Fatty oils are very good wetting agents for metallic surfaces, displacing water, which tends to be emulsified. Natural fatty oils may be mixed with mineral oils to act as friction modifier additives. Mixing natural fatty oils with mineral oils to increase lubricity is called *compounding* of lubricants.

14.6.2 Mineral Oils

Mineral oils are a mixture of liquid hydrocarbons obtained from crude oil by different methods of distillation and refining. The proportion of mineral oils compared to total crude oil production is quite low, i.e., about 1.5% and only about 15% out of this proportion represents mineral oils used as process fluids.[39]

The structure of mineral oils is complex. In addition to liquid hydrocarbons with a wide range of molecular and specific weights, there are impurities of sulphur, oxygen, and nitrogen compounds. Thousands of different organic compounds have been identified in the composition of crude oil. Nevertheless, the percentage composition of most types of oil is quite uniform (Table 14.2).

According to the nature of the oil fields, the prevailing properties of the hydrocarbon mixture belong to one of three main subgroups:

- C_P – linear or ramified saturated hydrocarbons, that is, paraffinic and isoparaffinic.

- C_N – cycloparaffinic hydrocarbons, that is, naphthenes.

- C_A – nonsaturated hydrocarbons with cyclo-aromatic and mixed structure.

Table 14.2. Elementary Composition of Mineral Oils

CONSTITUENT ELEMENTS	% - WEIGHT	COMPOUND TYPE
Carbon	82–87	Paraffins, naphthenes, aromatics, olefins (traces)
Hydrogen	10–14	
Oxygen	0–7	Acids, phenols
Sulfur	0.01–7	Disulfides, mercaptans, tiophenol, H_2S (traces)
Nitrogen	0.01–2.2	Amines, pyridine, quinoline
Metals (organo-metallic compounds)	< 0.05	Si, Fe, Al, Co, Mg, Ni, Na, V, Cu, etc.

These subgroups are all found in any base oil. The composition of an oil may be expressed by the proportions of the constituent subgroups:

Eq. (14.9) $C_P + C_N + C_A = 100\%$

The composition of the base oil determines important properties of a resulting process fluid, including viscosity, lubricity, and service life.* The most valuable stock oils are predominantly paraffinic (proportion up to 65%).[17] These oils are relatively stable from a thermal and chemical point of view and have optimum viscosity properties (high viscosity indexes, i.e., VI $\geq$ 70).

Naphthenic hydrocarbons, especially the aromatic ones, are more sensitive to chemical reactions and to pyrolitic decomposition. These chemical and tribochemical degradation processes reduce the efficiency and the service life of process fluids.

Mineral oils must be carefully hydrogenated to destroy carcinogenic polycyclic aromatics.

14.6.3 Synthetic Neat Oils

Synthetic neat oils consist of structural groups and are discussed in the following sections.

Synthetic Neat Oils with Hydrocarbon Structure. Synthetic hydrocarbon oils have similar tribological properties to those of super-refined mineral oils. This type of synthetic oil is a mixture of pure hydrocarbons, which have good viscosity-temperature properties and are sufficiently stable against oxidation and high temperatures. Due to the purity of synthetic hydrocarbons, the base oils have rather low solubility with additives.

Synthetic Neat Oils with a Complex Structure. In general terms, synthetic base oils may be considered as new products aimed at applications where natural oils are not entirely satisfactory. Many liquid organic synthetic compounds have been tested to establish the possibility of use as lubricants. Research on synthetic oils has focused mainly on the possibility of exceeding the thermal limit of 200°C for mineral oils. For this

---

* Nevertheless, the petrochemical manufacturing industry does not assign the same importance to the nature of the oils on which process fluids are based, as it does for the production of lubricants for other application domains.

reason, some compounds such as polyphenilesters have been synthesized for which the threshold exceeds 350°C.[19] Attempts have also aimed to improve viscosity properties for synthetic fluids. Besides the floride alkylesters, a group of oils have been synthesized with P, O, N, and S in their complex structure, that yielded very good results.[40] Representative groups of synthetic base oils include:

- *Organic esters*: A large family comprising esters of dibasic organic acids, esters of polyalcohols, polyphenilesters, and esters of phosphoric acid. Most of these compounds have good thermal and oxidation stability. The lubrication stability is also acceptable.[41]

- *Silicium-based compounds*: A family comprising silicones, silans, and silicates. These compounds have excellent viscosity properties and good thermal and oxidation stability up to 300°C. The lubrication quality in the absence of additives is weak.

- *Halogenate organic compounds*: These compounds include halogenated polyaryls and fluorocarbons (low VI indexes, high thermal stability, and oxidation resistance) and fluorochlorocarbons (very good boundary lubricating qualities).

Table 14.3 compares the application characteristics of synthetic oils and mineral oils. The use of synthetic oils is limited by cost, as these fluids are at least ten times more expensive than mineral oils.[19] Development of these fluids is aimed at synthesizing biodegradable oils to overcome environmental problems.

14.6.4 Classification of Neat Oils by Additives

Finally, neat oils can be classified by the use of additives,[42] and they are divided into twelve groups according to the nature of the additives. Additives determine the applications suitable for these fluids (Table 14.4). Additives are obtained by chemical treatment of base natural oils, either mineral or fatty. Tribological additives are supplemented with other essential agents with anticorrosion, detergent, anti-oxidant, and antifoam properties. In Secs. 14.9 through 14.13, a comprehensive classification of additives is presented, illustrating the diversity of active substances.

Table 14.3. Comparison Between the Properties and Prices of Synthetic and Mineral Oils

OIL SORT	A	B	C	D	E	F	G	H	I	J	K	L	M
Paraffinic mineral oil	o	o	o	o	o	o	o	o	o	o	o	o	1
Polyolephine	+	+	o	o	o	o	+	o	+	o	o	+	5
Alkyl-benzene	o	+	-	+	o	o	o	o	o	-	o	-	5
Polyphenyl	-	+	-	+	o	o	o	o	-	-	o	-	10
Typical organic ester	+	+	-	-	-	+	+	o	+	-	o	+	10
Silicone-oil	+	+	+	+	-	-	o	o	o	o	-	+	25
Silicate-ester	+	+	o	o	-	o	o	o	o	o	-	o	20
Phosphate-ester	-	o	o	-	-	+	o	+	o	-	o	o	10
Polyglycol	+	+	-	-	-	+	+	+	-	+	-	+	7
Fluoride-chloride hydrocarbons	-	o	o	+	-	+	o	+	-	-	o	o	250

Signs: + better than, - worse than, o equal to paraffinic mineral oils.

Note: A - viscosity-temperature relation
         B - behavior at cold
         C - oxidation stability
         D - thermal stability
         E - hydrolysis stability
         F - wear protection
         G - volatility
         H - fire resistance
         I - additive response
         J - lubricant compatibility/elastomers
         K - lubricant compatibility/mineral oils
         L - toxicity
         M - relative cost

Table 14.4. Classification of Neat Oils According to EP-Additives Formulation and Application

GROUPS	TRIBOLOGICAL-ADDITIVE FORMULATION	APPLICATION
A	Mineral oils without polar- or EP-additives	Superfinishing of non-ferrous metals
B	Mineral oils compounded with fatty- or synthetic-oils (effective friction modifiers)	Superfinishing of non-ferrous metals
C	Mineral oils with sulphurized fatty oils (good EP-properties and low Cu corrosivity)	Finishing of non-ferrous metals
D	Mineral oils with sulphurized mineral oils (good EP-properties and high Cu corrosivity).	Inexpensive process fluids for machining ferrous metals
E	Mineral oils with sulphurized mineral and fatty oils (good EP-properties and high Cu and its alloys corrosivity).	Cheap fluids for machining ferrous metals
F	Mineral oils with chlorinated paraffin (good EP-properties at relatively low work-temperatures)	Fluids for abrasive machining of nickel-alloys
G	Mineral oils with chlorinated paraffin and sulphurized fatty oils (good synergistic effect of additive combination)	Process fluids for a large application field
H	Mineral oils with chlorinated fatty oils (good EP- and lubricity-properties)	Fine machining of ferrous and non-ferrous metals, i.e., excellent honing and grinding oils
I	Mineral oils with chlorinated fatty oils and suphurized fatty oils (very good corrosion inhibition and deactivation properties)	A very large application field, a long life at relative high temperatures, without chemical corrosion
J	Mineral oils with sulphochlorinated fatty oils (excellent EP-properties, but associated with mechano-chemical corrosion problems)	Fluids applied for difficult to machine materials, such as stainless steel
K	Mineral oils with sulphur and phosphorus compounds (AW-additives, low mechano-chemical corrosion)	Machining of different kinds of steel workpieces
L	Mineral oils with phosphorus compounds (relatively low EP-effects, no mechano-chemical corrosion)	Machining of titanium-alloys

14.7 WATER-BASED FLUIDS

14.7.1 Physical Aspects

Water is the first machining fluid used by man. It is still widely used for cut-off and drilling operations and even for grinding. Water has a better cooling ability than oil and from an environmental point of view, it is almost ideal. But, due to its low viscosity; inferior lubricity; and anticorrosive, antirusting, and antibacterial properties, water is less than ideal. Machines that use water without efficient rust inhibitors have to be protected against damage to slideways. Emulsions, i.e., oil plus water, combine some of the advantages of both water and oil,[43] but this symbiosis is never completely successful.

Water-based fluids are more widely employed than neat oils with a wider range of formulations. The only common element in this class is the water used to dilute the different substances, that make up the concentrate. The best results with water-based fluids are said to be obtained in centerless grinding and in diamond abrasive machining.[8][44] Water-based fluids may be subdivided into the following groups:

- Effective solutions, based on inorganic and/or organic (synthetic) concentrates, that are soluble in water.

- Emulsions, based on oil concentrations which may vary between large limits and on additives that form a steady dispersion phase in water, with different degrees of spreading.

This classification does not permit a clear distinction between the two groups. Most industrial emulsions have a high degree of homogeneity, due to the use of very efficient emulsifying additives. The only case of a true water solution is where the concentrates used are a mixture of electrolytes. In other situations, uniform dispersions are achieved that lie somewhere between rough emulsions and a molecular solution (Table 14.5).

14.7.2 Properties of Water as a Base Liquid

Although water lacks the complex structure of mineral oils, water properties have a significant influence on the properties of process fluids. Access to a quality water supply with a volume of hundreds of cubic meters can be a major problem. It may, therefore, be necessary to install a chemical section in a laboratory for analysis and control of water quality.

Table 14.6 presents minimum quality standards for water used as process fluids.[45] Direct use of tap water without softening and without analysis presents a serious risk of undermining performance and the service life of a process fluid. De-ionized water is recommended for grinding sensitive and difficult-to-grind materials. By ensuring consistently high water quality, it is possible to avoid or reduce the incidence of parasitic chemical and tribochemical processes during abrasive machining.

Table 14.5. Dispersion: Degrees of Different Emulsions

DISPERSION SCALE (μm)	EMULSION TYPE	MACHINING APPLICATION
1000…1	Microscopic emulsion	Plastic deformation, such as: fine rolling, etc.
1…0.1	Rough colloidal emulsion	Rough cutting
0.1…0.01	Middle colloidal emulsion	Fine cutting, medium grinding
0.01…0.001	Fine colloidal emulsion	Fine grinding
< 0.001	Molecular emulsion	Superfinishing

Table 14.6. Standard of Quality for Water as Base Liquid

FEATURES	UNITS	RANGE
Hardness	mmol/l	1.8 to 10.8
pH-value	-	6 to 7
Ions of Cl⁻ (concentration)	mg/l	< 100
Ions of SO_4^{--}	mg/l	< 100
Residue by evaporation	mg/l	< 500
Bacterial density	1/ml	< 10

14.8 WATER SOLUTIONS

A water solution is a homogeneous fluid. Water solution process fluids include mineral salt solutions and water solutions of synthetic organic compounds.

14.8.1 Mineral Salt Solutions

Inorganic salt solutions consist of dilute solutions of simple electrolyte compounds. Mineral salt solutions are used in abrasive machining processes for cooling the process and for preventing rusting and corrosion of machine tools and of processed pieces. The soluble concentrate includes substances such as sodium carbonate to raise alkalinity (pH > 8.0). Alkalinity is the main requirement for reducing the tendency for rusting. The concentrate also contains halogens, nitrites, nitrates, phosphates, and/or borates of metals from the first three groups of the periodic table. The role of these substances is to generate an adherent and thin reaction layer on the workpiece surface. The very fine thickness, compactness, and low resistance to shearing of these layers facilitate abrasive machining and ensure chemical protection of the workpiece by passivation of the surface. The total concentration of both groups of electrolyte, in the water solution does not usually exceed 3.5%. Concerning the environmental aspects of the chemical protective additive, sodium nitrite ($NaNO_2$), this salt can react with a secondary amine and generate a carcinogen, *nitrosamin*.[45]

14.8.2 Water Solutions of Synthetic Organic Compounds

In contrast to emulsifying concentrates, soluble synthetic fluids do not include an insoluble oily-phase in the composition. Synthetic organic compounds are used instead of oils. These compounds are derived from liquids, that are more viscous than water and completely miscible in water, e.g., glycols and polyglycols. The absence of mineral oils causes problems for the inclusion of additives. Most organic additives are soluble only in hydrocarbons and, in the machining process, act exclusively in the presence of hydrocarbons. The development of synthetic fluids is an important theme of research and development activities in the petrochemical industry, although the results obtained so far are relatively modest.

14.9 WATER-OIL EMULSIONS

A water-oil emulsion is the scattered distribution of insoluble oil into water. The distributed oil represents the scattered stage (interior) and the surrounding water, that preserves its continuity, represents the exterior stage. On a broader scale, industrial dispersions for abrasive machining may include not only liquid mixtures, but also foams (gas/liquid) and mist (liquid/gas).[38]

Water-oil emulsions for abrasive machining can be classified into three subgroups and are discussed in the following sections.

14.9.1 Oil-concentrate Emulsions

Oil-concentrate emulsions are called *soluble oils*. The dispersion consisting of mineral or synthetic oils may represent up to 85% of the concentrate. It also contains various additives which are mostly soluble only in oil.

The exterior stage may be either water (oil in water emulsion) or oil (water in oil emulsion) depending on the proportion of the concentrate in the emulsion. Soluble oils are typically general-purpose products, which offer both cooling and lubricating properties.

14.9.2 Semisynthetic Emulsions

The oil content in semisynthetic emulsions is partially replaced by synthetic products which are soluble in water. The percentage of oil in the concentrate is less than 60%; the additives are also partially replaced. the VI-improver, anticorrosive, rust-inhibitor, and other additives are water-soluble. Because they contain less mineral oil, the semisynthetics allow better control of rancidity and microbiological activity. The distinction between the oil-concentrate emulsions and semisynthetic emulsions is rather conventional as the semisynthetic emulsions represent, above all, a tendency towards modernization of the whole group.

14.9.3 Synthetic Emulsions

Synthetic emulsions contain water-soluble organic compounds such as esters. These are in addtion to the additives found in complex synthetic emulsions. Synthetic emulsions have better lubricating properties than other types of emulsions, especially for heavy- or high-speed machining. The concentrate used to obtain the water emulsions mainly contains oil emulsifiers, detergents, anticorrosives, rust-inhibitors, EP-properties, antifoam and antimicrobial additives, and friction modifiers. The most important additive is the emulsifier.

Most process fluids are water based and in the 1980s, water emulsifiable oils represented about 75% of water-based fluids. Emulsifiable semisynthetic oils represented about 20% and synthetic fluids only 5%. Since then, the last two subgroups experienced a slow growth, and the proportion of oil emulsions decreased accordingly.

14.10 THE INFLUENCE OF ADDITIVES

Most industrial additives have an organic chemical structure and additives are required to have the following effects:[42]

- Endow new properties compared with those of the base liquid.

- Enhance the essential and necessary properties of the base liquid.

- Cancel some properties of the base liquid, which could detract from the machining process efficiency.

The concentration of additives in the base liquid may vary from a few parts per million up to several percent, according to the nature and the purpose of the additives. The nature and the proportion of the additives have a decisive influence on the efficiency of the machining process and particularly on the cost. Additives are introduced into the base liquid by intense mixing until the fluid becomes perfectly homogeneous. The simultaneous presence in this uniform chemical blend of so many different substances, i.e., the "right formulation" of the process fluid must permit the harmonization of all component properties, in order to obtain synergetic effects.[46] *Synergy* means that the effect of each component in the additive mixture must be superior to the individual action of each componenet. The components that have antagonistic effects towards the other additives, base oil, or workpiece materials are gradually eliminated by testing.[47]

The additives used in abrasive machining can be classified according to the nature of their functions in relation to the fluid ensemble (Table 14.7).

Table 14.7. Functions of Additives

PROPERTIES	ADDITIVES	EFFECTS
1. Physical		
Viscosity	VI-Improvers	Reduction of temperature dependence
Detergency	Detergents	Flush away fines
Dispersivity	Dispersants	Dispersion of impurities
Foaming	Antifoam	Foam inhibition
Emulsion stability	Emulsifiers	Reduction of separation rate
2. Chemical		
Thermal stability	Thermal stability-improvers	Increase temperature threshold
Oxidational stability	Anti-oxidation	Fluid oxidation prevention
Metal deactivation	Deactivators	Metal catalytic effect inhibition
Corrosivity	Anticorrosion	Corrosion prevention
Rusting	Antirust	Rust prevention
Alkalinity	Alkaline	Corrosion prevention
Ash-content	Ashless	Improve lubricity
3. Tribochemical		
Friction behavior	Friction modifiers	Improve oiliness
Wear behavior	Antiwear	Reduction of adhesive wear of tool
EP-properties	Extreme pressure	Prevention of seizure and scuffing
4. Biological		
Microbial behavior	Biocides	Long time preservation of fluid
Toxicity	Strong clinical control	Non-toxic fluids

14.11 PHYSICAL PROPERTIES OF PROCESS FLUIDS

14.11.1 Density

The density is expressed as the mass of unit volume of fluid at a temperature of 20°C. The values of density can be used to calculate cooling-lubricating circuit characteristics for neat oils only. For water-based fluids, the data refers only to the concentrate. The density of a water solution or emulsion must be evaluated, taking into account the dilution, or preferably establishing it experimentally. Comparing products of different companies, neat oils are found to have densities ranging between $d_{20} = 815$–890 Kg/m^3. Concentrates for water-based fluids have greater values of density, i.e., d_{20} = 940–1150 Kg/m^3.

14.11.2 Viscosity and Viscosity Index

Basic definitions of viscosity and viscosity index were given in Sec. 14.3. Base oil characteristics particularly influence the viscosity dependence on temperature and pressure. For example, a preponderance of paraffinic fraction in a base oil leads to VI improvement for the process fluid. The viscosity characteristics can also be improved by additives of liquid polymers or co-polymers, with the help of different soaps. The values of kinematic or dynamic viscosity are normally given for 20°C (in certain circumstances also for 40°C or 100°C). Thus, for typical neat oils used in honing, the kinematic viscosity $v_{20} = 11$ m^2/s, while for those used in grinding, $v_{20} = 20$–60 m^2/s. The concentrates for water-based fluids have viscosities ranging from $v_{20} = 50$–600 m^2/s.

14.11.3 Color

A light hue demonstrates a well-refined neat oil and possibly a high degree of additivation. Poor refining negatively affects quality and service life of the oil. Neat oil is normally additivated for a prolonged service life of months or years.

.

14.11.4 Transparency

Transparency is an indication of the extent of impurities in an oil. Solid suspensions in a process fluid are usually due mainly to the machining process, but may also be a segregation of paraffinic crystals.

14.11.5 Fluorescence

Fluorescence refers to the specific property of a mineral oil, which changes color according to the angle it is viewed by, and according to the light source. Fatty- and synthetic-oil groups do not exhibit fluorescence.

14.11.6 Detergency

Detergency relates to the cleaning effect of an oil. Detergency relies on specific properties that maintain the contact surfaces of the abrasive tool and keep the workpiece as clean as possible. The efficiency of abrasive machining requires cleaning properties, as well as, preventing the emergence of polymers (varnishes) on the working surfaces. Thus, besides facilitating removal of wear particles and swarf from the workpiece surface, it is also necessary to avoid loading the active surface of the abrasive tool. The most efficient detergents are sulphonates-, phenates-, and phosphonium salts of calcium, barium, or magnesium.

14.11.7 Dispersive Ability

Dispersing additives aim to prevent accumulation of mud deposits in the machining area. Dispersing additives have a particular affinity to loose solid particles, that they surround, maintaining them in suspension. Particular liquid polymers, as well as amines or amides, of high molecular weight are used as dispersing additives.

14.11.8 Foam Depressing

Control of fluid foaming is much more important in grinding than in other machining processes. The foam depresser acts on small air bubbles from the liquid and makes them coalesce into larger bubbles. Large bubbles can be eliminated more easily, without generating foam. Silicone polymers and tributylphosphates are used for foam-depressing additives.

14.11.9 Flash Point

The *flash point* is the minimum temperature at which the volatile part of a mineral oil-based fluid can be ignited on the surface of the liquid. This important characteristic represents the lower limit for the explosive range of air + oil vapors. Commercial fluids used for superfinishing processes have flash points ranging between 120°–145°C, and those for grinding processes range between 145°–210°C. The flash point of neat oils depends primarily on the composition of the stock oil.

14.11.10 Emulsion Stability

Emulsion stability is the ability of the emulsion to resist separating, either at rest or in use. Emulsifiers are surface-active substances that act through adsorption on the separation surfaces of the two phases. Stabilizing additives for the emulsion can be classified as either ionogenic (anionic or cationic) or nonionogenic (nonionic) agents.[42] The stability of the emulsion can be quantitatively expressed by the simplified Stokes relationship. It is represented by the value of the inverse separating speed ($1/v$). The inverse separating speed is directly proportional to the absolute viscosity of the external phase (η), and is inversely proportional to the square of the droplet radius ($1/r^2$) and to the difference in densities of the two phases ($d_1 - d_2$):

Eq. (14.10) $1/v = \eta/[r^2 \cdot g \cdot (d_1 - d_2)]$

The degree of dispersion of the internal phase influences both the emulsion stability and the quality of the finished surface (Table 14.5). The quality of the emulsifiers can be characterized by the value of hydrophile lipophile balance (HLB).[45]

The main emulsifiers are sodium-based soaps, naphthenes or amines, sodium sulphonates or sulphates, and the amine salts of esterphosphoric acids, etc.

14.11.11 Cooling Properties

Ideally, a process fluid ensures rapid convection of heat generated during a machining process, in order to cool the abrasive tool, the workpiece, and the swarf.[48] The cooling properties are mainly determined by the attributes of the base liquid and by the degree of dilution. The convection of heat depends also on the rate of fluid supply, the shape and position of the nozzles, and so forth.

14.11.12 Boiling Point

The *boiling point* is the temperature at which the liquid phase of a process fluid rapidly evaporates. When the temperature of the liquid in the contact area exceeds the boiling point, it is assumed that the cooling function of the process fluid ceases. Afterwards, the grinding process and its thermal stability are seriously disturbed.[49] Neat oils have a boiling point above 300°C, but for water-based fluids the boiling point reduces to almost a third of this value.[50]

Most physical properties of process fluids cannot be influenced by additivation. Physical properties are mainly influenced by control of the structure of the stock oil and by the refining process.

14.12 CHEMICAL PROPERTIES OF PROCESS FLUIDS

14.12.1 Thermal Stability (TS)

Thermal stability (TS) is the ability to withstand machining temperatures and maintain the chemical composition of the process fluid components. Thermal instability of a fluid leads to catalytically and tribologically accelerated chemical decomposition under the application of heat. For more stable base liquids from a thermal point of view (see Sec. 14.5.2), the stability limit may be increased up to 300°C. This superior threshold can be extended to around 650°C by using special TS additives. The most efficient substances are the compounds of type α-phthalonitril and tetrafluoro-phthalonitrile.[51] These special compounds generate surface layers by tribochemical reactions with the workpiece material during the machining process. The boundary layers are stable from the thermal point of view and must have very good tribological properties.

14.12.2 Oxidation Stability

The resistance to oxidation is one of an ensemble of chemical stability properties of the process fluid. Anti-oxidant (AO) additives are combinations that prevent self-oxidation chain-reactions by blocking organic peroxides and by preventing catalytic reactions. The presence of AO additives increases service life of a process fluid and decreases the volume of resins and corrosive substances generated by the machining process. Various AO additives are used to increase oxidation stability including phenol or amine type, organic sulphurs, and various compounds of type zinc-dithiophosphate (Zn-DTP).

14.12.3 Catalytic Effects of Metals

Some metals with a newly formed contact surface, or which are finely dispersed in the fluid, have a strong catalytic effect leading to increased reaction speed and, consequently, to the decomposition of the additives and to oxidation of the whole fluid. The greatest problems are caused by the presence of copper and its alloys. The participation of passivators (metal deactivators) in the fluid additivation formula allows the formation of screen films on the metal surface. The films produced by adsorption on the metal surface prevent catalytic action. Passivators include diamines, organic combinations with sulphur, triarylphosphite, etc.

14.12.4 Fluid Corrosivity

Corrosive properties are due to the chemical action of the additives and of the decomposition products. Neat oil is also corrosive, depending on the materials with which it comes into contact. The aggressiveness of the liquid increases as the process fluid gets older. As with deactivators, the anticorrosive (AC) additives produce a chemical protective film on the metallic surface. Sulphurized and/or phosphorized olefins (terpenes) are used for this purpose.

The problems of corrosive attack are even greater with water-based emulsions. Anticorrosive additives include water-soluble inorganic salts, particularly sodium nitride or sodium tetraboride. Some organic compounds, partially soluble in water, include diethanolamin, esters or amides of boric acid, amides of mono- or di-carbonic organic acids, etc., may also be used. For water-based fluids, it is also necessary to ensure a high alkalinity (pH buffering) for corrosivity control of the liquid medium, i.e., pH = 8 to 9.

14.12.5 Rusting

The presence of water in a process fluid facilitates the formation of galvanic microcells in the pores of the thin oxide layer on a metal workpiece surface. A ring of rust appears around these galvanic microcells and the attack extends below the surface. Antirust (AR) additives build up adsorption barriers of polar-compounds that prevent access of water to the metal surface. As with corrosivity, a high pH of the process fluid in the alkaline domain is required. For most commercial fluids, a value of pH $\geq$ 8.5, is recommended for a dilution of 3–5%. A value of pH > 9.0 for a resin-bonded grinding wheel should be avoided, because higher values reduce tool life.[52] Antirust additives may be fatty acids, ammonium phosphates, sodium (calcium or magnesium) sulphonates, or amino-phosphates (the first two types are ashless additives).

14.12.6 Ash Content

Ash content is the content of noncombustible salts in the process fluid. It may be determined by gravimetric method after calcination of a fluid sample in a melting pot. Additivated neat oils have a greater percentage of ash than nonadditivated oils. The lubricating properties of the fluid are positively affected when the ash content is lower than 2%.

14.13 TRIBOLOGICAL PROPERTIES OF PROCESS FLUIDS

The tribological characteristics of process fluids are introduced here but explained in detail in the next chapter.

14.13.1 Friction Properties

A *friction modifier* (FM) is an active substance with improved wettability that forms films of polar substances with solid lubricating effects on the workpiece surface. The emergence of these tribophysical films during the machining process reduces the cutting forces. For this purpose, animal or vegetable natural fatty oils, fatty acids, or their derivatives,[42][53] and/or fine suspensions of solid lubricants such as graphite and molybdenum disulfide are used.[54]

14.13.2 Wear Resistance

An antiwear (AW) additive is a tribochemical active substance. The term *tribochemical* refers to the chemical action that takes place in a rubbing/abrasive contact and which may affect and be affected by the rubbing/abrasive action. The AW additive develops a reaction layer on the workpiece surface during the machining process. The reaction products are easily deformed and removed, reducing the loading of the abrasive tool. The effect of the tribochemical action of the AW additives is to make the machining process easier and to reduce abrasive tool wear. The tribochemical mechanism is particularly important for optimizing abrasive machining of difficult-to-grind materials and for prolonging the microprofile of the cutting edges and the macroprofile of the abrasive tool. The main AW substances are tricresyl-phosphate (TCP) and different types of Zn-DTP.[55]

14.13.3 Extreme Pressure (EP) Properties

Extreme pressure (EP) additives react with metallic workpiece asperities above a high-energy threshold. It prevents the transfer of microchips onto the tool surface and the formation of a built-up edge on the abrasive grains. These welding processes reflect a tendency to seizure of the friction pair.[18] The efficiency of EP additives lies, mainly, in the prevention of loading and scoring of the tool abrasive layer. Extreme pressure additives are organic compounds with P: Cl, or S, including chlorinated paraffins or aromatics; di- or polysulphurized organic compounds; sulphurized mineral oils; esters of aryl(alkyl)phosphoric acid; chlorinated and/or sulphurized fatty acids or olefins; esters of phosphoric acid; polyalkylene glycol; etc.[56] Extreme pressure additives with various properties are used both for neat oils and for water-based fluids.

Extreme pressure additives are generally soluble only in oil and act directly from this phase. In aqueous media, the inorganic layers of chlorides and sulfates act as mild EP agents.[57] For water-soluble additives, e.g., esters of phosphoric acid or polyalkylene glycol, the solubility into the thin layer of lubricant is reduced at the high temperatures generated on the workpiece surface.[58] Water-soluble EP additives exhibit weak activity that explains the low efficiency of soluble fluids in deep grinding or in honing processes.

14.14 BIOLOGICAL PROPERTIES OF PROCESS FLUIDS

Process fluids are required to have a long service life in the factory environment. Consideration must be given to fluid tank cleaning, cooling, and transport, and it is important to take microbiological stability into account.[59] Various physical and chemical methods have been employed for fluid preservation.[60] Only a few antibacterial (AB) substances, i.e., chemicals with biocide/fungicide properties, have proven to be sufficiently safe and inexpensive. Antibacterial additives have to fulfill a large range of techno-economic conditions:

- Quick action.
- Wide application domain.
- Long service life.
- Chemical inertness towards other additives or metals.
- Thermal stability.
- Constancy of the tribological and cooling properties.
- Stable pH value.
- Good solubility in the base liquid.
- Tolerability from the ecological and environmental point of view.
- Cost savings.

Most AB additives did not fulfill all these conditions so new additives have been introduced, i.e., derivatives of formaldehydes (O-Formale and

N-Formale), nitroderivates, organic compounds with sulphur and chlorine, and phenols and their derivatives. The efficiency of AB additives is monitored by periodic measurement of the minimum inhibitory concentration (MIC) values,[45] and by using DIP-slides (bacterium number/ml).[40] Together with the analysis of the microbial content, the toxic effects of additives on the human body are also controlled (i.e., the acceptability),[58] as well as the damaging effect of waste fluids on the environment.

14.15 DEGRADATION OF FLUID PROPERTIES DURING OPERATION

During the machining process, the fluid undergoes a slow, but continuous, alteration of its composition and properties due to tribomechanical and thermal actions, as well as the effects of the environment. In this respect, a process fluid is subject to a wear process and has a service life determined by the working conditions.[48] The main modifications to a process fluid are:

- Increase of fluid density and viscosity due to oxidation, cracking, pyrolysis, and hydrolytic processes suffered by its constituents; consequently, the diminution of the viscosity index.

- Darkening color and diminishing transparency due to the growth of impurities (wear particles, chips, oxidation products, tramp oil, decomposition products, impurities of the environment, etc.).

- Increase in response time of the fluid constituents, owing to a decrease in the concentration and chemical activity of the additives.

- Deterioration of the emulsion qualities (stability, concentration, alkalinity) and increase of its electric conductivity; the decrease of its AC protection.

- Decrease of fluid lubricity, as well as of its antiseizure EP properties.

- Increase of bacterial content, which is a sign of bulk biological contamination.

For longer life of the process fluid in operation, it is necessary to carry out periodic controls by analytical and functional methods. Laboratory investigation starts by collecting fluid samples at the outlet from the fines clarification unit. Where possible, an analysis and control laboratory should be managed by a lubricants department; this is only possible for medium or large companies.

Ideally, laboratory equipment should be setup with two distinct aims:

- To analyze the physico-chemical and biological properties of the fluids. These properties do not depend on the tribosystem.

- To determine the fluid tribochemical and tribomechanical characteristics, which ensure optimum functioning of the machining process by modeling the characteristics of the process.

14.16 ANALYSIS OF PHYSICO-CHEMICAL AND BIOLOGICAL PROPERTIES

The properties of the fluid determine its performance under machining conditions. In Table 14.8, some testing methods are enumerated for determining fluid properties and for carrying out the essential corrections.

Because there are so many variables, any variations in the fluid properties may be critical. An important criterion for a good fluid is reliability.[38] Fluids that quickly deteriorate should be sampled and analyzed more frequently. On the basis of these analyses, the composition of the process fluid can be corrected and restored to the initial values of fresh fluids.

14.16.1 Water-based Emulsion and/or Solution Characteristics

Due to their dispersed structure, water-based fluids are much more sensitive to chemical degradation (decomposition, oxidation, and/or hydrolysis) in machining than the immiscible oils. In Table 14.8, the first five points refer to physico-chemical characteristics of these fluids, including Point 4 with regard to fluid electro-conductivity. By increasing salt concentration as the fluid degrades, the electrical conductivity increases. Exceeding the upper limit for conductivity requires partial evacuation of the degraded fluid from the cooling-lubricating circuit and its replacement with fresh fluid. Water-based fluids also require periodic analysis of hardness and of other characteristics (Table 14.6).

Table 14.8. Process Fluids Physico-Chemical and Biological Properties

POINT	PROCESS FLUID PROPERTIES	LABORATORY TESTING METHODS	OBSER-VATIONS
1	Emulsion concentration	Measuring flask for soluble oils Titration kit Refractometry method	Water-based fluid
2	Emulsion stability	Separatory funnel Centrifuging method	Water-based fluid
3	Alkalinity	Test paper indicator pH metering	Water-based fluid
4	Specific electrical conductivity	Electrical conductivity cell	Water-based fluid
5	Fluid cleanliness: Color Limpidity Filterability.	Colorimetry (8 degrees) Transparency Fluid drop on filter paper Insoluble content (including ash content)	All process fluid types
6	Hardness of water	Water analysis	Water-based fluids

(Cont'd.)

Table 14.8. *(Cont'd.)*

POINT	PROCESS FLUID PROPERTIES	LABORATORY TESTING METHODS	OBSER-VATIONS
7	Corrosion inhibition	Copper strip test Steel strip test Chips on cast iron plate Chips on filter paper Humidity cabinet Chem. gravimetric method Electro-chemical method Radiotracer method	All process fluid types
8	Heat transfer to fluid	Measuring device of heat transfer rate	Water-based fluid
9	Thermal reactivity	Thermographic differential analysis (TDA) Hot wire test apparatus	All process fluid types
10	Bacterial level and resistance	DIP-slide method Corrosion resistance	Water-based fluid
11	Dermatological- and physiological- properties (toxicity)	Clinical investigation on tissue	All process fluid types

14.16.2 Corrosion Inhibition

The initial specific corrosivity should be determined as a reference, as this is the first indicator of the EP and AW additive activity. The increase of corrosivity with time is also due to gradual decomposition of the AC additives. Due to the complexity of the phenomena, most of the methods of analysis have a highly qualitative character and are based on visual estimations (comparison with standards). Quantitative methods of measuring corrosivity (see Point 7) include chemical or electro-chemical methods, as well as the method of measuring the process fluid corrosivity with the help of a radioactive tracer.[61]

14.16.3 Heat Transfer Rate

The convection parameter "h-value" (W/m^2K) is an important characteristic of a process fluid, related to its capacity for cooling (see Point 8). The determination of the magnitude of the convection parameter for the investigated fluid, in contact with a particular material (heated at a temperature, θ_w) can be affected with the help of equipment designed and built at Okayama University (Japan).[62] For emulsion-type fluids, the h-value decreases with the increase in the proportion of oily-concentrate within the emulsion. This means that increase in the concentration of active substances reduces the dissipation of thermal energy during the machining process. This tendency is contrary to the beneficial tribological influence of the oil and additives.

14.16.4 Thermal Reactivity of the Tribosystem

Thermal reactivity of the tribosystem includes the thermal physico-chemical processes, which take place at the interaction of the cutting edges on the workpiece, the fluid, and the environment. These phenomena can be investigated by thermographic differential analysis (TDA) with a device called a derivatograph (see Point 9).

The TDA method is based on the determination of mass-variations of the system components contained in a closed volume. Physico-chemical interactions take place between the system components, as a consequence of constant rising of temperature (T) with a gradient of 20°C/min. These changes are accompanied by alterations of the system enthalpy; the variations are recorded and compared to a standard gauge. As a result, the differential thermal analysis curve is obtained and this allows identification of the exo- and endothermal processes for the tribosystem (Fig. 14.3).

By using the TDA method, the reciprocal action of the gaseous environment and of the other tribosystem elements can be studied. By combining the TDA method with other physical procedures, the thermo-chemical behavior of the process fluid in contact with the elements of the friction couple and the environment can be studied.[63]

Another method for investigating thermal reactivity is the use of the hot wire test apparatus.[64] This method determines the variations of electrical conductivity of a heated metal wire submitted to chemical attack by the process fluid.

A criticism of such methods is that the test cannot precisely reproduce the development of the actual machining process, particularly of the abrasive contact. However, such methods do provide comparative information for different fluids.

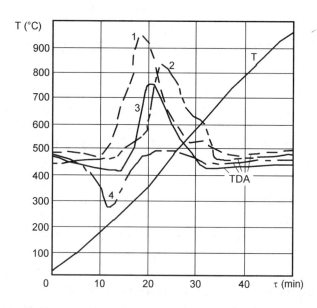

Figure 14.3. Thermodiagrams obtained by TDA-method for some process fluid types: (1) fatty alcohol; (2) petroleum; (3) neat oil; (4) dimethylsulphoxide; (T) temperature increase (°C); (τ) heating time (min).[11]

14.16.5 Biological Characteristics

Aerobic and/or anaerobic bacteria radically affect and even destroy the stability and other technological properties of all water-miscible process fluids. The most damaging are the aerobic bacteria, which seriously degrade the tribological and chemical properties. Periodic inspection of fluids is necessary to monitor growth of bacteria and the need to correct the biocide concentration (see Point 10, Table 14.8). If the correct levels are not maintained, the process fluid becomes highly infected and so corrosive that disposal of the entire fluid in the system becomes necessary. The quality control laboratory should also examine the potential effects of fluid specific toxicity for human beings from a health point of view[65] and also have consideration for the wider environment (see Point 11, Table 14.8).

14.17 TRIBOLOGICAL AND APPLICATION CHARACTERISTICS

The main functions of a fluid in the machining process are tribological, i.e., ensuring efficient process lubrication. Fluid ageing and contaminating processes during application diminish the lubricating quality, which is why fluids must be periodically inspected and corrected. The control of the tribological properties may be affected with different levels of complexity.[66][67] In order to carry out tests for the first level of control, a tribometer is recommended. A *tribometer* is an instrument which emulates the friction couple with regard to the material combination and the contact type (see also Ch. 2, "Tribosystems of Abrasive Machining Processes"). Ideally, it should replicate all relevant parameters of the tribosystem including the speed, contact pressure, and rate of fluid supply. This is not usually possible; therefore, test results using a tribometer must be considered as comparative for the test conditions employed. Results obtained using a tribometer are valuable for comparing particular tribological characteristics of a series of fluids.

Using a tribometer, it is possible to quickly compare the lubricity and load-carrying properties of the friction couple with different process fluids. Figure 14.4 shows a tribometer intended for comparative testing of process fluids, developed at the Technical University in Aachen, Germany.[68]

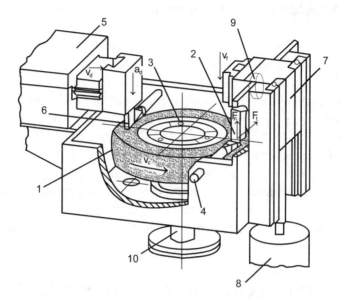

Figure 14.4. Tribometrical equipment for testing process fluid quality: (1) grinding wheel; (2) workpiece; (3) balancing washer; (4) fluid inlet; (5) dressing device; (6) dressing tool; (7) infeed slide; (8) stepping motor; (9) two D-load cell; (10) main spindle with pulley; (F_n and F_t) cutting forces; (v_c) peripheral speed; and (v_d and a_d) dressing parameters.[68]

Table 14.9 presents a classification of the main types of tribometers, indicating literature sources for essential details. With the help of the table, the most suitable tribometer can be selected to model a friction couple appropriate for a particular abrasive machining application. Another possible criterion is that the tribometer chosen should be similar to the tribometer used by the process fluid supplier. This allows the correlation of control-laboratory results for possible complaints. When the tribometer results indicate a level of wear that is too low to be reliably measured by conventional methods, a radioactive-tracer method offers higher sensitivity.[69]

When testing a new fluid or considering a change in machining technology, it is necessary to take the testing to a higher level of modeling (Sec. 2.4, Ch. 2, "Tribosystems of Abrasive Machining Processes"). It is evident that tribometers presented in Table 14.9 do not allow the whole complexity of the machining process to be modeled. Tribometer test results have only a preliminary and comparative character; they can rarely be used to obtain definitive predictions for machining performance of a real process set-up such as grinding.

Table14.9. Modeling the Tribological Processes of Abrasive Machining[36] (See also Ch. 2, Table 2.8)

FRICTION-PAIR CONTACT TYPE	NO. OF CONTACTS	TRIBOMETER TYPES	NOTATION	REFERENCES
Type "D" (dot contact)	1	Single grain on flat	D-1	Prins[70] and Koenig[71]
	1	Rider on flat	D-1	Rabinowicz[72]
	1	Hemisphere/ cylinder	D-1	Kirk[73]
	1	Crossing cylinders	D-1	FALEX (ASTM 2670)
	3	Four balls	D-3	Deckmann,[74] and Cholakov-Rowe[75]
Type "L" (line contact)	1	Cylinder on flat	L-1	Timken, i.e., (ASTM-2782)[46]
	1	Parallel cylinders	L-1	SAE[76]
Type "F" (flat area contact)	1	Pin on disk	F-1	Koenig and Vits[68] and Hrulkov[11]
	1	Pin on flat	F-1	Hong[77]
	1	Pin on band	F-1	Nakayama[78]
	2	Two flats on flat	F-2	Tannert[45]
	3	Three pins on disk	F-3	Thorp[79]
Type "C" (curved area contact)	1	Shoe on cylinder	C-1	Amsler[36] and Nakajima[62]
	1	Honing blade on cylinder	C-1	Salije and See[80] and Chepovitzki[63]
	1	Band on cylinder	C-1	Hrulkov[11]
	2	Two half-bearings on cylinder	C-2	Almen-Wieland[55]

The next stage of testing should be undertaken directly on a machine tool, having a technical specification similar to the machines used in production. The machine tool used for testing should be endowed with adequate measuring devices, as well as having an adequate fluid circuit.[8] The test methodology will be designed to allow optimization of the abrasive machining process including optimization of the process fluid specification. Machining processes that take place in the black box of the abrasive machining tribosystem can be greatly influenced by the fluid properties. The best fluid group may be chosen initially from comparative tests with a tribometer. The fluid specification will be selected and confirmed by further tests on the production type of equipment, with elements of the friction pair at the scale 1:1 with the real machining parameters.

We can also make use of complementary investigations, which may consist of analyzing the sizes, form, and composition of chips and wear particles.[82] The worn surfaces of the abrasive tool may also be investigated. Processed workpieces can be examined for residual stresses or other forms of damage.[83][84] Such methods can throw valuable light on the severity of the abrasive process and sources of problems.

Finally, it is important to consider waste treatment tests; these are the first tests required for a new fluid approval. The tests depend on the waste treatment technology[85] and on government regulations.[86] Today, environmental conservation is of essential importance, affecting individual, industrial, and also national concerns. The main goal should be to assure clean industrial production.[87]

14.18 ADJUSTMENT OF FLUID PROPERTIES IN OPERATION

A trend in larger modern factories is to provide a centralized fluid delivery system. Optimized fluid management can improve control and maintenance of quality of the fluid supply. It can also bring about an important reduction in costs. A possible implication of centralized supply is that there will be a reduction in the variety of fluids employed. This can reduce costs, although it may be decided that some processes may have to operate with a nonoptimal fluid. Centralized fluid delivery implies a more centralized process of decision-making concerning the physico-chemical, biological, and behavioral analysis for selection of fluids.

Measures for controlling quality of the fluid supply may include the adjustment of flow and pressure, the improvement of purification methods, and the total or partial replacement of the fluid. Also, measures for maintenance, i.e., adjustment of the fluid composition and properties, can be taken. By introducing carefully determined quantities of appropriate concentrates into the feed tank, the balance of the constituents can be altered. These constituents supplied by the manufacturers of process fluids can be added based on results from the laboratory, in the correct quantities to reduce the deviation from the standard properties. After performing an adjustment, laboratory analysis should be repeated in order to check the operation. Adjusting the fluid properties is the key to achieving an efficient delivery system.[88] Very often, careful control of fluid quality can obviate the requirement for total replacement of the process fluid.

14.19 SELECTION OF PROCESS FLUIDS

Although there are plenty of multipurpose fluids, most customers prefer fluids proven to be best for particular materials. Productivity, tool life, and workpiece quality are the most important criteria for process fluid selection.[89] These criteria form the main basis for techno-economic optimization of an abrasive machining process (Ch. 2, "Tribosystems of Abrasive Machining Processes").

The first decision is the class of fluid to be chosen. The longest tool life and the lowest operating temperature can be achieved using additivated mineral oils as evident from the literature.[90][91] Except for dry machining, there is evidence that the worst performance is offered by mineral salt solutions (Figs. 14.5 and 14.6).

Increased specific removal rate (Q_w') for a prescribed surface roughness after machining can be achieved by a more suitable fluid application (Fig. 14.7).

Residual stress is an important property of a machined workpiece.[83] In spite of advances in synthetic and semisynthetic fluids, the lowest residual stress level is achieved with EP-additivated, soluble mineral oils (Fig. 14.8).

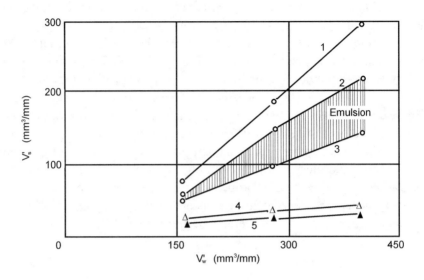

Figure14.5. Influence of fluid characteristics upon the wheel wear, for a grinding couple SiC/alloyed steel: (1) mineral salt solution; (2) emulsion without EP additives; (3) emulsion with EP additives; (4 and 5) neat oils with EP additives. Machining conditions: v_c = 60 m/s; v_f = 30 m/min; Q'_w = 9 mm³/mm · s.[90]

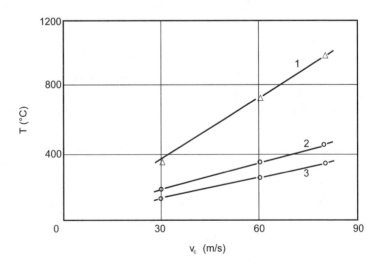

Figure 14.6. Influence of the cooling-lubrication regime on the thermal effects, for a grinding couple Al_2O_3/carbon steel: (1) dry grinding; (2) water emulsion; (3) neat oil. Machining conditions: v_c = 25–80 m/s; v_f = 6 m/min; a = 0.1 mm; Q'_w = 10 mm³/mm · s.[90]

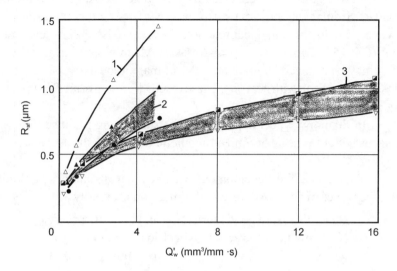

Figure 14.7. Influence of fluid specification and of specific removal rate on the roughness of processed workpiece of chromium steel: (1) AC-salt in water solution (3%); (2) mineral oil emulsion (3–8%); (3) neat oils + additives (0–15%). Machining conditions: $v_c = 45$ m/s; $Q'_w = 0.5$–16 mm³/mm · s.[92]

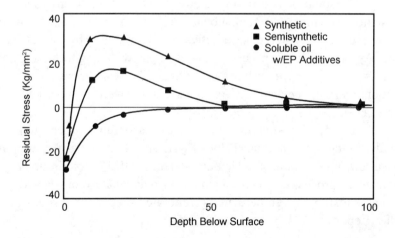

Figure 14.8. Influence of fluid specification on residual stress generated in a workpiece by grinding.[83]

Most of the above conclusions were reached using conventional abrasives. Similar conclusions tend to apply when using superabrasives.[82] However, due to tribochemical reactions between CBN grains and water vapor (Ch. 15, "Tribochemistry of Abrasive Machining"), grinding ratio values may be reduced by up to an order of magnitude using water-based fluids in comparison with neat oils. These results were confirmed[92] in cutter grinding of HSS tools (Fig. 14.9).

In diamond wheel applications for sharpening SC-insert cutting tools, it is better to use an optimized synthetic water solution as the process fluid (Fig. 14.10).

There is ongoing development to replace neat oils and/or soluble oils with other types of fluid. At present, the situation is broadly as follows:

- Neat oils assure good lubrication and long life in service. Neat oils help to prevent wheel loading and corrosion, wheel wear is low, and the quality of the finished workpiece surface is the best possible.[95] Neat oils are preferred for profile grinding, high-speed grinding, CBN grinding, honing, and some lapping operations.

- Water-based fluids offer better cooling of the bulk workpiece material and reduced grinding cost. These fluids are used for most grinding operations and sometimes also for super-finishing processes.

The situation is evolving rapidly due to restrictive environmental legislation at national and international levels.[66] For example, development has been undertaken to replace additivated oil for honing by a water-based fluid. As a result, it was necessary to develop new honing technology and devices compatible with water-based fluids. The new abrasive tool specifications and the special machine settings required were found to double the honing cost.[95]

A study of the situation and implications for future process fluids for metalworking was undertaken by PERA, an international research and consulting organization based in Leicestershire, UK.[96] For the majority of companies, process fluid management is inadequately developed. The recommendations given here are corroborated by the conclusions of the PERA report.

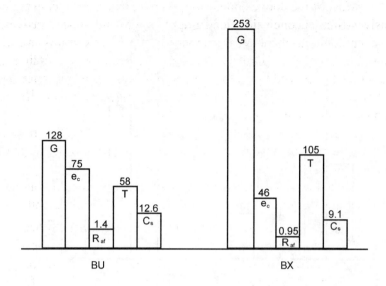

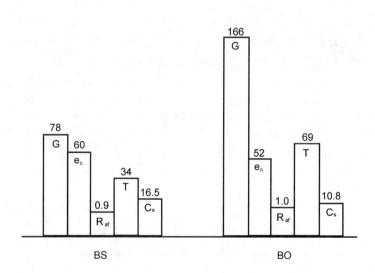

Figure 14.9. Influence of process fluids upon the techno-economic parameters in cutter grinding with couple CBN/HSS, where: BU - dry regime; BX - additivated mineral oil; BS - soluble oil (5%); BO - stock oil. Machining parameters: $v_c = 24$ m/s; $v_f = 2.4$ m/min; $a = 0.03$ mm.[93]

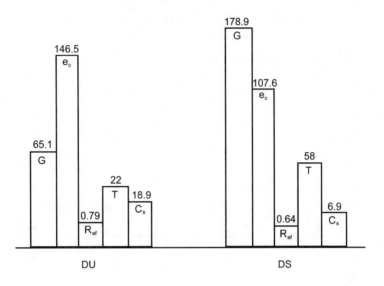

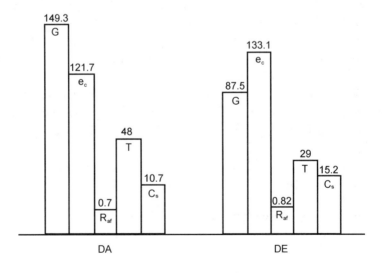

Figure 14.10. Influence of process fluids upon the techno-economic parameters in cutter grinding with couple SD/SC, where: DU - dry regime; DS - synthetic water solution (5%); DA - mineral salt in water solution (1.5%); DE - soluble oil (5%). Machining parameters: v_c = 16.5 m/s; v_f = 1.6 m/min; a = 0.03 mm.[93]

14.20 CONCLUSIONS AND RECOMMENDATIONS

The process fluid fulfills the roles of lubricating, process cooling, bulk cooling, flushing, and cleaning in the abrasive machining process. Surface roughness reduction, tool life increase, and grinding power reduction depend on the lubricating properties of the process fluid. For this reason, the description of the process fluid as a cooling-lubrication liquid used in most European countries is partially justified.

The behavior of the abrasive tool/workpiece pair and its functional regime has not yet been completely elucidated. The idea of mixed or of elasto-plasto-hydrodynamic lubrication has been proposed.

Optimum specification of the process fluid may be based both on manufacturer recommendations, as well as on optimization trials. The fluid specification influences technological and tribological aspects of abrasive machining.

Dry abrasive machining processes should be avoided if possible. This recommendation is particularly important for workpiece quality.

For most abrasive machining processes, neat oils give best results. However, in spite of excellent results, neat oils tend to be used only when other fluids prove inadequate. Water-based fluids, in spite of poorer results, tend to be chosen for economic and ecological reasons.

New process fluids are required for machining advanced materials such as ceramics, composites, and minerals.

Where possible, centralized fluid delivery systems, whose functioning can be submitted to strict quality control, are recommended. Periodic adjustment of fluid composition is required to ensure the initial quality is maintained.

Detailed consideration should be given to biological and ecological aspects concerning the use of process fluids to meet internal, national, and international obligations.

The main goal of process fluid management is the extension of the working life of the tool.

REFERENCES

1. Inasaki, I., Toenshoff, H. K., and Howes, T. D., Abrasive Machining in the Future, *Annals of the CIRP,* 42(2):723–732 (1993)

2. Uhlmann, E., Laufer, J., and Spur, G., Technological and Ecological Aspects of Cooling Lubrication During Grinding Ultrahard Materials, *Proc. Suprabrasives & CVD-Diamond: Theory & Application,* Windsor, Ontario (1998)

3. Blau, P. J., Friction and Wear Transitions of Materials: Break-in, Run-in, Wear-in, Noyes Publ., Park Ridge, NJ (1989)

4. Schey, J. A., Tribology of Metalworking, *ASM Metals,* Park, OH (1983)

5. Brinksmeier, E., Heinzel, C., and Wittman, M., Friction, Cooling and Lubrication, *Annals of the CIRP,* 48(2):581–598 (1999)

6. Brumgard, J. W., Fundamentals of Honing Fluids, *Proc. Symp. Understanding the Basics of Honing,* Dearborn, MI (1993)

7. Buckley, D. H., Definition and Effect of Chemical Properties of Surfaces in Friction, Wear, and Lubrication, *Proc. Int. Conf. Fundamentals of Tribology,* MIT, Cambridge, MA (1978)

8. Dimitrov, B., Grigorescu, A., Gheorghiu, Z., and Dutza, R., Tribologische Bewertung der Fertigungshilfstoffen (FZH) beim Werkzeugschleifen, Schmierungstechnik, 18(5):146–150, (In German) (1987)

9. Hutchings, I. M., Tribology: Friction and Wear of Engineering Materials, Ed Arnold, Sevenoaks, England (1992)

10. Hsu, S. M., Advanced Lubrication Concepts and Lubricants, *Proc. Int. ASM-Conf.,* Gaithersburg, MD (1988)

11. Hrulkov, V. A., Matveev, V. S., and Volkov, V. V., New Cooling-Lubricants for Grinding of Hardprocessing Metals, Mashinostroenije, Moskow, (In Russian) (1982)

12. Komanduri, R., Larsen-Basse, J., Tribology in Manufacturing, *Proc. Int. ASM-Conf.,* Gaithersburg, MD (1988)

13. Lucca, D. A., Brinksmeier, E., and Goch, G., Progress in Assessing Supraface and Subsurface Integrity, *Annals of the CIRP,* 47(2):669–693 (1998)

14. Stribeck, R., Die Wesentlichen Eigenschaften der Gleit- und Rollenlager, VDI-Z, 46:1341–1342, (In German) (1902)

15. Dowson, D., and Higginson, G. R., Elastohydrodynamic Lubrication SI Edition, Pergamon Press, *Int'l. Ser. Materials Science and Technology,* Vol. 23 (1977)

16. Winer, W. O., and Cheng, H. S., Film Thickness, Contact Stress, and Surface Temperatures, Wear Control Handbook, (M. B. Peterson and W. O. Winer, eds.), *ASME*, NY (1980)

17. Luo, J., Wen, S., and Huang, P., Thin Film Lubrication. Part I: Study on the Transition Between EHL and Thin Film Lubrication Using a Relative Optical Interference Intensity Technique, *Wear*, 194(1):107–115 (1996)

18. Dowson, D., Elastohydrodynamic and Microelastohydrodynamic Lubrication, *Wear*, 190(1):125–133 (1994)

19. Wiliams, J. A., Engineering Tribology, Oxford Univ. Press, NY (1994)

20. Anon, Le Systeme International d'Unites (SI), Editee par le Bureau International des Poids et Mesures, Paris, (In French) (1981)

21. Fuller, D. D., Theory and Practice of Lubrication of Engineers, John Wiley & Sons, NY (1956)

22. Crouch, R. F., and Cameron, A., Viscosity-Temperature Equations for Lubricants, *J. Inst. Petroleum*, 47(9):307–3131 (1961)

23. Chu, S. Y., and Cameron, A., Pressure-Viscosity Characteristics of Lubricating Oils, *J. Inst. Petroleum*, 48(5):147–150 (1962)

24. Chepovetski, I. H., Interactions in Mechanic Contacts by Diamond Processing, *Naukova Dumka*, Kiev, (In Russian) (1978)

25. Blaendel, J. S., Taylor, J. S., and Piscotti, M. A., Summary Session Precision Grinding of Brittle Materials, *ASPE Spring Topical Meeting*, Annapolis, MD (1996)

26. Anon, Glossary of Terms and Definitions in the Friction, Wear, and Lubrication – Tribology, Research Group on Wear of Engineering Materials – OECD (1969)

27. Kajdas, C., Harvey, S. S. K., and Wilusz, E., Encyclopedia of Tribology, Elsevier Sci. Publ., Amsterdam (1990)

28. Matveevski, R. M., Problems of Boundary Lubrication, *Tribology Int.*, 28(1):51–54 (1995)

29. Iliuc, I., Tribology of Thin Layers, Elsevier, Amsterdam (1980)

30. Childers, J. C., The Chemistry of Metalworking Fluids, *Metal-working Fluids*, (J. P. Byers, ed.), pp. 165–189, Marcel Dekker Inc., NY (1994)

31. Armarego, E. J. A., and Brown, R. H., The Machining of Metals, Prentice Hall Inc., NJ (1969)

32. Leep, H. R., Metal Cutting Processes, *Metal-working Fluids,* (J. P. Byers, ed.), Marcel Dekker Inc., NY (1994)

33. Moore, F. D., Principles and Applications of Tribology, Pergamon Press, Oxford (1975)

34. Guo, C., and Chand, R. H., Application Adaptive Ceramic Machining, *Proc. Suprabrasives & CVD-Diamond: Theory & Application,* Windsor, Ontario (1998)

35. Campbell, J., Modeling and Analysis of Fluid Dynamics in the Grinding Zone, *Proc. 3rd Int. Machining & Grinding Conf.,* Cincinnati, OH (1999)

36. Pavelescu, D., Tribotechnics, Ed. Tehnica, Bucharest, (In Romanian) (1983)

37. Kalpakijan, S., Manufacturing Processes for Engineering Materials, Addison-Wesley Publ. Co., NY (1991)

38. Siliman, J. D., Cutting and Grinding Fluids: Selection and Application, One SME Drive, Dearborn, MI (1992)

39. Szeri, A. Z., Tribology: Friction, Lubrication and Wear, Hemisphere Publ. Corp., London (1979)

40. Mang, T., Die Schmierung in der Metallbearbeitung, Vogel Buchverlag, Wuerzburg, (In German) (1983)

41. Minke, E., Contribution to the Role of Coolants on Grinding Process and Work Results, *Proc. 3rd Int. Machining & Grinding Conf.,* pp. 13–32, Cincinnati, OH (1999)

42. Bartz, W. J., Additive Fuer Schmierstoffe, Curt R. Vincentz Vlg., Hannover, (In German) (1983)

43. Boothroyd, G., and Knight, W. A., Fundamentals of Machining and Machining Tools, pp. 155–166, 307–309, Marcel Dekker Inc., NY (1989)

44. Sluhan, C. A., Grinding with Water Miscible Grinding Fluids, *Lubrication Eng.,* 26(10):352–374 (1970)

45. Mang, T., et al., Wassermischbare Kuehlschmierstoffe Fuer die Zerspannung, *Expert Vlg.,* Grafenau, (In German) (1980)

46. Dimitrov, B., Aspects of Corrosive Wear in Lubricating Media, PhD-Thesis, Polytechnic Univ. Bucharest, (In Romanian) (1974)

47. Spickes, H. A., Additive-Additive and Additive-Surface Interactions in Lubrication, *Proc. 6th Int. Coll.,* Esslingen (1988)

48. Saint-Yves, M., La Coupe et les Emulsions, *Machine-Outil,* pp. 322–334, (In French) (1976)

49. Jannone da Silva, E., Bianchi, E. C., de Aguiar, P. R., and Gomez de Oliveira, J. F., Optimales Schleifverhalten Durch Verbessertes Kuehlschmierstoff-Management, *Ind. Diamanten Rundschau,* 33(4):350–368, (In German) (1999)

50. Ju, Y., Farris, T. N., and Chandrasekar, S., Effects of Grinding Conditions on Heat Partition and Temperatures in Grinding, *Proc. Conf. STLE/ASME Tribology Surveillance* (1999)

51. Lawton, E. A., Yavrouian, A. H., and Repar, J., Additives for High-Temperature Liquid Lubricants, *Proc. Int. ASM-Conf.,* Gaithersburg, MD (1988)

52. Elendman, M., How Coolants Affect the Performance of Resin-Bonded Abrasive Wheels, *Machinery,* 12:86–90 (1968)

53. Christakudis, D., Friction Modifier – eine neue Klasse von Schmieroeladditive zur Senkung des Energieverbrauches, *Schmierungstechnik,* 16(6):164–168, (In German) (1985)

54. Peterson, M. B., Kanakis, M., Solid Lubricants, *Proc. Int. ASM-Conf.,* Gaithersburg, MD (1988)

55. Dimitrov, B., Lueb, H. A., and Schoon, T. G. F., Eine Verbesserte und Automatisierte Ausfuehrung der Almen-Wieland Maschine, *Schmiertech + Tribologie,* 24(2):40–42, (In German) (1977)

56. Borsoff, V. N., and Wagner, C. D., Studies of Formation and Behavior on an EP-Lubrication, *Lubrication Eng.,* 13(2):91–99 (1975)

57. Jain, V. K., and Shukla, D. S., Study of the EP Activity of Water-Soluble Inorganic Metallic Salts for Aqueous Cutting Fluids, *Wear,* 193(2):226–234 (1996)

58. Sluhan, W. A., Extending of Water Miscible Cutting and Grinding Fluids, *Proc. JSLE-ASLE, Int. Lubr. Conf.,* Tokyo (1975)

59. Bartz, W. J., Tribologie, Schmierung und Schmierstoffe, *VDI-Z,* 121(9):445–455, (In German) (1979)

60. Grisey, R., and Sachoux, P., Conservation des Liquides de Coupe, Machine-Outil, 402:42–43, (In French) (1982)

61. Dimitrov, B., Equipment for Determination of the Corrosivity of Neat Oils and Their Additives, Romanian Licence No. 68637, (In Romanian) (1971)

62. Nakajima, T., and Tsukamoto, S., A Standard for Proper Selection of Water Soluble Type Grinding Fluids, SME Technical Paper MR 88-617 (1987)

63. Chepovitski, I. H., Kizikov, E. D., and Rydzov, J. E., Almaznoe Honingovanije Termo-Obrobotanyh Stalej, *Izd Nauka,* Kiev, (In Russian) (1988)

64. Sakurai, T., and Sato, K., Study of Corrosivity and Correlation Between Chemical Reactivity and Load Carrying Capacity of Oils Containing Additives, *ASLE Trans.,* 9(1):77–87 (1966)

65. Koenig, W., et al., Schadstoffe Beim Schleifvorgang, Schriftreihe der Bundesanstalt Fuer Arbeitsschutz, *Dortmund,* (In German) (1985)

66. Kipp, E. M., and Riddle, B. L., A Guide of the Development of Advanced Metalworking Lubricants, *Proc. 8th Int. Coll. Tribology 2000,* Esslingen (1992)

67. Cholakov, G. S., and Devenski, P., Correlation Between Different Levels of Evaluation of Lubricants, *Proc. 8th Int. Coll. Tribology 2000*, Esslingen, (In German) (1992)

68. Koenig, W., and Vits, R., Pruefung von Kuehlschmierstoffen Fuer die Schleifbearbeitung, Industrie Anzeiger, 72(9):26–28, (In German) (1983)

69. Wagner, K., Die Anwendug von Radionukliden Fuer Verschleiss- und Schmierungs-Untersuchungen, *Isotopen Praxis,* 4(3):85–94, (In German) (1968)

70. Prins, J. F., Wechselwirkung Zwischen Diamanten und Stahl in Einkornversuchen, de Beers Ind. Diam. Div., *Diamant Inf.*, M 24, (In German) (1978)

71. Koenig, W., Steffens, K., and Ludewig, T., Single Grit Tests to Reveal the Fundamental Mechanisms in Grinding, *Proc. Shaw-ASME Grinding Symp.* (1985)

72. Rabinowicz, E., Friction and Wear of Materials, John Wiley & Sons, NY (1994)

73. Kirk, J. A., Cardenas-Garcia, J. F., and Allison, C. R., Evaluation of Grinding Lubricants – Simulation Testing and Grinding Performance, *ASLE Trans.,* 20(4):333–339 (1977)

74. Deckmann, D. E., Jahanmir, S., Hsu, S. M., and Gates, R. S., Friction and Wear Measurements for New Materials and Lubricants, *Proc. Int. ASM-Conf.,* Gaithersburg, MD (1988)

75. Cholakov, G. S., and Rowe, G. W., Lubricating Properties of Grinding Fluids, Part II, Comparison of Fluids in Four-Ball Tribometer Tests, *Wear,* 155:331–342 (1992)

76. Blouet, J., Tribometrie, Le Frottement et l'usure, GAMI Paris, pp. 59–74 (In French) (1967)

77. Hong, H., and Standnik, N. M., Evaluation of High Temperature Lubricants in Ceramic/Metal and Metal/Metal Contacts, *ASLE Trans.,* 36(4):791–735 (1993)

78. Nakayama, M., Kudo, K., Hirose, T., and Iino, M., Experimental Study of Grinding Fluids for Abrasive-Belt Grinding of Stainless Steel, *Tribology Int.*, 20(3):133–143 (1987)

79. Thorpe, J. M., The Novel Tri-Pin-on-Disc Tribometer Designed to Retain Lubricants, *Tribology Int.,* 4:121–125 (1981)

80. Salije, E., and von See, M., Pruefverfahren Fuer Honleisten und Kuehlschmierstoffe Zum Honen, *Industrie Anzeiger,* 79/80(4):46–47, (In German) (1985)

81. Byers, J. P., Laboratory Evaluation of Metalworking Fluids, *Metalworking Fluids,* pp. 191–221, Marcel Dekker Inc., NY (1994)

82. Tripathi, K. C., Nicol, A. W., and Rowe, G. W., Observation of Wheel-Metal-Coolant Interactions in Grinding, *ASLE Trans.,* 20(3):249–256 (1977)

83. Carius, A. C., Effects of Grinding Fluid Type on CBN-Wheel Performance, *Abrasive Eng. Soc. Magazine,* 29(2):22–27 (1990)

84. Mukai, R., and Imai, T., The Effect of Coolant onto CBN Wheel Grinding Performance, *Toyoda Koki Technol. Rev,* 29:3 (1989)

85. Sutton, P. M., and Mishra, P. N., Waste Treatment, *Metalworking Fluids,* pp. 367–393, Marcel Dekker Inc., NY (1994)

86. Lucke, W. E., Government Regulations Affecting Metalworking Fluids, *Metalworking Fluids,* pp. 423–461, Marcel Dekker Inc., NY (1994)

87. Spur, G., and Niewelt, W., Gesundheits- und Umweltvertraegliche Kuehlschmierstoffe Fuer das Schleifen, *ZWF-CIM,* 89:70–73, (In German) (1994)

88. Blenkowski, K., Coolants & Lubricants – the Truth, *Mfg. Eng.,* pp. 3–4, 90–96 (1993)

89. Smith, C. A., Performance of Metalworking Fluids in a Grinding System, *Metalworking Fluids,* pp. 99–134, (J. P. Byers, ed.), Marcel Dekker Inc, NY (1994)

90. Koenig, W., and Vits, R., Einsatz von Kuehlschmierstoffen bei Spanenden Metallbearbeitung, *Proc. 3rd Int. Coll. Schmierstoffe in der Metallbearbeitung,* Esslingen, Germany, (In German) (1982)

91. Koenig, W., Lung, D., and Hennig, B., Kuehlschmierstoffe Fuer die Zerspanung Metallischer Werkstoffe, *Proc. Int. Coll. Tribology 2000,* Esslingen, (In German) (1992)

92. Vits, R., Technologische Aspekte der Kuehlschmierung beim Schleifen, Dr-Ing Thesis, T. H. Aachen, (In German) (1985)

93. Dimitrov, B., Ratziu, G., and Gheorghiu, Z., Kuehlschmierstoffe Fuer Hochleistungsschleifstoffe, *Ind. Diamanten Rundschau,* 28(3):156–160 (In German) (1994)

94. Malkin, S., Grinding Technology: Theory and Applications of Machining with Abrasives, Ellis Horwood, Chichester (1989)

95. Korn, C., Honing Process Change – Oil to Water Base Fluid, *Proc. Conf. Advanced Honing and Superfinishing Clinic,* Cleveland, OH (1999)

96. Mosley, S. E., Archibald, L. C., and Bowes, C., Strategic Developments and Future Implications for the Metalworking Lubricant Market – a Contract Research Perspective, *Proc. Int. Coll. Tribology 2000,* Esslingen, Germany (1992)

15

Tribochemistry of Abrasive Machining

15.1 DEFINITION OF TRIBOCHEMISTRY

Tribochemistry concerns interacting chemical reactions and wear processes. Tribochemical wear is the most complex type of wear.[1] Chemical processes, due to active agents in the process fluid and/or the surrounding atmosphere, belong to the chemical corrosion domain. The friction couple is submitted simultaneously to mechanical action and chemical action. Mechanical action arises due to abrasive movements of the couple elements under the action of the applied forces. Chemical action results from the active substances in the environment and in the process fluid. All these elements are components of the tribosystem (see Ch. 2 "Tribosystems of Abrasive Machining Processes"). Corrosion wear is manifested by formation of reaction products as a result of chemical interactions between the elements of a tribosystem by tribological action.* When resting from mechanical action, interaction between the couple and other elements of the tribosystem is limited to chemical corrosion.

In the following sections, the prefix *tribo* is used to denote the special nature of a mechanical, physical, and chemical process when subject to intense surface deformation in abrasive contact.

---

* German standard DIN 50320

Corrosion generates conversion layers on the surfaces of a friction couple. The mechanical process simultaneously activates micro-areas of contact intensifying reaction processes and progressively removes the reaction layer.[2] The main type of corrosive wear is a mild process in which chemical interaction of the couple elements with the atmosphere and the process fluid is predominant.[3] The relative weight of chemical and mechanical processes, may be different for each particular case.[4]

Corrosion wear is a subject ot tribophysics and tribochemistry. Tribophysics[5] and tribochemistry[6] merge into one another. In the *Encyclopedia of Tribology*,[7] the following definitions are proposed for the two domains:

- "*Tribophysics* is the part of physics that deals with interacting surfaces in relative motion."

- "*Tribochemistry* is the branch of chemistry that deals with the chemical reactions in the friction zone. These reactions cause mechanical and physico-chemical changes on the surface layer of the mating parts. The reactions involved are the result of different types of energy and of catalysis."

Corrosion is a function of reaction time when pressure and temperature are kept constant.

Eq. (15.1) $G_c = f(t_c)$

There are two main types of corrosion.

A *Type C corrosive reaction* is a continuous chemical process, lacking any clear passivating tendency at the surface of the materials. The reaction process has a linear development and tends to proceed unhindered to complete destruction of the workpiece and/or of the abrasive tool (Fig. 15.1a). This type of evolution may be explained by:

- Weak adherence of corrosion products to the underlayer.

- A porous structure.

- Dissolution of the converted layer in contact with the reactive medium.

The corrosion rate (Q_c) remains almost constant. Type C corrosion is considered undesirable in abrasive machining.

A *Type P corrosive reaction* represents a chemical process with a passivating tendency.[8] Corrosion of this type slows down with time and even stops with sufficient compactness and thickness of the conversion layer (Fig. 15.1b). The reaction layer is generated *in situ* on the workpiece surface and plays a protective role. The insoluble surface layer does not permit diffusion of active agents into the underlayer. Consequently, corrosion rate (Q_c) decreases gradually over time, while its value has an asymptotic tendency towards zero. The generation of a passivating layer on the workpiece surface through limited chemical corrosion during rest periods represents an important property of a process fluid and surrounding atmosphere in machining.

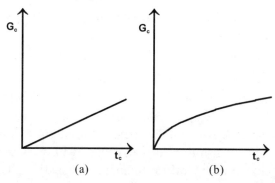

Figure 15.1. Evolution of the quantity of corroded material (G_c) from the workpiece surface vs reaction time (t_c) for: *(a)* corrosion reaction Type C, and *(b)* corrosion reaction Type P.

The kinetics of chemical reaction at a machined surface undergo important changes during dynamic periods of abrasion. The dynamic period is during the loading and relative movements of the couple necessary for material removal. Tribomechanical strains inherent to abrasive processing greatly intensify chemical reactions. The rate of tribocorrosion (Q_t) is, therefore, much faster than the rate of static corrosion (Q_c):

Eq. (15.2) $Q_t \gg Q_c$

Initially, the progression of a tribochemical reaction with time is similar to that inherent to chemical corrosion with passivation.[1] In abrasive machining, the thickness of the conversion layer is drastically reduced.[9]

This reduces passivation. Therefore, the rate of wear depends on the thickness of the conversion layer. The evolution of tribochemical wear changes with time according to the Type C corrosion reaction (Fig. 15.2).

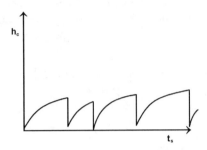

Figure 15.2. Evolution of conversion layer effective thickness(h_c) on the workpiece surface vs machining time (t_s).

A consequence of this evolution is a rapid increase in volume of material removed and rate of material removal. In a study[2] of the behavior of the friction couple SC/Co, the rate of tribochemical reaction was three orders of magnitude higher than the chemical reaction. In this study, the process fluid used for finish abrasion was neat oil with an extreme pressure (EP) additive, chlorinated paraffin. In another study,[10] laboratory tests showed the tribochemical rate increased by five orders of magnitude compared with the chemical reaction rate. Finally, one study[11] found an increase of up to seven orders of magnitude.

Chemo-mechanical technology may be used to increase productivity of abrasive machining.[12] Increased wear of stainless steel pumps used to pump watery slurries is an example of chemo-mechanical effects.[13] The phenomenon of activation by friction of slow chemical reactions was confirmed in work on industrial tribosynthesis of metal-organic compounds.[2]

It must be emphasized that the combined rate of wear in tribochemical reactions is greater than the sum of the rates of wear due to mechanical and chemical processes acting separately. Unusually good removal rates in chemo-mechanical machining may be simplistically explained through weak mechanical resistance of most conversion layers. Conversion layers provide chemical protection for workpieces under static conditions, but are easily removed by abrasive machining. Careful analysis shows that tribomechanical energy greatly accelerates chemical reactions.[14] Judicious use of additives in process fluids (see Ch. 14, "Process Fluids for Abrasive Machining," Sec. 14.4.4) increases productivity and quality in abrasive machining and offers cost savings.

Chemo-mechanical polishing (CMP) is an example of a chemo-mechanical process that may be applied to advanced ceramics. Careful control of tribochemical properties of a process fluid constitutes a main avenue for optimizing a CMP process[15] (see Sec. 15.6.3).

15.2 MODELING A TRIBOCHEMICAL PROCESS

15.2.1 Special Factors in Abrasive Machining

A tribochemical process cannot easily be explained through classical theories of kinetics and thermodynamics intended for static chemical reactions. First, as previously mentioned, a large majority of triboreactions are characterized by very high speeds, some of them even reaching an explosive state.[6] Second, due to tribochemical reaction, products involving less active agents or even inert gases from the environment may be present in the contact zone. From a purely thermodynamic point of view, such chemical reactions would be improbable. A most notable example is tribooxidation of noble metals, in spite of extremely negative values for chemical affinity (A). (Affinity is a criterion in thermodynamic chemistry concerning the possibility of reaction.)[16] This is a physico-chemical parameter considered for the whole temperature range within a triboreaction zone.[11] An example of a tribochemical reaction for a noble metal is

Eq. (15.3) $Au + \frac{3}{4} CO_2 \rightarrow \frac{1}{2} Au_2O_3 + \frac{3}{4} C$

The differences between triboreactions and the usual chemical reactions may be better understood by considering the peculiarities of an abrasive contact.

A real contact area is almost infinitesimal compared with a nominal contact area. The ratio of the two areas may be two to five orders of magnitude.[17] As a result, nominal pressure of a few tens of bars generates much higher real contact pressures in abrasive machining.

Deformation processes occur mainly in the workpiece, due to the rigid crystalline structure of the abrasive grains and their very high hardness. Workpiece strains occur mainly in the plastic part of the elasto-plastic domain. Elastic strains are extremely small compared with plastic strains.

Abrasive machining is characterized by high-energy consumption and relatively low mechanical efficiency.

The short duration contact of abrasive particles with the machined surface leads to quasi-adiabatic exchange of energy at the microcontacts between grains and workpiece.[6] Energy dissipation is analyzed in Ch. 6, "Thermal Design of Processes." Typical durations of energy exchange are given in Table 6.3.

Depending on materials and process conditions, mechanical removal of workmaterial consists of microcutting by plastic flow, fracture, and detachment. These are strongly influenced by friction.

Most of the energy consumed in abrasive machining involves redundant rubbing and ploughing energy. This tribo-energetic component of the total machining energy generates a high excitation state in the contact microzones which activate the subsurface of the workmaterial. Elevated temperatures at the contact interface imply structural transformations resulting from enormous energy concentration in that area. At the same time, material removal causes tribomechanical treatment of the machined surface.

All of the above leads to the conclusion that energetic efficiency of abrasive machining depends not only on mechanical processes, but also on complex effects of temperature and tribochemistry.

15.2.2 The Magma-Plasma Model for Abrasive Machining

Until the mid-20th century, the discipline of chemical kinetics was dominated by the concept that tribomechanical energy influences chemical reactions only by thermal effects. Gradually, based on evidence for tribomechanical structural transformations, a purely thermal excitation model had to be discarded.[18] With some correction, the old model remains valid for corrosion reactions of a static couple or when approaching zero wear rate.[19]

Materials may be highly mechanically activated by machining, in spite of relatively low temperatures measured in the working area. For example, chemical reactivity of cold-rolled nickel is increased by two orders of magnitude in synthesis of nickel-carbonyl:

Eq. (15.4) $Ni + 4 CO \rightarrow Ni(CO)_4$

Tribomechanically activated silicon carbide reacts relatively easily with hydrogen,[20] despite chemical inertness of this abrasive material:

Eq. (15.5) $SiC + 2H_2 \rightarrow Si + CH_4$

The rate of a triboreaction between a mechanically activated solid and a gas depends very little on the pressure of the surrounding medium.[6] This is quite different from the situation with usual chemical reactions.[16] According to the Arrhenius equation,[21] chemical reaction rate increases exponentially with temperature.[8] This is not the case in tribochemical reactions. The influence of temperature on tribochemical reactions is considerably reduced and is sometimes less significant than in usual chemical reactions.[20]

The magma-plasma model was conceived in order to explain the kinetics of tribochemical reaction with a freshly generated surface, taking place simultaneously with abrasive action or immediately after cessation (Fig. 15.3). The key to the model is triboplasma. Triboplasma in abrasive machining has the following characteristics:

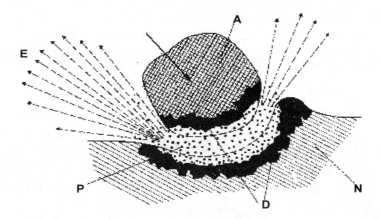

Figure 15.3. Magma-plasma model[6] representing the main phenomena that occur at the workpiece surface in contact with an abrasive grain: N is the normal structure of processed material, D is the modified material structure, A is the abrasive grain, P is the triboplasma, and E is the exo-emission.

- A very short duration, circa 10^{-7} s.

- The possibility of triboplasma is limited locally to microvolumes of contact between the couple elements. The excitation of inner faults, i.e., pores, cracks, etc., and of outer microcontact surfaces, has a significant influence on a tribochemical process.

- A tribochemical process is accelerated by generation of new workpiece surfaces, by abrasive wear. One such process is the so-called Russell Effect,[22] where hydrogen peroxide (H_2O_2) is produced when a freshly machined surface is exposed to moist air. A thin surface layer on a machined surface significantly influences dynamic properties of the workpiece.[6]

- Flash temperature on the "hot spots" of the contact area has questionable importance.[23] Friction force plays a greater activation role in an abrasive process due to a very high level of tribo-energy in the material crystalline lattice. This activation energy of the tribochemical process has to be taken into account. The tribological process causes structural transformation of the workmaterial, such as amorphization, recrystallization, lattice distortion, etc. Each of these material modifications influences reactivity.

- A high level of energetic excitation of the workmaterial in abrasive machining can explain the so-called Kramer Effect.[22] This refers to exo-emission of photons and of fast electrons accompanied by generation of OH^- ions and O^{2-} radicals. The kinetics of triboreaction are affected both by impact effects of exo-emitted particles and by structural transformations at the surface and in the subsurface due to mechanical strains.

- Activity of the surrounding atmosphere and of the process fluid also determines the type, rate, and equilibrium of tribochemical reactions at the friction interface.

15.3 TRIBOCHEMICAL BEHAVIOR OF ABRASIVE TOOLS

15.3.1 General Aspects

Abrasive machining is often considered simply as a microcutting process. Abrasive wheels are likened to milling tools with an aleatory cutting geometry.

Rayleigh and Beilby, around the year 1900, started to consider tribochemical aspects.[24] A systematic study includes not only geometric and kinematic parameters, but also tribochemistry. Since then, the number of unexplained questions have increased:

- Abrasives containing Al_2O_3 or CBN have proved to be techno-economically superior for grinding hard steels into tools containing abrasives with higher toughness, i.e., SiC and SD. This is reversed for nitrided carbon steels.

- Optimal machining of sintered carbides (SC) is assured only by using superabrasive tools containing a friable type of SD and not other tougher sorts.

- Tools including SiC or the toughest types of SD offer the best results for grinding softer materials, such as hard cast iron, nonferrous metals, and a wide variety of nonmetallic materials.

15.3.2 Triboreactions Between Tool and Workpiece

For better understanding of tribochemical processes in the contact between abrasive and workmaterial, it is necessary to consider the nature of the components of an abrasive (Table 15.1).

Abrasive Powder. An abrasive tool consists of abrasive powder together with bonds and other ingredients. The abrasive powders are hard, pulverized materials having a crystalline structure. Abrasives can be grouped according to hardness (see Chs. 11, "Abrasives and Abrasive Tools" and 16, "Processed Materials"). Abrasives can also be grouped by chemical composition according to the presence or absence of carbon atoms in the molecular structure. Machining difficulty is increased for most ferrous metals when using abrasives with a structural carbon content such as SiC, B_4C, or SD. This is due to triboreaction between carbon from the abrasive and ferrous atoms from the workmaterial. A triboreaction results in carbides of the cementite type and of afferent solid solutions. Diffusion of carbon atoms from the abrasive causes rapid tool wear. In order to prevent destructive wear, it is necessary to avoid tribochemical affinity (see Sec. 15.2.1 and relevant literature[11]).

Table 15.1. Main Components of the Tool Active Layer

TYPE OF COMPONENT	CLASSIFICATION	EXAMPLES
Abrasive powder	Conventional abrasives	Al_2O_3, SiC, B_4C, etc.
	Superabrasives	SD, CBN
Bonds	Polymers (resin)	Elastomers and thermosetting materials
	Metals	Sintered or electro-deposited metallic layers, etc.
	Vitrified	Advanced ceramics
Ingredients	Wetting agents	Organic agents, alloys, silicates
	Chemically active agents	Sulphur, sulphides, chlorides, fluorides, tellurium, etc.
	Solid lubricants	Graphite, MoS_2, etc.
	Thermal conductivity improvers	Fine powders of metals and alloys: iron, copper, silver, bronze, etc.
	Padding	Al_2O_3, SiC
	Materials with secondary functions	Drying agents, mineral coloring agents

In addition to avoiding tribochemical affinity of the material pair, it is essential to determine optimal working conditions, which can be different from case to case.

Dry grinding in air of hardened steel with SD wheels should be avoided at high cutting speeds and working temperatures higher than 800°C. These conditions lead to accelerated tribochemical wear.[25] The result is accelerated wear of diamond grains and fast oxidation of iron particles. As a consequence, the microchips consist of an abrasive micropowder containing $Fe_2O_3 + Fe_3O_4$. Simultaneously, nonoxidized microchips are transferred to the active surface of the tool; a fact that interferes with its proper function. Replacing oxygen from the environment with methane significantly reduces tool wear by more than an order of magnitude.[26] In the case of superabrasive finishing of gears through *coroning* technology[27] abnormal wear of the galvanic tool is not observed. An explanation for this different behavior can be found in the fact that the machining operation takes place at relatively low cutting speeds and contact intervals. Diamond grinding constitutes a high-quality machining operation of outstanding productivity when used for machining nonferrous materials.

Corrosion wear of the active layer of a SD tool is a mild, gradual, and continuous process. When tribochemical wear is dominant, there is a reduced frequency of wear traces on the surface of the couple elements, i.e., by adhesion and abrasion. But, there is increased wear of the tips of the grains and thermal degradation of a resin bond in dry grinding (Fig. 15.4a). This behavior causes "flatting" of the active layer and reduces removal rate due to reduced tool sharpness.

Tool dressing must be carried out in good time to maintain process efficiency. Dressing is also necessary to avoid intensifying unwanted wear processes, as a result of intensive friction on the machined surface, (see Ch. 12, "Conditioning of Abrasive Wheels"). Optimum grinding conditions achieve the simultaneous action of all four types of wear process on the tool surface.[28] Sometimes, under tribochemical attack by water,[29] corrosion wear of a CBN tool can be more intensive than other wear processes (Fig. 15.4b).

CBN tools withstand corrosion wear better than SD tools when grinding HSS. This behavior is explained by tribochemical inertness of CBN in contact with ferrous material and the environment. It also helps that CBN has a higher threshold of thermal instability (1,000°C for CBN compared with 800°C for diamond).[30]

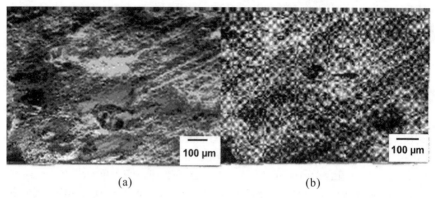

(a)   (b)

Figure 15.4. Scanning electron microscopy aspects of the action of superabrasive active layer, after subjection to an intensive tribochemical wear:[29] *(a)* SD grinding of SC (dry), and *(b)* CBN grinding of HSS (synthetic water solution). The mechanical parameters for machining are optimal for wet grinding.

The cost of Al_2O_3 abrasives is considerably lower than the cost of superabrasives. From a techno-economic viewpoint, Al_2O_3 abrasives may be justified for grinding hard steels. However, redress life of Al_2O_3 decreases when machining heavily oxidized workpieces. Oxide layers may result from previous machining or treatment stages. The rapid wear of an Al_2O_3 wheel is a consequence of tribochemical reaction between metallic oxides and the abrasive grains that results in a chemical complex of spinel-type:[31]

Eq. (15.6) $FeO + Al_2O_3 \rightarrow FeAl_2O_4$

Ferrous materials may be machined with abrasives containing structural carbon if the worksurface is protected by an efficient conversion layer. The result is reduced wear of the abrasive grains as the grains interact with the coated workmaterial. The protective layer forms an atomic or ionic barrier against tribochemical reaction with the grains. Very efficient machining may be carried out by prior carbo-nitriding or cementation of the workpieces.

Increasing silica content of cast iron results in increased hardness. Abrasive machining of this material with Al_2O_3 wheels leads to rapid tool wear, owing to a tribochemical reaction that generates a complex "mullite" molecule.[24] In this situation, SiC tools are preferred for grinding. The reaction is:

Eq. (15.7) $2Si + 2O_2 + 3Al_2O_3 \rightarrow 3Al_2O_3 \cdot 2SiO_2 + Q$

For machining nonmetallic materials, SiC or SD abrasive tools are considered to be best. The grinding of many nonmetals does not raise problems with tribochemical activity, and energy consumption may be reduced. Tools with tougher abrasives give reduced tool wear and high productivity for nonferrous materials.

SD grinding of cutting tools with SC inserts is interesting. Best results are obtained using superabrasives of low toughness rather than high toughness as might be expected, i.e., using a friable type of SD. Friable superabrasives maintain cutting-edge microgeometry by a self-sharpening process of the diamond grains.[32]

Bonds. Bonds consist of a variety of inorganic materials (metals or ceramics) or of organic ones (polymers). Bond materials have a relatively low hardness compared with abrasive grains. An important requirement of a bond material is a capacity for wetting and fixing abrasive grains into a structure. Optimizing the specification of an abrasive includes obtaining a low wear rate for all components of the composite in grinding.

Auxiliary Ingredients. Auxiliary ingredients include other components of the abrasive layer. The choice and proportions of auxiliary ingredients should take into account the behavior with bond, abrasive, workmaterial, and environment. Auxiliary materials can contribute to overcoming particular problems that may be experienced. A discussion of the groups of ingredients follows.

Wetting agents promote physical binding of the components during manufacture of the tool. These agents change the surface properties of abrasive grains and of other solid constituents in contact with the bond material. The nature of the wetting agents depends on the nature of the bond, i.e., metal alloying elements, surfactants or organic monomers, silica complexes, etc.

Chemically active agents modify tribochemical properties of a tool when in contact with the workmaterial and the environment. During or after machining, a triboreaction layer is generated on the worksurface. Tribolayers are important for chip removal and for workpiece protection. For example:

- Sulphur is one of the most widely used active agents. Sulphur can be introduced as a constituent in the abrasive composite layer by mixing or by impregnation. A tribogenerated surface layer is produced on the workpiece and inhibits adhesion of chips to the tool.

- Various metallic sulphide powders may be introduced into the bond mixture to similar effect.

- The mineral cryolite, a double fluoride of Na and Al, has a similar tribochemical behavior to sulphur. Abrasive tools with a cryolite content are used exclusively for machining HSS.

- Similar effects can be achieved with some halides, vinylidine chloride, fluoroborates, etc. The halogens in the chemical structure of these materials produce the required conversion layer on a worksurface during machining.

- From a scientific and technical viewpoint, an interesting method is brought about by dispersion of metallic tellurium powder within the active layer. This agent is very active in a conversion process on the workpiece. While machining, very fast intergranular tribocorrosion takes place on the worksurface. The triboreaction facilitates micro-cutting.[24]

Solid lubricants include lamellar materials, such as graphite and molybdenum disulphide. These materials produce a tribochemical layer on the workpiece that lubricates the process.

Thermal conductivity improvers are metallic powders finely dispersed into the abrasive composite during tool manufacture. For superabrasive tools, a significant augmentation of thermal conductivity is obtained by covering SD or CBN grains with mono- or multicoatings of metallic films. The metals used as thermal conductivity improvers have very high conductivity coefficients, good corrosion resistance, and acceptable tribochemical compatibility with abrasive grains, bond, and workmaterials. All these factors promote cool grinding.

Padding consists of less reactive solids designed to increase the hardness of an abrasive layer. These materials possess a high degree of dispersion and can include conventional abrasives such as Al_2O_3 or SiC added to superabrasive tools. Padding may also include oxides or carbonates of metals from the second group, in the case of conventional abrasive tools. Because of relatively low hardness and small grain size of these ingredients compared to the basic abrasives, padding decreases the effect of hardness differences between the various constituents of the composite. These padding substances, therefore, augment the hardness specification of the abrasive layer.

Materials with secondary functions can include sicativators (drying agents, i.e., CaO) to absorb humidity from the bond materials and inorganic coloring matter used for marking tools with tool specification data.

All of the components of an abrasive tool must work synergistically with the additives in the process fluid.

15.3.3 Triboreactions Between Tool and Environment

Tribochemical processes in abrasive machining work in the following ways.

During intense tribophysical processes in abrasive machining, water is absorbed on the contact surfaces of the friction couple.[33] The abrasive Al_2O_3 is affected by the surrounding humidity leading to increased tool wear compared with machining in a dry atmosphere. Condensed water on an abrasive surface acts as a catalyst for the following tribochemical reaction:

Eq. (15.8) $$2\,Fe + O_2 + 2\,Al_2O_3 \xrightarrow{\;(+H_2O)\;} 2\,FeAl_2O_4$$

Applying liquid hydrocarbons as a film on a tool surface reduces the effect of humidity on alumina. A film may be generated using an immiscible process fluid or a water emulsion.[34] Additivation of these process fluids greatly improves the machining process.

Using SD tools without process fluid causes detrimental tribochemical reactions, such as graphitization and oxidation. These reactions are mainly due to the enormous energy concentrated at the microcontact and to the effect of oxygen on superabrasive grains.[35] The intensity of tribochemical reactions in diamond grinding tends to make such machining processes economically infeasible.[26]

High contact temperatures in diamond grinding cause a chemo-sorption and tribochemical reaction of SD grains with oxygen. Decreasing the partial pressure of oxygen in the surrounding atmosphere reduces the rate of tribochemical reaction.[36] Tool wear can also be reduced by replacing air with an inert gas. However, total absence of oxygen in the surrounding medium makes it difficult to reach a steady-state level of wear and material removal.[37]

Tribochemical reaction of oxygen with SD grains in grinding is accelerated by the presence of water vapor in the atmosphere.[36]

When grinding temperature exceeds 800°C, wear rate of a diamond abrasive increases due to thermal destabilization of the diamond compact structure.[38] Under these conditions, tribo-oxidation combines with allotropic transformation of the crystallized carbon [see Eq. (15.9)]. Graphitization and tribochemical wear are accelerated in a tribosystem by presence of heavy metals such as Fe and Ni. Tribochemical processes proceed at a lower rate under vacuum conditions or in an inert atmosphere.

Eq. (15.9)
$$\text{Diamond (cube - octahedral system)} \xrightarrow{\ T,\,(Fe,\,Ni)\ } \text{Graphite (hexagonal system)}$$

SD grinding is best performed wet, i.e., in the presence of process fluids selected as optimal after experimental research. In grinding, it is accepted that: "The higher the toughness of the workmaterial the more intense are the tribochemical reactions."[32] For this reason, a neat mineral oil, even with a complete formulation of additives, cannot ensure an optimum abrasive process when grinding very tough materials from the SC group. The active surface becomes loaded with pyrolysis products. The SD grains become blunt or are destroyed, and tool life becomes short (Fig. 15.5a). However, good results can be obtained by changing to a synthetic oil-in-water solution. This group of process fluids provides better cleaning of the active surface of the abrasive and has excellent cooling properties. The active surface of an SD grain exhibits a desirable self-sharpening action (Fig. 15.5b). Finally, it is inadvisable to carry out dry SD grinding. In dry grinding, the active surface becomes rapidly worn and loaded with transfer workmaterial (Fig. 15.5c). Dry grinding produces poor quality work-surfaces.[29]

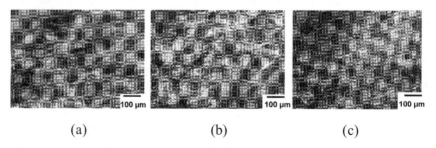

(a)   (b)   (c)

Figure 15.5. Scanning electron microscopy aspects on the SD active layer after SC grinding:[29] *(a)* dry, *(b)* synthetic water solution, and *(c)* additivated immiscible mineral oil. Machining parameters: v_c = 16.5 m/s; v_f = 1.6 m/min; a = 0.03 mm.

Very hard steels can be machined well using CBN grinding wheels. Excellent behavior of CBN can be explained by tribochemical inertness of CBN grains in contact with ferrous workmaterial. The only justification for replacing CBN by a conventional abrasive for repeated batch machining of hard ferrous materials may be cost when working with outmoded processing technology.

In spite of superior chemical stability, CBN superabrasive material exhibits tribochemical affinity with water. For this reason, water-based process fluids are not optimal with CBN (Fig. 15.6a). Water can also result from the machining operation itself or from vapor condensation. Regardless of the water source, the result is reduced process productivity, an increase of specific grinding energy, and a shorter tool life.[28] In practice, the effect of water is comparatively small, although oil is better. Good grinding results have been obtained using water-based fluids and it is far better to use water for grinding hard ferrous workpieces than to grind dry. Wear of a CBN active layer in the presence of water is a tribochemical reaction[39][40] that develops according to the following equations:

Eq. (15.10) $2BN + 3H_2O \rightarrow B_2O_3 + NH_3$

Eq. (15.11) $4BN + 3O_2 \rightarrow 2B_2O_3 + N_2$

Eq. (15.12) $B_2O_3 + 3H_2 \rightarrow 2H_3BO_3$

The SEM image presented in Fig. 15.6b is of an active surface used to grind HSS in the presence of additivated immiscible oil. This view illustrates the excellent condition and grinding ability of the CBN. Figure 15.6c shows the bad effect on an active layer of dry grinding. Dry grinding results in tribochemical oxidation, pyrolysis of the resin bond, and also in metal transfer to the active layer. Dry metal-to-metal friction results in low productivity and reduced tool life.

15.4 TRIBOCHEMICAL ASPECTS OF THE WORKMATERIAL STRUCTURE

Abrasive machining depends not only on the organic nature of additives and of the basic oil in the process fluid, but also on the inorganic components of these complex molecular structures. The process is also affected by positively charged cations originating in the particles of free wear of the couple elements. The inorganic part of the surrounding medium also contributes to the generation of thin surface layers on machined workpieces and influences the process (see Sec. 15.6.2).[42]

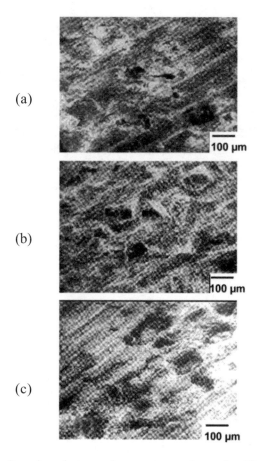

(a)

(b)

(c)

Figure 15.6. Scanning electron microscopy aspects on the CBN active layer after HSS grinding:[41] *(a)* synthetic water-solution, *(b)* additivated immiscible mineral oil, and *(c)* dry. Machining parameters: v_c = 24 m/s; v_f = 2.4 m/min; a = 0.03 mm.

15.4.1 Initial Structure of a Rough-machined Workpiece

Sectioning and structural analysis of a metal workmaterial may reveal a complex multilayer at the surface. Contamination and reaction layers are found next to the irregularities of the surface due to previous machining. Often, there is a so-called "white layer" that may be approximately 3 μm thick.[43] This white layer included in the plastically deformed layer is a consequence of previous mechanical working of the surface.[44] The overall thickness of a work-hardened layer may be over 10 μm, while its structure is very fine by comparison with the underlayer of the main workmaterial (Fig. 15.7).

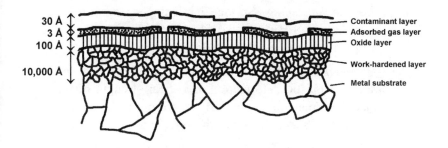

Figure 15.7. Representation of a workmaterial complex structure on the cross section of the workpiece.[1]

The thickness of a plastically deformed layer depends on the machining parameters and also on the efficiency of the process fluid used during previous treatments and machining processes.[45] The surface multi-layer usually has a fine and relatively hard structure. It adheres strongly to the underlayer. These surface properties may offer wear and corrosion resistance. However, a white layer is best removed during abrasive machining to ensure reliability of the machined part. The literature reveals that mechanical and chemical properties of a machined part can be negatively affected by the presence of a white layer. The best way to remove the unwanted layer is by careful abrasive finishing, in the presence of an efficient process fluid. Abrasive machining generates a new surface layer, but its very small thickness has less effect than the previous thick layer.

15.4.2 The Rehbinder Effect and Tribological Implications

The Rehbinder Effect leads to a reduction in body strength caused by adsorption of active agents from the surrounding medium as it comes into contact with the couple elements. The Rehbinder Effect can be interpreted thermodynamically, based on the premise that the surface energy decreases by physico-chemical contact between a solid body and an active fluid.[46] From a structural point of view, the Rehbinder Effect is explained by fracture of interatomic and intermolecular bonds and the establishing of links to molecules, atoms, or ions from the surrounding fluid. The strength of the effect depends on the nature of existing structural bonds, on temperature, on stress within the workmaterial, and on the activity of the external medium. Two distinct manifestations of the Rehbinder Effect[47] are:

- *Surface hardening.* Surface hardening as a result of machining and increased friability, both of the surface and within the sublayer of the workmaterial.

- *Surface plasticizing.* Surface plasticizing reduces resistance to plastic deformation and shear strength of the machined material. These effects are produced by surface-active agents from the environment.

Both the above manifestations may be explained by a decrease of mobility of the dislocations in the underlayer, due to the breaking and blocking action of the thin surface layer generated by adsorption or reaction processes.[48] These descriptive aspects can be quantified by determination of the creep stresses of the modified layer (σ_{ss}) and of the base workmaterial (σ_s). The ratio of these stresses illustrates the interaction between the couple element and the environment:

$$\text{Eq. (15.13)} \quad \sigma_{ss}/\sigma_s \begin{cases} = 1, & \text{inactive environment;} \\ < 1, & \text{plasticizing environment;} \\ > 1, & \text{hardening environment.} \end{cases}$$

15.4.3 Other Tribochemical Interactions Between Workmaterial and Surround

The action of the surrounding environment on a machining process is not limited to the effects of oxygen and water (see Sec. 15.3.3).

In grinding of martensitic steel, a process of nitrogen tribosorption from the surrounding atmosphere takes place at the machined surface. A triboreaction layer Type N-martensite results. The main outcome is an increase in surface hardness by almost five times.

A process of metal hardening can occur during wet grinding, when carbon from the tribodecomposition of mineral oils starts a cementing process at the worksurface. In the case of stainless steel, a similar process leads to formation of chromium carbide, causing weakening of the material structure and generation of fissures in the workpiece.[2]

15.4.4 Influence of the Chemical Composition of Workmaterial

Before an abrasive process, machinability of metal workpieces may be influenced by ions of certain elements implanted into the worksurface. For these machining operations, it is important to use AW and/or EP additivated process fluids (see also Ch. 14, "Process Fluids for Abrasive Machining"). The modification of the composition and structure of a workmaterial by implantation of ions has importance for industrial application. For example, increased grinding removal rate and G-ratio is gained by the presence of implanted Mo ions in the surface of steel workpieces.[49] Additivated process fluids with sulphurated paraffin are used for grinding. Similar results have been obtained by implanting Ni ions into steel workpieces and by using mineral oils additivated with ZDTP.[50]

The idea of additivation of a workmaterial for increased machinability is gaining ground. Thus, in the initial manufacture of HSS by powder metallurgy, additives are introduced into the powder composition, i.e., TiC (as a bonding agent and enhancement phase) and the mixture CaF_2 + MnS (as a solid lubricant). As a result, very low values of specific energy (e_c) are obtained. The processed workpiece has a stable and fine structure.[51]

Grinding is more efficient when a fluid additivated with sulphurated paraffin is used. This is due to *in situ* generation of a thick layer containing iron sulphide (FeS).[52] The presence of iron sulphide in the couple interface prevents scuffing or seizure of the elements. Thermo-chemically generated conversion layers also reduce microcutting forces during abrasive machining. A similar effect may be obtained with other types of preliminary coating based on chemical or electro-chemical reaction layers and plasma layers (see Ch. 16, "Processed Materials").

Preparation of advanced ceramics for abrasive machining requires solving complex problems. A ceramic structure is exceedingly sensitive to thermo-mechanical strains in abrasive machining. If abrasive machining conditions are unsatisfactory, visible and hidden defects cause rejection of finished parts. In general terms, the problem is the brittleness of the structure. The brittleness of advanced ceramics can be reduced by previous application of a structural stabilization treatment. For example, one of the most interesting materials for manufacturing forming tools is tetragonal zircon (3Y-TZP). This advanced ceramic must be surface modified before grinding by infiltration of $Y(NO_3)_3$ solution to reduce structural friability during abrasive finishing.[53]

15.4.5 Workmaterial Selection by Tribosimulation

Chapter 2, "Tribosystems of Abrasive Machining Processes," made a case for modeling and simulation of tribological processes in abrasive machining, especially for sensitive and less well-known workmaterials. Selection of optimal materials for the tool-workpiece couple and process fluid required for a particular machining application can be assisted by tribosimulation.[54] If there is active tribochemical attack by the environment on metallic couple elements, the tribometer used must be an advanced device. The tribometer must provide strict control over the physico-mechanical parameters and on the surrounding medium of the friction couple (Fig. 15.8). A radio-tracer may be used as a continuous and highly sensitive method for measuring tribochemical wear.[55]

Abrasive machining of sensitive composite workmaterials can be particularly difficult with insufficient data for structure and other physico-mechanical and tribochemical properties. For special composites (see Ch. 16, "Processed Materials"), tribosimulation is essential.[43] Some examples are:

- Selection of the best workmaterial from a group of composites, having almost similar structures, using the criteria of abrasive machinability and machined workpiece quality.

- Determination of the optimum chemical treatment to achieve a ductile machining regime during finishing for an advanced ceramic material.

- Determination of optimum machining conditions for profiling, sharpening, and polishing polycrystalline diamond (PCD) inserts for cutting tools.

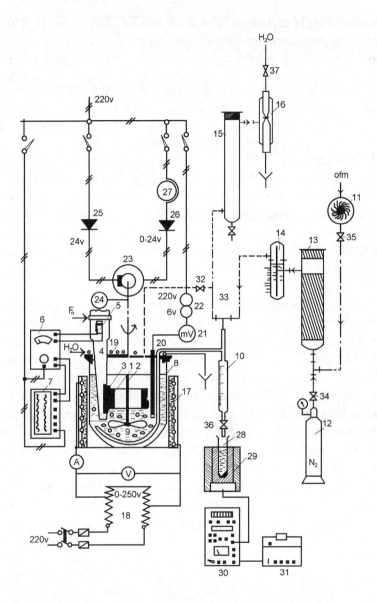

Figure 15.8. Laboratory device for tribosimulation of corrosion wear on metallic friction pair:[55] (1) cylindrical sample; (2) main shaft; (3) workpiece sample; (4) loading lever; (5) strain gauge; (6) strain-meter; (7) recording potentiometer; (8) glass tank for triboreaction; (9) stirrer; (10) glass pipe for bubbling and collecting; (11) compressed air; (12) inert gas cylinder; (13) drying tower; (14) flowmeter; (15) drop-separator; (16) vacuomation; (17) cylindrical furnace; (18) transformer for supply; (19) thermal barrier; (20) thermocouple; (21) c.c electrical motor; (22) tahymeter; (23) sample for process fluid collecting; (24) gammaray counter; (25 and 26) radiometer with recorder.

15.5 TRIBOCHEMICAL ASPECTS OF DRY ABRASIVE MACHINING

Abrasive machining in air leads to tribo-oxidation. Tribo-oxidation during abrasive machining is affected by process conditions including material removal rate; the nature, grain size, and concentration of abrasives in the active layer of the tool; and by the composition, structure, and hardness of the workmaterial.

A damaging tribo-oxidation process may be experienced by the presence of free wear and transfer-wear particles in the contact zone. The existence of metal and oxide particles causes imperfections in the surface quality of the workpiece and loading of the active layer; sometimes termed *oxidation wear*.[56] This mechanism is better considered as a damaging type of corrosion wear experienced in dry abrasive machining (see also Sec. 15.3.2).

It may be possible to achieve mild dry abrasive machining at a low processing temperature. A mild process is distinguished by the generation of a ductile layer of oxides. In this case, the oxide layer behaves almost as a solid lubricant.

With a more intense machining process, fragmentation and even exfoliation of the surface layer may occur. A fragmenting surface layer no longer exercises a protective function. In consequence, removal rate of the protective layer surpasses the rate of restoration. The result is increased adhesive wear of the abrasive tool.

However, if free-wear oxide particles are not removed continuously, both couple elements will be subject to a destructive abrasive process. It follows that negative effects of dry machining can be reduced by air lubrication, i.e., by blowing cold air on the processed workpiece and by vacuum pumping the resulting abrasive powder from the machine tool. The wear process can be moderated in dry grinding by limiting the material removal rate. Intensive dry grinding leads to extreme types of wear, i.e., seizure and delamination. At the limit of control over the wear process, tribosystem behavior is greatly influenced by tribochemical stability of the material pair.[57]

Other developments in gas lubrication include loading the process atmosphere with saturated hydrocarbons. Studies have been performed using SD abrasive on Si_3N_4 ceramics and using Al_2O_3 on Si_3N_4 ceramics operating at different pressure increments of the n-butane environment.[58]

Under these conditions, evidence of the magma-plasma model is found. This evidence includes photon and ion emission, as well as friction polymer-like stick deposits (see also Sec. 15.2.2). A conclusion arrived at by classical methods is that gas pressure for the maximum tribo-emission is not influenced by the material combination of the sliding pair.[59]

15.6 TRIBOCHEMICAL ASPECTS OF WET ABRASIVE MACHINING

As previously stated in Ch. 14, "Process Fluids for Abrasive Machining," wet abrasive machining takes place in the realms of mixed, boundary, and plasto-hydrodynamic lubrication. An efficient lubricating film under these conditions may be ensured by directly generating thin layers of a tribochemical nature in the contact interface. The boundary layers are spontaneously generated by tribomechanical energy following complex mechanisms.

15.6.1 Lubrication by a Tribosorption Layer

This lubrication technology is based on tribosorption of active substances.[14]

Physi-sorption Layer. A mild working regime is characterized by relatively low temperature and loading. Under these conditions, contact surfaces of the friction couple are protected by physi-sorbed layers. These lubricating films produce a viscosity enhancement in the contact interface.[60] Physi-sorbed layers are bonded to the surface of the couple elements by weak van der Waal forces. Binding occurs without exchange of electrons with the materials in contact.[42] The heat resulting from an adsorption process is small, typically in the range of 1,000 to 2,000 cal/mol.[14] The adsorbate molecules are mainly constituted by linear paraffin hydrocarbons, the remainder from saturated fatty acids, and/or other active substances in the base liquid. Physi-sorbed molecules stick to the friction surface.

Wear resistance of an adsorption layer depends both on the nature of the active substances and on that of the workmaterial. The contact surface of the workpiece must be as clean as possible (see Sec. 15.6.1). For example, the wear resistance of adsorbed layers[61] measured by the number of strokes of the tool element is in direct proportion to the length of adsorbed polar molecules (Fig. 15.9).

Physi-sorbed layers are monomolecular, directed perpendicularly to the surface. The length of the adsorbed molecules determines the thickness of the layer. Due to a strong, ordered quasicrystalline structure, these layers offer good mechanical resistance under dynamic conditions at elevated temperatures. Increasing temperatures, usually up to 100°C, increases molecular movement of the lubricant layer. This internal agitation leads to lower mechanical resistance of the layer, caused by increased probability of molecular detachment from the surface.[62] Return to a disordered state at the contact interface has the effect of rapidly increasing friction and increased adhesion wear of the unprotected surfaces.[60]

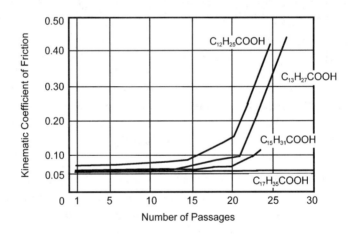

Figure 15.9. Influence of the catenate length of the molecules of fatty acids on the wear resistance of adsorbed layer (illustrated by the evolution of friction coefficient).[61]

Chemi-sorption Layer. Compared to physi-sorption, chemi-sorption produces a stronger bond to the surface. Chemi-sorption involves transfer of electrons between the solid body and the adsorbate. An electronic bond possesses a higher energy level, closer to that of a chemical bond.[63] The energy level of chemi-sorption usually lies in the range of 10 to 100 Kcal/mol; at least one order of magnitude greater than physi-sorption.

Activation of the surfaces of the couple elements is essential to adsorption. Surface activation can be achieved through friction, based on two distinct physical mechanisms: the tribo-energetic effect[64][65] and tribo-emission of exo-electrons.[6] Tribo-emission of low-energy electrons is made possible by pre-existence of a surface oxide layer on the workpiece. The reaction of electrons with various components of the process fluid results in the formation of negative ions and/or free radicals. The presence of these unstable and/or electrically charged particles accelerates the creation of a chemi-sorbed layer on the contact surface.[66]

Heating a workpiece beyond the desorption point of a layer leads to the release of the initial chemi-sorbed substance without chemical change at the worksurface.[67] Nevertheless, a chemi-sorbed layer is more stable than a physi-sorbed layer when heated, as well as from the action of organic solvents.[17]

The large majority of substances that generate chemi-sorbed layers belong to the group of friction modifier (FM) additives (see Ch. 14, "Process Fluids for Abrasive Machining," Sec. 14.12.1). Well known FM additives include compounded acids and fatty alcohols having a linear and saturated structure. Friction modifier additives provide good lubrication of friction, only if the molecular length exceeds that of the C_9 catenary, i.e., pelargonic acid.[68] The presence of a branched structure visibly worsens a lubrication process. This is because deviation from catenary linearity gives way to internal stresses in the chemi-sorbed layer.[69]

Friction modifier additives also include silicon-oxygen compounds from the silane group. These compounds are classified as synthetic process fluids (see Ch. 14, Sec. 14.3.2). Surface layers generated in the presence of silanes have a tri-dimensional structure and exhibit resistance to inter-mediate and cyclical strains.[70] Consequently, the use of silanes is recommended for FM additivation of process fluids for finishing.

Water-based process fluids, including synthetic fluids and emulsions, require FM additives for use in the presence of water and oxygen. Freshly generated metallic surfaces in abrasive machining react spontaneously with oxygen from the atmosphere and/or from a liquid. The resulting film of oxides constitutes the first protection of the worksurface against adhesion wear. In the case of wet grinding, oxygen can access the contact interface. Oxygen is carried by the process fluid, which has a good solubility for oxygen. The most efficient role of oxygen in mineral oil is with a concentration of approximately 0.1 to 0.5% (Fig. 15.10). Dissolved oxygen takes effect, especially at high loadings, when it reduces wear intensity of the friction pair.

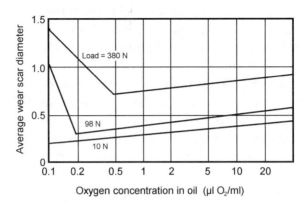

Figure 15.10. Influence of oxygen concentration in lubricating oil on the volumetric wear on the couple.[71]

Chemi-sorption of active substances from the process fluid takes place on oxidized parts of a machined workpiece.[71] The clean and nonoxidized microsurfaces release decomposition triboreactions of the catalytic type of adsorbed active substances.[68][72] The existence of these selective reaction mechanisms is well established from tribometer testing.

The simultaneous action of oxygen and water ensures relatively low values of friction coefficient.[73] The effects of water in the working environment are described in a technical review.[34] Hydrolytic dissociation of the surface-protective oxide layer was observed resulting in damage to the machined workpiece. The acceleration of chemical corrosion and of tribo-electrolytic reactions have also been observed with a film of liquid water on the contact surfaces of the couple. These secondary tribochemical processes are bad for surface quality of the machined workpiece.[74] Effective AC additives and AR additives should be introduced into emulsion and solution concentrates when using water-based fluid. (See Ch. 14, "Process Fluids for Abrasive Machining," Secs. 14.11.4 and 14.11.5.) The abrasive machining process and the nature of the workmaterials have direct impact on the characteristics of the worksurface after machining (i.e., μ coefficient).[75]

15.6.2 Lubrication by Chemical Triboreaction Layers

Antiwear (AW) additives (see Ch. 14, Sec. 14.12.2) are used for moderate machining temperatures, i.e., mild finishing processes. Lubricating layers are generated directly on the contact surface of the workpiece.[76] The thickness of the lubrication layer is approximately 0.010 to 0.015 mg/cm^2.[77] The triboreaction layer is often characterized by the structural features mentioned in the following paragraphs.

Friction-Polymer Layers. A friction-polymer layer has a tri-dimensional structure generated by tribocatalytic polymerization of oxidized hydrocarbons under the influence of pressure and temperature.[78] Nascent metal surfaces resulting from the machining process can exert a strong catalytic effect on polymerization reactions. The surface layer produced consists of friction-polymers generated at nonoxidated micro-areas of the machined surface and can provide good protection against adhesive wear.[79]

Soap Films. When contact energy surpasses a critical level, fatty acids from the process fluid attack hydrated oxides from the surface layer. A soap of metals results from this neutralization triboreaction.[64][80] The thickness and viscosity of the soap film provides lubrication and tribo-chemical protection of a workpiece surface.

Other Kinds of Tribochemical Layers. The structure of a thin layer resulting from a tribochemical reaction between AW additives and metals is difficult to characterize.[81] Due to the friction process, the thin layer undergoes a loosening process and becomes amorphous.[82] Its thickness exceeds the monomolecular level, thus ensuring good wear resistance for the workpiece.[83] Phosphates and sulphur compounds result-ing from tribochemical reactions may be included in a surface film of friction-polymers on the processed workpiece.[78] Consequently, the elements of the friction couple are separated by a soft lamellar composite film. The structure of the film is specific to the particular solid lubricant layer, generated *in situ*.[84] Due to its compactness, this layer provides good protection against corrosive action from the environment. A very interesting additive group consists of structural variants of the zinc-dialkyl-dithio-phosphates (ZDTP). The most stable structure corresponds to the basic form of ZDTP, followed by the neutral form of this additive.[85] This product has a number of useful physico-chemical and tribological properties.[75] In addition to good stability and reactivity, a third useful property of ZDTP is its capacity for adsorption on the surface of machined steel workpieces.[86] The adsorption of AW agents is assisted by the presence of FM additives and of mineral oils in the process fluid.[75] Also, the presence of oxygen, by itself or as an oxide layer, on the processed surface increases the efficiency of ZDTP. Abrasive machining in an inert gas confirms the comparatively low chemical activity of ZDTP in the process fluid.[55] Synthetic oils cause a decrease of adsorption of AW additives (see Ch. 14, "Process Fluids for Abrasive

Machining"). Combinaing adsorption and reaction processes with antiwear (AW) additives is essential for optimizing abrasive machining.[86]

The kinetics of tribochemical decomposition of immiscible mineral oils with ZDTP additive is complex. Many tribochemical mechanisms have been proposed.[87][88] The only certain information has been obtained from chemical micro-analysis of products of triboreactions (Table 15.2).

As previously mentioned, the nature of metal constituents (cations) in AW and AC additives has a special influence on process fluid efficiency. The nature of cations determines the stability of additive molecules and the structure of thin layers. Different metals can be found within the composition of thin layers. This can lead to new additive formulations of rare earth lubricants for process fluids.[91] *Rare earth additives* are products of synthesis reaction with lesser-known metals, such as cerium, lanthanum, neodymium, and gadolinium.[92][93]

Table 15.2. Constituent Products of Immiscible Process Fluids[89][90]

CLASS OF CONSTITUENTS	NATURE OF CONSTITUENTS
Volatile products	H_2S, olefin
Products soluble in mineral oil	RS-SR; RSR; RSH; $SP(SP)(OR_2)$; $SP(SR)_2(OR)$, etc.
Products insoluble in mineral oil	Zn-polyphosphate

15.6.3 Lubrication in Extreme Pressure (EP) Conditions

Theoretical Aspects. The lubrication regime in heavy abrasive machining, such as deep grinding, profiling, high-speed grinding, and cut-off grinding is boundary and/or plasto-hydrodynamic lubrication (see Ch. 14, "Process Fluids for Abrasive Machining," Sec. 14.2). All machining operations are characterized by very high values of real contact pressure and by elevated temperatures. Under such conditions, if the thin reaction layer is destroyed or is missing, there is a real danger of destructive tribological processes, such as seizure and scuffing.[94] It is also possible that tribochemical reactions characteristic for the additive formulation will contribute to an acceleration of other destructive chemo-mechanical processes, such as corrosion fatigue.[95]

Under high removal rate conditions, the only way to achieve effective lubrication is to create a triboreaction sacrificial layer on the working surface.[14][96] The generation of a sacrificial layer is achieved by means of an extreme pressure (EP) additive. An EP chemical reaction can be considered as controlled tribocorrosion, acting where and when a threshold energetic level is exceeded in the working interface of the friction couple.[97] In moderate machining conditions, most EP additives have low tribochemical activity. This selective energetic behavior in the absence of extreme mechanical strains reduces corrosion attack against the tool-workpiece couple, as well as the entire machine tool.

EP Lubrication of Metal Machining Processes. An EP tribomechanism can be considered as a controlled corrosion of metal workpieces.[10] The layer created by an EP triboreaction under hard abrasive machining conditions is removed during the next passage of the tool. The cutting forces with a sacrificial layer are much smaller than without a sacrificial layer. The effectiveness of an EP additive relies on the nature of the active groups within the complex molecular structure. The main chemical and tribochemical attributes of an additive are relative chemical activity and relative load-carrying capacity.[96] These characteristics can be experimentally determined for a process fluid by comparing test results to results for a reference fluid. Comparative tests are conducted using two specialized instruments: a hot-wire corrosion apparatus and a four-ball tribometer (see Ch. 14, Secs. 14.15.4 and 14.16).

In Fig. 15.11, three main EP process fluids groups are compared on the basis of changes in tribological characteristics. There is a direct proportionality between the tribomechanical characteristics and the chemical properties. The best tribological efficiency is achieved using a sulphur additive, while the worst is achieved using a chloride additive;[98] the load-carrying capacity of phosphorus additives is situated between them. The effectiveness and applications domain for the Cl additive is limited due to the relatively low melting point of the iron chloride surface-reaction layer (circa 680°C). Above this temperature on the worksurface, there is a danger of seizure for the friction couple. Even in the absence of a seizure tendency, there is a strong corrosive action on the workpiece. This tribochemical attack spreads rapidly into the sublayer.[2] However, due to low cost, process fluids containing various chloride additives are widely used in abrasive machining with some essential precautions.

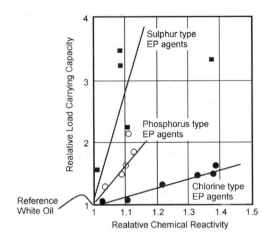

Figure 15.11. Influence of the tribochemical corrosivity of the main EP-additives on their load-carrying capacity.[96]

Tribological efficiency of EP additives and, consequently, the optimal concentration in process fluids depends on the chemical composition of these substances. The properties of chlor-type EP additives are enhanced by the presence of polar groups such as -CH$_3$, -OH, and -COOH in their structure (Fig. 15.12). The optimal concentration of additives depends, not only on the nature of the polar groups, but also on the nature of the material machined.

The application of EP process fluids at a wider range of temperature is achieved by decreasing the lower limit for the normal working temperature, which is achieved by including FM additives (i.e., fatty acids) in the EP fluid formulation (Fig. 15.13).

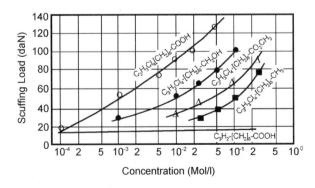

Figure 15.12. Influence of the structure of additives on the load-carying capacity of several EP-process fluids.[84]

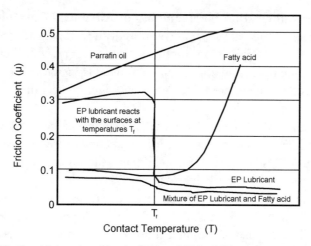

Figure 15.13. Combining the effects of EP and FM additives for a wider application doamin of the process fluids.[17]

An EP additive reacts with a clean, newly generated surface. Usually, a high oxygen concentration in EP fluids is undesirable because oxygen creates a reaction additional to that of the additive itself. The effect of an oxygen reaction is to screen the processed surface. However, a low oxygen concentration is beneficial, since some oxidation of the work-surface reduces corrosion by EP additives and increases efficiency of FM additives.[10] A sacrificial layer may contain products of tribochemical reactions involving sulphates, phosphates and/or chlorides, and also a mixture of base-metal oxides.[99]

A triboreaction layer produced *in situ* can also be controlled by addition of some inorganic additives (e.g., graphite, titanium oxide and zinc borate) to water-immiscible process fluids. The powders of submicron grain size are insoluble in the base mineral oil. Uniform dispersion in the process fluid is made possible by the nanometric size of the powders. This is known as *nano-additivation*.[100] A good dispersivity for these false solutions is assured by dispersivity additivation. Older, well-known process fluids have been rehabilitated using nano-additivation; these include fluids containing solid lubricants, such as hexagonal boron nitride (HBN) and/or molybdenum disulfide (MoS$_2$). Extreme pressure nano-additives from the borate additive class have been introduced[101][102] as additives with a complex structure such as Mo-S-alkyldithiobiurets.[103] The new nano-additives have very good EP and AW attributes contributing to lower friction, reduced abrasive tool wear, and a higher seizure load in abrasive machining.[103]

EP Lubrication in Machining of Advanced Ceramics. Abrasive machining of advanced ceramics is rapidly developing as a technology, and involves very small values for the total stock removed. When it comes to the finishing operation, the volume removed is insignificant. The most important requirements are achieving a high-quality machined surface and absence of structural defects in finished parts.

In Ch. 16, "Processed Materials," extreme stiffness of the sintered structure of ceramics explains very high real contact pressures in abrasive machining. A SD tool and ceramic workpiece should, therefore, be considered as a candidate for EP lubrication. The employment of traditional finishing methods for such brittle materials leads to the generation of numerous structural defects in the machined workpieces, leading to rejection in quality control. Costs of machining ceramics are extremely high and may constitute 50–90% of total cost.[15] Consequently, there is an urgent need for new technologies for finishing ceramics.

The importance of advanced ceramics in modern technology has led to the development of ductile finishing technology.[104] In particular, nanogrinding is used for ceramics. This method requires special machine tools of high stiffness (approximately 100 MN/m) and excellent accuracy allowing control of critical depth of cut to 2 nm/rev resolution.[105] It is also necessary to use a new generation of superabrasive wheels containing SD micropowders. These powders are diamonds with a grain size of 4 to 8 μm that require a special dressing and profiling process[105] (see also Ch. 12, "Conditioning of Abrasive Wheels").

An unconventional development proposed for finishing ceramics is rather similar to a method used for metals under EP conditions. The proposal is to achieve controlled tribocorrosion on the contact microsurfaces of the process couple. Extreme pressure additives from the process fluid are designed to cause tribodissolution of the asperities of the worksurface. As result, an increase of the bearing area of the couple reduces real contact pressure. A sacrificial layer is generated by triboreaction. The properties of the changed worksurface provide a more elastic contact in the friction pair interface. The result is a so-called "softening action" due to the EP additives. Consequently, it is possible to obtain a significant reduction in the variations of couple-loading pressure, variations caused by mechanical shocks and vibrations.

A new technology involves chemo-mechanical machining (CMM).[15] A particular case of CMM is chemo-mechanical polishing (CMP).[106]

There has been considerable research on SD finishing of Si_3N_4 ceramics.[107] Very good results are obtained using an additive from the C_1 to C_9 alcohol group. A thin surface film results from a condensation reaction. The film consists of silicon alkoxide and a support material of long chain hydrocarbons.[108]

Eq. (15.14) $Si_3N_4 + n\text{-ROH} \rightarrow SiO_2 \text{ gel} + \text{long chain hydrocarbons}$

Chemo-mechanical polishing has started to use a suspension of conventional abrasives instead of superabrasives. Conventional abrasives have lower hardness than diamond. The use of insoluble inorganic powders having reduced abrasive properties is based on the idea of applying gentle pressures on the ceramic structure. The aim is to achieve EP tribochemical reactions between the abrasive, the environment, and the ceramic workmaterial, using a process fluid consisting of a watery suspension (slurry) of nanometric Cr_2O_3.[109][110] The machining of Si_3N_4 ceramics with a chrome tri-oxide slurry is considered to be based on the following tribochemical reactions:

Eq. (15.15) $Si_3N_4 + Cr_2O_3 \xrightarrow{(+3H_2O)} 3SiO_2 + 2CrN + 4NH_3$

Eq. (15.16) $Cr_2O_3 + SiO_2 \longrightarrow Cr_2SiO_4 + \frac{1}{2} O_2$

Eq. (15.17) $3SiO_2 + 6H_2O \longrightarrow 3Si(OH)_4$

The slurries consist of abrasives and water, plus EP additives, detergent-dispersants, pH-adjusters, etc. The main tribochemical role of the slurry components is to ensure a transition from brittle to a ductile regime of machining. Other conventional abrasives used for finishing ceramics (i.e., FeO, CeO_2, HBN, $SrCO_3$, $SiCO_3$, SiC, $BaCO_3$, Si, Al_2O_3, Si_3N_4, etc.) act in a similar way to chromium tri-oxide. A few examples are given below:

- Polishing of silica wafers by colloidal silica water slurries.[106]

- Polishing of Si_3O_4 by HBN water slurry.[111]

- Fine abrasive machining of Si_3N_4 balls for bearings using CeO_2 slurries.[112]

- Chemo-mechanical polishing of Cu or Ta wafers using 3% Al_2O_3 slurry.[113]

For all types of water slurry, a basic pH value in the range between 8 to 11 is needed. The optimum pH value for each individual application must be established.[112][114] A review of efforts underway to depart from the exclusive use of water as a CMP-bases liquid and to replace it with ethyl-alcohol is given in Ref. 113 or for mineral oils in Refs. 114 and 115.

15.6.4 Combined Effects of Tribochemical Processes Induced by Additivation

Tribochemical additives with different tribological functions act according to distinct mechanisms.[14][87] Thin layers on a worksurface are the result of combined activity of different tribochemical processes. A tribochemical reaction is specific to each active substance from the environment coming into contact with the couple elements and depends also on machining conditions.[14] The thin layers generated are a weighted combination of each of three essential tribochemical processes, as previously described. Based on the individual characteristics of a tribosystem, one of the tribochemical processes will become dominant (Table 15.3).

Different aspects of contact surfaces are observed when closely examined. The SEM images in Fig. 15.14 show different effects of tribochemical processes caused by additives on a standard steel sample using an Almen-Wieland tribometer.[99] The clear definition of the test sample at the beginning gradually becomes blurred. This change is due to tribochemical effects of FM, AW, and EP additives and also due to deposition of friction polymers resulting from decomposition of the base liquid (light paraffin mineral oil).

The effects of active substances on a friction couple can be quickly tested by measuring friction coefficients (see Ch. 14, "Process Fluids for Abrasive Machining").[116] This method determines the evolution of the friction force with an increase of load, up to the point when seizure effects begin[99] (Fig. 15.15).

There are advantages of using a tribometer fitted with tangential and normal force measurement in order to control both the tribosystem elements and the machining conditions. The behavior of the abrasive tool/workpiece couple is closely linked to the physico-chemical nature of the process of generating the surface layer, as well as to the strength of the layer during machining and wear processes.

Table 15.3. Fundamental Tribochemical Processes of Boundary Lubrication

LUBRICATION REGIME	THE PREVAILING PROCESS	RESULTS OF THE LUBRICATION PROCESS
Boundary and plasto-hydrodynamic lubrication	Lubrication by physical and chemical adsorption	Ensures thin mono-molecular films at the interface of friction couple, through physical adsorption and chemi-sorption processes.
	Lubrication by tribochemical reaction	Causes a modification of the contact surfaces by generation of triboreaction layers, which limit adhesion wear intensity.
	Extreme pressure lubrication	Creates a sacrificial layer by a controlled tribocorrosion of EP additives against the highly strained contact microsurfaces of the workpiece. The sacrificial films protect the friction couple against seizure. Attention must be paid to the possibility of uncontrolled corrosion of the machined surface.

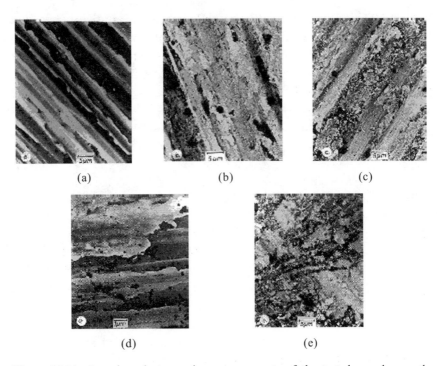

Figure 15.14. Scanning electron microscopy aspects of the tested samples on the Almen-Wieland tribometer using various process fluids:[99] *(a)* initial surface (as reference), *(b)* base paraffin oil (PO), *(c)* PO + 1% FM additive, *(d)* PO + 2% AW additive, *(e)* PO + 2% EP additive.

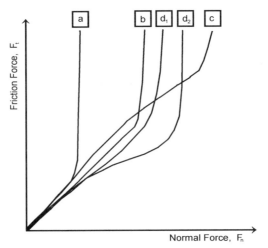

Figure 15.15. Influence of process fluid additivation on friction force evolution: *(a)* base paraffin oil (PO), *(b)* PO + 1% FM-additive, *(c)* PO + 2% (AW + EP) additive, *(d1)* PO + 2% (EP)$_1$ additive, *(d2)* PO + 2% (EP)$_2$ additive.

Values of friction coefficient and wear rate are always represented by average values; these results represent the influence of each of the two or three main tribochemical processes. The kinetics of the thin layer generated has been studied using radio-tracer S-35 dibenzyl disulphide (DBDS).[83][98] The wear-resistant layer is composed of both metallic sulphide and of adsorbed DBDS. Investigation of kinetics of surface-layer generation in the presence of ZDTP additives shows a sequence of physi-sorption, chemi-sorption, and tribochemical reaction in the development of the thin layer.[89]

A final warning is that although controlled corrosion generates protective and lubricating tribochemical layers, there is also a risk of damaging secondary effects. For example, using process fluids with chlor additives brings a danger of hydrochloric acid in the couple environment. This strong acid diluted in the remnant water can cause corrosion of oxide layers on the processed surface.

Intensifying the tribochemical reaction using new additives brings risks still to be determined. There are risks in machining of creating new layers with outstanding catalytic properties. Catalytic effects, when stimulating corrosive processes, are totally unpredictable. Also, there is the possibility of stress corrosion caused by residual stresses from machining. Thus, by intensification of corrosion, the tool life of an abrasive wheel and the life of finished parts could be significantly shortened.

15.7 CONCLUSIONS

Type C corrosion must be avoided in abrasive machining. The selection of additives should ensure the prevalence of Type P corrosion in the working zone. Chemical passivation provides temporary protection for the machined surface, for the abrasive tool, and for the machine tool.

Under favorable machining conditions, active substances from the process fluid provide rapid and permanent creation of thin layers on a worksurface. These layers assist machining and avoid ongoing chemical corrosion of the surfaces.

Synergetic abrasive machining and tribocorrosion assist material removal from the workpiece. This phenomena increases process efficiency and improves the quality of the machined surface.

In CMP finishing of advanced ceramics, use of special EP additives increases tribochemical activity of the friction couple environment and can help to reduce real contact pressures below the dangerous level for microcracking.

Tribochemical processes in abrasive machining are interdependent on the nature of the surfaces of the couple elements.

Deep grain penetration must be avoided in machining advanced ceramics. Abrasive tool wear is accelerated in dry machining and can lead to irreversible workpiece damage.

Additivated process fluids are used to obtain surface layers with specific tribological functions. The main selection criteria for a fluid specification are the nature of the couple elements and the intensity of the machining process. Laboratory simulation and semi-industrial testing are essential requirements for testing suitability of process fluids for particular material combinations and machining conditions.

REFERENCES

1. Rabinowicz, E., Friction & Wear of Materials, J. Wiley & Sons, NY (1994)

2. Rowe, G. W., The Chemistry of Tribology, *RIC-Rev*, 1(2):135–204 (1968)

3. Zumgahr, K. H., Microstructure and Wear of Materials, Elsevier, Amsterdam (1987)

4. Dimitrov, B., On the Nature of the Corrosion Wear, *Rev. Roum. Mec. Appl.*, 18(8):571–585 (1973)

5. Suh, N. P., Tribophysics, Prentice-Hall Inc, NJ (1986)

6. Heinicke, G., Tribochemistry, *Akademie-Vlg,* Berlin (1984)

7. Kajdas, C., Harvey, S. K., and Wilusz, E., *Encyclopedia of Tribology*, Elsevier, Amsterdam (1990)

8. Nenitzescu, C., *General Chemistry*, Ed Did & Ped, Bucharest (in Romanian) (1975)

9. Gutman, E. M., Interaction of Corrosion Processes with Metal Mechanical Processing, *Physical-Chemistry Mechanics of Mater.*, 3(6):548–559 (in Russian) (1967)

10. Dimitrov, B., Aspects of Corrosion Wear in Lubricating Media, Ph.D. Thesis, Polyth. Univ., Bucharest (in Romanian) (1974)

11. Thiessen, P. A., Meyer, K., and Heinicke, G., Grundlagen der Tribochemie, *Akademie Vlg*, Berlin (in German) (1967)

12. Batchelor, A. W., and Stachowiak, G. W., Predicting Synergism Between Corrosion and Abrasive Wear, *Wear*, 123(2):281–291 (1998)

13. Fan, A., Long, J., and Tao, Z., An Investigation of the Corrosive Wear of Stainless Steels in Aqueous Slurries, *Wear*, 193(1):93–97 (1996)

14. Stachowiak, G. W., and Batchelor, A. W., *Engineering Tribology*, Elsevier, Amsterdam (1993)

15. Gates, R. S., and Hsu, S. M., Chemo-Mechanical Machining of Ceramics, *Ceramic Trans.*, 102:67–76 (1999)

16. Sandulescu, D., Physical-Chemistry, Ed. Sci. & Encicl., Bucharest (in Romanian) (1979)

17. Bowden, F. P., and Tabor, D., *The Friction and Lubrication of Solids*, Clarendon Press, Oxford (1950)

18. Fink, M., Chemische Aktivierung Nicht Durch Temperatur, Sondern Durch Plastische Verformung als Ursache Reibchemischer Reaktionen bei Metallen, *Fortsch Ber* VDI-Z, 5(3):5–26 (in German) (1967)

19. Beyer, R. G., and Ku, T. G., *Handbook of Analytical Design for Wear*, Plenum Press, NY (1964)

20. Heinicke, G., and Sigrist, K., Zur Thermodynamik Tribochemischer Reaktionen, *Z-schft fuer Chemie*, 11(6):226–235 (in German) (1971)

21. Glasstone, S., and Lewis, D., *Elements of Physical-Chemistry*, McMilan & Co, London (1964)

22. Rowe, G. W., Smart, E. F., and Tripathi, K. C., Surface Adsorption Effects in Metal Cutting and Grinding, *ASLE Trans.*, 20(4):347–353 (1977)

23. Block, H., The Flash Temperature Concept, *Wear*, 6(3):483–494 (1963)

24. Coes, L., Jr., Abrasives, *Springer Vlg*, NY (1971)

25. Hitchiner, H. F., and Wilks, J., Some Remarks of the Chemical Wear of Diamond and CBN During Turning and Grinding, *Wear*, 114:327–338 (1987)

26. Wilks, E., and Wilks, J., *Properties and Applications of Diamond*, Butterworth-Heinemann, Oxford (1994)

27. Toenshoff, H. K., Karpuschewski, B., Mandrisch, T., and Inasaki, I., Grinding Process Achievements and their Consequences on Machine Tools Challenges and Opportunities, *Annals of the CIRP*, 47(2):651–668 (1998)

28. Marinescu, I. D., Dimitrov, B., Gheorghiu, Z., and Grigorescu, A., Some Aspects Concerning Wear and Tool Life of Diamond Wheels, *Annals of the CIRP*, 32(1):251–254 (1983)

29. Dimitrov, B., Ratziu, G., and Gheorghiu, Z., Kuehlschmierstoffe Fuer Hochleistungs-schleifstoffe, *IDR*, 28(3):156–160 (in German) (1994)

30. Tawakoli, T., High Efficiency Deep Grinding, *Mechanical Eng. Publ.*, London (1993)

31. Georges, J. M., Nature of Surface and its Effect on Solid-State Interactions in New Directions, *Lubrication, Materials, Wear and Surface Interactions*, (W. R. Loomis, ed.), Noyes Publ., Park Ridge, NJ (1985)

32. Semerad-Radulescu, M., Friction and Wear Processes by Dry Grinding with Resin-Bonded Diamond-Wheels, Ph.D. Thesis, Institut for Materials Technology, Bucharest (in Romanian) (1978)

33. Bowden, F. P., and Throssell, W. R., Adsorption of Water Vapour on Solid Surfaces, *Proc. Royal Soc. of London*, pp. 297–308 (1951)

34. Schazberg, P., Influence of Water and Oxygen in Lubricants on Sliding Wear, *J. of ASLE*, 3:301–305 (1970)

35. Field, J. E., Mechanical and Physical Properties of Diamond, *Proc. Int. Conf: Sci. of Hard Materials, Sep. 23–28, Rhodes*, pp. 181–205 (1984)

36. D'Evelyn, M. P., Surface Properties of Diamond, *Handbook of Ind. Diamonds and Diamond Films,* (M. A. Prelas, et al., ed.), Marcel Dekker Inc., NY (1998)

37. Lansdown, A. R., and Price, A. L., *Materials to Resist Wear*, Pergamon Press, Oxford (1986)

38. Metzger, J. L., *Superabrasive Grinding*, Butterworths, London (1987)

39. Smith, L. I., and Tsujigo, Y., An Analysis of CBN Grinding, *Cutting Tool Eng.*, 3(1):49–53 (1977)

40. Okada, S., and Taniguki, N., Rectification Par la Meule Borazon a Liant Ceramique, *Annals of the CIRP*, 28(1):219–224 (in French) (1979)

41. Dimitrov, B., Grigorescu, A., Gheorghiu, Z., and Dutza, R., Tribologische Bewertung der Fertigungshilfsstoffe (FHS) beim Werkzeugschleifen, *Schmierungstechnik*, 18(5):146–151 (in German) (1987)

42. Iliuc, I., *Tribology of Thin Layers*, Elsevier, Amsterdam (1980)

43. Habig, K. H., and Meier zu Koeker, G., Possibilities of Model Wear Testing for the Preselection of Hard Coatings for Cutting Tools, *Surface Coating Technol.*, 62:428–437 (1993)

44. Tarassov, S. Y., and Kolubaev, A. V., Effect of Friction on Subsurface Layer Microstructure in Austenitic and Martensitic Steels, *Wear*, 231(2):199–206 (1999)

45. Tomlinson, W. J., Blunt, L. A., and Sparggett, S., Running-in Wear of White Layers Formed on EN24-Steel by Centerless Grinding, *Wear*, 128(1):83–91 (1998)

46. Rehbinder, P. A., and Shchukin, E. D., Surface Phenomena in Solids During Deformation and Fracture Processes, *Progress in Surface Science*, (S. G. Davison, ed.), 3:97–188, Pergamon Press, Oxford (1972)

47. Savenko, V. I., and Shchukin, E. D., New Applications of the Rehbinder Effect in Tribology: A Review, *Wear*, 194(1):86–94 (1996)

48. Westwood, A. R. C., Ahern, J. S., and Mills, J. J., Development in the Theory and Application of Chemomechanical Effects, *Colloids Surf.*, 2(1):1–35 (1981)

49. Yang, D., Xue, D., Zhang, X., Ding, X., and Lin, W., Characteristics of the Chemical Reaction Between Pure Iron Implanted with Molybdenum Ion and S-containing Lubricating Additive, *Wear*, 193(1):66–72 (1996)

50. Yang, D., Zhou, J., and Xue, Q., Tribochemical Behavior of Ni Ion Implantated Pure Iron Lubricated with ZnDDP as an Additive, *Wear*, 199(1):60–65 (1996)

51. Liu, Z., and Childs, T. H. C., The Influence of TiC, CaF_2 and MnS Additives on Friction and Lubrication of Sintered High Speed Steels at Elevated Temperature, *Wear*, 193(1):31–37 (1996)

52. Kapsa, P., and Martin, J. M., Boundary Lubricant Films: A Review, *Tribology Int.*, 14(2):38–41 (1982)

53. Sato, T., Besshi, T., and Tada, Y., Effects of Surface-Finishing Condition and Annealing on Transformation Sensitivity of 3 mol-Y_2O_3 Stabilized Tetragonal Zirconia Surface Under Interaction of Lubricant, *Wear*, 194(2):204–211 (1996)

54. Santner, E., and Meier zu Koeker, G., Utility and Limitation for Quality Control and Material Preselection, *Wear*, 181/183:350–359 (1995)

55. Dimitrov, B., Aspects of Corrosion Wear into Lubricating Oils, *Chemistry & Technology of Fuels and Lubricants*, 11(2):139–142 (in Russian) (1975)

56. Quinn, T. F. J., The Origins of Oxidational Wear, Part I & II, *Tribology Int.*, 15(5):257–271; 15(6):305–315 (1983)

57. Wang, Y., Lei, T., and Liu, J., TriboMetallographic Behavior of High Carbon Steels in Dry Sliding, *Wear*, 231(1):1–37 (1999)

58. Nakayama, K., and Hashimoto, H., Tribo-emission, Tribochemical Reaction and Friction and Wear in Ceramics Under Various n-butane Gas Pressure, *Tribology Int.*, 29(5):383–393 (1996)

59. Thiessen, P. A., Physikalisch-chemische Untersuchungen Tribochemischer Vorgaenge, *Z-schft Chemie*, 5(5):162–171 (in German) (1965)

60. Campbell, W. E., Boundary Lubrication, Boundary Lubrication—An Appraisal of World Literature, pp. 87–117, ASME Eng. Center, New York (1971)

61. Zisman, W. A., Friction, Durability and Wettability Properties of Monomolecular Films on Solids, *Friction and Wear*, (R. Davies, ed.) Elsevier, Amsterdam (1959)

62. Dorinson, A., and Ludema, K. C., *Mechanics and Chemistry in Lubrication*, Elsevier, Amsterdam (1985)

63. Buckley, D. H., Definition and Effect of Chemical Properties of Surfaces in Friction, Wear and Lubrication, *Proc. Int. Conf. Fundamentals of Tribology*, MIT, Cambridge, MA (1978)

64. Okabe, H., and Kanno, T., Behavior of Polar Compounds in Lubricating-Oil Films, *ASLE Trans.*, 24(4):459–466(1981)

65. Barwell, F. T., Advances in Friction and Wear, *Tribology Int.*, 17(5):299–307 (1984)

66. Kajdas, C., Tribochemistry—A Review, *Proc. 5th Trib. Congr. '89*, Espoo (Finland) (1989)

67. Forbes, E. S., and Reid, A. J. D., Liquide Phase Adsorption/Reaction Studies of Organo-Sulfur Compounds and their Load Carrying Mechanism, *ASLE Trans.*, 161:50–60 (1973)

68. Morecroft, D. W., Reactions of Octanoic and Decanoic Acid with Clean Iron Surfaces, *Wear*, 18(3):333–339 (1971)

69. Klaus, E. E., and Bieber, H. E., Effect of Some Physical and Chemical Properties of Lubricants on Boundary Lubrication, *ASLE Trans.*, 7(1):1–10 (1964)

70. Ando, E., et al., Frictional Properties of Monomolecular Layers of Silane Compounds, *Thin Solid Films*, 180:287–291 (1989)

71. Goldman, I. B., Appeldoorn, J. K., and Tao, F. F., Scuffing as Influenced by Oxygen and Moisture, *ASLE Trans.*, 13(1):29–38 (1970)

72. Godfrey, D., Boundary Lubrication, *Proc. Int. Symp. Lubr. & Wear*, Houston, TX (1963)

73. Dokos, S. J., Sliding Friction Under Extreme Pressures, *Trans. ASME*, 68:148–156 (1946)

74. Staeger, H., Reibung und Schmierung, *Schweitzer Arch.*, 15(4):97–116 (in German) (1949)

75. Tingle, E. D., The Importance of Surface Oxide Films in the Friction and Lubrication of Metals, Part II, *Trans. Faraday Soc.*, 326:97–102 (1950)

76. Armstrong, D. R., Ferrari, E. S., Roberts, K. J., and Adas, D., An Investigation into the Molecular Stability of Zinc di-alkyl-dithiophosphates (ZDDP) in Relation to their Use as Antiwear and Anticorrosion Additives in Lubricating Oils, *Wear*, 208(1):138–146 (1997)

77. Palacios, J. J., Thickness and Chemical Composition of Films Formed by Antimony Dithiocarbamate and Zincdithiophosphate, *Tribology Int.*, 19(1):35–39 (1986)

78. Furey, M. J., The Formation of Polymeric Films Directly on Rubbing Surfaces to Reduce Wear, *Wear*, 26(3):369–392 (1973)

79. Tripathi, K. C., Correlation of Chemical and Mechanical Effects of Lubricants, *Tribology Int.*, 8:146–152 (1975)

80. Adler, K., and Schoon, T. G. F., Untersuchungen zur Kenntnis der Schmier- und Verschleissvorgaenge im Gebiet der Grenzreibung, *Schmiertech + Trib*, 15(3):257–271 (in German) (1968)

81. Okabe, H., Nishio, H., and Masuko, M., Tribochemical Surface Reaction and Lubricating Oil Film, *ASLE Trans.*, 22(1):65–70 (1979)

82. Ludema, K. L., Boundary Lubrication, *Characterization of Tribological Materials*, (W. A. Glaeser, ed.), pp. 98–115, Butterworth-Heinemann, Boston (1993)

83. Lacey, I. N., Kellsal, G. H., and Spickes, H. A., Thick Antiwear Films in EHD-contacts, Part I, *ASLE Trans.*, 29(3):299–305 (1986)

84. Studt, P., Zusammenhang zwischen Absorbierbarkeit und Wirksamkeit von Hochdruckzusaetzen in Schmieroelen, *Erdoel und Kohle*, 21:784–785 (in German) (1968)

85. Dimitrov, B., Einfluss Chemisch-aktiver Stoffe auf das Tribologische Verhalten eines Reibpaares, *Schmierungstechnik*, 8(6):192–195 (in German) (1977)

86. Studt, P., Boundary Lubrication: Adsorption of Oil Additives on Steel and Ceramic Surfaces and its Influence on Friction and Wear, *Tribology Int.*, 22(2):111–119 (1989)

87. McFadden, C., Soto, C., and Spencer, N. D., Adsorption and Surface Chemistry in Tribology, *Tribology Int.*, 30(12):881–888 (1997)

88. Wu, Y. L., and Dacre, B., Effects of Lubricant-Additives on Kinetics and Mechanisms of ZDDP Adsorption on Steel Surfaces, *Tribology Int.*, 30(6):445–453 (1997)

89. Bell, J. C., and Delargy, K. M., The Composition and Structure of Model Dialkyldithiophosphate Antiwear Films, *Proc. 6th Int. Congr. Tribology*, pp. 328–332, Budapest, Hungary (1992)

90. Georges, J. M., Nature of the Surface and its Effect on Solid-state Interactions, *Tribology in the 80s*, (W. R. Loomis, ed.), pp. 43–67, Noyes Publ., NJ (1985)

91. Yu, L., Nie, M., and Lian, Y., The Tribological Behavior and Application of Rare Earth Lubricants, *Wear*, 197(2):206–210 (1996)

92. Zhang, Z., Liu, W., and Xue, Q., Tribological Properties and Lubricating Mechanisms of the Rare Earth Complex as a Grease Additive, *Wear*, 194(1):80–85 (1996)

93. Chen, B., Dong, J., and Chen, G., Tribochemistry of Gadolinium Dialkyldithiophosphate, *Wear*, 196(1):16–20 (1996)

94. Tomaru, M., Hironaka, S., and Sakurai, T., Effects of Some Chemical Factors of Film Failure Under EP Conditions, *Wear*, 41(1):141–155 (1977)

95. Torrance, A. A., Morgan, J. E., and Wan, G. T. Y., An Additive's Influence on the Pitting and Wear of Ball Bearing Steel, *Wear*, 192(1):66–73 (1996)

96. Sakurai, T., and Sato, K., Study of Corrosivity and Correlation Between Chemical Reactivity and Load Carrying Capacity of Oils Containing Extreme Pressure Agents, *ASLE Trans.*, 91:77–87 (1966)

97. Borsoff, V. N., and Wagner, C. D., Studies of Formation and Behavior of an EP-lubrication, *Lubrication Eng.*, 13(2):91–99 (1975)

98. Kotvis, P. V., et al, The Surface Decomposition and Extreme Pressure Tribological Properties of Highly Chlorinated Methanes and Ethanes on Ferrous Surface, *Wear*, 147(3):401–419 (1991)

99. Dimitrov, B., Lueb, H. A., and Schoon, T. G. F., Tribologische Untersuchungen Gemischter Reibung auf einer Almen-Wieland Anlage, *Rev. Roum. Sc. Tech.*, 21(4):497–511 (in German) (1976)

100. Kimura, Y., et al., Boron Nitride as a Lubricant Additive, *Wear*, 232(2):199–206 (1999)

101. Dong, J. X., and Hu, Z. S., A Study of the Antiwear and Friction-reducing Properties of the Lubricant Additive, Nanometer Zinc Borate, *Tribology Int.*, 31:219–223 (1998)

102. Yao, J., and Dong, J., Tribocatalysis Reaction During Antiwear Synergism Between Borates and Sn (IV) Compounds in Boundary Lubrication, *Tribology Int.*, 29(5):429–432 (1996)

103. Tripathi, A. K., Bhattacharya, A., and Verma, V. A., A Study in the Tribochemistry of Alkyldithiobiurets and Their Mo-S Complexes as EP Additives, *Wear*, 209(1):134–139 (1997)

104. Bifano, T. G., Dow, T. A., and Scattergood, R. O., Ductile-regime Grinding: a New Technology for Machining Brittle Materials, *Trans. of the ASME*, 113(5):184–189 (1991)

105. Marinescu, I. D., Webster, J. A., and Dimitrov, B., Brittle/ductile Grinding Regimes for Brittle Materials, *Diamond & CBN Ultrahard Materials Symp. '93*, Windsor, Ontario (1993)

106. Moon, Y., Bevans, K., and Dornfeld, D. A., Identification of the Mechanical Aspects of Material Removal Mechanisms in Chemical Mechanical Polishing (CMP), *Ceramic Trans.*, 102:269–280 (1999)

107. Muratov, V. A., Luangvarnunt, T., and Fischer,T. E., The Tribochemistry of Silicon Nitride: Effects of Friction, Temperature and Sliding Velocity, *Tribology Int.*, 31(10):601–611 (1998)

108. Hibi, Y., and Enomoto, Y., Mechanochemical Reactions and Relationship to Tribological Response of Silicon Nitride in n-alcohol, *Wear*, 231(2):185–194 (1999)

109. Muratov, V. A., and Fischer, T. E., Tribochemical Reactions of Silicon Nitride in Aqueous Solutions, *Ceramic Trans.*, 102:245–258 (1999)

110. Bhagavatula, S. R., and Komanduri, R., On Chemomechanical Polishing of Si_3N_4 with Cr_2O_3, *Philosofical Mag.*, 74(4):1003–1017 (1996)

111. Saito, T., Imada, Y., and Honda, F., Chemical Influence on Wear of Si_3N_4 and HBN in Water, *Wear*, 236(1/2):153–158 (1999)

112. Komanduri, R., Lucca, D. A., and Tani, Y., Technological Advances in Fine Abrasive Processes, *Annals of the CIRP*, 46(2):545–596 (1997)

113. Her, Y. S., et al, The Role of the Physical Properties of Alumina Abrasive in the Chemical-Mechanical Polishing of Copper, *Ceramic Trans.*, 102:281–290 (1999)

114. Zhu, H., Tessarotto, L. A. B., Greenhut, V. A., Niesz, D. E., and Sabia,R., Solvent (pH) and Abrasive Effects in Chemical Mechanical Polishing of Si_3N_4, *Ceramic Trans.*, 102:259–268 (1999)

115. Winn, A. J., Dowson, D., and Bell, J. C., The Lubricated Wear of Ceramics, Part I – II, *Tribology Int.*, 28(6):383–402 (1995)

116. Dimitrov, B., Lueb, H. A., and Schoon, T. G. F., Eine Verbesserte und Automatisierte Ausfuehrung der Almen-Wieland Maschine, *Schmiertech + Trib.*, 24(2):40–42 (in German) (1977)

16

Processed Materials

16.1 IMPORTANCE OF MATERIAL PROPERTIES

The main requirement of abrasive machining is to produce workpieces of a specified quality and accuracy. A process specification normally includes:[1]

- Geometry of the workpieces.
- Dimensional tolerances.
- Maximum and minimum surface roughness.
- The integrity of the top and under layers of the finished surface.

Improvement of an abrasive finishing process can contribute to the durability of the finished workpiece and, consequently, to the durability of the whole assembly of which it forms a part. To meet quality requirements, machinining conditions should not be optimized in isolation from the characteristics of the material or from consideration of the effects of the process on the particular material. It is necessary to take into account the surface and volume properties of the materials to be processed, both before and after machining. These may conveniently be considered as external and internal properties depending on whether the properties act on the surface or just below the surface. The properties of a workpiece material, in relation to the abrasive process when considered as a tribosystem, were introduced in Ch. 2, "Tribosystems of Abrasive Machining Processes." In this chapter, structural aspects are considered, as are some implications for the process.

The surface properties of a solid cannot always be precisely defined. This is partly due to insufficient knowledge of the history of former processing and also because of the dynamic character of the atomic and electronic bonds at the surface. It is known from molecular physics that specific energies and tensions on and in the workpiece material play an important role in material behavior. Energetic aspects play a fundamental role in adhesion tendencies, or affinities, or energies of rejection, depending on the nature of the abrasive material and of the work environment.* These energies are proven by adhesion forces and constitute a partial explanation for the physico-chemical nature of the external friction process at a pair interface, i.e., the adhesion component of friction.

Volume properties of a solid depend on the energy of equilibrium and stability, which are much stronger under the surface than at the surface. Atomic and ionic forces produce structural order, evidenced in the degree of crystallinity of a material. Inner bonds contribute to the physico-mechanical properties. Cohesion forces inside the volume of material are manifested in internal friction during shear process, i.e., the deformation and ploughing components of friction. Due to differences in hardness of the friction pair at the asperities in sliding contact, elastic and plastic deformations take place. In an abrasive process, the deformations lead to material removal. It has been suggested that optimizing the abrasive machining process causes physico-mechanical aspects of friction to predominate over physico-chemical aspects.[3]

It follows that surface and volume properties (Tables 16.1 and 16.2) both affect the magnitude of friction forces. For the external friction process, the removal of an older top layer generates a new surface; accordingly, adhesion forces tend to increase since a new surface is more adhesive than an old surface.

Volume properties of a processed material influence wear of the tool and also the nature of the abrasive machining process. Although it may not be immediately apparent, under abrasive machining conditions, friction

---

* Regarding the differences between the unambiguous definition of the volume properties and the lack of accuracy in characterizing the surface state, the eminent scientist Wolfgang Pauli may be quoted: "God made solids, but the surfaces were made by the Devil."[2]

and wear processes are strongly interdependent, mutually influencing each other. Abrasive machining is strongly influenced by tribological considerations of friction, lubrication, and wear, and is influenced by the internal and external structure of the workpiece.

Table 16.1. Surface Properties of Workmaterials

SURFACE PROPERTIES OF WORKPIECES	NOTATIONS	COMMONLY USED UNITS
Workpiece roughness	R_a, R_z, R_t	μm
Hardness of surface microcomponents	ΔHv	$N \cdot mm^{-2}$
Workpiece surface tension	σ_s	$J \cdot mm^{-2}$
Proportion of surface chemical components	A_l, B_l, C_l, etc.	%
Surface heat treatment	Denomination and description	—
Thickness of surface layer	h_s	μm

Table 16.2. Bulk Properties of Workmaterials

BULK PROPERTIES OF WORKPIECES	NOTATION	COMMONLY USED UNITS
Volume density	δ	$Kg \cdot dm^{-3}$
Elastic modulus	E	GPa
Tensile strength	R_m	$N \cdot mm^{-2}$
Bulk hardness	Hv	$N \cdot mm^{-2}$
Fracture toughness	K_{1C}	$MN \cdot m^{3/2}$
Thermal conductivity	λ	$W \cdot m^{-1} \cdot K^{-1}$

The range of materials subject to abrasive machining processes is large. Materials can be classified as follows:

- *Ductile materials.* Most materials machined are ductile including basic metals and coatings. Ductile materials have greatly varying compositions and structural properties that affect choice of machining conditions and tribological behavior in processing.

- *Brittle materials.* Brittle materials are mostly nonmetals. The group can be subdivided into crystalline minerals, advanced ceramics, optical glasses, etc. Crystalline minerals include abrasive and superabrasive powders, semiconductors, and so on. Although superabrasives are typical nonmetals, some of their physical properties and structural characteristics bring them close to metals in nature.

- *Transitional materials.* Transitional materials lie between ductile and brittle materials. This class includes composites, polymers, and other materials. In particular aspects, transitional materials may be included in either the brittle or the ductile classes.

Examples of the three classes are illustrated in Table 16.3, with typical values of volume properties for comparison. Units are as listed in Tables 16.1 and 16.2.

16.2 STRUCTURAL ASPECTS OF METALS

Ductile materials have a compact structure distinguished by nondirectional electronic bonds. The electronic density at the Wigner-Seitz atomic cell boundary is an indicator of solid-state properties.[5] Consequently, this type of structure leads to high values of density, thermal conductivity, elastic modulus, elongation, and breaking point. The microstructures of metals can display various degrees of crystallinity. This characteristic can lie anywhere between the clearly defined properties of a crystalline lattice of a pure metal and the completely amorphous structure (absence of Bragg-type reflexes) of some alloys.[6] Ductile processes tend to predominate during abrasive machining of metal workpieces.[7] Due to low chemical and mechanical resistance, metals with a high degree of purity are likely to be unsuitable for many industrial applications.[8]

Table 16.3. Main Bulk Properties of Some Metals (Alloys), Advanced Ceramics, and Polymers[4]

MATERIALS	δ	E	R_m	K_{1C}	HV	λ
Steel	7.8– 7.9	210	440– 930	50– 214	100– 900	30– 60
Cast iron	7.1– 7.4	64– 181	140– 490	6– 20	100– 850	30– 60
Aluminum alloy	2.6– 2.9	60.8	300– 700	23– 45	25– 140	121– 237
Aluminum oxide	3.9	210– 380		3–5	1400- 1900	25– 35
Zirconium oxide	5.6	140– 210		8– 10	1200	2
Silicon nitride	3.2	170		3– 7	1600– 1800	25– 50
Silicon carbide	3.2	450		4.5	2500	90– 125
Polyamide (PA)	1.01– 1.14	2–4	40– 80	3	80– 100	0.25– 0.35
Polyimide (PI)	1.3	3–5	100– 300			0.27– 0.52
Polytetrafluoro-ethylene (PTFE)	2.1– 2.3	0.4	15– 25		12	0.25
Polyethylene (PE-HD)	0.92	0.2	14– 18	1– 2	13	0.33– 0.57

16.2.1 Control and Modification of Structure in Metals

Processes used to modify and control the bulk and surface properties of metals fall into three categories:

- *Bulk treatment.* These are processes that modify, where possible, properties of the whole mass. Methods mainly consist of alloying and/or heat treatment.[9] Sometimes, alloying is designed to produce wear-resisting fine structures. However, this can lead to the undesired consequence of structural segregation.[10]

- *Surface treatment.* Physical or chemical processes are used for surface treatment.[7] For example, surface layers may be modified by thermal hardening, atomic and ionic diffusion, or by chemical or electro-chemical conversion.[11] Surface treatments include flame hardening, induction hardening, diffusion processes, ionic implantation, and hard anodic coating. (Table 16.4).

- *Surface coating treatment.* Various surface coatings may be applied, having a different composition and structure than the base metal. Surface coating treatments include electroplating, flame sprayed coating, hard facing by welding, and physical coatings applied using a variety of techniques (Table 16.5).

A recent development in coatings is the generation of various carbon-type layers, especially for cutting tools. This broad group of hard, smooth, corrosion and wear resistant physical layers includes carbon, hydrocarbon, metal-doped hydrocarbon, diamondlike carbon, and diamond coatings.[12]

The main purpose of the majority of these processes is to increase resistance to mechanical and chemical wear of the parts produced.[13] However, the process of increasing service life of the parts produced may also lead to increased difficulties in abrasive finishing of the parts involved.[14]

Table 16.4. Conversion Technologies and Resulting Surface Layers Properties

MAIN TECHNO-LOGICAL GROUPS	CONVERSION METHODS	BASIC METALS	CONVERSION LAYER PROPERTIES	
			DEPTH (mm)	HARDNESS (HV)
Heat treatment	Flame hardening	Steels	3.0–6.0	500–700
	Induction hardening	Steels	0.5–4.0	600–700
Physico-chemical conversion methods	Carburizing	Steels	0.1–2.0	600–800
	Nitriding	Steels	0.1–0.8	600–750
	Cyaniding and carbonitriding	Steels	0.1–5.0	700–850
	Chroming	Steels	0.010–0.015	1200–1650
	Siliconizing	Steels	0.1–0.2	1600–1800
	Boronizing	Ferrous	0.1–0.3	150–160
Hard anodic coating	Electrolitical oxidation	Al, Ti, and Mg alloys	0.01–0.2	400–500

Table 16.5. Application of Wear Resistant Surface Layers

MAIN TECHNO-LOGICAL GROUPS	LAYER DEPOSI-TION METHODS	BASIC MATERIALS	PROPERTIES OF SURFACE LAYERS	
			DEPTH (mm)	HARDNESS (Hv)
Electro-plating	Hard chromium	Metallic	0.003 to 0.20	600 to 800 Hv
	Hard nickel		0.50 to 6.0	650 to 800 Hv
	Rhodium		0.002 to 0.005	540 to 650 Hv
Flame sprayed coatings	Aluminum alloy	Metallic	0.01 to 3.0	72 HRB
	Copper			32 HRB
	Iron			80 HRB
	Moly-bdenum			
	Monel			39 HRB
	Nickel			49 HRB
	SS 18-8			78 HRB
	High C-steel			39 HRC
	Tantalum			450 Hv
	Oxides (Os, Cr, Hf, Zr)			250 to 800 Hv
	Carbides (W, Hf, Ta).			750 to 1000 Hv

(Cont'd.)

Table 16.5. *(Cont'd.)*

MAIN TECHNO-LOGICAL GROUPS	LAYER DEPOSI-TION METHODS	BASIC MATERIALS	PROPERTIES OF SURFACE LAYERS	
			DEPTH (mm)	HARDNESS (Hv)
Hard facings	Austenitic steel	Ferrous	1.5 to 6.0	600 Hv
	Martensitic steel			650 Hv
	Cr-steel			550 Hv
	Mn-steel			200 Hv
	Co-alloys			550 Hv
	Ni-alloys			600 Hv
	CW			700 Hv
Physical boundary coatings	Carbides	Cemented carbides	0.001 to 0.003	
	Nitrides			
	Carbo-nitrides			
	Oxides			
	Carbon (Diamond) layers	(+ Si_3N_4)	0.001 to 0.002	

16.2.2 Machinability of Metals

Machining forces tend to increase with hardness. This has implications for the machinability of metals including alloys and coatings. In general:

- Cutting force and machining difficulty tend to increase with hardness of the base metal and/or its surface.

- Hardness of steels tends to increase with carbon content, therefore, increasing machining difficulty.

- In terms of metallurgical structure, machinability of steel tends to decrease in the following order:

Ferrite (low C) $\rightarrow$ Pearlite (Soft) $\rightarrow$ Bainite $\rightarrow$
Pearlite (high C) $\rightarrow$ Martensite (high C)

- The difficulty of machining increases where steel is alloyed with the following elements, in the order:

Cr $\rightarrow$ Cr-Mo $\rightarrow$ Ni-Cr-Mo

- A harder surface layer increases the resistance to abrasive machining. In addition to the direct effect of hardness, abrasivity of a protective coating on a workmaterial increases rate of wear of the abrasive tool.

Despite the above generalities, it is important to note that soft materials may introduce other problems into abrasive machining. Softness implies lower forces on the grains of the abrasive and soft ductile materials tend to imply generation of long chips. Long soft chips have an increased tendency to clog and load the pores of the abrasive, a process that quickly destroys the cutting ability of the abrasive. In addition, if the material has a tendency to adhere to the abrasive, surface roughness will rapidly increase. In this sense, a very soft material does not necessarily have good machinability. A free-machining iron or steel has graphite or other elements introduced into the structure to break up the chips. Free-machining materials exhibit better machinability than a soft homogeneous material.

At the other end of the spectrum, difficulties will be experienced with high strength alloys designed for high temperature operation in aerospace engines. In this case, alloying is designed to ensure the material maintains its hardness at high temperatures. In abrasive machining, the forces on the abrasive grains will be high leading to rapid wear of the abrasive. This may be combined with ductile behavior of the chips and a tendency for wheel loading. Machinability of such materials is poor, but the problems can be overcome by use of high machining speeds to minimize grain forces, use of superabrasives to minimize grain wear, use of increased porosity in the abrasive to facilitate chip removal, and use of high pressure fluid delivery.

16.3 STRUCTURAL ASPECTS OF NONMETALS

The main structural characteristic of brittle materials is the existence of directional inner bonds of an ionic/covalent nature. This type of structure leads to relatively low values for thermal conductivity, thermal expansion, density, and toughness. Exceedingly small values of fracture strain (under 0.3%) and of breaking energy (circa 10^{-2} J cm^{-2}) lead to a preponderance of brittle fracture as the failure mechanism.[15]

16.3.1 Advanced Ceramics

Advanced ceramics have a polycrystalline structure obtained by sintering. Inner bonds of atomic or ionic nature lead to relatively high hardness values. Elastic properties are comparable with metals. Ceramic materials can have unusually high hardness, typically 1,200 to 2,500 Hv. Other physico-chemical characteristics are low density and good resistance to wear and corrosion. These properties allow ceramics to be used as a possible alternative to metals. The ratio of elasticity to hardness (E/Hv) for ceramics is smaller than for metals by approximately an order of magnitude.[16] These characteristics make ceramic materials highly suitable for applications such as insert cutting tips for cutting tools and for other parts requiring excellent wear resistance. The main types of advanced ceramics are:[17]

- Silicon nitride (Si_3N_4).
- Silicon carbide (SiC).
- Silicon aluminum oxynitride (SiAlON).
- Silicon boride (SiB_4).
- Alumina (Al_2O_3).
- Zirconia (ZrO_2).

These groups can also be classified into oxide and nonoxide ceramics, while ceramics of the SiAlON (Sialon) type are an intermediary product situated between these two other groups.

16.3.2 Optical Glasses

Glass is a mixture of melted inorganic materials, undercooled at temperatures below the transitional thermal threshold. Glass is a homogeneous mixture of silicates, phosphates, and/or borate of Pb, Ba, Al, La, Th, Ta, Nb, etc. From the thermal point of view, the most important parameter that characterizes an optical glass is the softening temperature (T_g). Depending on the type of glass, the softening point lies in the range $T_g = 470°$ to $685°C$. From a mechanical viewpoint, optical glass may be defined by hardness, which lies in the range Hv = 410 to 850 N mm^{-2}. Finally, from an optical viewpoint, glasses can be distinguished by their refractive index (n) and by the Abbe number (v). Based on the values of these indexes, variants of optical glass may be defined.[18]

16.3.3 Crystallized Minerals

Abrasives and Abrasive Powders. Abrasives are hard and extrahard materials which include conventional abrasives (SiC, B$_4$C, ZrO$_2$, Al$_2$O$_3$, Cr$_2$O$_3$, etc.) and superabrasives (natural and synthetic diamond and cubic boron nitride). Abrasives fall into three main categories according to grain size. These are normal powders, micropowders, and nano-powders. With reference to the shape and structure of the grains, abrasive powders can be monocrystalline or polycrystalline powders. Most conventional abrasives are ceramic.

In the last decade, new polycrystalline powders have been developed for conventional abrasives and for superabrasives. For example, for finish grinding steel or honing microcrystalline aluminum oxide powders, also called seeded gel (SG), have established a position between Al$_2$O$_3$ and CBN. Development continues into new types of microcrystalline abrasive powder.[19] A recent development is high aspect-ratio abrasive grains that retain sharpness for extended periods of grinding.

Semiconductors. These materials are located within Subgroups IIIB to VIIB of the Periodic Table. The most important semiconductor materials are silicon and germanium, found in Subgroup IVB. The crystalline structure of these semimetals is similar to diamond. Small concentrations of elements from Subgroup VB can be introduced into the lattice of Si and Ge crystals (doping), to improve properties and to extend the range of application.

Superconductors. These are crystalline materials having a complex ceramic structure. Semiconductors allow very high values of electrical and/or thermal conductivity to be obtained. In order to achieve electrical superconductivity, the materials must generate extremely weak magnetic fields. Superconductivity is most useful when the material achieves this condition at higher temperatures away from $0°K$. Materials in this category include $TiBa_2Ca_3Cu_4O_{11}$ or $Yb_2Cu_3O_{7-x}$.

16.4 STRUCTURAL ASPECTS OF TRANSITIONAL MATERIALS

16.4.1 Composite Materials

Composite materials have been developed to achieve unusual combinations of properties (hardness, strength, thermal properties, corrosion resistance, etc.); properties unachievable with the component materials. Composite materials can be classified into the following sub-groups.

Particulate Composites. These are materials with isotropic structure and properties. The main components, the dispersoids, are evenly distributed within the mass of the material. A reticular structure called a matrix is developed according to the following main categories of materials:

- Metal matrix composites (type MMC).
- Polymer matrix composites (type PMC).
- Ceramic matrix composites (type CMC).
- Glass matrix composites (type GMC).

Theoretical properties of a composite can be computed using the rule of mixtures. To apply this rule, the proportions (f_i - %) must be known and the values of the properties (A_j) for each and every component:

Eq. (16.1) $\qquad A = \Sigma A_I \times f_j$

The rule of mixtures can also be applied to the friction values of the composite, demonstrating that at the start of testing, the friction coefficient of the compositor depends on the friction coefficient of each component, as well as on the proportions. Due to selective wear, the loading of the couple gradually becomes unevenly distributed. Components that are more wear resistant carry more of the load and dominate the friction process.[20]

Particulate composites machined by abrasive processes include:

- *Cemented carbides (CC).* Hard carbides of W, Ta, Ti, etc., dispersed in a sintered cobalt matrix; the main use of these composites is to insert cutting tips for cutting tools.

- *Abrasive materials.* As described in Ch. 7, "Molecular Dynamics for Abrasive Process Simulation" and Ch. 15, "Tribochemistry of Abrasive Machining," abrasive materials are made from conventional abrasive or superabrasive powders dispersed in a metallic, organic, or ceramic matrix. These composite materials constitute the active layer of an abrasive tool. The abrasive material can form the whole tool or the abrasive material can be deposited as an abrasive layer on the tool.

- *Extrahard polycrystalline composites.* This group includes synthetic diamond (SD) and cubic boron nitride (CBN) powders in-grown or dispersed in a cobalt metal matrix. The polycrystalline compacts can be mechanically attached or bonded on the surface of a carrier. Sometimes the composite may be directly generated on the surface. The principal feature of interest in these materials is exceptionally good wear resistance. Abrasive machining of these composites, where necessary, is a challenge involving slow wear processes such as polishing with diamond paste.

- *Dispersion-strengthened composites.* These are submicron oxide particles (CdO, BeO, PbO, ZrO_2, Al_2O_3, Y_2O_3, etc.) uniformly dispersed in a metal matrix (Ag, Al, Cr, Ni, Co, W, Pb, Pt, etc.) in which they are insoluble. This type of structure blocks movement of dislocations, reduces creep, and enhances strength. These materials are ideal for some high-technology applications and have found application as abrasives for machining brittle materials.

- *Cast-metal particulate composites.* These are metallic materials obtained by a composite casting method used for incorporating ceramic (SiC) and glass particles in aluminum and magnesium alloys.[12] This material subgroup may also include metal-ceramic materials (cermets).

Fiber-reinforced Composites. Fiber-reinforced composites result from incorporating stiff, strong, and brittle fibers (metals, glass, carbon, boron, and polymer fibers) into a softer and ductile matrix (metals, ceramics, and polymers). The rule of mixtures may be employed, unless fiber orientation leads to anisotropy. Strength of composites depends on the adherence between the matrix and the fibers. Adherence is influenced by the physico-chemical properties of the matrix material. Mechanical properties of the fiber-reinforced composites are determined by fiber length, diameter, concentration, strength, and direction.

Laminar Composite Materials. Laminar composites are anisotropic. The rule of mixtures can only be applied for characteristics such as density. For elastic modulus, and electrical and thermal conductivity, an appropriate rule of mixtures for the direction of consideration may be applied. Other directional properties can only be determined through testing. Laminar composites may be:

- *Coatings.* Surface layers designed to protect against corrosion and/or wear (see also Sec. 16.2).

- *Claddings.* Metal-to-metal composites having high strength and good anticorrosion properties.

- *Laminates.* Metal layers joined by adhesive.

- *Sandwich structures.* Thin and resistant layers held together in light spatial structures.

16.4.2 Polymers

Polymers are long chain macromolecular structures. The molecular weight can reach huge values of 10^6 g/mol. Polymers are based on chemical polymerization processes. The principal chain formations are generated as linear and/or spatial structures. Polymerization is based on addition of structural units or monomers through condensation mechanisms. The degree of polymerization is the number of units participating in the final macromolecular structure. The basic elements permitting the formation of macromolecular structures are carbon and silicium, found in the fourth group of the Periodic Table. Polymers can be categorized as:

- Thermoplastic materials consisting of flexible, linear chains.
- Elastomers consisting of linear cross-linked chains.
- Thermosetting materials consisting of rigid three-dimensional lattices.

Polymer materials have relatively low mechanical strength and poor resistance to high temperatures compared with most structural metals. These features restrict the range of application. However, polymers are also used in composites for machine construction. In these circumstances, polymers may make up the matrix and/or the fiber or layer components of a composite.[21]

16.5 ADVERSE TRIBOCHEMICAL EFFECTS IN ABRASIVE MACHINING

As previously stated, there is strong interdependence between the abrasive machining process and material resistance to corrosion; the mechanical and chemical processes are mutually conditioned. Also, the abrasive process influences the durability of the machined part.

16.5.1 Physico-Chemical Effects

Some physico-chemical effects of machining on parts produced may be as follows:

- Residual stresses in a machined part create a vulnerability to stress-corrosion in service.
- Exceeding allowable process temperature can result in:
  - Structural changes of the tool and workpiece materials.
  - Modification of the elemental composition of the workmaterial due to oxidation and decarburation.

- Disturbance of chemical equilibrium and initiation of corrosion caused by additive decomposition products from the process fluid.

- Tensile residual stresses, distortion of the part, and thermal stresses.

- Physico-chemical interactions between the workpiece material and degraded process fluid damage the workpiece surface. Unfortunately, this type of damage may be difficult to discover during routine quality control and can remain undetected.[1]

- Materials of the friction pair widely separated in the table for "galvanic series" (specific to the environment) may be subject to transfer wear. Due to transfer wear,[22] galvanic microcells form on the ground surface. These lead to accelerated corrosion in microzones that can become even more active by subsequent chemical (CMP) or electro-chemical machining processes.

- Microbial corrosion may contribute to intensification of other corrosion processes through the increased aggressiveness of the environment.

16.5.2 Prevention of Adverse Tribochemical Effects

In general, an aggressive environment accelerates tribochemical and electro-chemical corrosion. The intensity of corrosion processes can often be reduced by measures taken before or during abrasive machining. Possible preventative measures include:

- Careful choice of friction pair materials with regard to electro-chemical compatibility.

- Cathodic protection by using a "sacrificial electrode"[12] or by passivating the processed surface.

- Control of the process fluid and use of anticorrosion additives (see Ch. 14, "Process Fluids for Abrasive Machining").

- Careful attention to the water quality used in the fluid and use of antirust additives (see Ch. 14).

16.6 TRIBOLOGICAL ASPECTS OF ABRASIVE MACHINING

16.6.1 General Aspects

A wear process is often characterized by complexity. In such cases, it is necessary to consider all the working conditions of the friction pair. This implies establishing a large number of physical, chemical, and mechanical parameters.[23]

Archard[24] proposed that the volume (V) of material worn from a body in sliding/rubbing contact is

Eq. (16.2) $V = K \times F_n \times L / Hv$

where F_N is the normal force, which is the most important parameter,[25][26] L is the total sliding distance, Hv is the hardness of the body subject to the wear to be estimated, and K is a wear coefficient. Unfortunately, K has a very large range of possible values, $K = 10^{-8}$ to 10^{-1} mm^{3}N^{-1}m^{-1}, and has to be determined empirically. Archard's Law is valid for the "steady state" regime of machining.[27]

Dividing both sides of Eq. (16.2) by the duration of the abrasive process yields a relationship for work removal rate:

Eq. (16.3) $Q_w = K \times F_n \times v_w / Hv$

where Q_w is the stock removal rate, v_w is the workpiece speed, and Hv is the hardness of the workpiece material.

In the case of grinding, Eq. (16.3) is of more use for determining the coefficient K than for determining Q_w since the work-material removal rate is accurately known from the grinding conditions. However for other abrasive machining processes, the removal rate is not so easily determined.

A similar relationship may be applied for tool-wear rate:

Eq. (16.4) $Q_s = K \times F_n \times v_c / Hv$

where Q_s is the abrasive tool wear rate, v_c is the cutting (grinding) speed, and Hv is the abrasive hardness.

The friction couples intended for industrial applications must satisfy the condition of abrasive wear resistance[21] where

Eq. (16.5) $K < 10^{-6} \ mm^3N^{-1}m^{-1}$

This condition is to be satisfied by the active layer of the abrasive tool.

16.6.2 Tribology of Metals in Abrasive Machining

The following are some empirical comments relevant to abrasive machining of metals:

- The tool-wear coefficients for a particular workmaterial should be approximately of the same order for both two-body wear and three-body wear.

- The tool-wear coefficient is directly proportional to workpiece hardness, for similar processing conditions. However, the behavior of metals in abrasive machining is not determined by hardness and cold hardening considerations alone, but also by phenomena such as plastic deformation and fracture.[28]

- The magnitude of the tool-wear coefficient depends on the metallographic structure of the workmaterial.

- The tool-wear coefficient is influenced by mechanical process conditions, such as load and sliding speed. However, a gradual reduction of the tool-wear coefficient occurs in ceramics grinding.[29] A possible explanation is the self-generation of an optimal active microgeometry of the diamond grinding tool during grinding (Fig. 16.1).

- The tool-wear coefficient for metal-bond superabrasive tools depends not only on the nature, hardness, and grain size of the abrasive powders, but also on the metallic component of the bond (sintered or galvanic), as well as on the metal coatings of the superabrasive grains. These metal components must be tribologically compatible with the workpiece material.[30] This refers to the tribo-metallurgical reciprocal compatibility, i.e., the degree of solid solubility of the metallic friction-pair elements.

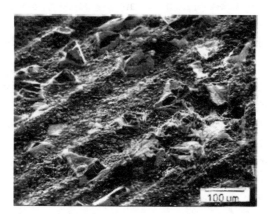

Figure 16.1. Self-sharpening process of the superabrasive layer; couple CBN/HSS. Optimized machining parameters: v_c = 24 m/s; v_w = 2.4 m/· min; and a = 0.02 mm.

- In processing of alloys, the dominant metal in the structure should be considered. The worst situation (from a compatibility point of view) is encountered when two metals in tribological contact are identical. A particular case of this extreme situation is constituted by the transfer onto the abrasive tool surface of workmaterial microchips generated by the machining process (Fig. 16.2).

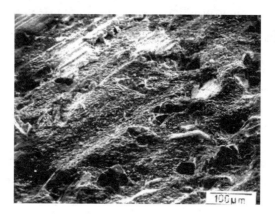

Figure 16.2. Beginning of the transfer process of the microchips generated by the couple CBN/HSS. Machining parameters: v_c = 30 m/s; v_w = 2.4 m/· min; and a = 0.03 mm.

- Compatibility difficulties are increased in dry abrasive machining (Fig. 16.3). Difficulties may also be experienced in wet processing when the shape of the abrasive tool is conformal with the workpiece profile.

- Some metallic materials have especially favorable tribological properties for abrasive processing from the viewpoint of low friction energy consumption. The following materials can be mentioned:

  - Noble metals (Ag, Au, Pt, Ir, Os, etc.) with a friction coefficient reduced by almost 70% compared to base metals.

  - Metals which crystallize in the hexagonal system such as Zr, Hf, Mg, Cd, Co, Ti, and their alloys. There are clear tribological advantages during the abrasive process provided that the process fluids are FM additivated to reduce adhesive tendencies.[31]

  - Alloys containing soft metal components facilitate the lubrication role of the process fluid during the machining process. The proportion of the soft components must be no less than 8% while the hardness should be a maximum of one-third of the hardest component (e.g., Cu-Pb alloy).

Figure 16.3. General distruction of the superabrasive active layer by its metallization process. Dry tool and cutter grinding CBN/HSS. Machining parameters: $v_c =$ 36 m/s, $v_w = 3.2$ m/· min, and $a = 0.05$ mm.

- Sintered metals or composites containing solid lubricants such as graphite, MoS_2, HBN. These sintered materials display a remarkable tendency for self-lubrication during abrasive machining.

16.6.3 Tribology of Abrasives in Abrasive Machining

Conventional Abrasives. The majority of conventional abrasives belong to the ceramic materials family, due to their similar methods of fabrication.

Differences are evident due to the diversity of elemental compositions. A variety of materials are introduced for the purpose of producing particular tribological properties specific to the abrasive grains such as brittleness, self-sharpening, and surface properties relating to the contact with different bonds, etc.

The tribological characteristics of conventional abrasives depend on the nature of the bond and the abrasive tool, on the nature of the workmaterial, and on the processing method.

Superabrasives. Diamond, both natural (ND) and synthetic (SD), is the hardest material on Earth, having a cubic and/or octagonal crystalline structure. Due to its structure, diamond has outstanding resistance to wear.[23] In the presence of air, diamond in contact with the majority of materials, or even with itself, exhibits low friction coefficients, $\mu = 0.05$ to 0.1. Under vacuum conditions after removal of the adsorbed layer, friction coefficients are increased at least seven-fold. Exceptional tribological properties make diamond a pre-eminent superabrasive. The tribological behavior of monocrystalline and polycrystalline diamond grains in a composite layer may be improved by light polishing with diamond paste.

Since diamond is anisotropic, the friction and wear properties depend on grain orientation in relation to the motion of the processed surface. For example, the crystal-plane {111} yields low friction coefficients and an easier polishing process. Using a tribometer of the Micro-Abrader type[15] differences in abrasion resistance of 15 to 20% were found for various types of natural diamond.

Abrasive wear of diamond grains is accompanied by propagation of microcracks in the stressed zone (peaks and/or edges of the crystals). Stress concentration leads to the breaking of atomic strong bonds that characterize the diamond compact structure. In exchange, the release of the dislocation from the stressed contact results in an increase in diamond resistance to abrasion.[32]

When it comes to diamond powders used in the manufacture of superabrasive tools, an important problem arises. The sharpness of the working micropoints and -edges of the grains need to be maintained as sharp as possible for a long period of processing. During the machining process, the diamond crystals gradually become dull or are pulled out of the abrasive layer. It is necessary to stimulate a sharpening and/or self-sharpening of superabrasive grains from the active layer during machining. Self-sharpening, depending on the friability of the monocrystals or the polycrystalline structure, is achieved in some synthetic diamonds.

16.6.4 Tribology of Ceramic and Glass Processing

Abrasive machining of brittle materials presents a high degree of difficulty due to the fact that both elements (i.e., tool and workpiece) have a rigid structure that make it difficult to operate in an elasto-plastic regime with conventional depths of grain penetration. With large grain penetration depths, large hertzian pressures are generated.[33] Glass and ceramic structures are especially sensitive to concentrated mechanical stresses. As a result, macro- and micromachining processes generate a whole network of cracks that compromise workpiece quality and may lead to complete failure.[34] It is possible to successfully achieve material removal by three possible mechanisms, i.e., flake formation, powder formation, and ploughing.[35] The problems of developing, machining, and industrial application of brittle materials depend on solving the following problems.

Improving Tribological Properties of Ceramics. During the development of ceramics, the following innovations may be applied to increase the toughness of these materials:

- The improvement of the quality of the component powders by the following:

  - Preparation of monodispersed oxidized powders.

  - Production of the ceramic powders by the sol-gel process.

  - Manufacturing the powders from the gaseous stage using the laser and/or plasma techniques.[4]

- The augmentation of toughness of sintered ceramics through the creation of a meta-stable state that results from the dispersion within the structure of additional materials in very small quantities; these materials include metallic oxides of Ti, Zn, Be, Y, Pb, etc. This "doping" assures the absorption or spreading of the crack energy during mechanical stressing and prevents the spread of microcracks.[36]

- The enhancement of toughness of ceramic materials by construction of fibrous composite structures (see Sec. 16.4.1). Many advanced ceramics are actually fiber-ceramic matrix composites.

Ductile Abrasive Machining of Ceramics and Glasses. Ductile machining of nonmetallic materials having a low toughness can be achieved under the following conditions:[37]

- Only workpieces checked for absence of structural defects should be selected for machining.

- The machining infeed must be extremely carefully controlled so that the grain depth of penetration is less than the critical grain depth. The critical grain depth of cut is specific to each material type; generally speaking, abrasive machining of brittle materials is a nanotechnology process.[38]

- Diamond grain size of the superabrasive tool should be reduced to the micropowder domain.

- Elastic bonds should be used to hold the diamond powder. The bond should have good thermal and electrical conductivity.

- ELID-dressing should be used[39] to achieve better control of the cutting microgeometry of the active abrasive layer.

- Dynamic stiffness of the machine tool should be very high in order to achieve a very low level of vibrations.

- The feed drives and positional control system should be capable of very high accuracy to avoid shock contact between the abrasive grains and the workpiece.

- Abrasive machining of brittle materials must be undertaken[40] under light plastic deformations and in the presence of a process fluid able to generate a resistant tribofilm in the contact interface.

16.6.5 Abrasive Machining Tribology of Polymeric Materials

The main advantage of polymeric materials lies in ease of manufacture by casting, injection molding, or forming into complex shapes. These parts may subsequently require abrasive machining.

In spite of lightness and chemical inertness, wider use of polymers for machine parts may be offset by high wear rate. When subjected to friction, resistance may be affected by gradual depolymerization in the contact area.[41] In the case of a metal/polymer couple, the extremely low values of surface energy of the polymer (i.e., low friction coefficient) means the wear rate can be very low at low contact temperatures.[42]

High wear resistance [see Eq. (16.5)] can be achieved by using polymers with fillers and reinforcements,[43] as well as by using adequate lubricants. Wear properties of multiphase materials are strongly dependent on the size of the reinforcing phase.[44] It is possible to achieve a quasi-crystalline polymeric structure by modification through copolymerization of different monomers, by inclusion in the network of "high-performance" blends, or by thermochemical treatment.[45] The toughness and strengths of these modified polymers, having a complex tri-dimensional structure, can reach very high values, close to those of metals.

16.7 CONCLUDING COMMENTS

Behavior in an abrasive machining process depends on the combined complex effects of the workmaterial, the abrasive, and the process fluid, as well as on the contact temperatures and mode of deformation. By careful study of the tribology of an abrasive machining system, it is possible to identify avenues for process and product improvement.

Based on tribological considerations and comparing behavior of workmaterials from the three main groups (Table 16.3) during abrasive processing, it is possible to reach the following general conclusions:

- Owing to high values of density and surface energy, as well as susceptibility to tribochemical attack, metals are beginning to be replaced by other materials for many applications.

- Polymers are convenient for ease of product development and machining based on the principles described. Shortcomings for many applications are lack of toughness and an insufficient range of operating temperatures that can be applied and sustained.

- Ceramic materials offer some important advantages because of their intermediate density, wide range of working temperatures, and chemical inertness. However, many ceramics are difficult to machine requiring microcutting due to their structural rigidity and very high hertzian contact pressures experienced in cutting.

- Plastics and ceramics are developed, processed and used as composite materials within which plastics or polymers constitute the matrix. There are numerous situations where a matrix can also be developed from metal. The specific structure of composite materials determines the achievement of previously mentioned advantages of matrix materials. Some new qualities can be achieved due to the specific role of the composite components.

- Metals continue to remain the basic workmaterials for machine construction due to their toughness, thermal conductivity, and superior machinability. For many applications in modern technology, abrasive machining processes have to be preceded by modification of the volume structure and/or the workpiece surface.

REFERENCES

1. Lucca, D. A., Brinksmeier, E., and Goch, G., Progress in Assessing Surface and Subsurface Integrity, *Annals of the CIRP*, 47(2):669–693 (1998)

2. Tabor, D., Status and Direction of Tribology as a Science in the 80s— Understanding and Prediction, *Tribology in the 80s,* (W. R. Loomis, ed.), pp. 1–16, Noyes Publ., NJ (1985)

3. Zum Gahr, K. H., Abrasiver Verschleiss Metallischer Werkstoffe Vortsch.Ber.VDI-Z, Duesseldorf (in German) (1981)

4. Czichos, H., Tribologie und Neue Werkstoffe, *Proc. 8th Int. Coll. Tribology 2000*, Esslingen (in German) (1992)

5. Subramanian, E. V., and Eliezer, Z., Abrasive Wear and Electronic Properties of Materials, *Wear*, 126(2):219–222 (1986)

6. Klinger, R., Ein Beitrag zur Korrelation Tribologischer und Metallphysikalischer Werkstoffeigenschaften, Ph.D. Thesis, T. U. Berlin (in German) (1984)

7. Mutton, P. J., and Watson, J. D., Some Effects of Microstructure on the Abrasion Resistance of Metals, *Wear,* 48(3):385–398 (1978)

8. Nakayama, K., Takagi, J., and Nakano, T., Peculiarity in the Grinding of Hardened Steel, *Annals of the CIRP*, 23(1):89–90 (1974)

9. Jang, J. W., Iwasaki, I., and Moore, J. J., Effect of Martensite and Austenite on Grinding Media Wear, *Wear*, 122(2):285–299 (1988)

10. Buckley, D. H., Importance and Definition of Materials in Tribology— Status of Understanding, *Tribology in the 80s*, (W. R. Loomis, ed.), pp. 18–42, Noyes Publ., NJ (1985)

11. Zum Gahr, K. H., Relation Between Abrasive Wear Rate and the Microstructure of Metals, *Proc. ASME Int. Conf. Wear of Materials,* pp. 266–274 (1979)

12. Askeland, D. R., *The Science and Engineering of Materials*, PWS Publ. Co., Boston (1994)

13. Chandrasekaran, T., Natarajan, K. A., and Kishore, Influence of Microstructure on the Wear of Grinding Media, *Wear*, 147(2):267–274 (1991)

14. Bhat, M. S., Zackay, V. F., and Parker, E. P., Alloy Design for Abrasive Wear, *Proc. ASME Int. Conf. Wear of Materials*, pp. 286–291 (1979)

15. Moore, M. A., and King, F. S., Abrasive Wear of Brittle Solids, *Proc. ASME Int. Conf. Wear of Materials,* pp. 275–285 (1979)

16. Gilman, J. J., Hardness—A Strength Microprobe in the Science of the Hardness Testing and its Research Applications, *ASM Int.*, Metals Park, OH (1973)

17. Komanduri, R., Lucca, D. A., and Tani, Y., Technological Advances in Fine Abrasive Process, *Annals of the CIRP*, 46(2):545–596 (1997)

18. Izumitami, T., *Polishing, Lapping and Diamond Grinding of Optical Glasses Treatise on Materials Science and Technology*, pp. 116–149, Academic Press, New York (1979)

19. Toenshoff, H. K., Karpuschewski, B., Mandrysch, T., and Inasaki, I., Grinding Process Achievements and Their Consequences on Machine Tools Challenges and Opportunities, *Annals of the CIRP*, 47(2):651–668 (1998)

20. Axen, N., Hutchings, I. M., and Jacobson, S., A Model for the Friction of Multiphase Materials in Abrasion, *Tribology Int.*, 29(6)467–475 (1996)

21. Czichos, H., Klaffke, D., Santner, E., and Woydt, M., Advances in Tribology: The Materials Point of View, *Wear*, 190(1):155–161 (1995)

22. Dimitrov, B., Aspects of Corrosive Wear in Lubricating Media, Ph.D. Thesis, Polith. University, Bucharest (in Romanian) (1974)

23. Hsu, S. M., Shen, M. C., and Ruff, A. W., Wear Prediction for Metals, *Tribology Int.*, 30(5):377–383 (1997)

24. Archard, J. F., Contact and Rubbing of Flat Surfaces, *J. Appl. Phys.*, 24:981–988 (1953)

25. Yang, C., and Bahadur, S., Friction and Wear Behavior of Alumina-based Ceramics in Dry and Lubricated Sliding Against Tool Steel, *Proc. ASME Int. Conf. Wear of Materials,* pp. 383–391 (1991)

26. Habig, K. H., Haerte und Verschleiss (in German) (1984)

27. Rigney, D. A., Comments on the Sliding Wear of Metals, *Tribology Int.*, 30(5):361–367 (1997)

28. Kato, K., Abrasive Wear of Metals, *Tribology Int.*, 30(5):333–338 (1997)

29. Hwang, T. W., and Evans, C. J., Diamond Wear in High Speed Grinding of Silicon Nitride, *Ceramic Trans.*, 102:27–37 (1999)

30. Rabinowicz, E., Wear Coefficients—Metals, *Wear Control Handbook*, ASME, pp. 475–506, New York (1980)

31. Rabinowicz, E., *Friction and Wear of Materials*, pp. 206–209, J. Wiley, New York (1995)

32. Wilks, E., and Wilks, J., Properties and Applications of Diamond, Butterworths-Heinemann, University Press, Cambridge (1991)

33. Woydt, M., Klaffke, D., Habig, K. H., and Czichos, H., Tribological Transition Phenomena of Ceramic Materials, *Wear*, 136(2):373–380 (1990)

34. Ajayi, O. O., and Ludema, K. C., The Effect of Microstructure on Wear Modes of Ceramic Materials, *Proc. ASME Int. Conf. Wear of Materials*, pp. 307–318 (1991)

35. Hokkirigawa, K., Wear Mode Map of Ceramics, *Proc. ASME Int. Conf. Wear of Materials*, pp. 353–358 (1991)

36. Leatherman, G. L., and Nathan-Katz, R., Structural Ceramics: Processing and Properties, pp. 671–696, Academic Press, New York (1989)

37. Shimada, S., and Ikawa, N., Scientific Approach to Machining Brittle Materials, *Jpn., Proc. Conf. Supertech,* Livermore, CA (1996)

38. Piscoty, M. A., Davis, P. J., Blaendel, K. L., Precision Ceramic Grinding Process Development for Brittle Materials, pp. 102:3–10, *Ceramic Trans.*, (1999)

39. Marinescu, I. D., Ohmori, H., and Li, W., ELID Grinding Ceramic Materials, pp. 102:47–55, *Ceramic Trans.*, 102:47–55 (1999)

40. Xiong, F., and Manory, R. R., The Effect of Test Parameters on Alumina Wear Under Low Contact Stress, *Wear*, 236(1/2):240–245 (1999)

41. Heinicke, G., Tribochemistry, *Akademie Vlg.*, Berlin (1986)

42. Czichos, H., Advanced Materials in Tribology, *Proc. Int. Conf. Nordtrib*, pp. 2–27 (1992)

43. Tewari, U. S., and Bijwe, J., On the Abrasive Wear of Some Polyimides and Their Composites, *Tribology Int.*, 24(4):247–254 (1991)

44. Simm, W., and Freti, S., Abrasive Wear of Multiphase Materials, *Wear*, 129(1):105–121 (1989)

45. Santner, E., and Czichos, H., Tribology of Polymers, *Tribology Int.*, 22(2):103–109 (1989)

Symbols and Units

SYMBOLS

a: Set depth of cut.

A_c: Contact area; overall contact area.

A_{cu}: Mean cross-sectional area of the uncut chips.

a_d: Dressing depth.

a_e: Real depth of cut.

A_r: Real area of contact.

a_{sw}: Depth of wheel wear.

a_t: Thermal expansion.

B: Brittleness index; cutting edge spacing in lateral direction, mean grain spacing; Peclet equivalent for inclined band source.

b_{cu}: Uncut chip width.

b_d: Active width of dressing tool.

b_s: Abrasive tool width.

b_w: Width of cut.

C, C_a: Cutting edge density; number of active cutting edges per unit area.

C: Factor for temperature solution taking into account the Peclet number, flux distribution, and geometry; specific heat capacity.

c: Specific heat capacity.

C_a: Area contraction coefficient for fluid jet.

C_d: Orifice discharge coefficient.

C_m: Machine and labor cost per unit time.

c_p: Specific heat at constant pressure.

C_s: Abrasive tool cost.

C_{stat}, C_{dyn}: Static and dynamic cutting edge density.

c_v: Specific heat at constant volume.

C_v: Velocity coefficient for orifice or nozzle jet.

d_{cap}: Diameter of capillary.

d_{cu}: Unloaded cut diameter.

d_e: Equivalent wheel diameter.

d_{ef}: Effective diameter when cutting.

d_{eg}: Effective grain diameter.

d_g: Grain diameter; mean grain diameter.

d_{jet}: Fluid jet diameter.

d_s: Diameter of grinding wheel.

d_w: Diameter of workpiece.

E^*: Equivalent elastic modulus for two bodies in contact.

E: Young's Modulus of elasticity.

E: Energy.

e_c: Specific cutting energy; energy per unit volume.

e_{ch}: Energy per unit volume carried away by chips.

E_f: Fracture energy.

E_p: Plastic flow energy.

$erf[u]$: Error function for argument u.

$erfc[u]$: Complementary error function for argument u.

$f = t/k$: Interface friction ratio at a friction contact.

F^*: Critical load to initiate a crack.

F: Force; resultant of normal, tangential, and axial forces.

F_a: Force parallel to wheel axis.

f_d: Dressing feed per wheel revolution.

F_{ij}: Notation used to represent forces on atom "i" from atoms "j."

F_n: Force normal to wheel surface.

f_n: Normal force on a grain.

F_n': Normal force per unit width.

F_t: Force tangential to wheel surface.

f_t: Tangential force on a grain.

F_t': Tangential force per unit width.

G: Grinding ratio; volume of workmaterial removed divided by tool wear volume.

H: Hardness; Brinell hardness.

h: Heat convection coefficient for moving surface; heat per unit area per degree temperature difference.

h_{air}: Thickness of the boundary layer of air surrounding a wheel.

H_b: Hardness of material bulk.

$\overline{h}_{cu}$: Mean thickness of uncut chips.

h_{cu}: Uncut chip thickness.

h_{eq}: Equivalent chip thickness.

h_f: Convection coefficient for fluid cooling in contact zone.

h_{fu}: Mean thickness of the useful process fluid layer that passes through the contact.

h_g: Convection coefficient for an abrasive grain.

h_{jet}: Thickness of a uniform jet of process fluid.

H_m: Momentum power of fluid at wheelspeed.

h_p: Depth of fluid penetration into the wheel.

H_p: Fluid pumping power.

h_{pores}: Mean depth of pores at surface of wheel.

H_{RC}: Rockwell hardness.

H_s: Hardness of material at surface.

h_{slot}: Slot nozzle gap thickness.

H_t: Total pumping and momentum power required to deliver fluid through grinding contact.

H_v: Vickers hardness.

h_w: Convection coefficient for workpiece.

h_{wg}: Convection coefficient for the workpiece at a grain contact.

ierf[u]: Integral error function for argument u.

ierfc[u]: Integral complementary error function for argument u.

K: Archard-Preston wear coefficient; yield shear stress, flow stress; permeability of a wheel.

k: Shear flow stress; thermal conductivity.

K_{1C}: Fracture toughness.

k_m: Stiffness of the machine-tool-workpiece system.

$K_0[u]$: Bessel function of second kind order zero for an argument of value u.

k_w: Thermal conductivity of workmaterial.

L: Cutting edge spacing in cutting direction; Peclet number for moving band source.

l_c: Contact length of tool and workpiece; uncut chip length.

l_{cap}: Capillary length.

l_f: Contact length due to force arising from deflection.

l_{fr}: Contact length due to force for a real surface with roughness.

l_{fs}: Contact length due to force for a smooth body.

l_g: Geometric contact length.

l_{mc}: Length of median crack.

M: Mesh number; measure of grain size based on number of wires used in sieve.

$M_{air}{'}$: Momentum of air boundary layer around wheel per unit wheel width.

M_f: Momentum of process fluid.

M_f':   Momentum of process fluid per unit wheel width.

n_s: Tool rotational speed.

n_w: Workpiece rotational speed.

P, P_c: Cutting or grinding power.

p: Pressure; normal stress.

$P_c(x)$:   Probable number of active cutting edges per unit length in cutting direction.

$P_c(z)$:   Probable number of active cutting edges per unit length in lateral direction.

P_c':   Cutting power per unit width.

P_e: Peclet number for moving band source.

p_p: Pumped pressure.

q: Tangential stress; speed ratio, tool-speed over workspeed; heat flux, heat per unit area per unit time.

q_{ch}:   Heat flux to chips.

Q_f: Flowrate of process fluid.

q_f: Heat flux to process fluid.

Q_{fu}:   Useful flowrate; flowrate that passes through contact zone.

q_o, q_t: Mean heat flux; total heat flux.

q_s: Heat flux to abrasive.

Q_s: Rate of tool wear.

Q_s':   Rate of tool wear per unit width.

q_w: Heat flux to workpiece.

Q_w: Rate of workmaterial removal.

Q_w':   Rate of workmaterial removal per unit width of contact.

r: Ratio of chip groove width to groove thickness.

R_a: Measure of average surface roughness.

R_{ch}:   Proportion of total heat taken by chips.

r_{cu}:   Uncut chip aspect ratio.

R_e: Reynolds number.

R_f: Proportion of total heat taken by fluid.

r_g: Effective radius of grain.

R_{ij}:   Notation used to represent 2-D positions of atoms.

R_L: Ratio of real contact length to geometric contact length.

r_o: Effective radius of wear flat on tip of grain.

R_r: Roughness factor; ratio of contact lengths due to deflection for rough and smooth surfaces.

R_s: Proportion of total heat taken by abrasive.

R_t, R_z: Measures of peak-to-valley roughness.

R_w: Partition ratio; proportion of total heat taken by workpiece.

R_{ws}:   Workpiece-abrasive subsystem partition ratio.

s: Feed of workpiece per cutting edge.

S_{cu}:   Uncut chip surface area.

S_o: Sommerfeld number.

t: Time.

T: Tool life or tool redress life; temperature.

t_c: Time of contact of point on workpiece based on real contact length.

T_d: Maximum wear spot on diamond dressing tool.

t_g: Time of contact of point on workpiece based on geometric contact length.

t_{gc}:   Time for passage of one grain through the contact zone.

T_{max}:   Maximum value of workpiece background temperature.

t_s: Machining time.

T_w: Workpiece background temperature in contact zone.

T_{wg}, T_g: Spike temperature at a contact between grain and workpiece; flash temperature.

U_d: Overlap ratio in dressing; active width of dressing tool divided by dressing feed.

v: Speed of a moving heat source.

V_a: Useful volume of abrasive layer.

V_b: Volume percentage of bond material in an abrasive structure.

v_c: Cutting speed.

v_{cap}:   Mean fluid velocity through capillary.

V_{cu}:   Uncut chip volume.

v_d: Velocity of roll dresser.

$v_{f.wear}$:   Rate of reduction of wheel radius due to wear.

v_f: Feedrate.

v_{fd}:   Dressing feedrate.

v_{fe}:   Effective feedrate in centerless grinding.

V_g: Volume percentage of abrasive grain in a wheel structure.

VI: Viscosity index.

v_{jet}:   Fluid jet velocity.

$v_{orifice}$:   Fluid velocity from orifice.

V_p: Volume percentage of air in an abrasive structure; porosity.

V_{pw}:   Effective porosity ratio at wheel surface; pore volume/ wheel volume at surface.

v_r: Regulating wheelspeed in centerless grinding.

v_s: Speed of the wheel or tool.

V_s: Total wear volume of wheel or tool material removed.

v_w: Speed of the workpiece.

V_w: Total volume of workpiece material removed.

W_d: Diamond size (carats).

w_{slot}: Width of slot nozzle.

$x, y,$ and z: Tangential, axial, and radial position coordinates.

$\{X\}$: A set of input parameters.

$\{Y\}$: A set of output parameters.

y_b: Thermal boundary layer thickness of the process fluid in the contact zone.

$\Phi\ (r)$: Potential energy at atomic distance r.

Ψ: Indenter angle.

α: Feed angle in angle-approach grinding; thermal diffusivity.

α_p: Workplate angle in centerless grinding.

β: Included tangents angle in centerless grinding; thermal property for transient heat conduction; wheel angle in angled-wheel grinding; thermal property for transient heat conduction in a material.

δ: Deflection.

ϕ: Angle between plane of motion and inclined plane of heat source.

ϕ_{pores}: Porosity of wheel at surface.

γ: Dullness ratio of abrasive grain; friction angle of slip line; sharpness ratio of dressing tool.

γ_r: Elevation angle of workpiece from regulating wheel center in centerless grinding.

γ_s: Elevation angle of workpiece from grinding wheel center in centerless grinding.

η: Dynamic viscosity.

η_p: Dynamic viscosity at elevated pressure.

μ: Friction coefficient; grinding force ratio.

v: Kinematic viscosity; Poisson ratio.

θ_{mp}: Melting temperature of chips.

ρ: Density.

σ_{hs}: Hydrostatic stress.

σ_n: Direct stress.

τ: Shear stress.

SI UNITS

QUANTITY	UNIT	SYMBOL	EQUIVALENCE
Base Units			
Length	meter	m	
Mass	kilogram	kg	
Time	second	s	
Electric current	ampere	A	
Thermodynamic temperature	kelvin	K	
Amount of substance	mole	mol	
Supplementary Units			
Celsius temperature	celsius	°C	
Angle	radian	rad	$m.m^{-1}$
Frequency	hertz	Hz	s^{-1}
Force	newton	N	$m.kg.s^{-2}$
Pressure and stress	pascal	Pa	$N.m^{-2}$ or $m^{-1}.kg.s^{-2}$
Energy, work, heat	joule	J	$m^2.kg.s^{-2}$
Power	watt	W	$J.s^{-1}$ or $m^2.kg.s^{-3}$
Kinematic viscosity	stoke	St	$10^{-4}\ m^2.s^{-1}$
Dynamic viscosity	poise	P	$1\ N.s.m^{-2} = 1\ Pa.s = 10\ P$
Atomic distances	ångström	Å	$10^{-10}\ m$
Thermal conductivity	—	—	$W.m^{-1}.K^{-1}$ or $m.kg.s^{-3}.K^{-1}$

CONSISTENCY OF UNITS IN EQUATIONS

Equations are always stated without reference to a set of units and do not contain conversion factors. When values are inserted into an equation in order to calculate the value of an unknown, it is important that a consistent set of units is employed. The SI system was developed in order to ensure consistency when using base force, length, mass, and time units. Some examples of consistent sets used in engineering are listed below. Other engineering quantities can be expressed in terms of these units, thus ensuring consistency. Some examples of equivalence in SI units are given in the previous table. Some conversion factors between SI and British units are given below.

SYSTEM	FORCE	LENGTH	MASS	TIME
SI	N	m	kg	s
British Engineering Units	lbf	ft	slug	s
Alternative British Units	lbf	in	$lbf\,in^{-1}\,s^2$	s
British Physical Units	pdl	ft	lb	s

SI – BRITISH CONVERSION FACTORS

Values are rounded to 4 significant figures.

Mass: $1\ kg = 2.205\ lb = 0.06848\ slug = 0.8218\ lbf\,in^{-1}s^2$

Length: $1\ m = 3.281\ ft = 39.37\ in$

Temperature rise: $1 \text{ K} = 1°\text{C} = 1.8°\text{F} = 1.8 \text{ R}$

Force: $1 \text{ N} = 7.233 \text{ pdl} = 0.2248 \text{ lbf}$

Volume: $1 \text{ m}^3 = 1000 \text{ liters} = 61020 \text{ in}^3 = 35.32 \text{ ft}^3$

Pressure: $1 \text{ bar} = 0.1 \text{ MPa} = 14.5 \text{ lbf.in}^{-2}$. $1 \text{ N.m}^{-2} = 0.000145 \text{ lbf.in}^{-2}$

Density:
$$1 \text{ kg.m}^{-3} = 0.06243 \text{ lb.ft}^{-3} = 0.001939 \text{ slug.ft}^{-3} = 0.00001347 \text{ lbf.in}^{-4}.\text{s}^2$$

Energy: $1 \text{ J} = 0.7376 \text{ ft.lbf} = 8.851 \text{ in.lbf}$

Power: $1 \text{ W} = 0.7376 \text{ ft.lbf.s}^{-1} = 8.851 \text{in.lbf.s}^{-1}$

Thermal conductivity: $1 \text{ W.m}^{-1}.\text{K}^{-1} = 0.0141 \text{ lbf.s}^{-1}.\text{R}^{-1}$

Heat transfer coefficient: $1 \text{W.m}^{-2}.\text{K}^{-1} = 0.003172 \text{ lbf.in}^{-1}.\text{s}^{-1}.\text{R}^{-1}$

Kinematic viscosity: $1 \text{ m}^2.\text{s}^{-1} = 10^4 \text{ St} = 10.76 \text{ ft}^2.\text{s}^{-1} = 1550 \text{ in}^2.\text{s}^{-1}$

Dynamic viscosity:
$$1 \text{ N.s.m}^{-2} = 10 \text{ P} = 1000 \text{ cP} = 0.000145 \text{ lbf.s.in}^{-2} = 0.000145 \text{ reyns}$$

Gem size: $1 \text{ metric carat} = 2 \times 10^{-4} \text{ kg}$

Glossary

A

AE sensor: Acoustic emission sensor used to detect wheel contact.

AFM: Atomic force microscopy.

Absorption: Assimilation of matter from a surround into a surface and into the body of the material.

Abrasion: Wearing or machining by rubbing, scratching, or friction.

Abrasion, 2-body: Wear of one body abraded by another.

Abrasion, 3-body: Wear of one body acted on by another with free abrasive in the contact.

Abrasive: Tendency to cause wear; a hard material which tends to wear a softer material.

Abrasive belt: Abrasive tool formed from abrasive layer glued to a flexible belt.

Abrasive composite: Material consisting of abrasive grains and bond structure.

Abrasive, conventional: Hard abrasive mineral grit such as alumina and silicon carbide.

Abrasive grains: Hard particles in an abrasive tool that provide the cutting edges.

Abrasive machining: Machining by an abrasive process such as grinding, honing, lapping, or polishing.

Abrasive medium: Suspension of abrasive grit in a liquid as used in lapping and polishing.

Abrasive tool: Tool used for abrasive grinding, lapping, honing, or polishing.

Abusive grinding: Grinding under conditions which produce severe damage, cracks, or burn.

AC additive: Anticorrosion additive.

Accuracy: Difference between intended value and achieved value.

Active cutting depth: Depth of abrasive surface that actively engages the workpiece surface.

Active layer: The layer of the abrasive tool that contains the abrasive.

Active surface: The surface of the active layer that makes abrasive contact.

Additivation: The adding of additives to a lubricant or process fluid.

Additive: Substances added in small quantities to modify properties of a lubricant or process fluid.

Adhesive loading: Workmaterial adhering to tips of abrasive grains of active layers.

Adhesion: Sticking together of two surfaces.

Adiabatic: Process takes place without heat loss or gain within an immediate volume.

Adsorption: Assimilation of material into close physical chemical contact with a surface.

Affinity: Measure of possibility of chemical reaction between two materials.

Agglomeration: Tendency of particles to group together in clumps within a masss.

Air barrier: Boundary layer of air around a high-speed wheel tending to deflect grinding fluid.

Algorithm: Series of statements constituting a calculation routine.

Aloxide, Alox: Terms sometimes used for aluminum oxide abrasive.

Alumina, Al_2O_3: Aluminium oxide abrasive.

Ambient: Conditions in the atmosphere surrounding the process.

Amorphous layer: A thin layer lacking organized structure due to abrasive deformation.

Analytical: Based on mathematical or logical consideration.

Angle grinding: The axis of the wheel or the feed motion is angled to the workpiece axis.

Anisotropic: Material having properties dependent on direction of measurement.

Annulus: Space or shape between two concentric diameters.

AO additive: Anti-oxidation additive used to increase service life of a process fluid.

Apparent contact area: Total area of contact, including spaces between contacts.

Apparent contact pressure: Normal force divided by apparent contact.

AR additive: Antirust additive.

Archard Constant: Coefficient used in Archard-Preston's Law for wear rate.

Archard-Preston's Law: Relationship between wear, normal force, hardness, and sliding distance.

Aromatic oils: Cyclo-aromatic and mixed structure nonsaturated hydrocarbons.

Arrhenius Law: Relationship for rate of chemical action based on temperature and time.

Asperity: Sharp edge.

Atmosphere: Air surrounding a process.

Atp: Standard atmospheric temperature and pressure.

Auxiliary processes: Processes additional to the main abrasive process, e.g., fluid delivery, dressing.

AW additive: Antiwear additive.

Axial force: Component of force axial to the grinding wheel surface.

B

Background temperature: Mean temperature at a point in a region of many flash contacts.

Bactericide: Additive that kills bacteria.

Balancing: Adding or removing weight to improve balance from a wheel.

Band heat source: Wide heat source having finite length.

Barite: Barium sulphate, sulphide mineral used for polishing.

Bauxite: Impure ore of aluminum and alumina.

bcc: Body-centered cubic lattice structure.

Bearing area curve: Area of solid phase increasing with depth into surface expressed as a fraction.

Bearing steel: Group of steels used for manufacture of rolling bearing elements.

Bessel functions: Mathematical functions used in solving moving heat source problems.

Binder: Medium for suspension of abrasive particles.

Biocide: Additive to kill bacteria and improve life of an oil.

Black box: Unknown system characterized by measuring inputs and outputs.

Blunt cutting edge: Cutting edge worn to a flat.

Boiling temperature: Temperature at which grinding fluid rapidly vaporizes producing bubbles.

Bond material: Material that bonds the abrasive grains in a tool.

Bond post: Bond material joining one grain to another.

Bore grinding: Grinding an internal cylindrical surface.

Boundary layer: Region close to a surface, e.g., as in fluid.

Boundary lubrication: Sliding contact with molecular layer of lubricant.

Brake-dresser: A dressing tool driven by the wheel, speed is controlled by a brake.

Brinell hardness: Measure of hardness using a standard ball indentor.

Brinelling: Indentation due to compressive action of blunt grains.

Brittle: Tendency to fail by cracking fracture.

Brittleness index: Measure of brittleness based on fracture toughness and Vickers hardness.

BUE: Built-up edge, material piled up at leading edge of tool while cutting.

Bulk temperature: Mean temperature of the whole workpiece.

Burn: Action of oxidation or material damage due to high maching temperature.

Burn-out, fluid: The complete drying out of grinding fluid in the contact zone at high temperature.

Burnishing: Smoothing by action of friction.

Burr: Lip of deformed material extending from edge of cut surface.

Bursting speed: Speed at which a wheel fails due to hoop stresses.

C

Carat: Measure of diamond grain size.

Carborundum: Early name for silicon carbide abrasive.

Carcinogenic: Tending to lead to cancer in humans.

Cation: Positively charged ion.

Cubic boron nitride (CBN): A extra-hard allotropic form of boron nitride.

CD: Continuous dressing.

Computational fluid dynamics (CFD): A type of computer software for fluid flow.

CFRP: Carbon fiber reinforced plastic.

C-factor: Temperature factor for heat conduction from a moving source into a workpiece.

Capillary tube: A tube of large length-to-diameter ratio.

Centerless grinding: Process for grinding without center-holes for workpiece support.

Ceramic: Materials made by firing clays or similar materials, e.g., silicon nitride ceramic.

Chalk: Fine white calcium carbonate powder used for polishing.

Chatter vibration: Regenerative vibration arising from an unstable machining process.

Chemi-sorption: Assimilation of a material on a surface by chemical action.

Chip aspect ratio: Ratio of length/width of uncut chip.

Chip, uncut chip: Undeformed workpiece material in the path of an oncoming abrasive grain.

Chip, cross-sectional area: Cross-sectional area of the uncut chip.

Chip length: Length of the uncut chip.

Chip thickness: Maximum or mean thickness of uncut chip.

Chip volume, mean: Volume of material removed divided by number of chips.

Chip width: Width of the uncut chip.

Chips: Pieces of material or swarf cut from workpiece.

CIRP: International College/Institution of Production Engineering Research.

Cleaning-up: Removing a layer of abrasive to remove loading and restore unworn surface.

Climb grinding: The grinding motion and the workpiece motion are in same direction.

CMC: Ceramic matrix composite.

CMM: Chemo-mechanical machining; Coordinate measuring machine.

CMP: Chemo-mechanical polishing.

CNC: Computer numerical control of a machining system.

Coarse dressing: Dressing with large dressing depth and large dressing feedrate.

Coated abrasive: Abrasive applied as a coating to a belt.

Compatibility: Suitability of two materials to form rubbing couple without severe damage.

Compliance: Inverse of stiffness, movement per unit force.

Composite: Body formed from mixture of two or more materials.

Concentration: Proportion of a material in a mixture or in a solution.

Conditioning: Process to prepare an abrasive surface for machining; see also *Dressing*.

Conduction: Transfer of energy through a body, e.g., heat conduction.

Conformal contact: Convex surface contacting within a concave surface.

Constant force process: Abrasive process controlled by application of constant force.

Contact: Touching between one body and another.

Contact area: Apparent area of contact between abrasive tool and workpiece.

Contact area, real: Sum of grain contact areas with workpiece.

Contact, grain: Contact between grain and workpiece.

Contact, wheel: Contact between wheel and workpiece.

Contact length: Length of tool and workpiece contact parallel to grinding direction.

Contact length, force: Contact length due to effect of force and deformation.

Contact length, geometric: Contact length predicted ignoring roughness, speeds, and deformation.

Contact length, kinematic: Geometric contact length modified for speed of wheel and workpiece.

Contact length ratio: Ratio of real contact length/geometric contact length.

Contact length, real: Contact length including all influences; also effective contact length.

Contact mechanics: Analysis of contact, particularly due to elastic/plastic deflections.

Contact pressure: Normal force divided by contact area; see *Real* and *Apparent contact pressure.*

Contact radius: Mean radius of the wear flat on an abrasive grain.

Contact surface: Surface in (apparent) contact area during abrasive machining.

Continuous dressing: Process of dressing concurrent with grinding.

Controlled feed process: Abrasive process controlled by application of feed motion.

Control wheel: Wheel used to control workpiece motion in centerless grinding.

Control wheelhead: Powered assembly to carry and rotate control wheel.

Convection: Process of carrying by physical movement, e.g., heat carried by motion of fluid.

Convection coefficient: Convective heat transfer per unit contact area per degree temperature difference.

Conventional: In accordance with traditional or widest practice.

Conventional speed: Wheelspeeds from 20 m/s to 45 m/s.

Coolant: Grinding fluid.

Corundum: Traditional name for aluminum oxide mineral.

Corrosion: Conversion of workmaterial surface by chemical or electro-chemical action.

Corrosion wear: Removal of material from a surface by corrosion.

Cost: Price to be paid.

Couette flow: Flow induced by proximity to a sliding surface; entrained flow.

Coulomb friction: Friction force is proportional to normal force.

Covalent: Electrons are shared by neighboring atoms.

Creep grinding: Grinding at very low work speeds and usually with large depth of cut.

Criterion: Standard for making a judgement; basis for making a decision.

Critical: Point at which an abrupt change takes place, e.g., critical temperature.

Critical temperature: Temperature for a change of physical condition, e.g., softening.

Cross section: View of section of body cut through to reveal internal structure.

Cross-sectional area: Area of a cross section.

Crush dressing: Dressing or truing by crushing. See *Crushing*.

Crushing: Removing a layer of abrasive by applying pressure with a block of softer material.

Crushing roll: Roller or disc used for crush dressing.

Crystal: Solid of regular atomic structure, such as quartz.

Crystalline: Having an ordered atomic lattice structure.

Curvature: Inverse of radius.

Cut, cutting: Process of shearing a material; abbreviation for depth of cut.

Cutting edge: That part of an abrasive particle that engages the workpiece.

Cutting edge depth: Depth of penetration of the grains into the workpiece.

Cutting edge density: Number of cutting edges per unit area; varies with depth.

Cutting edge spacing: Measure of spacing between cutting edges on an abrasive surface.

Cutting edge width: Measure of the width of the cutting edge in contact with the workpiece.

Cutting force: Resultant force in cutting process; vector sum of component forces.

C type corrosion: Corrosion with continuous progression.

CVD: Chemical vapor deposition process used to form diamond layer.

Cycle time: Mean time for the machining of a part.

Cylindricity: Measure of deviations from a cylindrical shape.

Cylindrical grinding: Grinding a cylindrical surface by rotating a workpiece.

D

D'Alembert force: The reaction experienced when accelerating a mass.

Damping force: Reaction force proportional to speed.

DBDS: Dibenzyl disulphide.

Debris: Small particles of abrasive and workpiece, e.g., swarf.

Deep grinding: Grinding depths of cut much in excess of 0.1 mm.

Delamination: Plate-shaped particles breaking out of surface.

Dendritic: Tree-like crystalline growth structure.

Density: Mass per unit volume.

Depth of cut: Instantaneous normal thickness of layer of material to be removed.

Detergent: Cleansing agent; additive to oil for cleansing surfaces.

Diamond tool: Cutting tool using diamond-cutting edges.

Dicing: Cutting into slices.

Diffusion: Migration of energy or atoms through a material.

Diffusivity: Thermal property; conductivity divided by density and specific heat.

DIP-slide: Used for measurement of bacterial concentration.

Direct stresses: Tensile or compressive stresses.

Dislocation: Vacancy in a structure, e.g., within a structure of atoms.

Dispersing additive: Additive to keep solid particles in suspension in the fluid.

Disturbance: Change to input value which causes system output to change.

Double-side grinding: Process for simultaneously grinding two sides of a workpiece; also known as duplex grinding.

Down grinding: The wheel motion and workpiece motion are in the same direction.

Dresser head: Powered assembly to carry and drive rotary dressing tool.

Dressing: Process to prepare an abrasive surface for machining; see also *Truing* and *Conditioning*.

Dressing, continuous: Process of dressing at the same time as grinding.

Dressing depth: Depth of cut in dressing operation.

Dressing feedrate: Traverse/feed rate of dressing tool in dressing operation.

Dressing increment: Dressing depth.

Dressing lead: Traverse/feed distance of dressing tool per wheel revolution.

Dressing plate: Plate dressing tool with diamond grit in the tool cutting surface.

Dressing, roll: Friction or power driven roll-shaped dressing tool.

Dressing sharpness ratio: Dressing depth/dressing width of wedge or cone-shaped tool.

Dressing speed: Peripheral speed of a dressing roll.

Dressing stick: A stick-shaped abrasive tool used to dress a wheel.

Dressing tool: Tool used for dressing, e.g., diamond, fliese, or rotary dressing tool.

Dressing tool-life: Tool-life of the dressing tool.

Dry machining: Machining without use of a process fluid.

Ductile: Tendency to deform by shear without cracking fracture.

Dull: Blunt; opposite of sharp.

Dynamics: Analysis of motions due to forces.

Dynamic balancing: Adding or removing weight in two planes to improve couple balance.

Dynamic stiffness: Stiffness at a particular frequency or at resonance.

E

EAM: Embedded atom method used in molecular dynamics.

Eccentricity: Displacement of a center of rotation relative to another.

ECG: Electro-chemical grinding.

ECM: Electro-chemical machining.

EDM: Electrical discharge machining.

Effective porosity ratio: Pore volume in active layer divided by total layer volume.

EHL: Elasto-hydrodynamic lubrication.

Elastic deformation: Recoverable linear deformation under load.

Elastic modulus: Material property; rate of increase of tensile stress with strain in tensile test.

Electrolysis: Material transfer from an electrode due to electric current and an ionized electrolyte.

Electrolyte: Electrically conducting liquid used in electrolysis.

Electroplated abrasive: Abrasive grains attached to tool by electroplated metal.

Element: A part of a system.

ELID: Electrolytic in-process dressing; a process for metal-bonded wheels.

Emery: Black abrasive based on corundum with magnetite or haematite.

Empirical: Deriving from a limited range of measurements, not based on physics.

Emulsion: Grinding fluid consisting of mixture of soluble oil in water.

Emulsion stabilizer: Surfactant added to resist separating out of the dispersed phase.

Energy: Capacity to do work; measure of work expended.

Energy dissipation: Transformation of work energy into heat.

Energy partition: Analysis of energy dissipation to particular heat sinks.

Entrained flow: Fluid flow induced by parallel sliding of surface; Couette flow.

Environment: Conditions in surround, e.g., atmosphere, noise, temperature.

EP additive: Extreme pressure additive for process fluid.

EPHL: Elasto-plasto hydrodynamic lubrication.

Equivalent diameter: Grinding wheel diameter modified to allow for workpiece diameter.

Equivalent chip thickness: Thickness of the layer of chips emerging from grinding action.

Erosion: Wear by series of small impacts from gas, liquid, or solid particles.

Error: Difference between measured value and ideal value, e.g., size error.

Error function: An integral function used in heat transfer calculations.

Esters: Compounds produced by acid-alcohol reactions with elimination of water.

Exo-emission: Radiation of photons or electrons.

External grinding: Grinding an external surface particularly for cylindrical grinding.

Extreme pressure (EP): Extreme pressures as in extreme pressure lubrication.

F

Face grinding: Grinding a flat surface including the flat surface on a cylindrical part.

Fatigue life: Expected life under a cyclic loading condition.

fcc: Face-centered cubic lattice structure.

Feed: An increment of grinding wheel position normal to machined surface.

Feed per cutting edge: Distance moved by workpiece in interval between succeeding cutting edges.

Feedrate: Speed of grinding wheel movement normal to machined surface.

FEM: Finite element modeling.

Ferrous material: Material containing mainly iron.

Fine dressing: Dressing with small dressing depth and small dressing feedrate.

Finish surface: Workpiece surface after abrasive machining.

Fixture: Fixed (holding) device.

Flash point: Lowest temperature at which vapor above a liquid may be ignited in air.

Flash temperature: Peak temperature at an individual cutting edge/grain.

Flatness: Measure of deviations from a flat plane.

Fliese: Diamond-coated wedge-shaped tool used for dressing.

Flood nozzle: Nozzle delivering a large volume of fluid at low velocity.

Flow utilization: Useful flow divided by total flow.

Fluid: Liquid or gas that can flow.

Flushing: Displacing swarf by the action of a fluid jet.

Flux: Flow; heat; rate of heat flow, e.g., heat per unit area per second.

Flux distribution: Distribution of flux in contact zone, e.g., triangular distribution.

FM additive: Friction modifier additive for process fluids.

Foam depressant: Additive to promote bubble coalescence.

Fog: Mist.

Forced vibration: Vibration due to application of an harmonic force.

Form grinding: Grinding a form profile with a form tool or by form generation.

Fracture toughness: Measure of resistance to fracture under impact loading.

Free radical: Reactive atom or molecule containing an unpaired electron.

Free abrasive: Loose abrasive grains used in free abrasive processes.

Free surface: Surface open to the environment.

Friable: Tending to fracture under compression.

Friction: Resistance to slip between two sliding surfaces.

Friction coefficient: Ratio of tangential force to normal force between two sliding bodies.

Friction couple: The pair of interacting body materials in abrasive contact.

Friction polymer: Tribocatalytically polymerized oxidized hydrocarbons.

Friction pair: Friction couple.

Friction polymer film: Polymer film on worksurface formed by friction process.

Friction power: Power required to overcome frictional drag.

Friction ratio: Ratio of interface shear stress/shear flow stress of the softer bulk material.

G

Garnet: Crystalline silicate abrasive.

Gaussian distribution: Normal distribution.

Gear grinding: Grinding a gear surface by hobbing or shaping motions.

Geometrics: Analysis of points, lengths, lines, curves, and shapes.

Glassy bond: Noncrystalline vitreous bond.

Glazing: Condition of large wear flats on the abrasive grains.

GMC: Glass matrix composite.

Grade, grade letter: System used to classify hardness of abrasive layer.

Grain boundary: Boundary of grains within a granular structure such as soft steels.

Grain, grit: Abrasive particle.

Grain contact: Touching between a grain and the workpiece.

Grain contact time: Time a grain is in contact with workpiece; time a point on workpiece is in contact with a grain.

Grain force: Resultant force on an abrasive grain; the sum of component forces on a grain.

Grain depth, penetration: Depth of penetration of a grain into the workpiece.

Grain protrusion: Measure of height of grain tips above surrounding bond.

Grain size: Measure of the sizes of abrasive grains.

G-ratio: Volume of material removed divided by volume of grinding wheel removed.

Grinder: Shop-floor term for grinding machine or operative.

Grinding: Removal of material from a surface by abrading with a hard rough surface.

Grind-hardening: Novel process for hardening steels from soft state by grinding at high temperatures.

Grinding fluid: Fluid used to lubricate, cool, and flush abrasive processes; see also *Process fluids*.

Grinding force: Resultant force in grinding process; a vector sum of component forces.

Grinding force ratio: Ratio of tangential force to normal force.

Grinding power: Product of tangential force and tangential grinding wheelspeed.

Grinding time: Part of time spent in grinding.

Grinding wheel: Cylindrical abrasive tool rotated at high speed in machining.

H

Hardness: Measure of resistance to penetration.

Hardness, hot: Hardness of a material at elevated temperature.

Harmonic: Integer multiple of basic frequency.

Heat: Thermal form of energy.

Heat conduction: Transfer of heat through a body due to temperature gradient.

Heat sink: Where heat goes: heat flows from a source to a sink.

HEDG: High efficiency deep grinding.

Hencky equations: Equations for rotation of slip lines.

Hertz: Unit of frequency: cycles per second.

Hertzian: Smooth elastic contact between a sphere and a plane or two spheres.

High speed: Wheelspeeds in excess of 45 m/s.

HLB: Hydrophile lipophile balance.

Homogeneous: Constant material composition throughout.

Honing: A cylindrical process using abrasive stones.

Honing stones: Abrasive blocks inserted in honing tools.

Horizontal grinder: Grinding wheel axis is horizontal.

HSS: High-speed steels used for cutting tools.

h-value: Heat convection value for oil under a standardized condition.

Hv: Vickers hardness measure.

Hydrodynamic: Usually refers to action of fluid due to surface movements; strict meaning: action due to fluid motion.

Hydrogenation: Hydrogen reaction used typically to saturate and stabilize a fatty oil.

Hydrostatic stress: Contribution to stress system for equal direct stresses, usually compressive.

I

Image processing: Techniques for extracting information from surface data.

Inclined heat source: Heat source moves in a plane inclined to the workpiece surface.

Internal grinding: Grinding an internal surface, particularly bore grinding.

Ion, ionic: Electrically charged atom or group of atoms; having electrically charged atoms or groups of atoms.

ISO: International Standards Organization.

Isotropic: Material with properties constant in all directions.

J

Jet: High velocity fluid stream; orifice for high velocity fluid.

Junction growth: Growth of an area of sticking contact.

Jewellers rouge: Red iron oxide powder used for polishing.

K

Kaolin: Fine white clay made into a paste for polishing.

Kinematics: Analysis of motions, ignoring forces.

Kinematic similarity: Uncut chip dimensions remain unchanged.

Kinetics: Effects or study of rates of action.

Kramer Effect: Exo-emission under abrasive conditions.

L

Laminar: Fluid tending to move along steady parallel paths; nonturbulent.

Lap, lapping: An abrasive tool used for lapping; process of improving form using a lap and abrasive paste/fluid.

Lattice: Structure of directional bonding of atoms.

LDA: Laser-Doppler Anemometry.

Limit: Maximum or minimum permissible value.

Limit chart: Chart showing maximum achievable values of process control variables.

Liquid: Intermediate phase on cooling between gas and solid, e.g., water or oil.

Log, diamond: Synthetic diamond coated on a prismatic log shape.

Longitudinal grinding: Grinding with a wheel traversing the workpiece length.

Losses: Difference between input and useful output, e.g., power losses.

Lower bound: Estimate known to be lower than real value.

Lubricant: Medium such as oil or graphite that eases sliding between surfaces.

Lubricity: Ability of a fluid to reduce friction other than by its viscosity.

M

M50: A tool steel specification.

Machinability: Qualitative measure for a material of machining ease.

Machine tool: A powered machine or system used in part production.

Machining: Production of shape by removal of material from a part.

Machining conditions: Process conditions; values of parameters employed for machining.

Magma-plasma: Material in energetic state due to intense abrasive deformation.

Magnetic abrasive machining: Process where magnetism applies force on the abrasive.

Martensite: Hard brittle phase of carbon dissolved in iron; white phase.

Material removal: Volume of material removed from workpiece.

Metal-bond wheels: Superabrasive wheels bonded with cast iron or other metal compositions.

Mean: Arithmetic average of a series of readings.

Mesh number: Measure of grit size based on number of wires in a sieve: high mesh number yields small grit size.

MIC value: Minimum inhibitory concentration of fluid.

Microscopy: Magnified visualization of a surface using one of several physical principles.

Microhardness: Hardness measured with an extremely small indentor.

Mineral oil: Natural hydrocarbon oils.

Misalignment: Deviation from parallelism between two elements.

Mist: An aerosol dispersion of particles/fluid in the atmosphere.

Mixed lubrication: Transition between hydrodynamic lubrication and boundary lubrication.

MMC: Metal matrix composite.

MNIR: Maximum normal infeed rate.

Model: Mathematical, physical, or conceptual representation of a structure or process.

Mohrs Circles: Construction relating direct and shear stresses.

Molecular dynamics (MD): Simulation of deformation of group of molecules or at a quasimolecular level.

Momentum power: Rate of kinetic energy.

Monocrystal: Grains constitute a single crystal.

Morphology: Shape, form, and structure of a body.

MQL: Minimum quantity lubrication.

N

Nano-: Refers to the nanometer order of magnitude; less than 0.02 microns.

Nano-additive: Nanosize powders used as additives.

Nanogrinding: Grinding with nanosize grain penetration.

Nanometer: 1 meter divided by 10 raised to the power of 9.

Nanotechnology: Technology involving machines and processes at the scale of a few nanometers.

Napthenes: Cycloparaffinic hydrocarbon oils.

Natural fatty oil: Animal, vegetable, or fish oil.

ND: Natural diamond.

Neat oil: Oil not mixed with water; immiscible oil.

Nip: Convergent gap at entry to grinding contact zone that creates coolant wedge.

Noise: Unwanted vibration.

No-load power: Power with grinding wheel rotating, but not grinding.

Normal: Perpendicular; usual.

Normal distribution: Continuous random distribution with same mean, median, mode, e.g., bell curve.

Normal force: Component of force perpendicular to grinding wheel surface.

Nozzle: The end of a pipe or hose shaped to direct grinding fluid.

Nozzle, jet: Nozzle shaped to intensify exit velocity of the grinding fluid.

Nozzle, shoe: Fluid delivery nozzle shaped to fit snugly around a wheel.

O

OOR: Out of roundness; roundness error.

Operator, machine: Person who operates a machine.

Organic: Having biological origins; based on or related to carbon compounds.

Oscillating: Motion to and fro.

Overlap ratio: Contact width of dressing tool/feed per revolution of wheel.

P

Padding: Solids added to bond mixture to modify the effective hardness of an abrasive layer.

Paraffinic oil: Linear or ramified saturated hydrocarbons.

Particle: Small part, fragment, grit, or grain.

Partition ratio: Energy to a heat sink/total heat energy.

Passivation: Slowing corrosion due to inhibitor or protective layer.

Passivator: Additive to prevent catalytic reaction.

Paste: Thick mixture of abrasive particles in liquid or wax.

PCBN: Polycrystalline cubic boron nitride.

PCD: Polycrystalline diamond.

Peclet number: Dimensionless speed parameter for moving heat sources.

Pendulum grinding: Shop-floor term for forward and back traverse grinding.

Permeability: Measure of the ability of fluid to diffuse through a material.

pH: Measure of acidity/alkalinity; acids have a pH less than 7.

pH-meter: Alkalinity meter.

Physi-sorption: Layers bonded by weak van der Waal forces.

Pick-up: Scuffing; a sensor.

Plane: A flat section of space.

Plasma: Hot ionized gas.

Plastic deformation: Nonrecoverable deformation by ductile shear within a material.

Plateau honing: Removing the peaks of the machining marks to form plateaus.

Ploughing: Grooving a surface without loss of material.

Ploughing energy: Cutting energy, less chip energy, and sliding energy.

Plunge grinding: The grinding wheel is fed directly into the workpiece.

PMC: Polymer matrix composite.

Poisson distribution: Discrete distribution; alternative to binomial distribution.

Poisseuille flow: Flow induced by a pressure gradient.

Polishing: Use of a conformable pad and an abrasive to smooth a surface.

Polycrystalline: Many small closely packed crystals form the grains.

Pore: A small hole or channel in a structure.

Pore loading: Abrasive layer with workmaterial loaded into pores.

Porosity: Measure of air/pores in a structure.

Potential energy: Energy having the potential to do work, e.g., pressure or height energy.

Potential function: Model of variation of potential energy, e.g., between atoms.

Preston Equation: Alternative source of Archard Equation.

Primary shear zone: Zone of primary shear between workpiece and chip.

Process: Action or sequence of actions or operations.

Process conditions: Specification of speeds, feeds, tools, fluid, and all conditions of the process.

Process fluid: Gas or liquid used to lubricate, flush, and cool an abrasive process.

Profile grinding: Grinding a form on a workpiece with a form tool or by generation.

Profilometry: Techniques for measuring shape or surface texture.

P type corrosion: Corrosion with passivated progression.

Pumice stone: Porous volcanic abrasive used for scouring and polishing.

Pumping power: Power required to deliver a flow from a pump.

Q

Qualification: Measurement/proof of quality.

Quality: Term used for attributes such as fitness for purpose, roughness, and durability.

Quartz: Hard crystalline silicon dioxide colorless rock.

Quasi-: Almost, but not exactly the real situation.

R

Rake angle: Angle between workpiece normal and leading cutting tool face.

Range: Difference between minimum and maximum values, e.g., cutting edge depth range.

Rare earth element: Elements of the lanthanide series.

Real contact pressure: Normal force divided by real contact area.

Reciprocal grinding: Successive forward and back traverse grinding.

Redundant energy: Energy consumed in nonuseful deformation.

Redress life: Machining time between dress and redress of a tool.

Rehardening: Heating of hardened surface to transformation followed by quenching.

Rehbinder Effect: Strength reduction due to adsorption of fluid molecules onto material surface.

Removal rate: Volume rate of material removal from a workpiece.

Replica, surface: A molded impression of a surface.

Residual stress: Stress that remains in a material after load is removed.

Resinoid wheel: Wheel having abrasive grains in resin-based bond.

Reynolds number, R_e: Dimensionless parameter increases with ratio of inertia forces to viscous forces.

Ringing: Acoustic test for flaws in a wheel from sound of a sharp tap.

RMS, rms: Square root of the mean of the sum of the squares of the differences from the mean.

Rolling: Motion without sliding, as of roller on plane.

Roll dresser: Dressing tool in form of roller or disk.

Roughness factor: Ratio of contact lengths due to normal force for rough and smooth contact.

Rounding: Process of improving roundness.

Roundness, error: Measure of deviations from a circle; out of roundness.

Roughness: Measure of micro-deviations in height of a surface.

Rubbing: Sliding of one surface on another with frictional contact.

Rules of mixture: Rules for properties of a mixture related to element properties.

Run-out: Error of circular motion.

S

SAE52100: A bearing steel specification.

Sanding: Sandpaper smoothing of wood or similar material.

Scratch: A groove made by dragging a sharp hard tool along a surface.

Screw grinding: Grinding a screw surface by rotation and axial feed of a workpiece.

Scuffing: Welded workpiece material on tool, pulls material out of the workpiece surface.

SD: Synthetic diamond.

Secondary shear zone: Zone of shear between chip and cutting face of tool.

Seeded gel: A tough commercial alumina abrasive that wears by microfracture.

Segmented wheel: Wheel with a number of pieces or segments of abrasive layer attached to a holder.

Seizure: General meaning: the welding together of the couple elements.

Seizure tendency: Similar meaning in abrasive machining to scuffing.

Self-dressing, self-sharpening: Tendency of grains to produce sharp edges when worn edges fracture.

SEM: Scanning electron microscopy.

Set-up: Geometry, speeds, and conditions for abrasive machining.

Shallow grinding: Grinding depths of cut less than 0.1 mm.

Sharp cutting edge: Pointed cutting edge.

Sharpness: Measure of angularity or narrowness of cutting tip.

Shear: Plastic sliding within a material acting like a pack of cards.

Shear strain rate: Rate of plastic sliding/thickness of sheared zone.

Shellac abrasive: Abrasive grains held together in a shellac resin bond.

Shoe nozzle: Nozzle that fits around the wheel periphery.

Side grinding: Grinding with the side of the wheel.

SI, Systeme International d'Unités: International system of units.

Siccative, siccativator: Drying agent.

Silica, SiO_2: Silicon dioxide; occurs naturally as quartz.

Silicon carbide, SiC: A hard bluish-black abrasive.

Simulation: Step by step imitation of a process over a short time period.

Single-point dressing: Dressing with a single-point tool such as a single diamond.

Sintering: Process of forming solid by partial fusing of compacted powder.

Size effect: Tendency for reduced specific energy with larger chip size.

Slide, slideway: Machine element for guiding sliding motion.

Sliding: Relative movement of two surfaces in tangential direction.

Sliding energy: Component of energy proportional to sliding area of grains.

Sliding heat source: Heat source that moves parallel to the workpiece surface.

Slip line field: Lines of maximum shear stress in a plastic field.

Slot nozzle: Nozzle having a large gap width to thickness ratio.

Slurry: Dense suspension of powder in a liquid.

Soap: Product of hydrated oxides and fatty acids; metallic salt of a fatty acid.

Solid lubricant: Lubricant in the form of a lamellar solid such as graphite.

Solution: Mixture formed of material dissolved in a liquid or solid.

Spacing length: Cutting edge spacing in direction of grinding.

Spacing width: Cutting edge spacing in lateral direction.

Spark-out: Period of dwell while depth of cut decreases.

Specific: Particular form of a parameter, e.g., specific force = force/unit width.

Specific energy: Energy per unit volume of material removed.

Specific energy in chips: Energy per unit volume of material removed, carried within chips.

Specific energy in fluid: Energy per unit volume of material removed, carried within fluid.

Specific energy in wheel: Energy per unit volume of material removed, carried within wheel.

Specific energy in workpiece: Energy per unit volume of material removed, carried within workpiece.

Specific force: Force per unit width of grinding wheel contact with workpiece.

Specific heat capacity: Specific heat: heat per degree temperature rise.

Specific power: Power per unit width of grinding wheel contact with workpiece; power per unit area of contact; power per unit volume removed.

Specific removal rate: Removal rate per unit width of contact.

Specific wear rate: Wear rate per unit width.

Specific wear resistance: Inverse of specific wear rate.

Speed ratio: Ratio of surface speed of wheel to surface speed of workpiece.

Spike temperature: Temperature at a grain contact; also flash temperature.

Spindle power: Power required to drive the main wheel spindle.

Spinel: Any of a group of hard glassy minerals.

Spray: Application of jet or shower of particles, usually liquid.

SSD: Single synthetic diamond.

Static deflection: Movement due to a steady applied force.

Stick-dressing, sticking: Use of a "soft" abrasive stick to open/clean a superabrasive wheel surface.

Sticking friction: Friction due to tangential shearing of a material.

Stiffness: Resistance to movement: rate of force divided by movement.

Stochastic: Acts randomly according to some statistical distribution.

Stock removal: Normal, radial, or diametral reduction of workpiece dimension.

Stone: A gem; an abrasive tool.

Stress: Local ratio of force per unit area, e.g., shear stress, tensile stress.

Stribeck curve: Graph of friction coefficient against Sommerfeld/Hersey number.

Structure number: Number used to indicate proportions of grit and bond volumes in an abrasive layer.

Stylus measurement: Profilometry using a stylus contact to sense shape or texture deviations.

Superabrasive: Extra-hard abrasive grit such as diamond or cubic boron nitride.

Superfinishing: Process for production of very low roughness.

Surface finish: Nonscientific term used to describe surface roughness.

Surface grinding: Grinding a flat surface.

Surface texture: Measure or nature of surface topography; see also *Roughness*.

Suspension: Dispersed particles in a liquid medium.

Swarf: Material debris machined from workpiece.

Synthetic oil: Oil produced chemically, e.g., silicone oils.

Synthetic emulsion: Emulsion of synthetic oil in water.

System: Process that transforms inputs to outputs.

T

Talc: Fine magnesium silicate powder used for polishing.

Tangential force: Component of force tangential to grinding wheel surface.

TDA: Thermographic differential analysis.

Temper: Diffusion process leading to softening and temper colors.

Temper colors: Colors produced on a surface by oxidation at high temperatures.

Temperature gradient: Rate of increase or reduction in temperature with increasing distance.

Tertiary shear zone: Zone of shear between machined workpiece surface and underside of tool.

Thermal boundary layer: A thermally affected layer of fluid near a surface.

Thermal conductivity: Heat transmitted per degree per unit length of transmission.

Thermal stress: Stress in a structure arising due to change of temperature.

Thermocouple: Temperature measurement device based on junction of dissimilar metals.

Thermal stability (TS): Temperature above which fluid chemical composition breaks down.

Thread grinding: Screw grinding of screw thread forms.

3-D: Three dimensional.

Threshold grinding force: Minimum force to achieve chip removal.

Throughfeed grinding: The workpiece is fed between centerless grinding wheel and control wheel.

Tilt: Deviation from vertical/horizontal plane, e.g., control wheelhead tilt.

Tolerance: Difference between maximum and minimum permissible limits.

Topography: Description of surface shape at macro or micro level.

Tool: A cutting part or implement.

Tool life: Machining time between dress and redress of tool.

Tool steel: A group of steels used for manufacturing cutting tools.

Touch dressing: Dressing process using 1–10 micron depth.

Transitional flow: The domain between fully laminar and fully turbulent flow.

Traverse: Linear movement; fast movement on a slideway.

Traverse grinding: Grinding with the wheel traversing the workpiece length.

Tresca: Criterion for plastic yield.

Triangular heat flux: Band heat source: flux varies linearly from maximum to zero.

Tribo-, tribocontact: Relating to conditions of abrasive deformation and surface generation in sliding contact.

Tribocatalytic: Catalytic effects due to abrasive contact conditions.

Tribochemistry: Tribological phenomena particularly related to chemical action.

Tribochemical wear: Chemical wear accelerated by abrasive action.

Tribology: The science of friction, lubrication, and wear of sliding contacts.

Tribomechanical: Quasitribophysical; mechanical aspects of tribology.

Tribometer: Machine to test friction and rate of wear for particular speed, force, and contact condition.

Tribo-oxidation: Oxidation mechanism under tribocontact conditions.

Tribophysics: Part of physics which deals with interacting surfaces in relative motion.

Triboplasma: Material in energetic state due to intensive tribocontact.

Triboreaction: Reaction under tribomechanical activation.

Tribosimulation: Testing on a rig replicating the real tribological contact conditions.

Tribosorption: Assimilation of a material on a surface under tribomechanical action.

Tribosystem structure: Elements of a system that effect transformation of inputs to outputs.

Tripoli: Soft powder of siliceous rock (originally from Tripoli, Libya).

Truing: Process to produce accurate form required on an abrasive tool.

Turbulent flow: Moving forward with agitated sideways buffeting motion.

2-D: Two dimensional.

Type C contact: Curved area contact.

Type C corrosion: Corrosion with continous progression.

Type D contact: Dot contact.

Type F contact: Flat area contact.

Type L contact: Line contact.

Type P corrosion: Corrosion with slowing progression due to passivation.

U

Ultrasonic machining: Machining by ultrasonic vibration of a tool against a workpiece.

Uniform heat flux: Band heat source of constant flux magnitude.

Universal machine: Machine capable of internal and external grinding.

Up grinding: The wheel motion and workpiece motion are in opposite directions.

UPM: Ulta-precision machining.

Upper bound: Estimate known to be higher than the real value.

Useful flowrate: Flowrate of grinding fluid passing through the contact zone.

V

Vertical grinder: Grinding wheel axis is vertical.

VI-Improver: Viscosity index additive used to reduce temperature dependence.

Vickers hardness: Hardness measured using standard diamond-shape indentor.

Vienna lime: Powder of calcium and magnesium oxides used for polishing.

Viscous friction: Friction force is proportional to sliding speed.

Vitreous, vitrified: Glassy phase produced by heating.

Vitrification: Conversion into a glassy form by heating.

Vitrified wheel: Wheel having abrasive grains held together by vitreous bond.

Von Mises: Criterion for plastic yield based on maximum deviatoric stress.

W

Waviness: Surface shape errors having a pattern of undulation.

Wear: Loss of material from a body by abrasive rubbing and sliding processes.

Wear, attritious: Slow wear evidenced by a polished surface.

Wear, fracture: Removal of parts of a body by cracking and breakage.

Wear, microfracture: Removal of very small parts of a body by small, localised cracks.

Wear rate: Wear volume per unit time; wear depth per unit time.

Wear ratio: Inverse of grinding ratio.

Workpiece: The part or body to be machined.

Wheel: Disc that rotates on a spindle; abbreviation for grinding wheel.

Wheelhead: Powered spindle assembly to carry and drive grinding wheel.

Wheel flange: Plate used to clamp wheel on hub.

Wheel hub: Adaptor; arbor used to mount wheel on a spindle.

Wheel loading: Tendency for softened workpiece material to adhere to the wheel surface.

Wheelspeed: Usual: tangential surface speed of wheel; sometimes: rotation speed.

Wheel-work partition: Division of energy at workpiece and grain contacts.

White layer: White layer of workpiece at surface due to mechanical or thermal effects.

Width of cut: Width of grinding wheel contact with workpiece.

Workblade, workplate, workrest: Plate to support workpiece in centerless grinding.

Workblade angle: Angle of top surface of workblade.

Work-hardening: Tendency for shear stress to increase with plastic shear strain.

Workhead: Powered spindle assembly to locate and drive workpiece.

Workmaterial: Workpiece material.

Work partition ratio: Proportion of heat conducted into workpiece in contact zone.

Workpiece: The part to be machined.

Work speed: Tangential speed of workpiece parallel to wheelspeed.

X

X-ray diffraction (XRD): Technique used for measuring residual stresses.

Z

Zinc-dialkyl-dithio-phosphates (ZDTP): AW additive for lubricating and process fluids.

Index